D1618496

REVERSE OSMOSIS

A PRACTICAL GUIDE FOR INDUSTRIAL USERS

WES BYRNE

TALL OAKS PUBLISHING INC.
P.O. Box 621669
Littleton, CO 80162

Library of Congress Catalog Card Number: 95-60061

ISBN 0-927188-03-1

FIRST EDITION

TALL OAKS PUBLISHING, INC.
P.O. Box 621669
Littleton, CO 80162
U.S.A.
303/973-6700

PRINTED IN THE UNITED STATES OF AMERICA

To Nathan

PREFACE

The superficially apparent simplicity of reverse osmosis (RO) as a technology has attracted numerous companies to the business of RO system manufacturing. Some companies have prospered, whereas many have fallen by the wayside. Frequently, this situation has resulted in the manufacturing of inferior RO equipment. Even more frequently, it has resulted in poor technical support being available for the operators of these systems.

The intent of this book is to assist RO system manufacturers and endusers in better understanding the technology of reverse osmosis. It is my hope that this will enable manufacturers to build more reliable equipment, system operators to run their systems more efficiently, and problems to be more quickly resolved.

As stated in the title, the emphasis of this book is to provide support for users and manufacturers as a practical guide to reverse osmosis. The discussion of RO theory will be limited. The book covers the basics so as to assist those new to RO. It also covers many of the subtleties of the design and maintenance of RO that are commonly overlooked even by experienced engineers and technicians.

This book is written about industrial systems in certain applications, chosen to cover common situations and basic concerns related to the application of reverse osmosis. The concepts discussed are comprehensive; therefore, the book should be of value to anyone who works with RO systems.

Acknowledgments

The overwhelming nature of writing a technical book is now apparent to me. I hope that it will contribute to the increased understanding of reverse osmosis and to its acceptance in new and demanding applications.

Although there may be certain nuances included in the book that may be new and creative for the water treatment industry, the vast majority of it is the result of work previously performed by numerous individuals and organizations. Of course, it would be impossible to name all parties who have contributed to the technology of reverse osmosis. The acknowledgments presented herein will be for individuals whose assistance had a direct impact on the book's material. I apologize to anyone whose contribution may have been overlooked.

Certain individuals provided a foundation for my understanding of reverse osmosis and deserve special recognition. These include W. Hobson McPherson of Process Scientific, Inc., in San Jose, California; Randy Truby of Spiral Composites in Santee, California; and Izadore Nusbaum of San Diego, California. I also want to offer special thanks for the technical and personal assistance given by Mark Roesner and Dr. David Blackford of American Fluid Technologies, Inc., in Wayzata, Minnesota; Steve Cappos of Fluid Systems in San Diego, California; Leland Comb of Osmonics, Inc., in Minnetonka, Minnesota; Ken Fulford of AT&T Analytical Services in Allentown, Pennsylvania; Chris Wise of Analytical Maintenance Services in Fremont, California; and Scott and Laurie Shekels of Scott Data Inc. in Crystal Bay, Minnesota.

I want to thank Dr. William Lite of Fluid Systems in San Diego, California, for his assistance in the discussion of reverse osmosis theory. Jerry Filteau and Bill Loy, also of Fluid Systems, also provided valuable assistance.

David Cohen of Flow Tech Industries of Glendale, California, assisted with RO instrumentation and control, as well as with peracetic acid sanitization. David Holland of Dalton Scientific in Eden Prairie, Minnesota, and Paul Osmundson of Osmonics, Inc. in Minnetonka, Minnesota, offered valuable insights for RO system and pressure vessel design.

In the chapter on RO pretreatment, Chub Michaud of Systematix in Brea, California, has been exceptional in his help on media filter design. Mel Hemp of Clack Corporation and Harold Aronovitch of Hungerford & Terry, Inc., also offered their assistance with media filtration. Mike Cloutier of Liquid Metronics of Acton, Massachusetts, assisted in the section on chemical injection pumps.

Dr. Gerold Luss of H.B. Fuller Company in St. Paul, Minnesota, provided much of the information that formed the foundation of the sections on scale inhibition, membrane cleaning, and analytical troubleshooting. Bob Burtleson of Osmonics, Inc., in Minnetonka, Minnesota, provided insights related to unusual applications of reverse osmosis.■

Wes Byrne
Wayzata, Minnesota

CONTENTS

Chapter 8: SPECIAL APPLICATIONS

CHAPTER 1

REVERSE OSMOSIS THEORY

Introduction

Reverse osmosis (RO) has become a popular water treatment technology in nearly every industry requiring separation of a dissolved solute from its solvent, the solvent usually being water. The most common application of RO is the purification of water, involving simply the removal of undesirable contaminants. Industry makes heavy use of this application of RO for producing highly purified process water, and for treating industrial wastewater. The RO process is used to make potable (drinkable) water by desalinization of seawater; and is even used residentially to improve the taste of water, as well as to remove potentially unhealthy contaminants.

Dissolved solutes can be salts, or they can be organics such as sugar or dissolved oils. The solutes are the species of lesser concentration held in solution by molecular attraction with the solvent, which is in greater concentration. The highest concentrations of these dissolved solutes in city water sources are typically salts.

Reverse osmosis is a fundamental component of treatment systems for the water used in the manufacturing of semiconductors, pharmaceuticals, and medical devices, as well as in purifying water used for dialysis and for power generation. The RO process is used in the concentration and reclamation of salty wastewaters, and is even used to concentrate sugar solutions in the food industry.

The increasing popularity of residential point-of-use (under-the-sink) RO systems has made *reverse osmosis* a household term. In this and many other applications, the RO also serves as an extremely fine filter. Its pores are believed to be less than 0.001 micron (μm) in diameter. (The diameter of a human hair is greater than 30 μm.) These pores give RO the ability to remove suspended solids, particulates, bacteria, and endotoxin (a very fine particulate of concern in the medical and pharmaceutical industries). (See Figure 1-1 and Table 1-1.)

The success of the reverse osmosis technology has been due mostly to the economics of its operation and to its simplicity. Compared to other salt-removal technologies, it is relatively inexpensive to purchase and to operate. It requires neither an energy-intensive phase change such as distillation nor a large volume of strongly acidic and caustic chemicals like those required for ion-exchange systems.

The basic process of reverse osmosis uses a pump and a semipermeable membrane. The pump provides the driving force. The semipermeable membrane passes water in preference to the solute that is dissolved in the water; thus the majority of solute is left behind to form a more concentrated stream. (See Figure 1-2.)

The process of reverse osmosis is not the same as filtration. Filtration (Figure 1-3) is the removal of particulates by size exclusion. Particulates are removed by filtration because they are too large to fit through physical pores (i.e., holes) in the filter media, whereas water molecules can readily fit through the pores.

The mechanism of reverse osmosis is different from filtration in that physical holes may not exist in an RO membrane. Such holes have never been found, even with microscopes of highest magnification. It is more likely that water and smaller-molecular-weight organics are able to diffuse through the membrane polymer by bonding between segments of the polymer's chemical structure. Dissolved salts and larger-molecular-weight organics will not permeate the membrane as readily, however, because of their size and charge characteristics. The RO membrane also is able to obtain nearly absolute removal of suspended particulates that have no means of traversing the membrane (Figure 1-4) except via mechanical lesions (such as a membrane tear).

Reverse osmosis is also a process of separation. The feedwater stream is separated into a stream of purified water and a stream of concentrated solutes and particulates. This is as compared to standard filtration where the entire feed stream passes through the membrane pores, leaving the particulates embedded in the filter media.

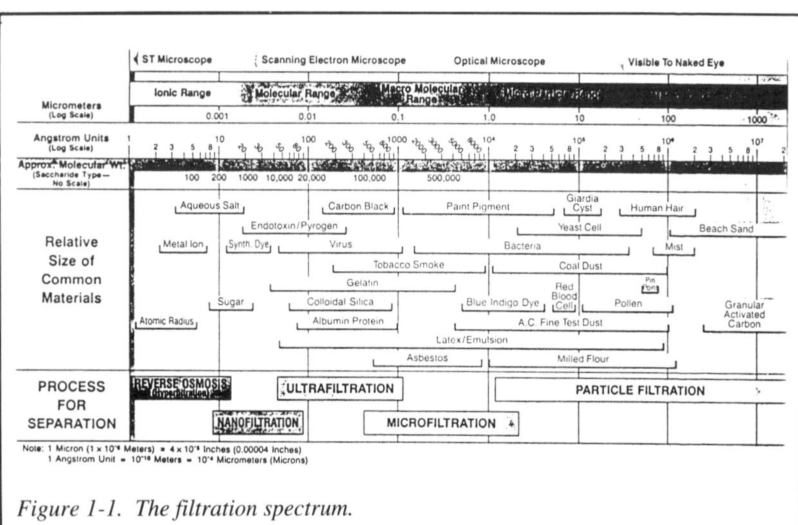

Figure 1-1. The filtration spectrum.
(Courtesy of Osmonics, Inc.)

Recent improvements in the semipermeable membranes used have caused dramatic improvements in the economics of reverse osmosis. Newer membranes reject more salts and pass more water at a particular pressure, thus resulting in greater energy efficiency. They also are more durable, able to handle harsher feedwater conditions and cleaning solutions.

These improvements will result in the continued growth of RO technology into new applications. This acceptance and growth of the RO industry will require the availability of more personnel educated in the various aspects and subtleties of reverse osmosis.

Reverse Osmosis Theory

Reverse osmosis systems separate dissolved solutes from water via a semipermeable membrane that passes water in preference to the solute. An RO membrane is very hydrophilic, meaning that water is attracted to its chemical structure. Water can bond with the ends of the polymer segments making up the membrane. This gives water the ability to readily diffuse into and out of the polymer structure of the membrane.

Water (H_2O) is a very polar molecule. There is a strong separation of charge that occurs across its molecule. When water is formed by the coming together of two hydrogen atoms with an oxygen atom, each of the two hydrogen atoms shares its single electron with the oxygen atom. This puts the hydrogen atoms into a more stable energy state. The oxygen atom is also in a more stable energy state when it borrows the electrons from the hydrogen atoms.

Table 1-1
Sizes of Some Common Particles

Substance	*Microns*
Table salt	100
Human hair	50-80
Sand	50 and up
Smallest visible to eye	30-50
Talcum powder	10
Cocoa	8-10
Iron rust	5
Carbon in oils	1-10
Clay	0.1-1.0
Pigments	0.2-0.4
Bacteria	0.4-2.0
Viruses	0.015-0.3

(Courtesy of Osmonics, Inc.)

The two hydrogen atoms will orient themselves on one side of the oxygen atom. Because it has given up its negatively charged electrons, the remaining positively charged proton of each hydrogen atom is able to exert its positive charge on its surroundings. The electrons that have migrated to the oxygen atom will tend to exert their negative charge on the far area of the oxygen atom. The end result is a separation of charge between the ends of the water molecule.

If this separation of charge is compounded by the large number of water molecules present in a typical solution, a substantial force is present that is capable of pulling apart the oppositely charged components of a salt. The hydrogen ends of the water molecules will attract the negatively charged anionic component of a salt. The oxygen end of water will attract the positively charged cationic component of the salt. As the salt molecules separate into their components, they go into solution (Figure 1-5).

While in solution, the charged ions will remain surrounded by water molecules that are attracted to the charge of the ions. In essence, an ion group is created. The size of this ion group will depend on the size of the ion at its center and the extent of the charge characteristics of the ion. How the size of this ion group compares to the spacing between the polymer segments has a lot to do with how well the ion will diffuse through the membrane.

Although separated by water molecules, the dissolved cations will still maintain an attraction to the dissolved anions in a solution. A hydrated cation is going to resist diffusing through the membrane if a corresponding hydrated anion cannot also pass through. Otherwise, a charge imbalance would be created between the feed side and the permeate side of the RO membrane.

The percent passage of any particular dissolved salt will depend mostly on the ionic component with greater size and charge characteristics. For example, if the hydrated sulfate anion of a sodium sulfate (Na_2SO_4) solution cannot diffuse through the area between polymer segments, the sodium cation will not permeate through either. If calcium chloride ($CaCl_2$) is added to that solution, the hydrated

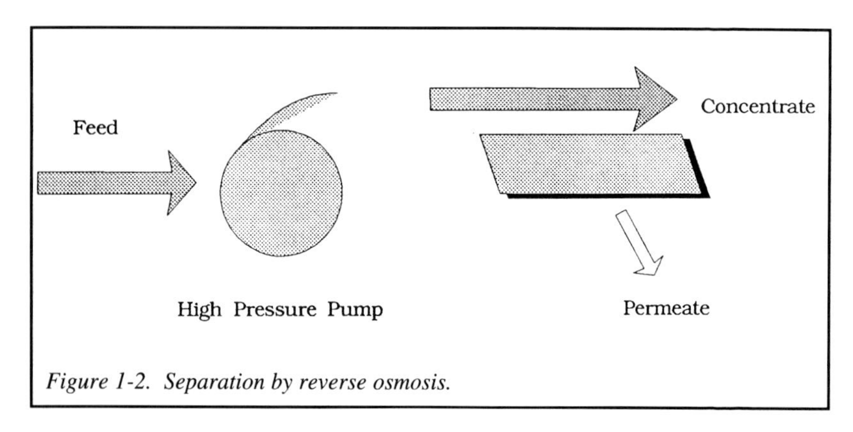

Figure 1-2. Separation by reverse osmosis.

sodium ion will then be able to diffuse more readily through the membrane because the hydrated chloride anion is better able to permeate the membrane.

The ability of an RO membrane to diffuse certain salts while rejecting others is not absolute. For any particular membrane, the percent passage of smaller ions, or lesser charged ions, will be relatively greater than that of larger ions, or ions with greater charge characteristics. Generally, cations or anions with greater valence numbers (greater charge) will be better rejected than ions with lower valence. A divalent cation, such as calcium with a valence of +2, will typically be rejected somewhere around three times better than the monovalent ion sodium (valence of +1). Approximate salt passage percentages for an RO membrane are given in Table 1-2.

How well an organic molecule permeates an RO membrane will depend on its physical size and shape and on its chemical characteristics. Generally, the closer an organic molecule is structurally to that of the membrane polymer, the more readily it will diffuse through the polymer. (Light, 1993).

Smaller organic molecules with polar characteristics will tend to diffuse better than larger, neutral ones. A very general rule of thumb is that organic molecules with a molecular weight less than 200 are more likely to permeate the membrane than ones with molecular weight greater than 200.

The physical structure of the molecular is critical as to whether it will permeate the RO membrane. An organic molecule with an elongated shape will not permeate the membrane as easily as will a round, compact molecule. Table 1-3 gives some of the rejection characteristics of certain organic molecules.

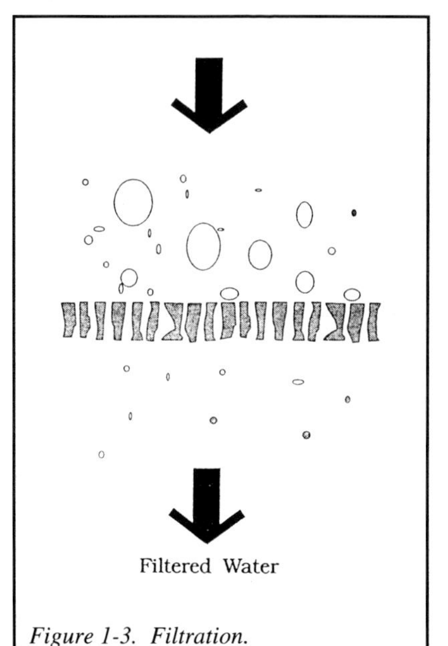

Figure 1-3. Filtration.

The ability of water to diffuse through a membrane gives RO a significant operating advantage over ultrafiltration technology. When water diffuses through the RO membrane, near-maximum use is made of the cross-sectional area of the membrane. The velocity of the bulk stream approaching the membrane surface is about the same as the local velocity of the water actually passing through the membrane (Figure 1-6).

This is in sharp contrast to the behavior of water as it passes through the pores of an ultrafiltration (UF) membrane. The flow through the UF membrane must pass through the pores

Table 1-2
Approximate Salt Passages for an RO Membrane

Salt	*Formula*	*Passage (%)*
Calcium sulfate	$CaSO_4$	1
Calcium bicarbonate	$Ca(HCO_3)_2$	2
Sodium chloride	$NaCl$	3
Sodium bicarbonate	$NaHCO_3$	3
Sodium sulfate	Na_2SO_4	1
Calcium chloride	$CaCl_2$	2
Magnesium bicarbonate	$Mg(HCO_3)_2$	2
Magnesium chloride	$MgCl_2$	2
Potassium chloride	KCl	3
Magnesium sulfate	$MgSO_4$	1
Potassium bicarbonate	$KHCO_3$	3
Potassium sulfate	K_2SO_4	1
Calcium nitrate	$Ca(NO_3)_2$	15
Sodium nitrate	$NaNO_3$	15
Magnesium nitrate	$Mg(NO_3)_2$	15
Potassium nitrate	KNO_3	15
Calcium fluoride	CaF_2	3
Sodium fluoride	NaF	6
Magnesium fluoride	MgF_2	3
Potassium fluoride	KF	6

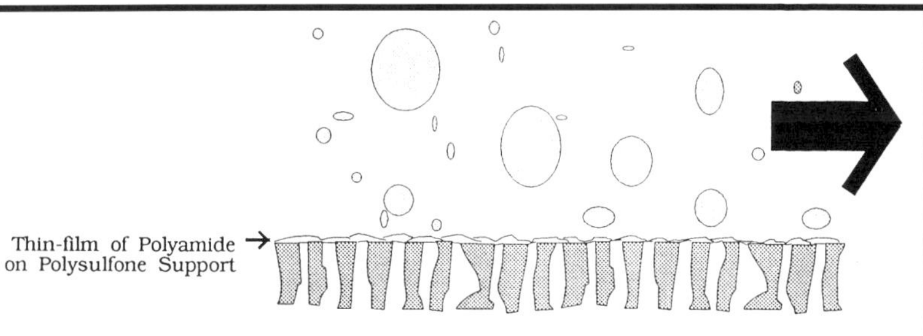

Figure 1-4. Crossflow reverse osmosis.

in the membrane. These pores will contain a minimal amount of cross-sectional area as compared to that of the membrane surface; therefore, the flow approaching the UF membrane surface will be forced through the open area of the membrane, which is a small percentage of the total membrane area. The result is that the local velocity of water going through the pores is dramatically greater than that of the bulk stream approaching the membrane surface (Figure 1-7).

With both RO and UF, particulates and suspended solids that are present in the feedwater are left behind on the membrane surface when water permeates the membrane. The continued flow of water through the membrane will place a force on these contaminants that can prevent them from returning to the bulk stream going across the membrane surface. In order for the contaminant to return to the

Table 1-3
Rejection of Organic Molecules by Polyamide Thin-Film Membrane

Organic	*Concentration*	*Rejection (%)*
Acetone	0.1 Molar	70
Ethanol	0.1 Molar	40
Phenol	0.1 Molar	40
Methylethyl ketone (MEK)	0.1 Molar	75
Isopropyl alcohol (IPA)	Independent	90
Sucrose	Independent	Almost 100
Dextrose	Independent	Almost 100

(Courtesy of Osmonics, Inc.)

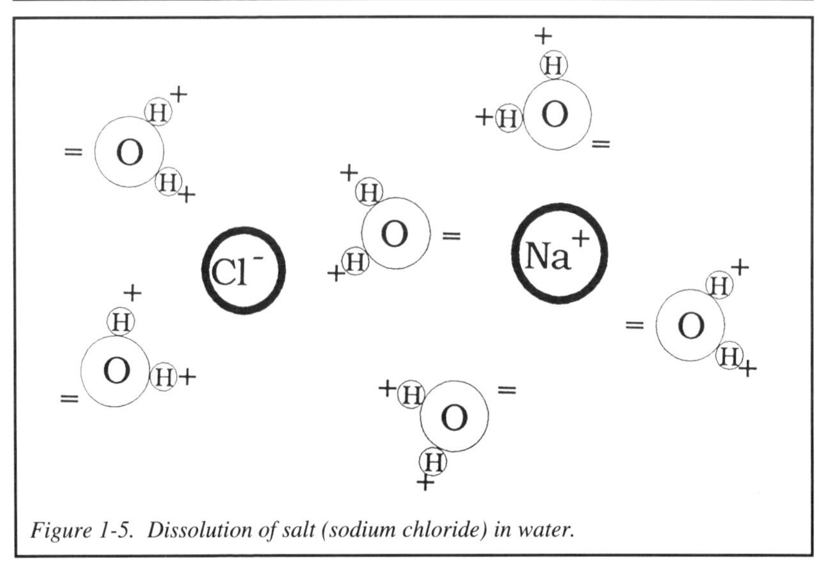

Figure 1-5. Dissolution of salt (sodium chloride) in water.

bulk stream, the shear force of the water going across the membrane surface has to overcome the shear force of the water going through the membrane. Because of the higher local velocity of water going through the UF pores, the shear forces of the water going across the membrane are less effective at preventing the contaminants from fouling the membrane.

In side-by-side studies, UF membrane systems have been known to foul faster than RO membrane systems by more than an order of magnitude (i.e., by more than ten times the rate of RO fouling). This is above and beyond the higher rate of UF fouling that can be attributed to its higher flux rate (overall permeate flowrate per area of membrane surface). It can be generalized that for most applications a UF membrane will foul faster from particulate and suspended solids contaminants than would an RO membrane.

Osmotic Pressure

Osmosis is a phenomenon occurring naturally in various biological processes. Cell walls will allow the passage of certain cell nutrients and waste products, while rejecting the passage of other materials. In these cases, the cell wall is considered a semipermeable membrane.

Through the particular semipermeable membrane used in reverse osmosis, water will permeate in preference to salts, particulates, suspended solids, and larger-molecular-weight organics. Reverse osmosis is a reversal of the standard flow of materials in osmosis, which is from the side of greater concentration to that of lesser concentration.

If a salt water solution is placed into a chamber and separated from a pure water solution by a semipermeable membrane, a change in the volume of each side of the chamber will occur. Water will pass through the membrane from the pure water side to the salt water side. The salt water side will thus increase in volume and water height as the pure water side decreases. This flow of water through the membrane to the salt water side is called *osmosis* (Figure 1-10).

In terms of physics, the water is diluting the salt solution as a means of increasing the system entropy. *Entropy* is defined as the state of randomness in nature. It is a law of thermodynamics that every spontaneous process that occurs

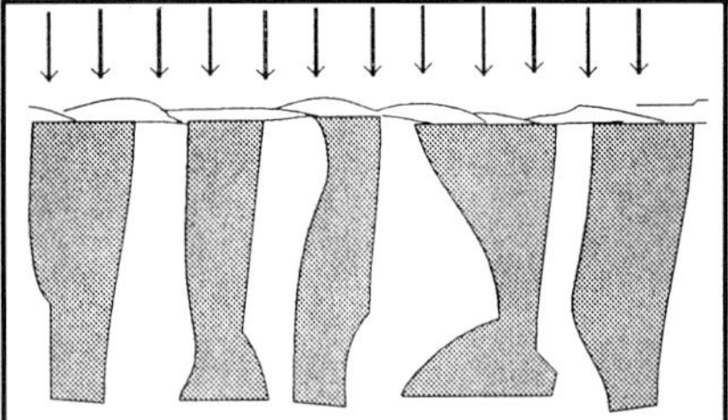

Figure 1-6. Diffusion velocity through an RO membrane.

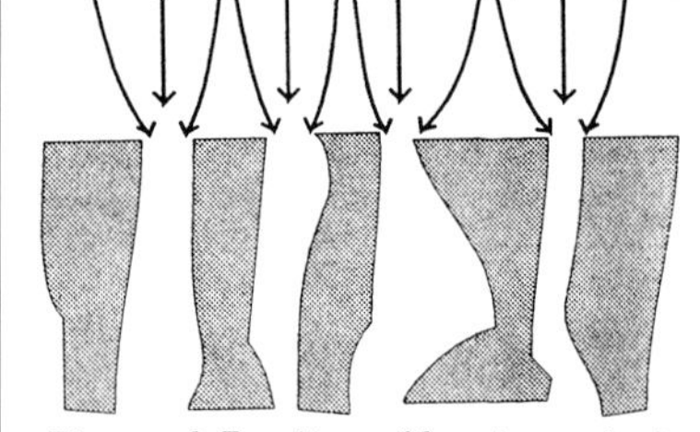

Figure 1-7. Pore filtration velocity through a UF membrane.

in nature will occur in a manner that increases the entropy of the system, unless energy is expended to force the process in the other direction (towards more order, or structure).

The difference in liquid height (called the *water head*) between the pure water and the salt water creates a difference in water pressure at the membrane surface. This pressure difference is called the *osmotic pressure* of the salt water solution and is a function of the particular salts and their concentration.

If the equilibrium established between the pure water and salt water is altered by applying physical pressure to the salt water side (or by raising the height of the salt water solution), pure water will permeate the membrane from the salt water to the pure water side. This permeation will continue until an equilibrium is again established where the osmotic pressure of the salt solution is equal to the pressure of the salt solution on the semipermeable membrane relative to the pressure of the pure water on the membrane.

For example, if additional salt water is poured into the salt water side of the chamber, part of the water will permeate the membrane. If the same salt concentration was used in this added solution, water will permeate to the pure water side until the difference in heights between the two sides returns to the original difference in height (Figure 1-9).

Pressure can be applied to the salt water side of the chamber with a pump. In this case, the pump supplies the force needed to force pure water through the membrane. The flow of water from the salty side of the membrane through to the pure water side is called *reverse osmosis*.

The rate at which water permeates the membrane is given by the following relationship:*

$$\text{permeate flux} = K_T \times (\text{membrane psid - osmotic psid})$$

where:

permeate flux is the flowrate through the membrane per unit of membrane area, K_T is some constant determined by the membrane and is a function of tempera-

* Fluid Systems Corporation, "Reverse Osmosis Principles and Applications", p. 13, October 1979.

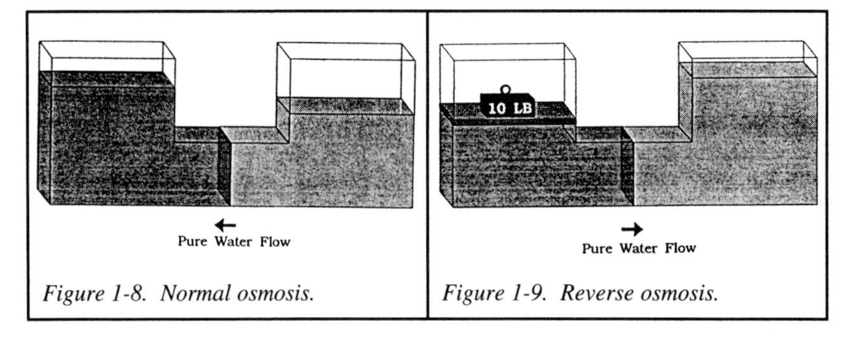

Figure 1-8. Normal osmosis. *Figure 1-9. Reverse osmosis.*

ture, *membrane psid* (pounds per square inch differential) is the difference between the pressure on each side of the membrane (i.e., the head pressure), and *osmotic psid* is the difference between the osmotic pressure of the solute on each side of the membrane. (When the solute concentration is negligible on the pure water side of the membrane, just the concentrated solute osmotic pressure needs to be considered.)

Not all the water can be forced through the membrane from the salt solution. As water leaves the salt solution, the solution becomes more concentrated. To overcome the increasing osmotic pressure of the saltier solution will require increasing hydraulic pressure. Otherwise, the flow of water will stop when the applied pressure equals the osmotic pressure of the salt solution.

Reverse osmosis is able only to separate a solution into a purer water component and a more concentrated component. Because of the nature of osmosis there must always be a concentrate stream (also called a *brine stream*). Usually, the amount of this brine stream is determined by the solubilities of the salts and/or by the osmotic pressure of the concentrate stream.

Although a semipermeable membrane used in reverse osmosis will preferentially pass water, a certain amount of salt (solute) will also diffuse through the membrane. This flow is proportional to the concentration gradient across the membrane (i.e., the difference in concentration between the feed side and the permeate side of the membrane).

$$\text{Salt flow} = K \times ([\text{solute}]_{\text{feed}} - [\text{solute}]_{\text{permeate}})^{*}$$

where:
K is a constant that is dictated by the membrane material and its thickness,
$[\text{solute}]_{\text{feed}}$ is the salt concentration on the feed side of the membrane, and
$[\text{solute}]_{\text{permeate}}$ is the salt concentration on the permeate side of the membrane.

* Fluid Systems Corporation, "Reverse Osmosis Principles and Applications", p. 13, October 1979.

The significance of the last two expressions is that the rate of salt passage through the membrane is not a function of pressure, whereas the water flowrate is. Therefore, a lower concentration of salts will result on the pure water side of the membrane if higher pressures are utilized on the concentrated solute side of the membrane. This is due to the greater dilution of the salt flow by the increased permeate water flowrate.

Reverse Osmosis Design

The basic process of reverse osmosis requires very little: a pump, a membrane element, and miscellaneous plumbing connections. In some cases, that is about all that goes into an RO design. In spite of their crude design, these systems are frequently successful at meeting the needs of their particular applications.

However, obtaining maximum life from the RO membrane, which is economically desirable for most larger applications, requires that greater effort be invested in the design subtleties of the system.

There are suggested ways of designing an RO system that will insure that the system will meet the needs of the application and will provide optimum membrane performance. Usually, the membrane manufacturer will provide this assistance for the system manufacturer. This chapter will outline some basic concepts that would also be covered by the membrane manufacturer.

Investigation of Design Conditions

Before an RO system is designed, it is important to understand the needs and limitations of the particular application. The RO system permeate water recovery and permeate quality, as well as maintenance requirements, will depend greatly on the quality of the water source. The quality of the potential RO feedwater will thus dictate the requirements of the RO pretreatment system for the particular application, and may affect the design of the RO itself. This in turn may affect the economics of the application. Given the wide variability in the quality of water sources, a thorough understanding of the water source and how it might impact the needs of the RO application will assist in designing successful RO pretreatment and RO systems.

Two parameters that first have to be determined are the desired RO permeate flowrate and the desired permeate quality. The desired permeate flow will depend on whether storage tanks are used in the system. If not, the RO system will have to be sized to meet the peak usage. If storage tanks are used, and sized properly, the RO can be sized to meet an average usage. In other words, if the storage tank is sized to handle any peak usage during any particular day, then the RO can be sized to handle the daily average usage (based upon a peak-usage day). (See "Storage Tanks" in Chapter 4.)

The water quality requirements may determine what membrane to use in the RO; or possibly limit the RO permeate water recovery (the percentage of feedwater that becomes purified water); or determine if another RO stage or some other means of further polishing the water is required. If ion exchange is to be used downstream of the RO to further polish the RO permeate water, a difference in RO membrane rejection between a 97.5%-rejecting membrane and a 95%-rejecting membrane will result in doubling the regeneration frequency of the downstream ion-exchange system.

Once a minimum value has been chosen for the permeate flow and a maximum value chosen for the permeate quality, other parameters can be considered. Perhaps there are size constraints as to how big the RO unit can be. Size restrictions have resulted in RO designs that are not necessarily easy to maintain. For example, if an RO system has its spiral-wound element housings mounted right next to a wall, or mounted vertically, then replacing membrane elements may require the removal of the entire housing. Maintenance is dramatically easier when technicians have ready access to instrument panels, pumps, and membrane elements.

The potential water source for the RO system should be completely evaluated for its contaminants, its variability, and even its availability. If the source is municipal water that is a blend of various well and surface waters whose ratios may vary at the whim of the municipality, it will be necessary to conservatively design the RO with each of the worst-case water source scenarios in mind.

If the water source pressure varies dramatically, this is a strong symptom that some component of the municipal or plant water distribution system is under-designed. Possibly the incoming pipeline or the backflow-preventer valve into the facility is undersized. (If there currently is no backflow preventer, it may be required by the municipality prior to installing the RO.) The RO should not have to compete with the building's yard-sprinkling systems for the availability of water.

Water temperature has a dramatic effect on the RO permeate flowrate. Winter conditions should be taken into account. A heat exchanger of some type may be desired to heat the RO feedwater as a means of increasing its output. This usually requires more energy and operating expense, but results in a reduced need for membrane elements in the RO system. A variable-speed drive may be desired to increase the high-pressure pump output during colder months.

The water source should be analyzed at the point of use for biocide concentration. Typical biocides used in municipal systems are free chlorine or chloramine. These have to be measured on-site because they are volatile chemicals, meaning that they will tend to degas to the air if transported to an off-site laboratory. Test kits are readily available from swimming pool chemicals suppliers, or from an industrial chemical test kit company. Typical free chlorine concentrations for municipal water sources are in the range of 0.3 to 0.8 ppm.

Chloramine concentration will analyze as "total chlorine" by the test kits. It should be noted if chloramine is being used by the municipality for biological control. Chloramine is not nearly as effective as free chlorine at killing and preventing bacteria growth, and may also require additional treatment equipment for its removal.

The water should be tested to confirm the absence of bacteria. If bacteria are present, or if the biocide concentration in the lines is low, it is likely that continuous injection of a biocide will be required as part of the RO pretreatment.

A comprehensive water analysis should be performed on the water source by an outside laboratory. This should be used to determine the potential for scale formation as the water is concentrated in the RO system. This may have a dramatic affect on the RO permeate recovery, or on the requirements for RO pretreatment.

Specific Analyses

Silt density index (SDI). Also known as the fouling index, this measurement has been commonly used in the water treatment industry as an indicator for the potential for membrane fouling. The test measures the rate of fouling of a 0.45-μm filter membrane by suspended solids. It is popular because the test is easy to perform and does not require any expensive instruments.

The test involves diverting water from the source through a 47-millimeter (mm) diameter filter under a constant pressure, usually 30 pounds per square inch gauge (psig). (See Figure 1-10.) The time for 500 milliliters (mL) of water to pass through the filter is measured initially, and again after the filter has been on-line for (typically) 15 minutes. The ratio between these flow measurements is used to determine the extent of filter fouling.

Unfortunately, the SDI measurement does not duplicate the same mechanisms by which reverse osmosis membrane elements operate. With the SDI test, all suspended solids are forced to foul out on the filter. With RO, some unknown percentage of these potential foulants will flow across the RO membrane, eventually ending up in the RO concentrate stream; therefore, a high SDI value does not necessarily mean that an RO is going to readily foul.

The SDI procedure:

1) Attach the SDI filter holder device to a sample valve.

2) Open the sample valve and set the pressure regulator for 30 psig. Shut the sample valve and open the filter housing.

3) Place a support pad in the filter housing with a 47-mm, 0.45-μm filter on top with the shiny side facing the incoming fluid. Place the housing O-ring on top of the filter. Wet the filter and close the housing.

4) Open the sample valve and immediately readjust the pressure regulator for 30 psig.

5) As soon as the pressure is set, measure the time to fill a 500-mL graduated cylinder. Allow the water to continue to flow, adjusting the pressure regulator as necessary to maintain 30 psig.

6) At 5-minute intervals, measure the time to fill a 500-mL graduated cylinder. The 15-minute measurement will be used to calculate the SDI unless the filter plugs in less than 15 minutes. If this is the case, perform the calculation using the 10-minute measurement, or the 5-minute measurement if the 10-minute test is not available.

7) Silt density index = $100 \times (1 - t_o/t_T) \div T$ where: T is the time before the second flow measurement (5, 10, or 15 minutes), t_o is the initial time to flow 500 mL, and t_T is the time to fill 500 mL after T minutes.

Typical SDI values:

Less than 3	Low membrane fouling rate
3 to 5	Normal fouling rate
Greater than 5	High fouling rate

(Aqua Media Ltd., Sunnyvale, California, 1981)

Example:
Initial time to fill cylinder: 23 seconds.
Test was run for 5 minutes because of excessive fouling of SDI filter.
Final time to fill cylinder: 62 seconds.

Answer: $SDI = 100 \times (1 - 23/62) \div 5 = 12.6$

With such a high SDI value, it is likely that significant fouling will occur in the RO membrane elements. It would be recommended that some means of reducing the foulant concentration in the water be considered. Some of these pretreatment methods will be discussed in Chapter 3.

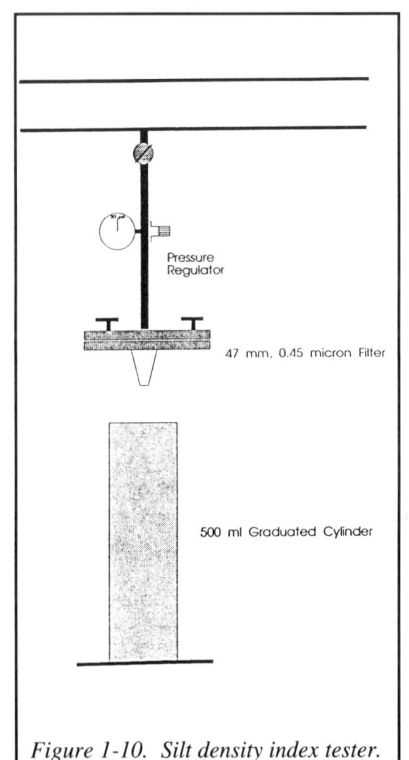

Figure 1-10. Silt density index tester.

Turbidity is another measurement that has been used as an indicator for the rate of RO membrane fouling. It is a measurement of the lack of clarity in a water sample. Turbidimeters (also called *nephelometers*) measure the scattering of light caused by various particles and suspended solids in the water sample.

As turbidity readings exceed 1.0, they are indicative of a greater tendency for membrane fouling. These readings are typically given in nephelometric turbidity units (NTU), or some other relative unit.

Like the SDI test, turbidity is only an indicator of fouling potential. High turbidity does not necessarily mean that what is causing the turbidity is going to fall out of suspension on the RO membrane, or vice versa. In fact, there are some membrane foulants that are clear to the passage of light, and that would not show up in a turbidity measurement. (W. H. McPherson, May 1981)

Although they are less than perfect as tools of analysis, turbidity and SDI measurements are useful for characterizing an RO feedwater. They are also useful in measuring the changing ability of pretreatment equipment to remove potential RO membrane foulants. It is therefore suggested that such tests be used on a regular basis to evaluate the performance of pretreatment media filters.

Zeta Potential. Organic suspended solids tend to carry negative charge characteristics on the outer surfaces of their groupings. Having a common charge, the various simple and colloidal groups will tend to repel each other. They resist coming into close proximity with each other, and so do not combine to form larger groups. These smaller groups are not as likely as larger groups to overcome the forces of the water molecules that keep the groups in suspension.

When such negatively charged solids are concentrated by a reverse osmosis system, they can be forced into close proximity. When this happens, the solids are more likely to come together and form larger colloidal groups. The large groups may overcome the water's charge and shear forces that keep them in suspension. They then can fall out of solution on the RO membrane.

The zeta potential is a measurement of the overall charge characteristics of the suspended particles in the water. A negative zeta potential indicates that the water contains free negatively charged suspended solids that are more likely to stay in solution. A neutral zeta potential indicates that the suspended solids do not carry a charge to assist in their repulsion of each other. They are more likely to coagulate into larger particulate groups and fall out of solution.

Zeta potential measurement is frequently used in conjunction with a cationic (positively charged) coagulant injection system. The coagulant concentration is adjusted in order to achieve a neutral zeta potential upstream of media filters; thus, the suspended solids are more likely to fall out of solution so that they can be filtered by the media filters. Coagulant addition is not necessarily recommended as part of the RO pretreatment. (See "Fouling Control" and "Coagulation Aids" in Chapter 3.)

Streaming current detector. Another type of monitor that is sometimes used in conjunction with the injection of coagulants is called a streaming current detector (SCD). This device uses a mechanical plunger to create a high water velocity, which causes movement of the charged ions that tend to surround oppositely charged suspended particles (usually negatively charged). It then measures the electrical current generated by the moving ions. If the charge of the suspended solids has been neutralized by the addition of a coagulant, there will be fewer free ions surrounding the particles, and a smaller current will be generated by the SCD (Carling, 1989).

Water Chemistry

It is important to understand the basics of water chemistry to understand the potential for scale formation in an RO system. If it is determined which salts have the potential for scale formation as a function of the RO permeate recovery, a method can then be devised to prevent and control the formation of scale.

When salts dissolve in water, they break up into their cationic (positively charged) and anionic (negatively charged) components. The most common ions appreciably present in raw water sources consist of the following:

Cations	*Anions*
Hardness	Alkalinity
Calcium - Ca^{+2}	Bicarbonate - HCO_3^-
Magnesium - Mg^{+2}	Carbonate - CO_3^{-2}
	Hydroxyl - OH^-
Sodium - Na^+	Sulfate - SO_4^{-2}
Potassium - K^+	Chloride - Cl^-
	Fluoride - F^-
	Nitrate - NO_3^-

Some of the other ions that can be of significance would include the following:

Cations	*Anions*
Iron - Fe^{++} or Fe^{+3}	Silica - SiO_2
Manganese - Mn^{+2} or other	Sulfide - S^{-2}
Aluminum - Al^{+3}	Phosphate - PO_4^{-3}
Barium - Ba^{+2}	
Strontium - Sr^{+2}	
Copper - Cu^{+2}	
Zinc - Zn^{+2}	

An analysis of potential RO feedwater should, as a minimum, include testing for the first set of ions as well as the pH. It is recommended that the analysis also include the second set of ions (if this is within the capabilities of the laboratory), and turbidity.

Concentrations are typically given as weight of contaminant per volume of water, usually as milligrams per liter (mg/L). Since the density of water is 1 gram per milliliter (g/L), in most aqueous solutions there are 1,000 grams of water in a liter. One milligram in a thousand grams of water is equal to 1 gram in one million grams of water. Therefore, this weight ratio is called a part per million; and, a concentration in mg/L is equal to that expressed in ppm if it is for a typical contaminant in a dilute aqueous solution.

Frequently, laboratory results are reported in milligrams per liter (mg/L) as calcium carbonate (as $CaCO_3$). This is a way of expressing the contaminants as a relative concentration of charged molecules. It is based upon the molecular weight and the valence charge of the ion versus that of calcium carbonate (100 molecular weight and an ionic charge of 2). A ratio is derived that is used to convert the concentration of the contaminant ion from mg/L as a weight-based concentration into a value better expressing the relative number of charged ions in the solution. If an ion has half the molecular weight of calcium carbonate (resulting in a conversion factor component of 2), and is trivalent (i.e., has three valence charges) versus the two present in calcium or carbonate ions, its concentration as $CaCO_3$ would be three times its weight-based concentration as ion ($2 \times 3 \div 2 = 3$). Expressing a concentration in its calcium carbonate equivalent concentration is useful because it better expresses the charge potential of an ion for reacting with other ions. (See Appendix for a table of calcium carbonate conversion factors.)

Molarity is another way of expressing concentration based on a quantity of molecules, but it does not take charge potential into account. In this case, the particular basis for the quantity is called a *mole*. When a substance is dissolved in a solvent, its molar quantity is the number of moles of molecules of that substance that are dissolved in a liter of solvent. A molar quantity is related mathematically to a weight-based concentration: it is the weight concentration of the substance divided by its molecular weight.

In a stable solution, the total of cationically charged ions has to equal the number of anionically charged ions when the ions involved are expressed as $CaCO_3$. Depending on the accuracy of the water analysis, the sum of the cations expressed as calcium carbonate will approximately equal the sum of the anions expressed as calcium carbonate. An accurate analysis will give total anion concentrations that are within 10% of the total cation concentrations.

Many laboratories do not have the capability of directly measuring the concentration of sodium in the water. They will derive the sodium concentration by subtracting the other cation concentrations from the sum of the anions (all expressed as calcium carbonate). Some laboratories will derive their alkalinity concentrations in this manner as a means of ensuring charge balance. Both of these practices are somewhat of a compromise on accuracy.

An easy way to check the accuracy of an analysis is to add the cations and the anions (expressed as calcium carbonate). If their totals are exactly equal, it is likely that the laboratory is using mathematics to derive the concentration of either sodium or alkalinity. If their totals differ by more than 10%, the accuracy of their analytical methods is less than ideal.

pH. A measurement of pH is important to water analysis, as the pH will dramatically affect the solubility of a number of slightly soluble salts.

The pH of water is a means of expressing its acidity or basicity. In water

(H_2O), a certain number of molecules will naturally dissociate into their cationic component, hydrogen ion (H^+), and their anionic component, hydroxyl (OH^-). At a temperature of 77 °F (25 °C), the mathematical product of the concentrations of these ions is roughly a constant with a value of 0.00000000000001. Another way to express this constant is 10 to the power of -14, or 10^{-14}. In equation form, this dissociation constant of water is expressed as

$$[H^+] \times [OH^-] = 10^{-14}$$

where brackets are used to designate the concentration of the particular ion in molarity.

If an acid or base is added to water, the specific hydrogen and hydroxyl ion concentrations will change by the amount of whatever is added. Some amount of hydrogen and hydroxyl ions will combine to form water as required in order to maintain the same mathematical product of hydrogen and hydroxyl concentrations. For example, if an acid (a chemical that dissociates to release hydrogen ions) is added to water, hydroxyl ions will combine with some of the hydrogen ions to form water, thus reducing the hydroxyl concentration as a means of maintaining the dissociation constant of 10^{-14}.

The pH is defined as the negative logarithm (base ten) of the hydrogen ion concentration expressed in molarity (moles/liter). A base ten logarithm is the value comparable to the number of times ten would have to be multiplied by itself to be equal to the other value in the equation. In mathematical terms, pH can be defined by either of the following two expressions:

$$[H^+] = 10^{-pH}$$
$$pH = -\log[H^+]$$

Example:
What is the pH of water if its hydrogen ion concentration is 0.01 mg/L?

Answer:
From the Appendix, the molecular weight of hydrogen can be found to be 1 g/mole, which is equal to 1,000 mg/mole. The molar concentration can be calculated as follows:

$$(0.01 \text{ mg/L}) \div (1{,}000 \text{ mg/mole}) = 0.00001 \text{ moles/L}$$
$$0.00001 \text{ moles/L} = 10^{-5} \text{ moles/L}$$
$$pH = -\log(10^{-5} \text{ moles/L});\ pH = 5$$

Example:
What is the pH of a solution containing 100 mg/L of sulfuric acid (H_2SO_4)?

Answer:
The molecular weight of sulfuric acid is determined by adding the molecular weights of its components, multiplied by their quantity of atoms in the molecule:

2 each H	1×2	= 2 g/mole
1 each S	32×1	= 32 g/mole
4 each O	16×4	= <u>64 g/mole</u>
		98 g/mole

$$100 \text{ mg/L} \div 98{,}000 \text{ mg/mole} = 0.001 \text{ moles/L } H_2SO_4$$

Sulfuric acid is a strong acid, meaning that it tends to dissociate completely in water. This dissociation releases two hydrogen ions; therefore, the molar concentration (moles/liter) of free hydrogen ions is equal to the sulfuric acid molar concentration times two.

$$pH = -\log[0.001 \text{ moles/L} \times 2] = 2.7$$

In this example, the original hydrogen ion concentration in the water of 10^{-7} (0.0000001) moles/L is overlooked. It is negligible when compared to the hydrogen ion concentration contributed by the sulfuric acid (0.002 moles/L).

Hardness. Calcium and magnesium make up the vast majority of what is called *water hardness*, and may be expressed as such in a water analysis. Technically, other multivalent cations are also included in what is considered hardness, although their concentrations are normally negligible when compared to those of calcium and magnesium.

Hardness and hard water get their names because of the way multivalent cations such as calcium and magnesium will attract and bind up natural skin oils. This tends to leave the skin feeling "hard" (Continental Water Systems Corporation).

Hardness in water is a concern for residential users. The hardness can fall out of solution, leaving behind scale formation on faucets and other outlets. The positive charge characteristics of hardness can cause it to bind up with phosphate ions (which are strongly anionic). Since most detergents rely on either phosphates or some other anionic surfactant, hardness will reduce the availability of the detergents for cleaning soils. Visible evidence of this is the sparseness of bubbles when detergent is mixed with hard water.

Hardness is a concern for RO systems because of its insolubility in the presence of high bicarbonate, carbonate, sulfate, fluoride, or silicate concentrations. Since hardness is present in most natural water sources, some means of controlling the precipitation of calcium or magnesium salts may be necessary to obtain satisfactory performance from an RO system.

Sodium salts are very soluble in water. Water softeners use resins that can exchange sodium for hardness in the water. When the resins are near depletion of available sodium, they are regenerated with sodium chloride, thus replacing the previously removed hardness cations with sodium.

Sodium may be a concern in certain high-purity water applications. Because sodium is monovalent (containing a single positive valence charge), it is not as well rejected by an RO membrane as many other ions. It is also not as well held by downstream ion-exchange equipment, and therefore tends to be a trace contaminant in high-purity water systems. In semiconductor manufacturing, trace sodium ions in process water can cause failures in integrated circuits.

Alkalinity consists of anions that remove hydrogen ions from solution. These anions include bicarbonate, carbonate, and hydroxyl. They remove hydrogen ions, thus shifting the ratio between hydrogen and hydroxyl ions. This gives water a higher pH (i.e., a more basic pH).

Most, if not all, of the alkalinity in naturally occurring water sources is in the form of bicarbonate alkalinity (HCO_3^-). Bicarbonate alkalinity is also known by the chemical used in titrating (measuring) its concentration, methyl orange (m.o.) alkalinity. Below a pH of about 8.3, the bicarbonate alkalinity will be in equilibrium with a certain concentration of dissolved carbon dioxide.

When water sources have pH values greater than 8.3, more of the alkalinity will be in the carbonate form (CO_3^{2-}). This is titrated using phenolphthalein, and is also known as phenolphthalein (p.) alkalinity. With water sources of pH greater than 11.3, hydroxyl alkalinity (OH^-) will be present.

Alkalinity acts as a natural pH buffer in most water sources. Natural conversions occur between the various forms of alkalinity and dissolved carbon dioxide to resist any potential change in pH that might be caused by a contaminant.

Water can dissolve carbon dioxide from the atmosphere, forming carbonic acid (H_2CO_3). Acidic water will tend to dissolve calcium carbonate from the ground as it passes over it or through it. This dissolution will occur until the pH of the water has increased, and is no longer aggressive toward calcium carbonate.

If the water supersaturates in bicarbonate and the water is still acidic, the excess bicarbonate will convert into carbon dioxide. If the water is supersaturated in carbon dioxide, the excess carbon dioxide will simply degas into the atmosphere.

If the water supersaturates in bicarbonate and the water is basic, however, the excess bicarbonate anion will convert into carbonate anion. If for some reason the water supersaturates in carbonate ion at a very basic pH, the excess carbonate will fall out of solution, with any residual alkalinity being present in the straight hydroxyl form.

Because of the ability of water to pick up alkalinity, most naturally occurring water sources are close to saturation in calcium carbonate. If the water is concentrated in an RO system, calcium carbonate salt is likely to precipitate. This tends to occur quickly, and prior to the formation of any other salts. Some means of preventing or controlling calcium carbonate precipitation is required in most RO systems.

Sulfates have limited solubility in water, depending on the concentrations of divalent cations also present. Such cations include calcium and magnesium, as well as barium and strontium.

Sulfates are present in relatively large concentrations in most raw waters. They are also added to water in the form of sulfuric acid when necessary for pH control. The prevention of sulfate scale formation in an RO system is usually performed by reducing or controlling the divalent cations in the raw water.

Chlorides pose little threat to an RO system from the standpoint of scale formation. Nearly all chloride salts are quite soluble in water.

High concentrations of chloride in an RO system can attack 304 grade stainless steel. If the chloride in the RO concentrate exceeds several thousand ppm, 316L grade stainless steel may be required for the high-pressure piping (or housings if those are stainless steel) to prevent corrosion.

Fluoride concentrations are usually low in most water sources. Calcium fluoride is fairly insoluble if fluoride happens to be present in an appreciable concentration.

Nitrates: The solubility of nitrates is not normally a concern. The presence of nitrates in a water is more a concern with respect to the ability of the water treatment system to remove them.

Nitrates in potable water, when ingested by mammals (including humans) can produce nitrites. These nitrites interfere with the body's ability to exchange oxygen in blood. This can cause serious problems for a fetus, for small children, or for other mammals. For this reason, it is desirable to maintain a nitrate concentration below 40 mg/L in drinking water. (L. Comb, February 1987).

Because of their poor charge characteristics, nitrates are not well removed by RO, and are lightly held by anion exchange resins. Depending on the membrane and the system recovery, nitrate removal by RO is in the range of 50% to 90%.

Iron and manganese are found in water either in a reduced state, which tends to be soluble, or in an oxidized state, which tends to be insoluble. Iron is usually the most prevalent of the two metals in naturally occurring water sources. Its divalent ferrous state (Fe^{+2}) is soluble. If it is oxidized to its ferric state (Fe^{+3}), it is insoluble and can foul an RO system.

The iron concentration can be substantial in a well water, particularly if its pump is starting to rust. Pretreatment equipment can be a source of iron if black iron or carbon steel has been used for piping, tanks, or tank internals. This problem will be compounded if acid is injected upstream of the equipment.

If the iron or manganese concentration is greater than 0.05 mg/L in an RO feedwater, a means for its removal should be considered. If the iron or manganese is in the reduced (soluble) state, it may not cause any fouling problem with the RO system. If any air gets into the system or if any oxidizing agent is introduced, however, the iron or manganese will oxidize into an insoluble state. It will then tend to foul the membrane elements. It can also catalyze the oxidative effects of residual oxidizing agents, possibly enabling the agent to attack and degrade the RO membrane. (See "Membrane Compatibility" in Chapter 2.)

Barium and strontium concentrations are not always included in a raw water analysis. However, if they are present in relatively small concentrations (0.01 mg/L or more) along with sulfates, they can easily fall out of solution as scale on the membrane surface. Barium and strontium sulfate scales are extremely difficult to redissolve; therefore it is critical that the precipitation of these salts be prevented.

Aluminum. Dissolved aluminum is usually not appreciably present in naturally occurring water sources. With its valence of +3, aluminum has high charge characteristics. It is also relatively small in size (molecular weight of 27), which results in aluminum metal ions having a particularly high charge density. This makes aluminum very reactive.

Aluminum is similar to iron in many of the ways it tends to react. Like iron, aluminum will combine with oxygen to form an insoluble oxide. Both metals tend to complex with negatively charged organic colloids, and they both will readily precipitate when silicates are present.

Because of aluminum's high charge characteristics, many municipalities use alum [aluminum sulfate, $Al_2(SO_4)_3$] or sodium aluminate [($NaAlO_2$)] addition in their treatment of surface waters. The aluminum acts as a coagulant to dissipate the negative charge characteristics of organic suspended solids in the water. This allows the suspended solids to group together and subsequently fall out of suspension.

Aluminum is more soluble when the water pH is extremely low, or when it is high; and it is less soluble over the normal pH operating range of an RO, 5.3 to 8.0. Because of this, aluminum fouling should be a concern if the municipal treatment plant ever uses alum or sodium aluminate addition and the RO uses acid injection. If this is the case, the location of the acid injection should be upstream of any media filters.

Trace aluminum is of particular concern in the dialysis industry. A concentration greater than 0.01 mg/L in the final dialysis water is a health concern for dialysis patients (AAMI Guidelines).

Copper and zinc are usually not present to any appreciable extent in natural water sources. However, it is possible to pick up trace concentrations from piping materials.

Copper and zinc will tend to fall out of solution with increasing pH, as might occur as salts are concentrated in an RO system. Because copper and zinc concentrations are always low in feedwaters of normal pH, their precipitants will affect the performance of an RO system only if allowed to precipitate over an extended period of time.

The primary reason for analyzing for copper and zinc in a potential RO feedwater would be if polyamide membrane is being considered. As transition metals, copper and zinc can increase the oxidation potential of oxidizing agents. This can be a concern with polyamide membrane systems, particularly when peracetic acid or hydrogen peroxide is used to sanitize the system.

Silica is a particularly interesting and dangerous ion for RO systems because of its stability once it falls out of solution. Below a pH of 9, silica is present mostly in the silicic acid form (H_4SiO_4, also written as $Si(OH)_4$). At low pH, silicic acid can polymerize to form a colloid (known as *colloidal silica*). As the water pH exceeds 8, silicic acid dissociates into the silicate anion (SiO_3^{-2}). At high pH, it can precipitate as a salt with calcium, magnesium, iron, or aluminum.

Silica and silicates can be difficult to redissolve. Ammonium bifluoride solutions are somewhat successful at cleaning silica if the scale formation is not too severe; however, ammonium bifluoride is considered a hazardous chemical that may pose problems for disposal.

The solubility of silica is dependent on temperature and pH. If silica is present in an RO feedwater at a concentration greater than 20 mg/L, the potential for silica precipitation should be evaluated.

Sulfides are frequently found in Florida in shallow well waters. They are present as a dissolved gas, hydrogen sulfide (H_2S). If oxidized by oxygen from the atmosphere, or by chlorine injected into the water for biological control, the sulfides will fall out of solution as elemental sulfur. Reverse osmosis systems have been known to operate well with sulfides present in the feedwater, only to have a sudden sulfur-fouling problem because of an air leak in the system. (W. H. McPherson, May 1981)

Hydrogen sulfide gas can be removed by running the water through a degasifier, which will give off the odor of rotten eggs. The usage of degasifiers for the removal of hydrogen sulfide may therefore be prohibited in some areas.

Another means of removing sulfides is to intentionally oxidize them by chlorine injection or some other means. Media filtration is then used to remove the precipitated sulfur.

Phosphates, with a valence of -3, have a strongly negative charge and a tendency to react with multivalent cations. Calcium phosphate has a very limited solubility at neutral pH, and an even lower solubility at higher pH.

If phosphates are present in an RO feedwater to any appreciable extent, unless the water is acidified, the phosphate will likely fall out of solution within the RO system. This is caused by the increasing concentration of phosphate as pure water permeates the membrane, and the tendency for the pH to increase within the RO system as the water becomes more concentrated.

Membrane Choice

For standard water purification applications, the two most common families of RO membranes are made using polymers of either cellulose acetate (CA) or polyamide (PA). A third membrane type, which has been introduced within the last few years, uses charged polysulfone (PS).

Cellulose Acetate. The CA membrane was the first to be used widely in a spiral-wound configuration. It offers stable performance (under its specified operating conditions), and is resistant to relatively high free-chlorine concentrations. It was the membrane of choice in RO applications until the polyamide thin-film membrane became available.

A CA membrane is considered an uncharged membrane, meaning that the functional groups on the ends of the polymer chains are not extremely polar. This gives it one particular advantage over charged membranes such as polyamide or charged polysulfone membranes, in that it is less likely to attract foulants to the membrane surface.

On a CA membrane, the surface exposed to the feedwater is an extremely thin, smooth layer known as a skin. From this skin, the membrane branches into a porous support structure. The smoothness of this skin layer also seems to help the resistance of CA membrane to fouling (Filteau, 1993). (See Figure 1-11.)

An important requirement for CA membrane is pH control. The CA membrane will hydrolyze with time. This is the natural break-off of the acetyl functional groups from the polymer. As this occurs, the membrane will lose its salt rejection characteristics. The rate of this breakdown is a function of pH and temperature. As shown in Figure 1-12, there is a dramatic difference in the rate of hydrolysis, depending on the operating conditions.

To reduce the rate of CA membrane hydrolysis, most CA membrane systems will use acid injection to reduce the feedwater pH. Although the optimum pH is around 4.8, maintaining a pH of 6.5 or less will usually minimize the CA hydrolysis rate sufficiently to obtain a minimum of 3 years of membrane life.

The original CA was specifically a cellulose diacetate. Later the performance characteristics of CA membranes were improved by blending in a cellulose triacetate. These CA-blend membranes offer greater permeate flowrates with better resistance to bacteria attack and hydrolysis. Nearly all CA membranes now being used in RO elements on the market today are made of these CA-blend polymers.

Cellulose acetate membranes are tolerant of normal concentrations of biocidal oxidizing agents such as free chlorine or chloramine. This makes the CA membrane a good choice for applications with a high potential for biofouling (fouling which results from the growth of biological material). In these applications, chlorine can be continuously injected in concentrations up to 1.0 mg/L into the RO feed stream to reduce biological growth without compromising membrane life expectancy.

A wide range of performance characteristics are available in CA membranes for reverse osmosis, depending on how the membrane was manufactured. The CA membranes made for higher-rejection applications will have a lower permeate flowrate than membranes made for lower-rejection applications.

The two most common CA membrane types have rejection rates for sodium chloride of typically about 95% for a membrane with standard rejection, and between 97% and 98% rejection for a high-rejection membrane. The higher-rejection membrane can be manufactured by changing subtleties in the extrusion and annealing of the CA membrane. Some of the CA membrane manufacturers will take standard rejecting membrane and improve its rejection characteristics

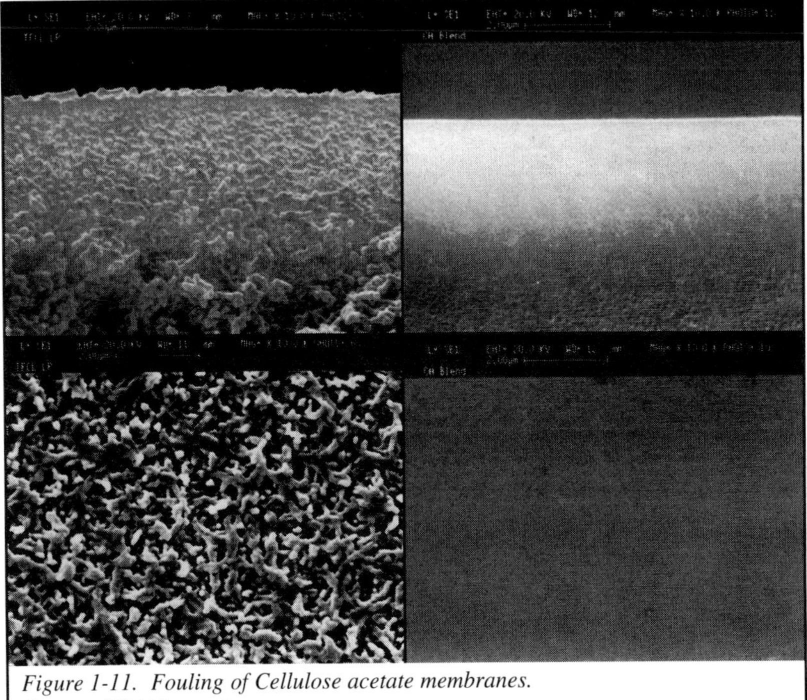

Figure 1-11. Fouling of Cellulose acetate membranes.
(Courtesy of Fluid Systems.)

by applying a surface treatment. This process is similar to applying a membrane rejuvenation chemical to a membrane surface. (See "Membrane Rejuvenation" in Chapter 5.) The surface treatment acts as a dynamic membrane that reduces the permeate flow as it improves the rejection. Like membrane rejuvenation, this process is somewhat unpredictable as to how long the surface treatment will stay attached, and a membrane so treated may not offer the stable performance of an untreated high-rejection membrane. (Comb, 1993)

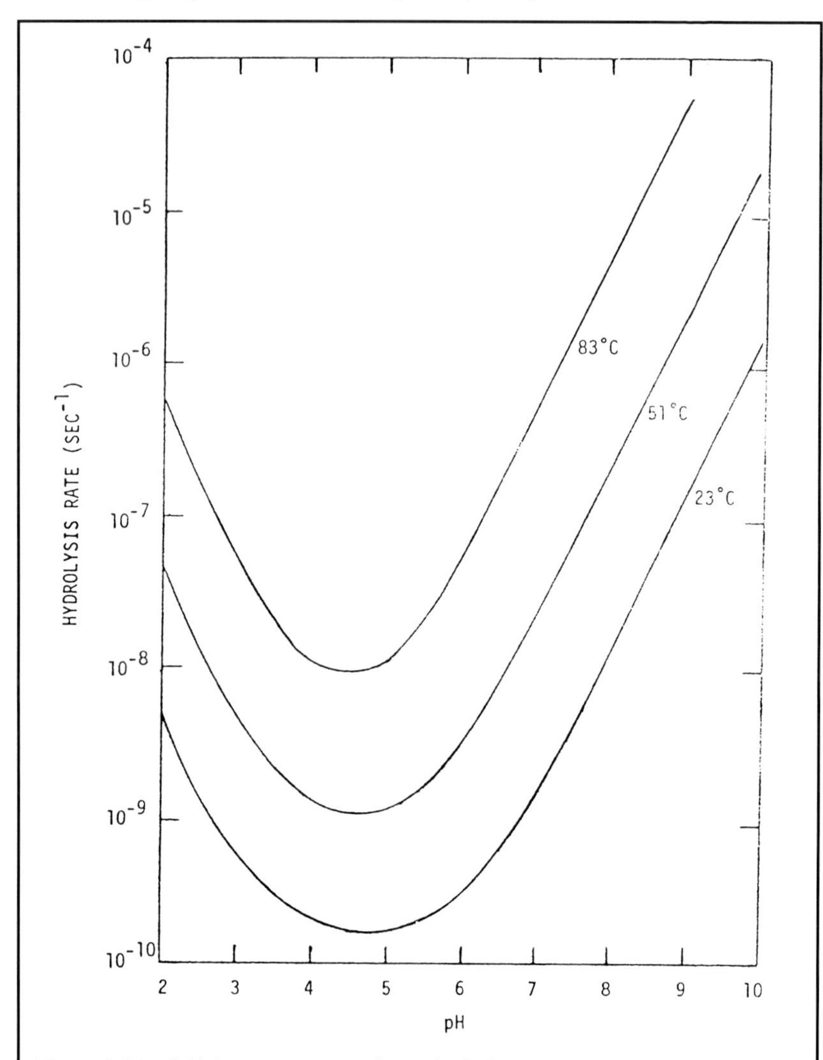

Figure 1-12. Cellulose acetate membrane hydrolysis rate constant versus pH. (Courtesy of Fluid Systems.)

In large RO systems, it is common to operate CA membrane RO systems at a feed pressure between 200 and 400 psig. When operating at 200 psig, the permeate flow will be about half of what it is at 400 psig. The salt passage into the permeate water will be almost double what it would be at 400 psig, which is about where the CA membrane element performance characteristics are tested.

When operated beyond 500 psig, CA membrane will tend to compact at an increased rate. Thus, CA membrane is not appropriate for applications at extremely high pressure. Seawater desalination is one example of an inappropriate application for CA membrane, because of its requirement for operating pressures in the 800- to 1,000-psig range.

Polyamide/thin film. Polyamide membranes are growing in popularity because of their high permeate flux (i.e., permeate flow per membrane area) characteristics. A PA membrane was first used in a hollow-fiber configuration (described in a later section). Its chemical structure is constructed of a linear aromatic polyamide that can be extruded into the fine hollow fibers (Figure 1-13).

The PA membranes used in a spiral-wound element configurations (to be discussed later) are constructed of a thin layer of an aromatic polyamide extruded onto a less dense polysulfone substrate. This particular type of spiral-wound membrane is called a thin-film. The common chemical structure of these membranes is cross-linked, which gives it greater tolerance to attack by oxidizing agents (Figure 1-14).

This thin-film PA membrane typically offers a flux rate significantly greater than that of a CA membrane. This enables PA membrane systems to be operated at half the pressure of a CA/RO system, and still obtain comparable permeate flowrates with slightly better salt rejection. This means that an RO system using a thin-film membrane will operate using substantially less energy than a CA/RO system operating at twice the pressure (400 versus 200 psig).

Capital equipment costs of thin-film RO systems are usually less than those of CA membrane systems for two reasons. A 200-psig system will need a pump with only half the pressure of that needed by a 400-psig system. Also, less expensive housings are available for operation at 200 psig.

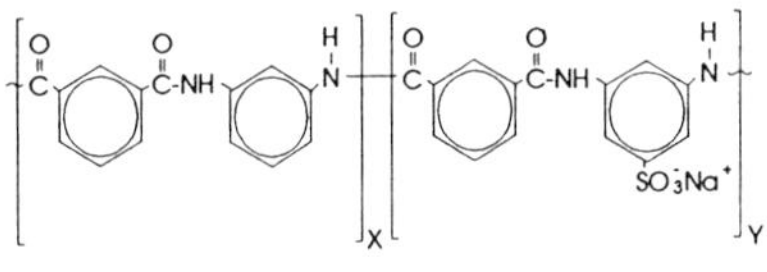

Figure 1-13. Basic structure of the DuPont B-9 and B-15 membranes.
(Courtesy of DuPont Corp.)

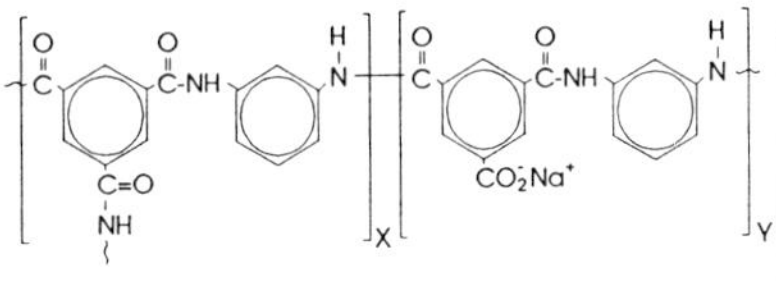

Figure 1-14. Basic structure of the FilmTec FT-30 membrane.
(Peterson, 1986, Courtesy of FilmTec Corp.)

As discussed, a CA membrane system can be designed for operation at 200 psig. However, it would require twice the number of membrane elements and housings needed by a 400-psig system, as well as additional interconnecting plumbing and a larger frame; and it would pass about twice the concentration of salts into the permeate water.

When the high-flux thin-film PA membranes were first introduced to the industrial market, they were operated at 400 psig, as were most CA systems at that time. Even though this allowed the thin-film systems to achieve twice the permeate flow of the CA systems with less than half the salt passage, the exceptionally high flux through the membrane resulted in a significant increase in the membrane fouling rate.

The membrane manufacturers chose instead to market the new membranes by promoting the energy savings that result from operation at lower pressures. The alternative of driving additional flux through the membrane by operating at 400 psig would have resulted in lower capital equipment costs (i.e., because less membrane and housings would be needed to achieve the same flowrate), but also would have meant higher maintenance costs. The increased rate of fouling would have necessitated more frequent cleaning and membrane replacement.

The PA membrane is very sensitive to oxidizing agents such as free chlorine or iodine. This requires that any chlorine present in the feedwater must be removed by injecting a reducing agent, such as sodium bisulfite, or by filtering the feedwater through an activated carbon filter. (Note: The membrane manufacturer should be consulted as to specific recommendations.)

Once the biocidal oxidizing agents are removed from the feedwater, the chances increase for biological activity and subsequent fouling of the RO. For many surface waters and some well waters, this biological growth can make PA systems impractical, particularly for systems that run intermittently.

The PA thin-film membranes most commonly used in the water purification industries have a negative charge characteristic. On some water sources, this charge has tended to attract certain foulants that can increase the fouling rate of the PA membrane. Only chemicals that are compatible with this charge should be allowed to come into contact with the membrane. For example, only anionic surfactants should be used to clean an anionic PA membrane system.

Seawater PA membrane. Special PA membranes are available for seawater desalination and other applications requiring extremely high pressures. These membrane elements use a thicker polyamide membrane layer. The permeate flux rate is lower with the thicker membrane, but the rejection characteristics are better.

Seawater PA membranes offer advantages desired in some non-seawater applications. The higher rejection rate may be needed to meet permeate quality requirements. Also, the lower permeate flux rate can be of assistance in reducing the potential for membrane fouling.

In applications requiring that the RO concentrate stream (brine stream) be highly concentrated, osmotic pressure will vary dramatically across the system as the feedwater salts become more concentrated. To overcome the high osmotic pressure in the latter stages of the system requires a high membrane pressure. However, the lower osmotic pressure in the lead end stage might result in overproduction of permeate water if standard brackish-water PA membrane elements were used. Seawater membrane elements can be used to keep the flux rates more balanced across the RO system, which should reduce the rate of membrane fouling in the lead end stage.

Charged Polysulfone. The PS membrane elements offer better chlorine tolerance than CA membranes, and chemical tolerance equal to or better than PA membranes. The permeate flux rate is comparable to PA membranes for economical energy usage. Rejection capabilities are comparable to standard-rejection CA membranes. (See Table 1-4 for a comparison of common RO membrane types.)

The charged PS membrane can handle a feedwater concentration of free chlorine of up to 5 ppm continuously. It can handle cleaning solutions with a pH between 2 and 12. (Consult with the membrane manufacturer.) This membrane is well suited for applications where the control of biological fouling is critical. It can also be applicable for waters containing certain hard-to-clean foulants or scale.

Normal PS membranes are used in ultrafiltration membranes because they contain physical pores that can pass salts. The PS membrane has to be chemically

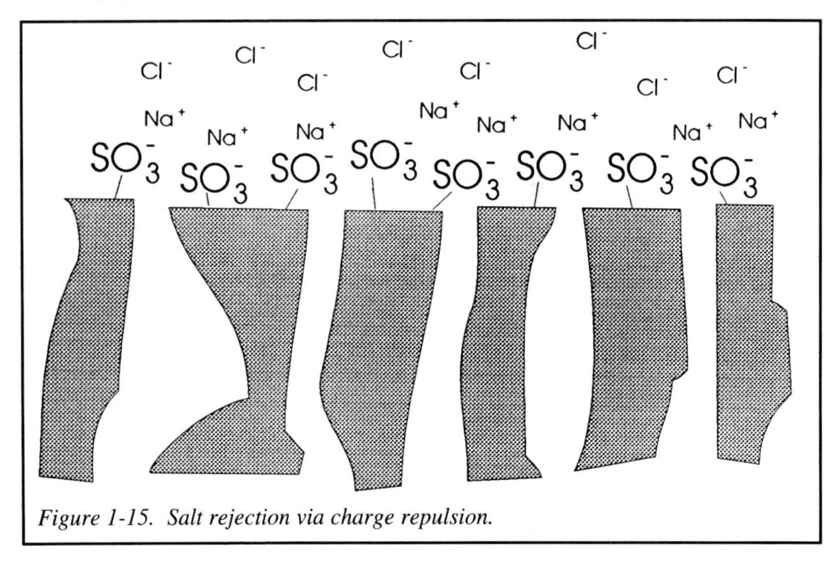

Figure 1-15. Salt rejection via charge repulsion.

sulfonated to get its ability to reject salts. The sulfonation process places sulfonate groups on the membrane surface. These negatively charged sites repel anions (which have a negative valence charge). The cations in the feed stream are indirectly repelled by the membrane due to their attraction to the anions in solution. Water solutions must maintain a balance between cationic and anionic charges. (See Figure 1-15.)

The sulfonation process is also used to give anion exchange resins their charge characteristics. The sulfonate functional groups that are attached to the basic resin bead have a strong negative charge. This allows the beads to attract and exchange cations (which have a positive valence charge).

Unfortunately, this ability of sulfonate groups to attract cations can result in the loss of performance of charged PS membranes. Should these membranes be exposed to any divalent or trivalent cations, such as calcium, magnesium, or iron in the feedwater, the membrane will hold onto the ions like an ion-exchange resin. This removes the charge characteristic of the membrane at that location, thus allowing salts to permeate the RO membrane.

Polysulfone RO systems require water-softening systems upstream to remove any calcium, magnesium, iron, or other divalent and trivalent cations in the feed stream. The breakthrough of any of these cations will foul the RO membrane. Should the membrane become fouled, it has to be regenerated like a water softener by running a concentrated salt solution (sodium chloride solution) through the system.

It is probably a good idea to use two water-softening systems upstream of an RO with the charged PS membrane. The second water softener can polish up any divalent cations that leak through the first softener.

Table 1-4
Performance Comparison of Common RO Membrane Types

	Thin-film	*Std CA*	*High-Rej CA*	*PS*
NaCl rejection (%)	98	95*	97.5*	95
Permeate flow @ 200 psig (in gpd)	1,800	1,000*	800*	2,000
Chlorine tolerant	No	Yes	Yes	Yes
Relative cost	Medium	Low	Low	High

*Note: Due to the lower flux rates of CA/RO membranes, they are more commonly operated around 400 psig, which results in a doubling of the permeate flowrate. Rejection rates are given for operation at 400 psig. Rejection of NaCl at 200-psig operation would be 90% and 95% for standard and high-rejecting CA membranes, respectively.

(Data courtesy of Dow Chemical/Filmtec, Osmonics, Inc., and Millipore)

Element Configurations

Reverse osmosis membrane is used in four configurations: Plate and frame, Tubular, Hollow fiber, and Spiral wound. These configurations are described below.

Plate-and-frame membrane systems are constructed using what is called flat sheet membrane, which is manufactured by extruding the membrane polymer onto a flat sheet of permeable backing material, usually a polyester fabric. Pieces of these sheets are clamped onto a plate backing device (held inside the unit frame) that allows the feedwater to flow across the membrane and allows the permeate water to be collected. Layers of these plates and frames are stacked such that the concentrated solution from one plate is fed to the next plate. The number of plates plumbed together in series will depend on the desired recovery ratio of the permeate flowrate to the feed flowrate (Figure 1-16).

Plate-and-frame systems are not commonly used for RO applications because they require a large amount of high-pressure hardware to construct a working system. This tends to be very expensive.

The advantage of the plate-and-frame unit is that the membrane is readily accessible by taking apart the frame. The unit is therefore desirable for membrane testing and pilot applications, since the membrane can be routinely inspected. It is also useful in applications with heavy fouling because the frame can be removed to gain the access needed to physically clean the membrane.

For most larger applications, however, plate-and-frame RO is not economically practical. The amount of hardware required to build a 350-square-foot (ft^2) system would be enormous compared to what is required to house a single 8-inch (diameter) by 40-inch (length) standard spiral-wound element containing about 350 ft^2 of membrane.

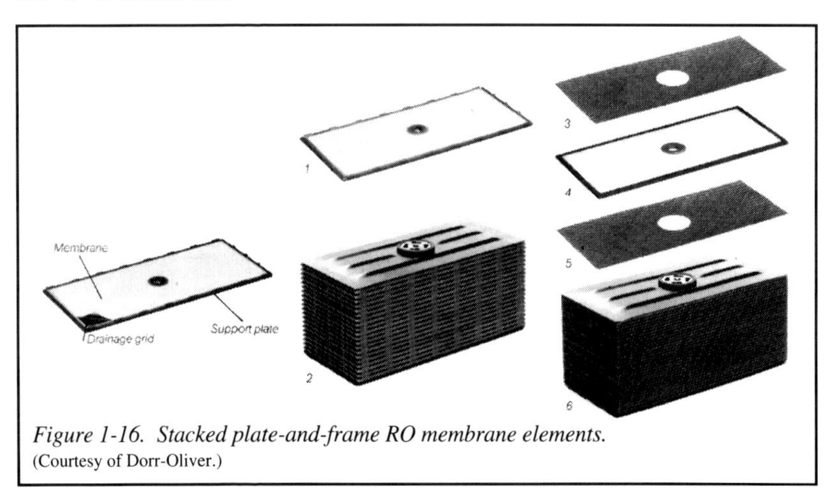

Figure 1-16. Stacked plate-and-frame RO membrane elements.
(Courtesy of Dorr-Oliver.)

Tubular. An infrequently used method of RO membrane configuration is the tubular membrane element. Here the membrane is cast onto the inside surface of a porous support tube that has a diameter between 0.125 and 1.0 inch. The tube, of fiberglass, ceramic, carbon, plastic, or stainless steel, acts as support against the RO operating pressure that will be present within it. Tubes are plumbed in series by connecting their ends with elbows, allowing them to achieve the desired permeate water recovery. The tubes are contained inside a low-pressure jacket that collects the permeate (Brandt et al., 1993). (See Figure 1-17.)

High flow velocities can be sent through the tubes during operation to reduce fouling by suspended solids. High velocities can also be used during cleaning. Because of this, tubular RO is sometimes used for heavily fouling applications like waste streams and food applications.

Tubular systems have minimal membrane area compared to the flow volume through the tubes, so that numerous tubes have to be connected in series to achieve a significant permeate recovery. Tubular RO tends therefore to be expensive to purchase and to operate, and it is not commonly used other than for heavily fouling applications.

Hollow Fiber. The first large-scale RO systems were built using hollow-fiber membrane elements. For these elements, fine hollow fibers of membrane material, either a cellulose triacetate or a polyamide, are extruded. The ends of

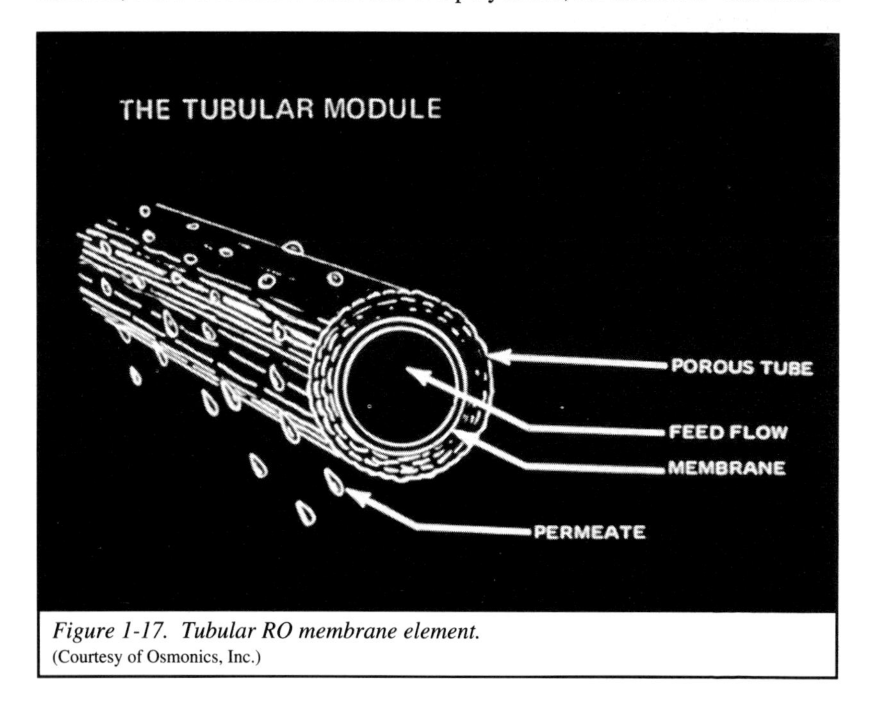

Figure 1-17. Tubular RO membrane element.
(Courtesy of Osmonics, Inc.)

the fibers are sealed in an epoxy block connected with the outside of the housing. The epoxy block is cut to allow the flow from the inside of the fine fibers to the other side of the epoxy block, where it is collected. The pressurized feedwater passes across the outside of the fibers. Pure water permeates the fibers and is collected at the end of the element (Figure 1-18).

The hollow-fiber housings are capable of holding a large quantity of fibers, thus allowing a single element to produce a large permeate flowrate. The capital cost of hollow-fiber systems tends, therefore, to be lower than plate-and-frame or tubular systems.

Hollow-fiber membrane systems are not as popular now as they were 10 years ago in brackish-water applications, due to growth in the popularity of spiral-wound membrane elements. However, hollow-fiber elements are still heavily used for seawater desalination, and for minimally fouling brackish-water applications such as well-water purification.

The hollow-fiber element design does not allow for turbulent or uniform flow across the fiber surface. In heavily fouling applications, the elements are more prone to fouling and scale formation than are other membrane configurations. A fouled or scaled hollow-fiber element can also be more difficult to clean because of the inability to get the cleaning solution into the fouled or scaled areas of the element.

Figure 1-18. Hollow-fiber RO membrane unit construction.
(Courtesy of DuPont.)

Manufacturers have improved on the layout of the hollow fibers within the element so that it is possible to create some turbulence and reduce the potential for fouling or scaling. There have also been improvements made in the permeate flux characteristics of the membrane fibers (Kates, 1993). This should help the hollow-fiber elements to compete with spiral-wound thin-film elements for the brackish-water treatment market.

Spiral wound. The most common membrane configuration used in brackish-water treatment is the spiral-wound element. It is relatively inexpensive to manufacture. It has fairly uniform flow velocities through the element and includes mechanisms to promote turbulence across the membrane surface, thus reducing the potential for membrane fouling or scale formation and facilitating the cleaning of fouled or scaled membrane elements.

Unlike hollow-fiber elements, spiral-wound elements are provided without a pressure housing. This housing needs to be provided separately for spiral-wound systems.

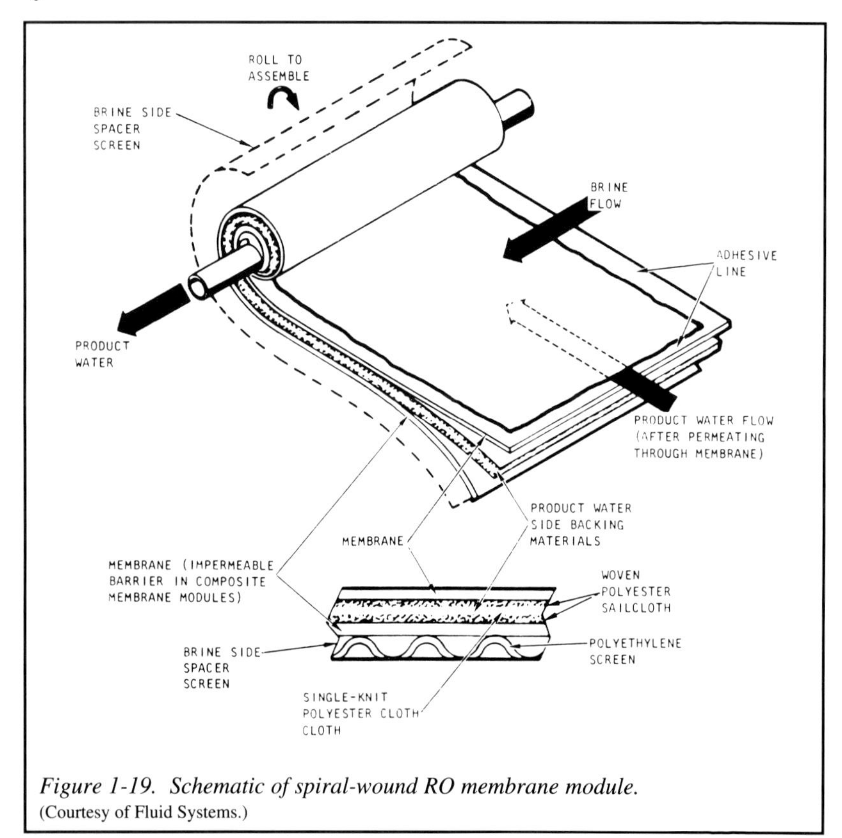

Figure 1-19. Schematic of spiral-wound RO membrane module.
(Courtesy of Fluid Systems.)

In the construction of a spiral-wound element (Figure 1-19), flat sheet membrane is folded and glued on three edges to create a membrane envelope (with the membrane on the outside). The open edge of the envelope is glued to a plastic (or sometimes metal) pipe with perforations to allow water from inside the envelope to pass into the pipe. Layered inside this envelope is a thin layer of tricot fabric that prevents the envelope from sealing itself off when the outside of the envelope is exposed to high pressure. This fabric still allows the passage of permeate water to the central collection tube.

Depending on the diameter of the desired element, a certain number of envelopes are wrapped around and glued to the central collection pipe (Figure 1-20). Between these envelopes, a plastic spacing material is placed to promote turbulence as the feedwater passes between the envelopes. The outside of the element is taped or fiberglassed to hold the element together. On the downstream end, the element's construction includes a cylindrical device that is held against the end of the membrane leaves. This prevents the downstream unraveling of the leaves, called *telescoping* (see Figure 1-21) as water pushes through the element.

Some type of gasket is placed around the outside of the element. This gasket seals the element to an external housing (provided independently of the membrane elements) and prevents water from bypassing the element. This gasket is sometimes called the *brine seal.*

Depending on the permeate flow needs of the RO system, the spiral-wound element used in industry may have a diameter as small as 2 inches (nominally),

Figure 1-20. Several membrane envelopes wrapped around the central collection pipe.
(Courtesy of Osmonics, Inc.)

Figure 1-21. The ends of the membrane leaves nearest the collection pipe have begun to telescope *on this unit.*
(Courtesy of Osmonics, Inc.)

or as large as 12 inches (nominally) (Figure 1-22). The 2-inch element would usually use just a single membrane envelope in its construction, whereas the 12-inch element would use a large number of envelopes. Besides the 2-inch element, other commonly used membrane element nominal diameters would include 2.5, 4, and 8 inches. Certain membrane manufacturers will also use 3 1/2, 6 and sometimes 8.5 inches. Although the typical industrial element length is either 40 or 60 inches, elements are also available that are about 12, 25, or 26 inches long.

Spiral-wound elements require an external housing, designed to withstand RO operating pressures (see Figure 1-23). A typical housing may contain up to six elements, each 40 inches in length. This size of housing could also handle four elements that are 60 inches in length. Using the longer elements offers a slight reduction in overall cost, and a slight increase in overall permeate flow.

The permeate collection tubes of adjacent elements are joined by what is called an *interconnector*, a small plastic tube with O-rings that seal against the collection tube. Each end of the housing will also have an adapter connecting the permeate collection tube of the end element with the permeate port on the end cap of the housing. (See Figure 1-24.)

During operation, the feedwater enters the end of the spiral-wound element and moves across the surface of the rolled-up membrane envelopes. The spacing material between the envelopes promotes turbulence so that as pure water

Figure 1-22. Spiral-wound elements are available in several diameters.
(Courtesy of Osmonics, Inc.)

permeates the envelopes, any salts left behind will diffuse back into the bulk solution. Inside the envelope, the pressure is typically near atmospheric, whereas the pressure on the feedwater side can be as high as 1,000 psig. This pressure differential drives the pure water to permeate into the membrane envelope. Here it passes through the tricot material and finds its way into the central collection pipe. The water in this collection pipe then travels to the end, where it either enters the collection tube of another element, or is transferred to the permeate port of the end cap of the housing.

Specialty spacing materials are sometimes used for applications with high suspended solids concentrations, or containing large particulates. Sometimes a corrugated spacing material is used that offers larger spacing between the membrane leaves and a free flowpath across the membrane surface. (See Figure 1-25.) Although an element with larger spacing is less prone to fouling, it has the disadvantage of a severely reduced amount of membrane surface area, and also requires much higher crossflow rates to achieve the same velocity at the membrane surface.

For some applications, it is desirable not to use an outer wrap on the membrane element. This is called the full-fit spiral-wound element (Figure 1-26), designed to expand to fill the inside area of its housing. In this manner, there are no stagnant areas around the membrane element, and there is crossflow through the

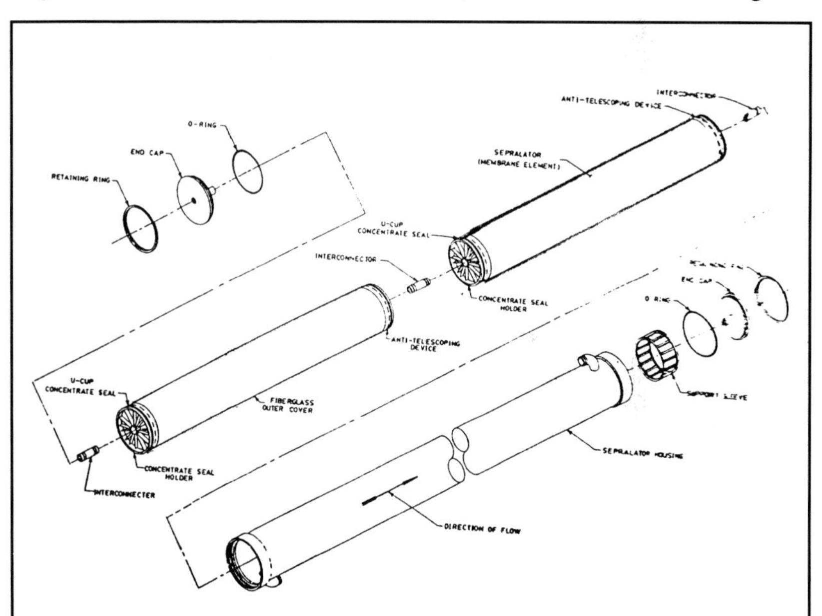

Figure 1-23. Schematic of spiral-wound membrane unit, including individual elements, anti-telescoping device, interconnectors, and internal housing.
(Courtesy of Osmonics, Inc.)

entire element. These elements are used in applications where biological control is a priority.

Full-fit elements contain more membrane than standard fiberglass or tape-wrapped elements; therefore, they will permeate more water, as much as 15% more in some cases.

The disadvantage of the full-fit design is that some relaxing of the membrane wrap will always occur within its housing. This means that the membrane leaves will not be as close together as with standard elements. The result is a lower pressure drop across the element due to the lower crossflow velocity. The lower velocity and turbulence through the element means that there is less shear placed on foulants at the membrane surface. This tends to subject the element to a greater fouling rate.

RO System Permeate Recovery

The RO system permeate recovery is defined as the ratio of permeate flow to feed flow, usually expressed as a percentage. It is an expression used to describe how efficiently the system is being operated, and is also used to determine the extent of concentration of the fluid solutes (in other words, the likelihood of scale formation).

$$\text{Permeate recovery} = \text{permeate flowrate} \div \text{feed flowrate}$$

The higher the system recovery, the less the concentrate and feedwater flowrates will be, relative to the particular permeate flowrate. Thus, higher RO

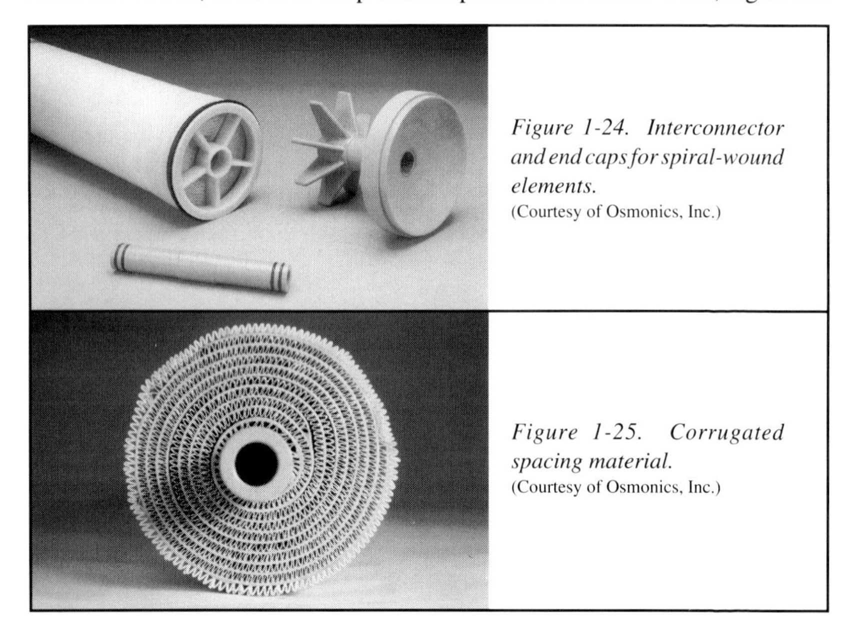

Figure 1-24. Interconnector and end caps for spiral-wound elements.
(Courtesy of Osmonics, Inc.)

Figure 1-25. Corrugated spacing material.
(Courtesy of Osmonics, Inc.)

permeate recoveries will result in using less pump electricity, pretreatment chemicals, or water; and in reduced sewer costs (for water purification applications). The size of pretreatment equipment required may be also reduced.

The system recovery also affects permeate water quality. Because of the higher concentration of salts present in the feedwater side of the downstream membrane stages of the RO, higher concentrations of salts will permeate the membrane. The higher the recovery, the more concentrated the salts will be in the tail end of the system, and subsequently in the permeate water of the tail end elements. For some applications, permeate quality can be critical enough to dictate a limit on the system recovery.

The recovery of larger RO systems is commonly chosen based upon the potential for scale formation. The higher the recovery, the greater the concentration of salts will be in the downstream membrane stages, thus the higher the potential for scale formation. If no means of changing the solubility of the salts is used (such as softening, acid injection, or injection of a scale inhibitor) there will exist an absolute limit to the recovery beyond which scale formation will almost definitely occur. Even with one or more methods of controlling salt solubilities, there will always be an upper limit to the recovery beyond which the risk of scale formation is too great.

Figure 1-26. Full-fit membrane elements.
(Courtesy of Osmonics, Inc.)

It is common for small under-the-sink RO units to operate at 30% recovery or less. With these small units, it is less expensive to waste water than it is to install a means of preventing scale formation. At very low recoveries, the units can operate safely without some means of controlling the solubility of dissolved salts.

Seawater systems will operate at approximately 30% recovery, limited by the high osmotic pressure of seawater. Small brackish-water RO systems may operate at 30% to 50% recovery simply because the cost of operation was not a priority in the system design. Achieving higher recoveries with smaller RO systems may require recycling some of the concentrate water back to the feed of the system. This will increase the concentration of salts in the system, subsequently reducing the permeate quality.

Brackish-water RO systems usually operate with recoveries from 50% to 80%. In areas where it is relatively easy to control scale formation with proper RO pretreatment, 75% has become somewhat of a standard system recovery for larger RO systems. This has occurred for the following two reasons:

- A 75% permeate recovery makes optimum use of the longest housings (the most economical), which contain six 40-inch-long membrane elements or four 60-inch elements. Using a minimum concentrate flowrate based upon a 6-to-1 concentrate-to-permeate ratio, the six-element housing would operate optimally at 50% recovery. With a 2 - 1 array (where the concentrate of the two lead housings are plumbed to feed a third housing), the overall recovery of the system ends up being 75%. A relatively high percentage of the feedwater is thus converted into permeate without using concentrate recycle.

- The overall RO permeate quality is not significantly affected at recoveries up to 75%. As the recovery exceeds 75%, water quality begins to drop dramatically because of the higher concentration of salts in the latter membrane stages (Figure 1-27).

In areas of water scarcity, RO systems are being designed to operate with permeate recoveries well over 75%. In these cases, the economics (or politics) particular to the area make it necessary to compromise on permeate quality in order to reduce water consumption. However, there are methods of achieving these higher recoveries without compromising water quality.

Optimizing RO permeate recovery. Because of the need to insure adequate turbulence within the RO elements, achieving higher recoveries is more difficult than just reducing the RO concentrate flowrate. However, two simple design modifications are available that make possible a reduction in RO wastewater without compromising on the minimum element concentrate flowrate.

Concentrate recycle. It is common for a small industrial RO system to recycle a portion of its concentrate back to the feed of the high-pressure pump. (See Figure 1-28.) This enables a system containing only a few elements to operate with a more practical recovery than that dictated by its minimum concentrate flowrate.

This same method can be applied to a larger RO system to increase its recovery beyond the 75% it may have already achieved. For example, by recycling half of the concentrate back to the feed, a 75%-recovery RO can be converted to an 85.7%-recovery RO, resulting in a 12.5% water savings.

A significant disadvantage of this method is that the permeate quality will decrease, depending on how much high-TDS (total dissolved solids) concentrate water is being recycled back to the feed stream. In the previous example, the permeate salt content will increase by roughly 75%. If the permeate is polished downstream by ion exchange, the regeneration frequency of the ion-exchange resins will increase by about 75%. Depending on how often the system currently regenerates, the additional water necessary to regenerate more frequently may nullify much of the water savings; therefore, the possible effects on the entire water treatment system should be evaluated before using concentrate recycle.

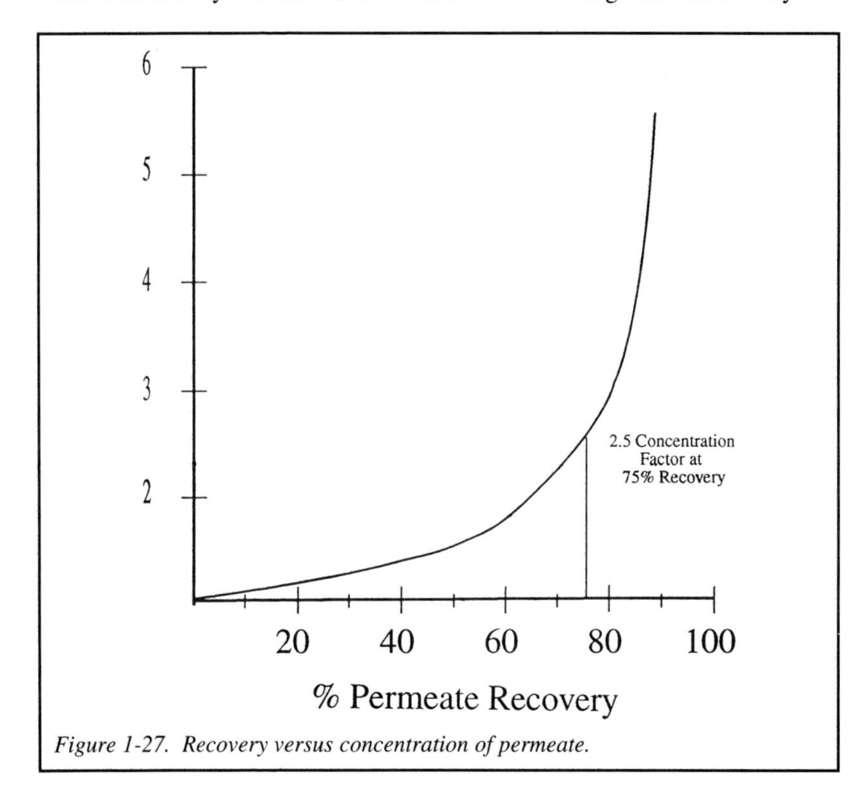

Figure 1-27. Recovery versus concentration of permeate.

Additional staging. Another method of achieving higher recoveries is to install smaller-diameter RO elements and housings in the concentrate line of the RO (upstream of the concentrate throttling valve). (See Figure 1-29.) The number of housings will depend on the amount of concentrate available. For example, an RO array that ends in a single 8-inch housing should have sufficient concentrate flow to feed two 4-inch housings. If scale formation is not a concern, there should be sufficient concentrate from the two 4-inch housings to feed a third housing.

The permeate water from the new membrane can be either plumbed into the existing permeate manifold or returned to a storage tank upstream. The advantage of supplementing the existing permeate is that more RO water is available for roughly the same operating costs. This is particularly handy if the current RO capacity is marginal.

However, since the new membrane is exposed to water containing a much greater concentration of salts, it will tend to pass to the same degree a proportionally greater percentage of salts. If ion exchange is used to polish the RO water, it will lead to a greater frequency of regeneration. For example, if an additional membrane stage is added to a 75%-recovery system to bring its recovery to 85.7%, its combined permeate quality will decline by about 15%. This will lead to approximately a 15% increase in ion-exchange regeneration frequency and regenerant water usage.

The alternative is to dilute the feedwater salt concentration by adding this new permeate water to a storage tank or degasifier upstream of the RO. The location needs to be one under atmospheric pressure to avoid back-pressuring the new membrane. Since many RO systems do not use a storage tank or degasifier upstream, this alternative may require that the water be repressurized, or it could just be used in some less critical application.

The maximum RO recovery should be limited by the scaling potential of the RO concentrate. If the solubilities of salts are to be exceeded, a means of

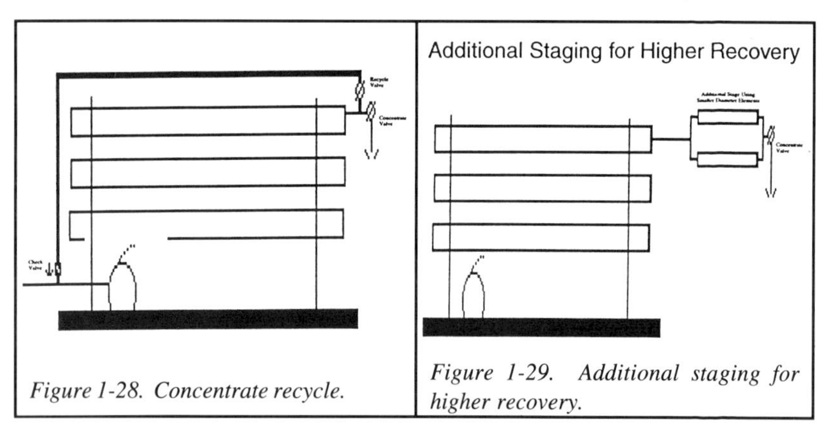

Figure 1-28. Concentrate recycle.

Figure 1-29. Additional staging for higher recovery.

preventing their precipitation must be considered. (See "Prevention of Scale Formation" in Chapter 3.) ■

CHAPTER 2

RO SYSTEM DESIGN

Component Designs

Membrane element housing (pressure vessel) length. With spiral-wound RO membrane systems, the element housing length is a variable that will have a major impact on other aspects of the RO design. It will affect the structural frame requirements, the flow hydraulics of the system, and the size of the high-pressure pump.

It is essential to know where the RO system will be located before choosing a housing length. Sufficient space must be available on each end of the housing to be able to remove and install membrane elements (unless it is intended that the housings be removed to change elements). Therefore, it is recommended that slightly more than 5 feet of space be allowed on each end of a horizontal system. (A minimum of 5 feet is required at the end of the housing in case 60-inch-long elements are ever used.) This may limit the length of housing that can be used for the system.

It is usually economical to use the longest housing possible if this results in the use of fewer housings. For aesthetic reasons, however, it may be desirable for small RO systems to use a larger number of smaller housings. For example, three 80-inch-long housings will look better on a small skid than one dangling 240-inch housing. It also will be easier to transport, and will take less space for installation.

Housings are designed for specific pressure applications. Seawater applications require heavy construction. If the housings are stainless steel, they will require 316L grade stainless because of the high chloride content of seawater. The housings will need to be built with a thick wall to safely handle the higher pressures. Most membrane housings, for safety reasons, are overdesigned. One manufacturer designs housings for a burst pressure that exceeds the operating pressure by a factor of six, and tests each housing at 1.5 times the rated operating pressure (Martinson, 1993).

Higher-pressure stainless steel housings have the advantage of structurally being able to handle side-entry ports. The high-pressure inlet and outlet pipe manifolding can then be plumbed into the side of the housing.

Fiberglass housings could not previously use side-entry ports without affecting the housing integrity. These housings usually use end-cap entry (Figure 2-

1). This made it necessary to remove the high-pressure plumbing when changing membrane elements. However, some fiberglass housings are now being manufactured with side-entry ports. The fiberglass also has the advantage of being relatively inert to corrosion.

For lower-pressure applications, the side-entry stainless steel housings can be significantly more expensive than fiberglass housings. For RO systems using 2-inch (nominal) diameter spiral-wound elements, polyvinyl chloride (PVC) housings are available as the least expensive of housings. For some lower-pressure applications, 4-inch PVC housings are also available.

RO structural frame. As with the structural framework of any piece of equipment, the RO frame must be sufficiently strong to handle transporting of the RO unit. The frame should be sufficiently rigid to maintain the alignment of the high-pressure pump and motor, as well as to protect the membrane housings and the pipe manifolding. It also should be sufficiently rigid to prevent the occurrence of harmonic vibration.

Welded steel (painted) is probably the most common frame material. In areas of high corrosion, such as in seawater desalination, fiberglass framework is sometimes used. Stainless steel is frequently used in food and pharmaceutical applications.

A smaller RO system, depending on its weight, can get by with 4- or 6-inch channel beam for the base of its main skid (Figure 2-2). A system larger than 180 inches in length should probably have I-beams for the base of its skid (Figure 2-3). It should also have a diagonal support across the entire frame to keep it square during shipping.

The frame section that supports the instrument and electrical panels, as well as the membrane housings, can be constructed of something like 2-inch-square

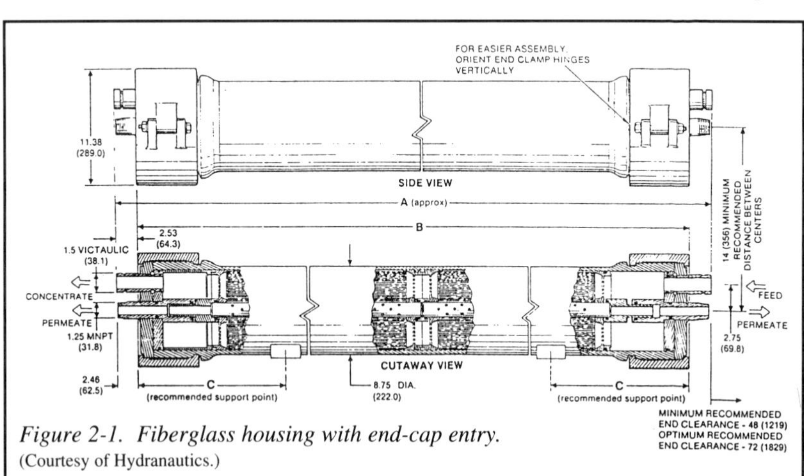

Figure 2-1. Fiberglass housing with end-cap entry.
(Courtesy of Hydranautics.)

tubing with a wall thickness of 3/16 of an inch. In the construction of the framework, care should be taken to avoid pockets that can collect water or debris. Such potential problem areas should be drilled with holes to provide drainage.

The frame base should be designed so that a fork-lift can easily move the system. A large unit may require a crane to lift the unit onto a flatbed truck. If possible, the RO should be lifted using a cradle under the frame base. Lifting eyes can be used, although heavier frame construction may be required for this. Lifting eyes should be located at the center of gravity of the RO unit.

If the high-pressure pump is mounted directly on the same skid as the RO instrument panel, care should be taken to insure that vibration does not affect the calibration of the instruments. Sometimes vibration dampeners are necessary if the frame is undersized.

Fiberglass membrane housings should be braced and cushioned against the frame. Pinhole leaks have been known to develop in fiberglass housings that are not adequately buffered against vibration from a steel frame, particularly if the high-pressure pump is mounted on the same frame. This is not as much a concern with stainless steel housings. One advantage of stainless steel housings is the additional rigidity they give to the RO skid (Osmundson, 1993).

An attempt should be made to balance the RO system on the base of the frame. It is not a particularly good design to put the high-pressure pump motors on a far end of the frame. When the system is moved, it will tend to tilt toward the end with the motors.

Figure 2-2. Skid for a smaller RO system.
(Courtesy of Osmonics, Inc.)

In areas prone to earthquakes, the system may need to meet special equipment codes designed to prevent equipment from falling or tipping during an earthquake. This should not be extremely difficult for an RO system if the high-pressure pumps are mounted directly on the base frame, and membrane housings are not mounted too far up in the air. (Meeting Seismic 4 calculations in California has a lot to do with keeping the equipment's center of gravity low and stable with respect to the supporting frame.)

The frame size will be limited by shipping restrictions and by accessibility at the installation site. Most RO systems will be limited to 72 inches in height and in depth, in order to fit through shipping docks and doorways. If this is not a concern, it is possible to build systems as large as 92 inches high and deep, and as long as 32 feet. This size may be manageable if shipment is by flatbed or by barge.

Carbon steel frames need to be treated to reduce corrosion, and for aesthetic reasons. They should be sandblasted, chemically cleaned, and primed with a rust inhibitor. Two layers of a high-quality epoxy or enamel paint are then suggested. Enamel paint has been reported to be easier to use than epoxy (Holland, 1993). Powder coating also provides a very durable finish.

Piping materials and design. High-pressure piping must be capable of conservatively handling the pressures potentially present in the system. Stainless

Figure 2-3. Skid for a larger RO system.
(Courtesy of Osmonics, Inc.)

steel is the most common piping material used for high-pressure piping and connections (Figure 2-4). Stainless steel tubing can be used for smaller flow systems. Tubing can be difficult to bend and fit when working with diameters larger than 3/4 inches.

For larger-diameter plumbing, stainless steel manifolding is manufactured by welding together straight pieces of pipe and various fittings. If this welding is performed using tungsten inert gas, called TIG welding, less carbon is introduced into the weld. The weld is then less susceptible to chemical attack. The manifolding is then passivated (this is also called *pickling*) in concentrated nitric acid at 120 °F. This strips free iron and other contaminants from the surface of any welds, down to about 1 millimeter; and prevents formation of the rust marks that are readily visible on unpassivated stainless steel welds.

It is possible to manufacture the piping manifolding using threaded fittings. Such a design typically has more leaks, and is not as structurally sound. It also is not as aesthetically pleasing.

Figure 2-4. High-pressure manifolding.
(Courtesy of Osmonics, Inc.)

Flexible, reinforced tubing is available that can handle the higher pressures. Flexible tubing may not have the longevity of stainless steel pipe, nor does it offer the structural rigidity. However, the flexible tubing can offer an advantage when used to connect end-entry housings to a central manifold. Since the connections are to the housing end caps, time and effort can be saved if the housing end caps can be removed without having to disconnect the high-pressure manifolding.

For water sources whose chloride concentration will exceed several thousand milligrams per liter in the RO concentrate, 316L grade stainless steel is suggested. With 316L, there will be reduced possibility of corrosion, particularly around welds. The cost of the 316L stainless will be slightly higher than standard 304 grade stainless steel.

In some applications, it is desirable to modify stainless steel to create a smoother surface that is less prone to attack or biological activity. This process is called electropolishing. The stainless steel (usually 316L grade) is made into an anode, and an electric current is generated, forming an anodic reaction process. The result is a surface so smooth that it shines (Combs, 1990).

Sufficient couplings (Figure 2-5) should be used between welded pieces of pipe so as to allow the disassembly of manifolding, should it be required to change out membrane elements or housings. These couplings will also make modifications easier, should the housing array ever need to be altered.

It is useful for cleaning purposes to have access to the interstage high-pressure manifolding. This can be a plugged Victaulic coupling connection, or the manifold can be teed into a high-pressure ball valve. This access can be used during cleaning to enable the recirculation of the cleaning solution around the individual membrane stage. This will make it possible to achieve maximize cleaning velocities in all of the membrane housings.

Occasionally, ball valves are used on the feed and concentrate streams of the individual housings. These can be used to isolate housings on larger RO systems (those with permeate flow in excess of 200 gallons per minute). The danger in using individual housing isolation valves is that if a valve leaks while a housing is isolated, the leaking water will leave virtually all of its salt content as scale on the membrane surface within that housing. Also, there is always the possibility that an operator will make a mistake in opening and closing valves that could result in the system being out of hydraulic balance. For these reasons, using isolation valves for the individual housings is not suggested.

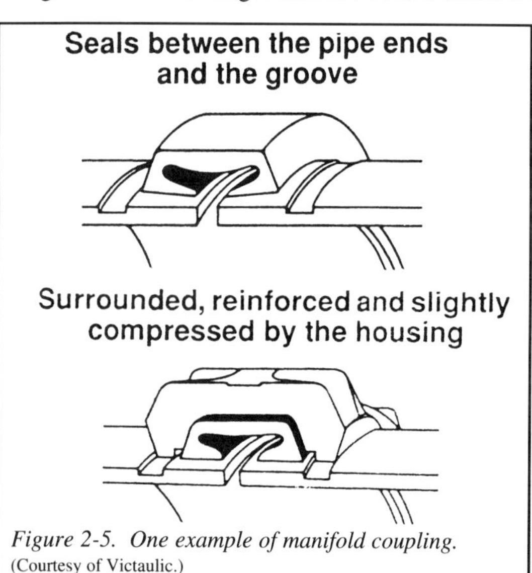

Figure 2-5. One example of manifold coupling. (Courtesy of Victaulic.)

Frequently PVC is used for the low-pressure

RO piping. For systems used in high-purity deionized (DI) water systems, polyvinylidene difluoride (PVDF) or some other "clean" piping material may be used. For the permeate connections to the housing end caps, using low-pressure flexible tubing will facilitate the removal of the end caps (Figure 2-6). If flexible tubing is exposed to direct sunlight (for systems installed outdoors), however, the ultraviolet (UV) radiation will tend to embrittle the tubing, and it will probably require replacement every couple of years.

Pipe Sizing. The pipe diameter should be chosen to avoid excessive pressure losses within the manifolding. This is particularly important with the permeate piping. Excessive permeate back pressure can cause flow reversal if this pressure should ever exceed the feed-side pressure, and can cause damage to the glue lines of spiral-wound membrane elements.

At comparable water velocities, pressure losses will be lower in piping with larger diameters because shear forces are lower at the pipe wall. Therefore,

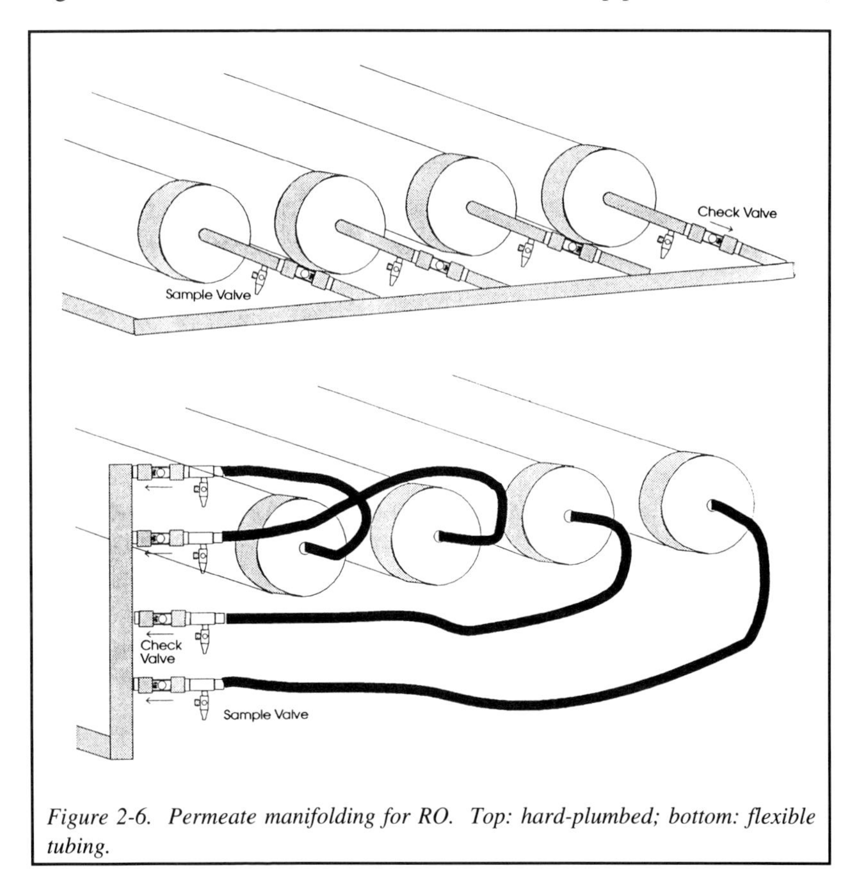

Figure 2-6. Permeate manifolding for RO. Top: hard-plumbed; bottom: flexible tubing.

higher velocities can be tolerated in larger-diameter piping without excessive pressure drops. The following guidelines for pipe sizing take this into account:

Flow (gpm	*Pipe Inside Diameter (inches)*	*Maximum Velocity (ft/sec)*
0-4	1/2	6.5
4-10	3/4	7.25
10-20	1	8.2
20-50	1-1/2	9.1
50-100	2	10.2
100-175	2-1/2	11.4
175-300	3	13.6

Velocity in piping is calculated as follows:

$$\text{Velocity} = \text{flow} \div 0.41\ \text{ID}^2$$

where:
velocity is the water velocity in feet per second (ft/sec),
flow is the flowrate in gallons per minute (gpm), and
ID is the inside diameter of the pipe in inches.

Partially filled membrane housings. Spiral-wound membrane elements are designed for their permeate collection tubes all to be connected together within a membrane housing; thus, the permeate needs to be removed from only one end of the housing. Either end can be used, although special permeate tube connectors, called *blanks* (see Figure 2-7), may be required within housings that are not completely filled with elements. The blanks connect the feed-end element's permeate tube with the end cap.

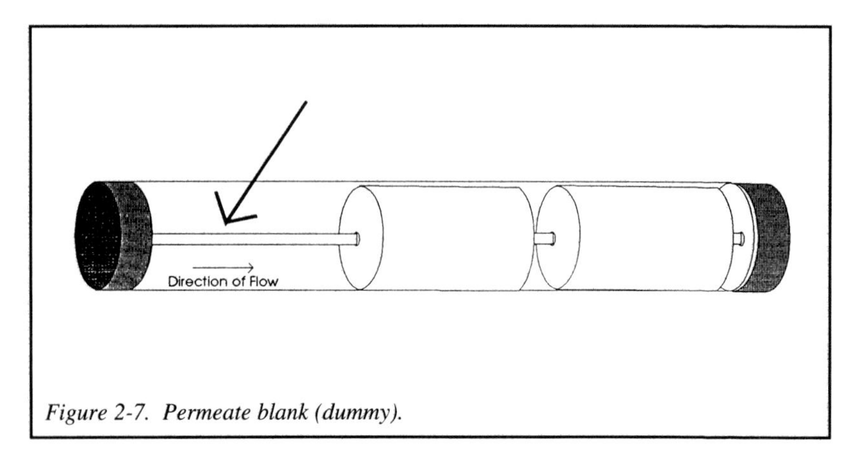

Figure 2-7. Permeate blank (dummy).

Permeate manifold design. To monitor the permeate water quality from each housing, a sampling valve should be teed into the permeate lines on the manifolding from each housing. A sampling valve can also be installed on the other end of the housing, although its water quality will indicate the performance of only the lead end element.

Check valves are recommended for each of the individual housing permeate lines. These serve two purposes. They reduce the possibility that water will backflow into a housing when a system is started up, should that housing have emptied during shutdown. (This backflow can blow apart the membrane leaves.) The check valves also ensure that water samples from the previously discussed sample valves are from only the desired housing, and do not get mixed with permeate that is feeding back from the permeate manifolding.

High-pressure throttling. A means of throttling the RO concentrate is required to finely control the flowrate while also creating a pressure loss as high as 800 psid (depending on the RO system pressure). It is always desirable if a valve or orifice can handle this task without creating excessive noise from the turbulence caused by the valve restriction.

Ball valves are usually not a good choice for this application. They are not designed for throttling flow. In fact, unless tightly constructed, the valve will tend to close during operation under high pressure drops. If the valve closes while the RO is operating, it can damage the high-pressure pump and provide insufficient flow for the RO membrane elements. Ball valves will create a lot of turbulence that can cause noise and possibly damage downstream piping.

Stainless steel needle valves have been successfully used for high-pressure throttling. These valves are designed specifically for this purpose. The stainless steel should be of a grade containing low concentrations of carbon, such as 316L.

Flow Coefficient. A value known as the flow coefficient is used to characterize valves. Valve manufacturers will supply a curve of their valve's coefficient graphed against the number of turns of the valve. The flow coefficient, C_v, is calculated by dividing the flowrate by the square root of the pressure drop across the valve.

$$C_v = \text{Flowrate} / (\text{psid})^{1/2}$$

where:
psid is the pressure drop across the valve at that particular flowrate.

In control situations, it is desirable to use a valve that offers very gradual variation in the flow coefficient when the valve is turned. Such a valve will better allow the operator to zero in on a desired flowrate. Over the desired flow and pressure range, the curve of the number of valve turns versus the coefficient should be steep. (See Figure 2-8.)

An orifice can be used to control a minimum concentrate flowrate. This orifice usually consists of a flat plate anchored in a stainless steel nipple. The plate has various-sized holes drilled through, depending on the flowrate and pressure drop desired. The advantage of using such a device is that it allows the concentrate flowrate to always exceed the minimum required value. With RO systems that use only a throttling valve to control the concentrate flowrate, it is possible for someone to inadvertently shut off the valve. When this occurs, it can cause severe damage to the membrane elements.

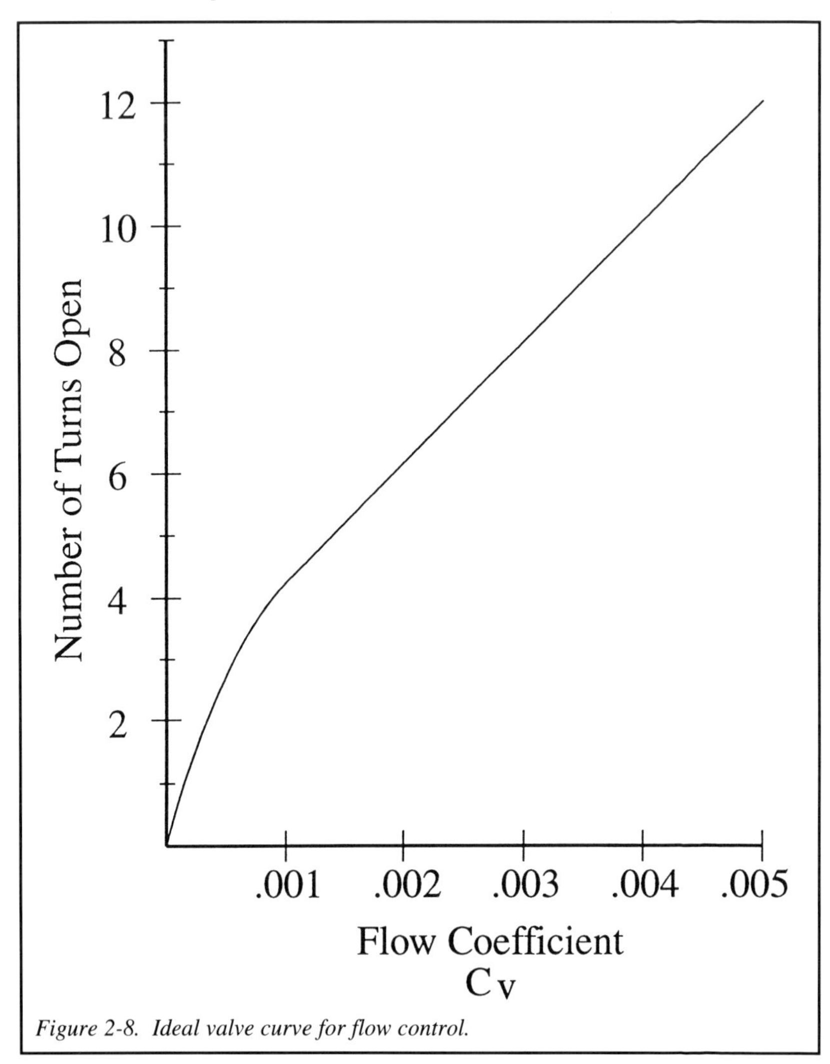

Figure 2-8. Ideal valve curve for flow control.

In high-recovery RO applications, like chrome and nickel recovery, it is useful to plumb two valves in parallel on the RO concentrate (Figure 2-9). One is designed for normal operation where the flow is quite small as compared to the system flowrate. The other larger valve can be used when larger concentrate flowrates are needed, such as during startup or cleaning.

With any throttling valve being used to create a large pressure drop, it is a good idea to plumb the valve with stainless steel pipe directly downstream of the valve. This will reduce the possibility of the valve turbulence gouging out the downstream piping.

High-pressure pump. The high-pressure pump is a fundamental component of an RO system. It must provide smooth, uninterrupted flow to the RO membrane elements. It must be sized to provide the necessary flowrate at the desired pressure at the various system operating conditions over the life of the RO membrane. Its energy consumption is one of the major expenses of RO system operation; therefore, energy efficiency is essential to the economics of the entire system.

Centrifugal pumps, which operate by spinning the fluid with the pump impellers, are most commonly used for RO applications requiring less than 600 psig of pressure. As the water reaches the outside of the impeller, it is under the greatest velocity. This velocity is then converted into pressure as the water leaves

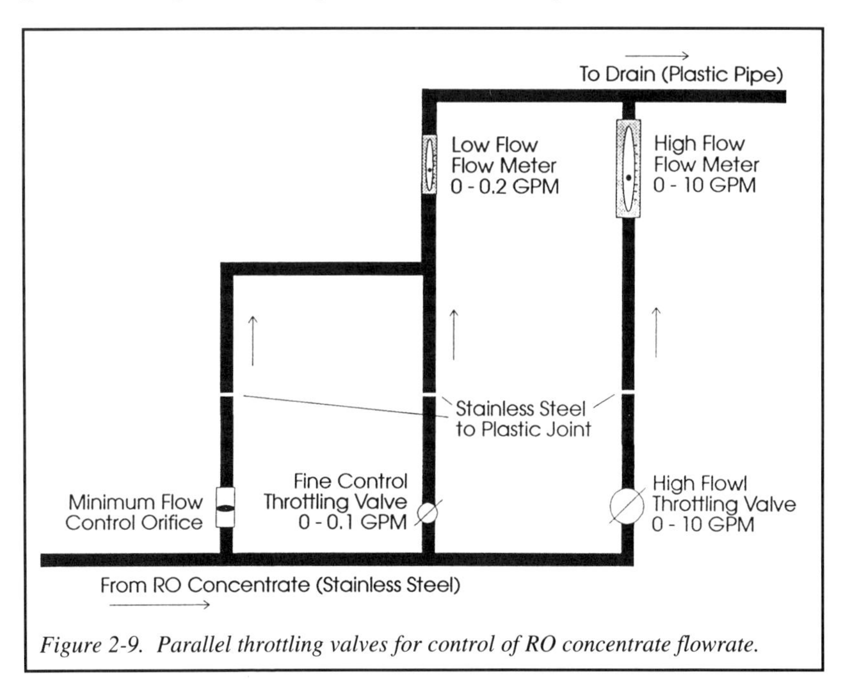

Figure 2-9. Parallel throttling valves for control of RO concentrate flowrate.

the pump (Figure 2-3).

Centrifugal pumps have the ability to provide smooth and continuous flow and pressure. Their efficiency will depend greatly on the shear of the impeller against the water. To provide the high pressures required for reverse osmosis with a single impeller necessitates either using an extremely large impeller, or using a standard size impeller at high speeds. The large impeller would tend to be impractical. However, at least one pump manufacturer does offer a pump with a standard size impeller that is spun at very high speeds.

A multistage centrifugal pump uses a number of impellers in series to build up the pressure, rather than relying on a single impeller. Each impeller takes the water and further pressurizes it. A diffuser bowl collects the water from the previous impeller and passes it on to the next impeller (Figures 2-11 and 2-12).

Most multistage centrifugal pumps are more energy-efficient than single-stage, high-speed centrifugal pumps. Each impeller of a multistage pump has to increase the pressure of the water passing through it by only 30 psid or less. There are reduced shear forces placed upon the water, and thus less energy is lost to friction and turbulence.

Piston pumps (Figure 2-13) are sometimes used for higher-pressure RO applications, such as those required for seawater desalination. Piston pumps use various numbers of mechanical plungers to create pressure. The pressure from piston pumps tends to pulsate, which can create velocity surges. These surges can promote telescoping and subsequent mechanical damage of RO membrane elements. It is therefore a good idea to use a dampening device downstream of a piston pump when it is being used for reverse osmosis.

Pump curves. If a particular pump model has already been chosen for use as the system high-pressure pump, its output pressure should be determined from its pump curve based upon the total flowrate at the pump. The total flow will be the sum of the permeate, concentrate, and recycle flowrates.

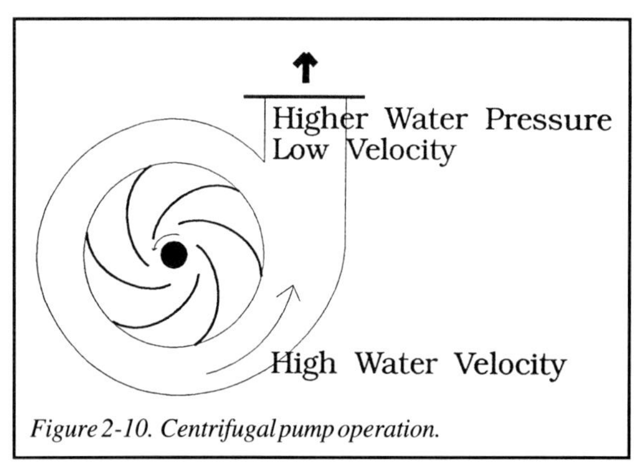

Figure 2-10. Centrifugal pump operation.

Pump curves can be used to choose the best pump for the required pressure and flowrate. As shown in Figure 2-14, if a 140-gpm flow is desired with a minimum pressure of 200 psig, the pump model of choice for this particular brand of pump would be the Model 12509. For that flow, it puts out a pressure of about 210 psig. In fact, the actual operating pressure will be greater than 210 psig since any incoming pressure is additive. For example, the city pressure may be 70 psig, which may drop to something like 55 psig after any media filters or cartridge prefilters upstream of the RO. The actual pressure at the high-pressure pump discharge in this example will thus be 265 psig.

In choosing a high-pressure pump, it is desirable to obtain the optimum energy efficiency; therefore, a pump should be chosen so that the desired flow falls on a point near the maximum efficiency of the pump. The horsepower of the motor should be chosen such that at no time will more flow be passed through the pump than what can be handled by that motor. The maximum horsepower of the motor is determined by multiplying its horsepower rating by its safety factor rating.

If there is a possibility that a line break or an incorrectly opened valve could cause the pump's motor to operate for an extended period of time at a horsepower beyond the maximum for the chosen motor, then a larger motor should be used. The motor draws power only for the horsepower that is actually used, called the *brake horsepower.* A slightly oversized motor does not necessarily mean that the system will be any less energy-efficient.

In the early years of RO it was common to grossly oversize the high-pressure pump and then restrict its output with a throttling valve. Given the uncertainties of membrane performance at that time, this offered RO system manufacturers a safety margin should the RO membrane compact or foul more than was anticipated.

For normal pure-water applications today, grossly oversizing a pump is usually not necessary, and results in a waste of energy. A pump should be correctly sized for the desired flowrate and pressure. The exception would be if a minor system expansion is anticipated in the near future.

If a major expansion is possible in the future, it might be more practical to build an independent RO array. This could be installed on the existing RO frame, if the frame is sized for a second array when first built. Many endusers like the

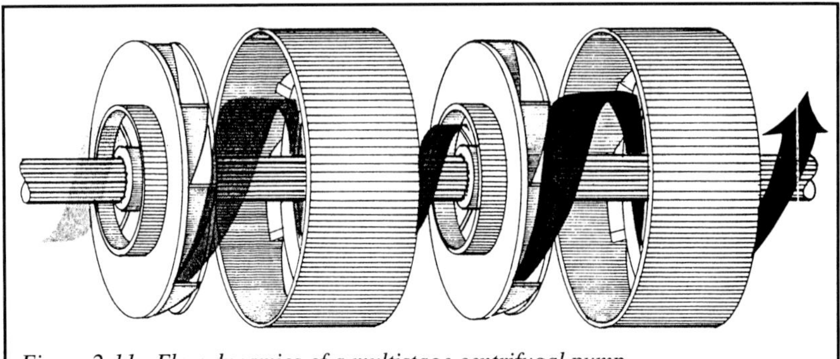

Figure 2-11. Flow dynamics of a multistage centrifugal pump.
(Courtesy of Osmonics, Inc.)

flexibility that comes from having two independent RO systems. If one is down for maintenance, the other is available to assist in keeping up with any water demands.

Motor efficiency. The efficiency of the motor is also a factor that should be considered. Some pump motors may have energy efficiencies as low as 92%. This can be compared to the 98% to 99% efficiency for motors designated for high efficiency. The higher efficiency can result in substantial energy savings. Also, some power companies offer special incentives to facilities for using high-efficiency motors.

Submersible high-pressure pumps, installed in a horizontally mounted pressure tube, are sometimes used in RO applications. Although these pumps are promoted for their energy efficiency, their submersible motors are actually less efficient than standard air-cooled motors, which tends to offset the higher efficiency of the pump wet end. Their primary advantage is their low noise level.

Pump materials of construction. The materials of construction of the pump's wetted parts should be evaluated for the application. Typically, pump impellers and diffuser bowls are made of either plastic, 304 grade stainless steel, or 316 grade stainless steel. The pumps with plastic parts are usually less expensive. However, the possibility for generating pump shavings is reduced if stainless steel parts are used.

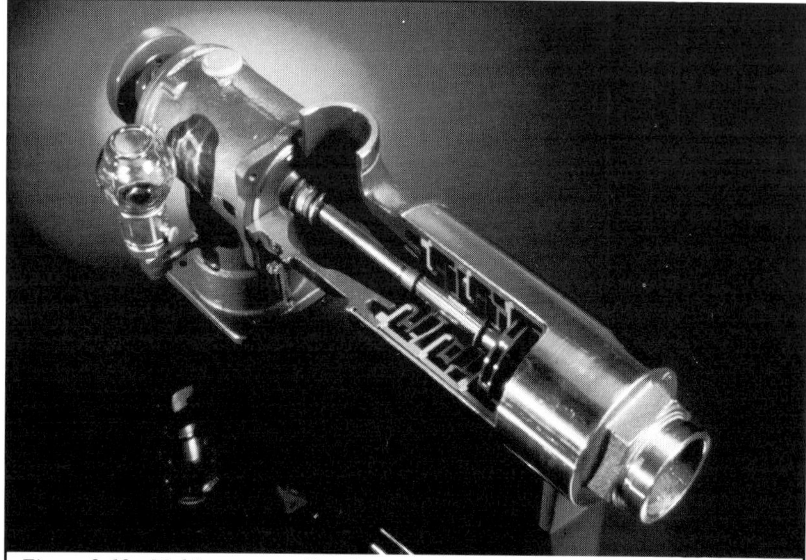

Figure 2-12. Multistage centrifugal pump.
(Courtesy of Osmonics, Inc.)

Pump shavings from plastic impellers have been known to foul the lead end elements of RO systems. The manufacturers of these types of pumps usually offer an option for a small discharge screen that is installed directly downstream of the pump to catch any of these shavings before they can get to the membrane elements. This screen (Figure 2-15) is recommended for RO pumps with plastic components.

Pump repair. Since the high-pressure pump is such a critical item in the operation of the RO system, its repair should be considered. If the pump were to fail, how quickly could it be repaired? Trained maintenance personnel can perform many repairs on-site. Many pump distribution companies offer repair services.

If the problem is severe, it may not be possible to repair the pump, or there may be insufficient time. In this case, some manufacturers will stock new or rebuilt pumps that could be couriered to the site. In critical applications, it may be desirable to keep a spare pump and motor on-site in case of problems. It may not be necessary to actually plumb in the extra pump on the RO skid if personnel are available who can quickly replace a pump.

Electrical Controls. Most RO systems require a means to control the operation of the system (Figure 2-16). The control system must handle functions such as opening an inlet water solenoid valve, initiating the high-pressure pump, and

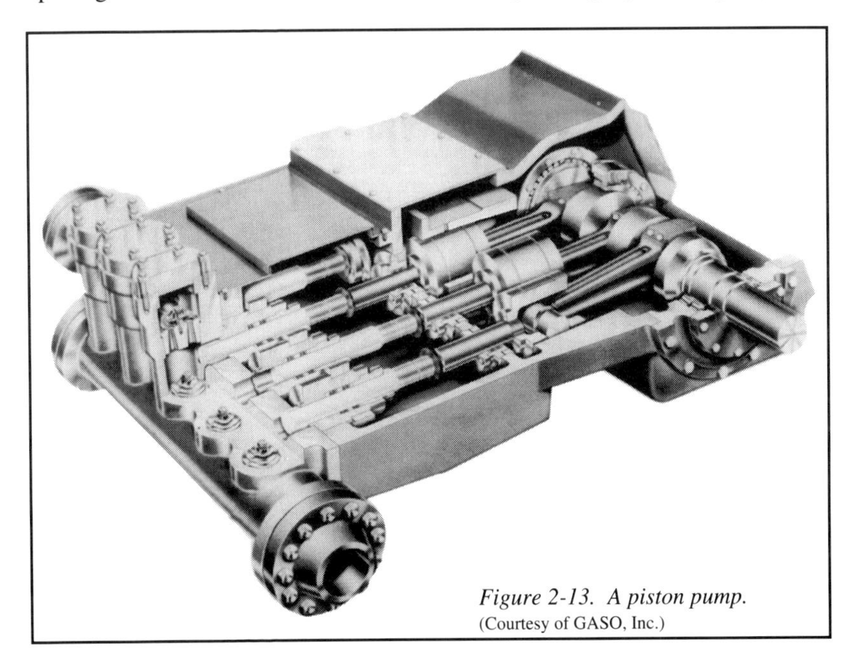

Figure 2-13. A piston pump.
(Courtesy of GASO, Inc.)

shutting down the system if a storage tank is full or if a monitoring instrument indicates a hazardous condition.

In early systems, this control process was done by mechanical relays in conjunction with timing devices. The relays used a low-voltage signal to initiate contact of a high-voltage device.

In recent times, the timing devices and many of the relays have been replaced with inexpensive microprocessor control units. These units can be programmed to take the signals from the monitoring instruments and provide the necessary control response. They can also provide output signals to another computer system for external monitoring. In some cases, the microprocessor can be programmed for monitoring and controlling the RO system from off-site via a telephone line.

Power. Most RO systems will need two types of electrical power: typically 220-

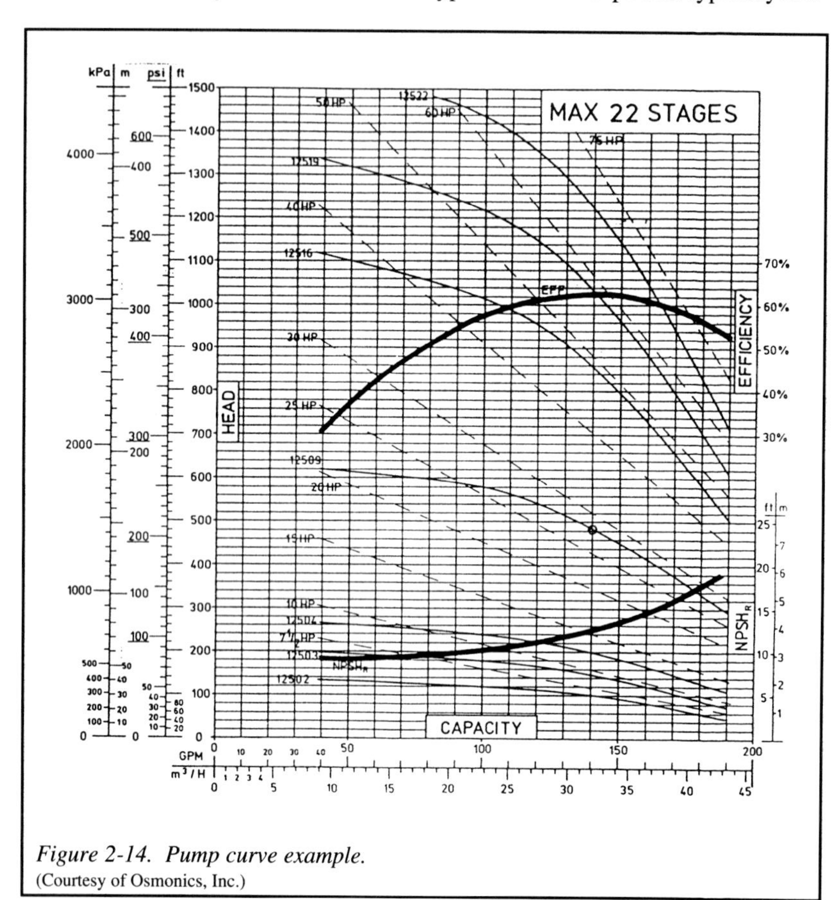

Figure 2-14. Pump curve example.
(Courtesy of Osmonics, Inc.)

volt or 480-volt alternating current (VAC) for the high-pressure pump motor; and lower-voltage power to control the motor starter and whatever instruments, alarms, and shutdown features are in use.

Small RO units use fractional horsepower motors for their high-pressure pumps. These may be designed to operate with standard voltage power, 110 VAC, possibly using a cord that can be plugged into a standard wall electrical outlet. It is suggested that a dedicated outlet and circuit be made available for the RO. Extension cords should be avoided.

Standard 110 VAC power in the United States is provided in single phase. This means there is only one run winding in a 110 VAC motor. When used to energize a motor, a capacitor is used to create an opposing magnetic field to the one created in the motor winding. The repulsion effect from the two windings spins the motor.

Higher-voltage power sources in the United States are available in three-phase power. Three-phase power motor systems use three current-carrying conductors, which power three separate windings in the motor's housing. The three alternating currents are 120 degrees out of phase with each other; thus, when the current in one conductor is at its low point, the currents in the other two conductors are each at 85% of their maximum value. This means that the motor shaft is pushed three times during its revolution, as opposed to one push with single-phase motors, making the three phase motor more energy-efficient than single-phase motors.

Power usage is equal to the voltage pulled, multiplied by the current draw. If 480 VAC power is available, there is an advantage in using 480 VAC versus 220 VAC. The current draw for the higher-voltage motor will be roughly half that of a lower-voltage motor for the particular power being required for the pump motor. Lower current results in lower electrical line resistance, and subsequently in the ability to use smaller wire, fuses, motor starters, and overloads. This can amount to substantial savings. (David Cohen, 1993).

High-pressure pump motor. The National Electrical Code requires that a motor disconnect be located "within sight" of the motor it controls. Some RO manufacturers will supply a disconnect and/or circuit breaker mounted inside the RO control panel. To meet the code requirements, the handle that disconnects the pump motor must be operable without opening the control panel. If a convenient disconnect is not supplied by the RO manufacturer, one will need to be installed on-site.

When a high-pressure pump first starts up, its motor will pull from five to twenty times as much current as its running current. Most circuit breakers are designed to handle this inrush current without tripping the breaker's overload elements. Motor starters are available with a soft-start feature that gradually brings the motor up to running speeds. This also will help to reduce the amount of water hammer that can occur when the high-pressure pump first starts. (Cohen, 1993).

In areas where the raw water temperature will vary seasonally, and a heat exchanger is not being used to raise the water temperature to a constant value, variable-speed drives can be used to control the high-pressure pump motors. A variable-speed drive will vary the speed of the pump motor according to an analog input. It can be used to soft-start the pump's motor. It also can be used to reduce the high-pressure pump output when less pressure is required for the RO to produce the desired permeate flowrate. Thus, when the water temperature is warmer, the pump output pressure is reduced to maintain the desired permeate flow (since the permeate flowrate is a function of pressure and temperature), and reduce the power consumption by the motor. This eliminates the need for an energy-inefficient throttling valve on the pump effluent line.

Another possible application for a variable-speed motor drive would be for RO applications that see a wide variation in the incoming solute concentration. To achieve the same permeate recovery when the solute concentration is low requires less pressure than when the concentration is high, because of the difference in the osmotic pressure of the solution. The variable-speed drive could be used to reduce the pump output pressure, thus reducing the permeate flowrate and the power consumption.

It is suggested that heaters be used with the motor starter. A heater is a bimetallic strip that will bend away from its electrical contacts if it heats past its design setting; thus, heaters will disconnect the motor if it draws too much current through its winding. This protects the motor from overheating.

An improvement on the heater device would be to use a motor with a thermistor on the motor winding. In this manner, the temperature of the motor can be monitored. If a problem is gradually occurring in the winding, a trend for increasing motor temperatures will become apparent prior to the motor overheating. (Cohen, 1993).

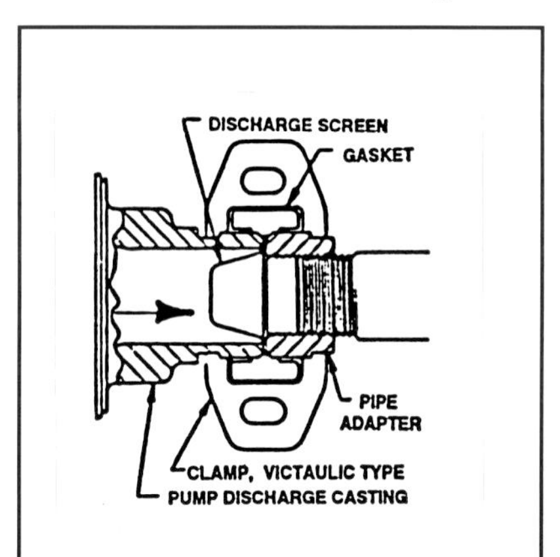

Figure 2-15. Installation of discharge screen.
(Courtesy of Osmonics, Inc.)

Instrument power. Since the RO monitoring instruments are in close proximity to water, it is safer to use low-voltage power such as 24 VAC for the control circuitry. The 24 VAC is a standard control voltage that is commonly used for analog instruments such as pH and temperature monitors, as well as for solenoid valves and motorized ball valves. Any voltage less than 50 VAC is classified by the National Electrical Code as Class II wiring (multiconductor sheathed cabling). Class II wiring between components is much less expensive to run than wire conduit.

The source of the low voltage can be obtained by bringing in separate control power to the RO control panel, or by using a control transformer to step down the voltage source from the high-pressure pump motor. (The low-voltage power should be grounded if taken off a transformer.) There are some advantages to using the control transformer. First, it saves the cost of bringing in another control voltage system. Secondly, it eliminates the risk that a technician servicing the equipment would de-energize only one of the two power sources, and possibly come into contact with high-voltage power. (Cohen, 1993).

Instrumentation. For setting flowrates and pressures, it is useful to have the monitoring instruments readily accessible on the RO system. Placing the monitors in one central panel also makes it easier to collect operating data. It is suggested that electronic instruments, as well as the system controls, be placed in a waterproof panel. Such panels are given ratings by the National Electrical Manufacturers Association (NEMA). A panel rated at NEMA 4 is designed to

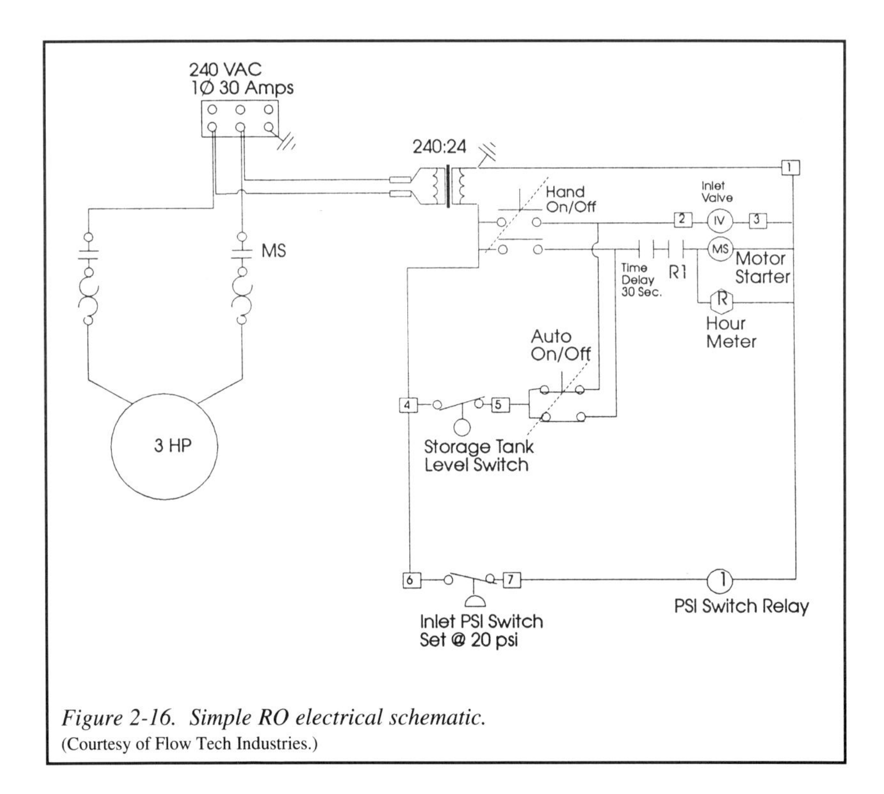

Figure 2-16. Simple RO electrical schematic.
(Courtesy of Flow Tech Industries.)

be watertight, even if water is sprayed directly on the panel. NEMA 12 panels (Drip-proof) are sometimes used, but should not be directly sprayed with water.

The signals from most transducers (also called *probes*) are typically not strong, usually in the millivolt range. Not more than 25 feet of wire should be used to connect the transducer with the monitor. Otherwise, the wires can pick up electromechanical interference (EMI) that can affect their accuracy.

Transducers should be grounded, but only in one location. Usually, they are grounded at the monitor. If the transducer is metallic and mounted into a metal pipe, however, it will be grounded at the pipe. It should not also be grounded at the monitor. A difference in potential between the two grounding points can affect the instrument reading. (David Cohen, March 1993).

Alarms and shutdowns. With any RO system, there is the possibility of a malfunction in the system or with the pretreatment equipment. Instruments (Figure 2-17) can be used to monitor the quality and other characteristics of the RO feedwater, as well as to monitor the performance of the RO system. These instruments can be connected with an audible alarm that will sound if some parameter is not within design specifications. Such an alarm must be loud enough to be heard over the noise in the area. The initiation of some of the alarm parameters may be significant enough to warrant automatic shutdown of the RO system. It is always useful if there is some means of identifying which parameter caused the alarm.

When performing maintenance on the RO, such as when calibrating the instruments, it is desirable to have the ability to silence audible alarms. There

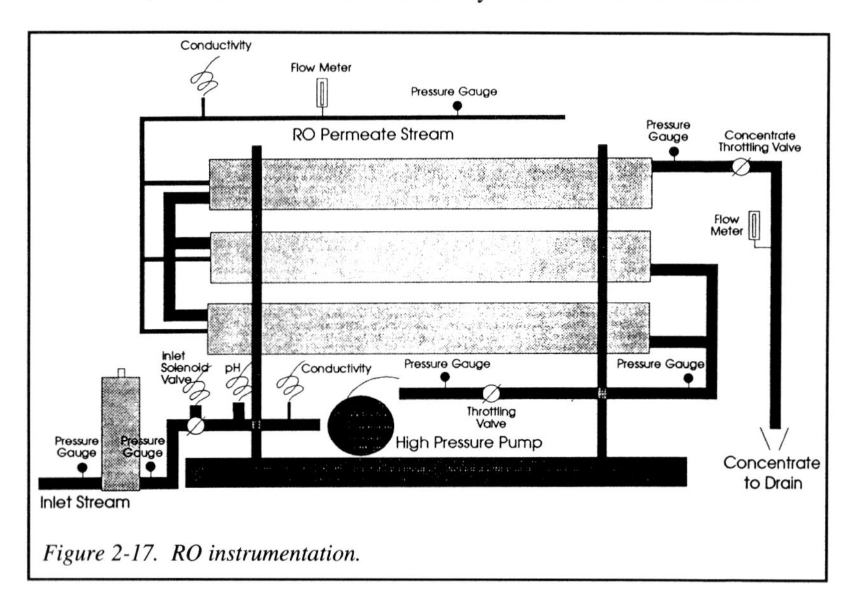

Figure 2-17. RO instrumentation.

should also be a means of temporarily bypassing alarm parameters that might otherwise shut down the RO system.

If it were practical, one might wish to install a means of protecting the RO against every possible thing that might go wrong in the system, particularly if the problem could result in RO membrane damage. Since system designs vary, based upon the endusers' needs, the necessary protective alarms or shutdown features will also vary. The practical number of alarms and shutdown features for any particular RO system will depend, however, on how much is invested in RO membrane, and the extent of the possibility for membrane damage. It would not make sense to put more money into alarms and shutdowns than what it would cost to frequently replace the membrane elements under worst-case conditions.

Larger RO systems should generally have more safety features than smaller systems. Suggested alarms and shutdowns might include the following:

Low inlet pressure. The high-pressure pump must be flooded to operate properly. If the pretreatment equipment is not properly sized, if the RO prefilters have fouled heavily, if there is insufficient city water flow available, or if an upstream valve has been inadvertently shut, insufficient flow may be available for the high-pressure pump. Many of these pumps use plastic impellers that can melt down within a few seconds if the pump is operated without flow. It is therefore suggested for both large and small RO systems that a low-pressure alarm and shutdown be used to protect the high-pressure pump.

The low inlet pressure alarm should be set conservatively if there is normally sufficient inlet pressure. It is common to use a setting between 10 and 20 psig. With a conservatively high setting, it is not necessary to put the alarm/shutdown on a time delay.

With some RO systems, there may be an initial drop in suction pressure as the high-pressure pump draws more flow in pressurizing the RO system. If this drop is severe, it probably indicates that insufficient time has been allowed for the RO system to fill with low-pressure water, prior to the high-pressure pump coming on line. This problem should be resolved immediately, since high water velocities that could be damaging to the membrane elements may be occurring at system start-up.

In cases where there is not enough pressure available for a conservative setting, it may be necessary to reduce the setting in order to keep the RO operating under normal conditions. For most high-pressure pumps, the absolutely critical parameter is that the pump is always under positive pressure.

High/low pH. Protecting RO membrane against high or low pH extremes is more critical for cellulose acetate membranes than for polyamide or charged polysulfone membrane systems. The CA membrane can hydrolyze fairly quickly if subjected to adverse pH conditions. It can also hydrolyze more rapidly if calcium carbonate precipitates on the membrane surface. (See "Membrane Compatibility" in Chapter 3.) Scale formation is also a concern for PA membrane systems, but can be cleaned from the membrane elements if discovered before the scale formation is severe.

Shutting down the RO unit when the pH goes out of specification may not be sufficient to protect the membrane, as this will result in the extreme pH water being left in contact with the membrane. This can possibly do more damage than if the water was allowed to rinse out of the system under normal operation.

For well-attended water treatment systems, it is suggested that an alarm system be relied upon to attract the operator's attention if the pH goes out of specification. The audible alarm must be loud enough to be heard when the RO is operating. A time delay of a few seconds should be used to allow the pH to stabilize within its desired range during RO system start-up.

High temperature. Heat exchangers used in the pretreatment for RO systems are notorious for causing membrane element problems. If the heat exchanger uses a heat source that would be hot enough to damage the membrane elements, the potential exists for the RO feedwater to equilibrate with this high temperature during a system shutdown. When the RO starts up, it can be exposed to the high-temperature water.

Excessively high fluid temperatures can also occur while an RO system is being cleaned. If the cleaning solution was heated to a warm cleaning temperature, the heat added by the recirculation pumps can cause its temperature to increase even further during the recirculation of the solution.

For these reasons, a high-temperature alarm is suggested. As with pH alarms, it is not necessarily a good idea to shut down the RO if the alarm is actuated. This would just leave the high-temperature water in contact with the membrane elements. If the system is allowed to continue operating, the high-temperature water can rinse out of the system. Of course, this is not the case when recirculating a cleaning solution.

High feedwater hardness. A high feedwater hardness alarm is useful for RO systems that rely on water softeners to remove the calcium and magnesium hardness from their feedwater. The hardness can precipitate on the membrane as carbonate or sulfate scale. It is common for softeners to occasionally leak hardness into the RO feedwater. If the softeners are not sufficiently regenerated, the shedding of hardness can be significant.

On-line hardness monitoring instruments are available with an alarm output signal. These instruments are an excellent safety measure for RO systems with upstream softeners, and are particularly recommended for CA membrane systems, because of the potential for membrane deterioration caused by carbonate scale.

High oxidation potential. For PA systems, membrane deterioration caused by low concentrations of oxidizers is a great concern. Oxidation-reduction potential (ORP) meters are available that can measure whether the RO feedwater is in an oxidative state. The meter should have an output that can be tied into an alarm that will shut down the RO system.

This meter would be particularly recommended if the injection of a reducing agent is being relied upon to chemically reduce any oxidizing agents. The meter will give an indication as to whether sufficient concentrations of reducing agent are being injected.

Low day-tank fluid level. Chemical injection systems are used for various critical purposes in the pretreatment of RO systems. The effectiveness of the chemical injection is defeated if the chemical day tank is allowed to run dry. Level switches are available to send a signal to an alarm if the fluid level drops to its setpoint. This alarm should be used to shut down the RO system. Be sure to use a level switch that is compatible with any chemical it might contact.

High permeate pressure. Conditions that create high pressure in an RO system are a threat to the structure of spiral-wound membrane elements. Should the RO shut down before relieving this pressure, the permeate may flow backward into the membrane envelopes. This can cause the membrane leaves to blow apart at the envelope glue lines.

A high-pressure alarm can be tied into a system shutdown if the shutdown is quick-acting and the setting is conservative. The unit must shut down before the permeate pressure becomes excessive; otherwise, shutting down the RO can actually be the cause of element damage. It can release the pressure on the feed and concentrate streams without necessarily releasing the pressure on the permeate, promoting the potential for element damage. It is suggested that the pressure alarm/shutdown be set at 40 psig or less. (Consult the membrane manufacturer.)

Some systems will use a pressure-relief valve either in addition to or in place of an alarm/shutdown. Pressure-relief valves use a spring set to prevent flow unless the water pressure exceeds the set value. The valve would be plumbed into a line that tees off the permeate line and would then be plumbed to drain (Figure 2-18).

If the RO permeate is plumbed into an open tank, it is best to avoid the possibility for high permeate pressure by not plumbing any valves into the permeate line. If the permeate flows directly into another piece of water treatment equipment (such as an ion-exchange unit or a filter), the possibility for high permeate pressure is present and precautions should be taken.

High membrane pressure. Although not as commonly used as some of the previously discussed alarms and shutdowns, a high system pressure alarm and shutdown can offer protection against the closing of the concentrate valve. If the RO concentrate valve is closed, the system pressure will increase, as the lack of concentrate flow will cause the high-pressure pump to operate higher on its curve. A high-pressure alarm/shutdown may also be a good idea if a valve is being used to throttle the high-pressure pump. To be of assistance, the setting of this alarm/shutdown will need to be within 40 to 80 psig of the normal operating pressure, depending on the particular pump curve.

Low concentrate flow. A more sensitive means of protecting against the inadvertent closing of the RO concentrate valve would be a low concentrate flow alarm/shutdown. This can be used to shut down the RO if the system recovery is too high because the concentrate flow has been adjusted too low. If the system shuts down on this alarm, the operator should rinse out the system at low recovery to remove any supersaturated salts. This alarm/shutdown should be wired with a time delay of a few seconds to prevent the alarm from actuating during system start-up.

High permeate conductivity. A high permeate conductivity alarm is useful if necessary to protect downstream equipment or processes. If this is extremely critical, the alarm can be tied into an RO shutdown. A high permeate conductivity alarm or shutdown is not particularly useful in protecting the RO itself, since the RO membrane rejection is really more of an after-the-fact indicator of system performance. A time delay on such an alarm will be necessary during RO system start-up to allow the salt rejection to return to normal.

Parameter	*Suggested Alarms*	*Suggested Shutdowns*
Low inlet pressure	X	X
High/low pH	X	
High temperature	X	
High water hardness	X	X
High oxidation potential	X	X
Low day-tank level	X	X
High permeate pressure	X	X
High membrane pressure	X	X
Low concentrate flow	X	X
High permeate conductivity	X	

Array design. As the RO feedwater passes across the RO membrane and through the system, permeate water is being drawn off, leaving behind a reduced flowrate. Reverse osmosis systems are designed to stage the hollow-fiber elements or the spiral-wound membrane housings to achieve a crossflow rate within a desirable range. This range is dictated by the following needs:

- To achieve sufficient crossflow turbulence to minimize any fouling/scaling potential.
- To minimize the pressure differential across the system. This differential takes away from the driving pressure at the tail end of the system, and thus reduces the permeate flowrate. Also, excessive pressure differentials can mechanically damage the membrane elements.
- To obtain the desired concentrate flowrate coming off the tail end of the system, as dictated by the desired system permeate recovery (i.e., the percentage of feed flowrate that becomes permeate).

The membrane manufacturers recommend minimum concentrate flowrates, and/or maximum individual element recoveries, as a means of insuring that adequate crossflow turbulence is obtained. They may also recommend a maximum feed flowrate to prevent the occurrence of excessive pressure drops that might damage the element.

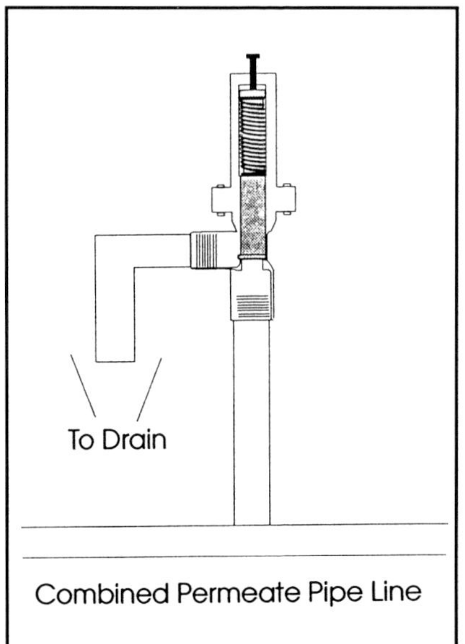

Figure 2-18. Permeate line pressure relief.

For a particular array design, meeting the first two needs (minimum concentrate flowrates and maximum element recoveries) may result in a concentrate flowrate greater than that required to meet the desired system recovery. This is not necessarily a problem, but will result in additional pump and energy requirements for the system. If a system of recycling part of this extra concentrate back to the RO feed stream is not used, it will also result in additional water usage, as well as in additional sewer and pretreatment requirements.

A commonly used array designed to operate at 75% permeate recovery without the need for concentrate recycle is a 2-1 array. The concentrate lines from two housings, each containing six 40-inch long elements, are plumbed to feed a third housing. Large systems may use multiples of this arrangement, such as a 4-2 or a 6-3 array. Each stage of the array recovers 50% of its feedwater as permeate, for the overall system recovery of 75%. This is based on an individual element concentrate-to-permeate ratio of 6:1.

The 2-1 array will probably not be the best design for RO systems using shorter housings (i.e., housings that contain less than six 40-inch elements), or for systems that do not operate at 75% permeate recovery. These systems will require custom arrays based upon the parameters of their applications.

In the design of a custom array, certain parameters will be treated as independent variables. Values will be chosen or found for them based upon the demands of the application. These parameters will include the membrane type and configuration, the desired permeate flowrate, the membrane driving pressure, the minimum concentrate-to-permeate flow ratio, and the water temperature.

Desired permeate flowrate. A number of parameters can affect the RO permeate flowrate of a reverse osmosis system. These include water temperature, salt concentration, and the membrane pressure as the feedwater flows through the system. In the design of an RO system, these factors need to be taken into account to insure that the desired permeate output is achieved.

For existing RO systems, it may not be particularly easy to upgrade the RO for additional permeate flow, unless the design initially included this flexibility. It is therefore suggested that the desired permeate flow be chosen somewhat conservatively. Then some means of controlling the system operation can be designed in to compensate for any RO overproduction. The most common means of controlling the RO system production to meet the specific permeate demands of the application is to interlock the RO system operation with a storage tank located downstream. The RO control circuitry is wired into the level controls in the storage tank. When the storage tank reaches a high level the RO automatically shuts off. When the tank empties to a low level, the RO automatically starts up again.

Although designing the RO to produce more than the required permeate flow may be desired, it is suggested that the system not be overdesigned by more than 50% of the required permeate flowrate. If the system is interlocked with a storage tank, the RO may sit idle for an extended period of time if the storage tank water is not being consumed. This can contribute to biological growth within the system, which can foul the RO membrane elements and possibly reduce their life expectancy.

In some applications, such as those where biological activity must be minimized, it is desirable for the RO to operate continuously. In this type of system, it is common to recirculate the RO permeate through a distribution loop and back to the feed of the RO. It should be noted that this will result in additional water losses, as well as higher power and pretreatment costs due to the continuous loss of RO concentrate water.

Desired membrane driving pressure. The membrane driving pressure is the pressure that is actively available to produce permeate. As part of an RO system design, a desired membrane driving pressure should be chosen based on the particular membrane type and application. Higher pressures result in higher permeate flowrates and better salt rejection characteristics. Higher pressures also require more power and can result in higher membrane fouling rates and reduced membrane life expectancy. In addition, they may require special membrane housings or piping materials.

Cellulose acetate membrane elements are frequently operated at pressures from less than 100 psig up to 500 psig. Above 500 psig, compaction will usually result in reduced membrane life. Typical pressures for larger CA/RO systems are 200 psig for applications requiring high energy efficiencies, or 400 psig for applications that require high salt rejection.

Larger PA and charged PS systems are typically operated at approximately 200 psig for high energy efficiency and low fouling rates. This typically allows these RO systems to use less expensive membrane housings than those used by systems operating at 400 psig.

The driving pressure will not necessarily be the actual membrane pressure. It is simply a number that is used to take into account membrane fouling when estimating an individual element permeate flowrate. Also, if a particularly low permeate flux rate is desired (as would be suggested for a heavily fouling application), adjusting the value for the driving pressure can take this into account. The actual membrane pressure will be calculated later when better values have been determined for other variables that affect the membrane permeate flowrate. The driving pressure, as defined, is used to estimate an approximate individual membrane element permeate flowrate.

$$\text{Individual element permeate flowrate} = \frac{\text{driving pressure} \times \text{fouling factor} \times \text{design permeate flowrate}}{\text{design pressure}}$$

where:

driving pressure is the desired membrane driving pressure;

design pressure is the membrane manufacturer's design driving pressure (usually about 205 psig for PA, and 400 psig for CA brackish-water membrane elements);

fouling factor is a number less than 1, used to estimate how much permeate flow will be lost due to membrane fouling; and *design permeate flowrate* is the membrane manufacturer's design permeate flowrate for the chosen membrane element.

Number of membrane elements. The individual element permeate flowrate can then be used to estimate the number of membrane elements required to produce the desired system permeate flowrate. With spiral-wound systems, this number should be divided by the element capacity of the chosen housing in order to determine the number of required housings for the system. Fractions should be rounded up to the next housing (i.e., if the calculation results in a requirement for 4.3 housings, 5 housings should be used).

Number of elements = desired permeate flow/individual element permeate rate
Number of housings = number of elements/elements per housing

Actual individual element permeate flowrate. The individual element permeate flowrate can now be better defined based upon the desired total permeate flowrate and the number of elements that will actually be used in the system array. The required number of elements will be based upon the number of housings in the system.

It should be noted that sometimes it may be desirable to not fully load housings for spiral-wound membrane elements. This is more common when longer housings are used. Possibly the elements are not required to meet the total permeate requirements for the system, or possibly the array crossflow rates balance better if the housings in a stage are not fully loaded.

It is critical that all the housings in any particular stage have the same number of elements. If there are three parallel housings in a stage and it makes sense to short them one element, each of the three housings in the stage should be shorted an element. This is necessary to keep the crossflow rates balanced between the three housings.

The missing element should be in the lead end of the housing. *Blanks* (also called *dummies*) are available from the membrane or housing manufacturer to connect the lead end permeate connection on the end cap of the housing to the first element of the housing. (See Figure 1-36.)

Simple plugs can also be used to block the permeate tube on the first element. However, using plugs will not prevent movement within the housing when the system starts up or shuts down. This movement can lead to O-ring breakage on the element interconnectors. If the system permeate is collected only from the lead end of the housing (i.e., the end where the feed flow enters the housing), a blank will have to be used to transport the permeate flow from the housing elements to the RO system permeate manifolding.

Number of elements = number of housings × elements/housing

$$\text{Element permeate flow} = \frac{\text{desired total permeate}}{\text{number of elements}}$$

Minimum concentrate flowrate. Membrane manufacturers specify a minimum concentrate flowrate for their elements, or they specify a minimum concentrate-to-permeate ratio (or both). The concentrate-to-permeate ratio is the ratio of the concentrate flowrate exiting an individual element, to the permeate flowrate of that element. It is also referred to as the minimum brine-to-permeate ratio, or it can be written as the maximum individual element permeate recovery.

Typical values for the manufacturer's recommended minimum concentrate-to-permeate ratio can range from 4:1 for 60-inch-long elements, to a 9:1 ratio for 40-inch elements in a heavily fouling application. In terms of individual element recoveries these values could be translated, realizing that:

$$\text{feed flow} = \text{permeate flow} + \text{concentrate flow}$$

and

$$\text{recovery} = \text{permeate flow} \div \text{feed flow}$$

In translating a 4:1 minimum ratio, we will assume a basis of one unit of permeate flow per four units of concentrate flow:

$$\text{recovery} = \frac{\text{1 unit permeate}}{\text{1 unit permeate + 4 units concentrate}} = 0.2 \text{ (20\% recovery)}$$

For a 9:1 ratio:

$$\text{recovery} = \frac{\text{1 unit permeate}}{\text{1 unit permeate + 9 units concentrate}} = 0.1 \text{ (10\% recovery)}$$

With the previously calculated individual element permeate flowrate, the minimum concentrate flowrate can be calculated using the minimum concentrate-to-permeate ratio.

The system concentrate flowrate can be calculated using the desired system permeate recovery and the desired system permeate flowrate. For example, if the desired recovery is 75% and the permeate flow is 15 gpm, the system concentrate flowrate is found using the definition for recovery:

$$\text{recovery} = \text{permeate flow} \div \text{feed flow}$$

where:

$$\text{feed flow} = \text{permeate flow} + \text{concentrate flow}$$

Thus:

$$0.75 = 15 \text{ gpm} \div (15 \text{ gpm} + \text{concentrate flow})$$

and this equation can be solved algebraically:

$$11.25 \text{ gpm} + (0.75 \times \text{concentrate flow}) = 15 \text{ gpm}$$
$$0.75 \times \text{concentrate flow} = 3.75 \text{ gpm}$$
$$\text{concentrate flow} = 5 \text{ gpm}$$

Allocation of housings. Most of the values are now available to enable a determination of the best staging for the application. This process requires allocating housings (or elements in the case of hollow-fiber systems) for the last stage (concentrate end) of the system, and working upstream to the next stage.

The initial allocation is based on the maximum number of housings (or hollow-fiber elements) through which the available concentrate flow can be split, while still meeting the minimum concentrate flow requirements for the individual elements. The feed flowrate to the last stage can then be calculated. This will be the same flowrate as the concentrate flowrate of the second-to-last stage. Housings can then be allocated in the same manner for the second-to-last stage.

If this allocation process happens to be able to allocate all housings without any left over, the resulting array would have all flowrates minimized. This array would operate with a minimal system pressure drop.

However, it is likely that there will be housings remaining, such that there are insufficient housings left to be allocated according to the available feed flow from the previously allocated stage. There are a few options available for what to do about the residual housings.

If there are only one or two housings left to be allocated for a system that uses a large number of housings, the following options should be considered:

- An array could be used that uses only the housings already allocated. The system recovery would be less than originally intended. The permeate flowrate could be increased to meet the desired rate by slightly increasing the pressure to the system.

- The residual housings could be added one at a time to the particular stage that has the lowest individual housing feed and concentrate flowrate. This also will result in a lower system recovery, but pressures will not have to be increased to meet the desired overall permeate flowrate.

In either of the two previous options, the system recovery can be set at the desired ratio by recycling water from the concentrate back to the feed. However, this will result in a greater concentration of solutes entering the feed of the membrane system, and subsequently a lower permeate quality.

If there are a number of housings left to be allocated (or if the system does not use a large number of housings), it usually results in better overall RO performance to reduce the number of housings used in one of the allocated stages. It should be determined which stage can best handle the increased individual housing flowrate that would result from having the total stage flow dispersed through one less housing. One housing would be removed from this stage and the flowrate recalculated.

Housings would then be allocated again for the upstream stages. If residual housings are still left over, another housing would again be removed from one of the stages using the same process. Eventually, an array will result that has another stage on the lead end of the system, and there will be no residual housings.

An array designed in this manner will meet the desired system permeate flowrate and recovery without the use of concentrate recycle. However, such an array design may result in a substantial system pressure drop.

In applications where there is a high potential for fouling or scale formation, it is common to design in higher crossflow rates in order to reduce this potential (although usually not exceeding the membrane manufacturer's guidelines for maximum flowrates or pressure drops). Sometimes this may be a concern in just the downstream stages of the array. In this case, these stages are designed with fewer housings than what might be optimal in order to obtain higher crossflow rates.

Actual membrane pressure. To calculate the actual membrane pressure needed to produce the desired system permeate flowrate, approximate values for the following are required:

- The temperature correction factor
- The average osmotic pressure of the solution as it passes through the RO system
- The hydraulic differential pressure across the membrane elements
- The permeate pressure

Temperature correction factor. The RO permeate flow is strongly dependent on the temperature of the feedwater. Higher temperatures result in higher permeate flowrates. There is some variation in the effect of temperature between different membrane types, although it does not become significant unless the temperature swing is significant.

Temperature correction factors are typically available from charts, graphs, or calculations supplied by the membrane manufacturer. It is a simple matter to look up or calculate the factor (see Appendix).

Average osmotic pressure. The osmotic pressure is a function of the concentration of total dissolved solids. The TDS is a measurement of the concentration of dissolved salts and dissolved organics in the water. (i.e. the non-filterable solids). In most city water systems, the osmotic pressure of the dissolved organic solids will be negligible as compared to the concentration of dissolved salts that are typically present.

A standard water analysis will usually include a value for the TDS. Otherwise, the concentrations of the individual ions (as ion) from the analysis can be added to obtain the TDS (if it is a complete analysis). Otherwise, the TDS can be estimated on-site by using a meter that measures conductivity of the water. This reading can be converted into a TDS measurement. Some conductivity meters include their own internal conversion mechanism to obtain a reading for TDS, based upon the conductivity of a common salt (usually sodium chloride or calcium carbonate).

For most brackish-water RO systems with a low feedwater TDS and a standard permeate water recovery, the osmotic pressure will have a negligible affect on the membrane permeate flowrate. However, for systems having feedwaters with a TDS greater than 1,000, or for systems operating at greater than 75% permeate water recovery, there will be a sufficient osmotic pressure to make it worth evaluating. If the TDS is not taken into account in the RO system design, the permeate flow will probably be less than anticipated.

Osmotic pressure is a variable that must be taken into account when designing RO systems for concentrating most organic solutes. For example, RO systems that concentrate apple juice will perform dramatically differently depending on how concentrated the fructose (sugar) is in the feed stream. It is common in juice concentration applications for the sugar concentration to be as high as 24%.

As discussed earlier in the theory of reverse osmosis, the RO permeate flow is proportional to the pressure across the membrane after subtracting out the effects of the solution osmotic pressure. The specific osmotic pressure for a solution will depend on the sum of the osmotic pressure contributions from each of the solutes being removed by the RO membrane.

For example, seawater contains approximately 35,000 mg/L (3.5%) of sodium chloride. The osmotic pressure of sodium chloride can be estimated by multiplying its concentration by the factor 0.0115 psig/ppm. Thus, the osmotic pressure of the sodium chloride component is 35,000 ppm times 0.0115 psig/ppm, which equals 402 psig.

If 10,000 ppm of sodium sulfate is added to the water, the sodium sulfate conversion factor is approximately 0.006 psig/ppm, and thus its osmotic pressure contribution is 10,000 ppm times 0.006, or 60 psig. The total osmotic pressure of the solution is then 402 plus 60, which equals 462 psig.

For most pure water applications the osmotic pressure can be estimated for the combined salt content of the water by multiplying the TDS by 0.01 psig/ppm.

Since the TDS increases in concentration as the feedwater is concentrated through the system, the osmotic pressure also increases. For normal permeate water recoveries, an average osmotic pressure for the system can be approximated by first calculating an arithmetic average of the RO system TDS. This is accomplished by adding the feed and concentrate TDS values and dividing by 2. The average TDS is used then to calculate the average osmotic pressure.

If the concentrate TDS is not available, it can be estimated for most high-rejecting membrane systems by dividing the feed TDS by:

$$1 - \text{recovery (as a fraction)}$$

For example, given an RO operating at 75% recovery, with a feedwater with 1,000 ppm of TDS, it is possible to get a close approximation for the concentrate TDS using the following calculation:

$$1{,}000 \text{ ppm} \div (1 - .75) = 4{,}000 \text{ ppm}$$

The average TDS would then be

$$(1{,}000 \text{ ppm} + 4{,}000 \text{ ppm}) \div 2 = 2{,}500 \text{ ppm}$$

and an average osmotic pressure would be

$$2{,}500 \text{ ppm} \times 0.01 \text{ psig/ppm} = 25 \text{ psig}$$

Hydraulic differential pressure. As the feedwater passes through the spacing material between the membrane leaves of a spiral-wound element or between the fibers of a hollow-fiber element, there is a loss of pressure due to the turbulence of the fluid. This means that a membrane element at the lead end of an RO system will experience a higher pressure than an element at the tail end of the system. The effects on the permeate flowrate by this loss in pressure can be estimated.

Membrane manufacturers usually provide graphs or some other means of determining a single element's hydraulic pressure differential as a function of feed flowrate, or based upon an arithmetic feed and concentrate flow average (i.e., feed flowrate plus concentrate flow divided by 2). Hollow-fiber elements are designed for operation with permeate recoveries from 35% to 75%. It is common for a single parallel banking of elements to be used in a hollow-fiber RO system. Estimating the pressure drop across the system is simply a matter of estimating the pressure drop across a single element based upon its individual-element flowrates.

Estimating the system pressure differential for a spiral-wound system with more than one element per housing is slightly more involved. Each housing of a stage can contain between one and six membrane elements, each element contributing to the pressure differential of that stage. Since the flow going through each element within the housing is reduced by the permeate flow that was drawn off by the previous element, the pressure differential provided by each subsequent element in the housing declines along with the crossflow rate.

The stage pressure differential can be estimated by calculating an arithmetic flow average of the individual housing feed flowrate plus the concentrate rate divided by 2. Using the manufacturer-supplied pressure differential chart or calculation, the individual-element pressure differential can be found for that average flowrate. This pressure differential is then multiplied by the number of elements in the housing to find the approximate pressure differential for the entire housing.

Permeate pressure. If it is anticipated that the permeate stream from the RO system is going to be plumbed a significant distance, or into another treatment device, permeate pressure must be considered in the RO design. Permeate pressure acts as back pressure to the membrane driving pressure. It uniformly subtracts from the driving pressure throughout the system.

For example, if the RO permeate is plumbed directly into a pressure tank (also called a *bladder tank*), the tank pressure at any particular moment will be working against the driving pressure, and will result in a reduced permeate flowrate. The maximum tank pressure should be used as a potential permeate pressure in the RO design.

If the RO is plumbed directly into a distribution loop, ion-exchange unit, or a cartridge filter, there will be an appreciable permeate back pressure placed on the RO system. If the RO feeds a piping loop, the desired loop pressure should be used as the value for RO permeate pressure. For water treatment equipment like ion-exchange units or cartridge filters, the manufacturer can assist in estimating the pressure drops as a function of the flowrate passing through their equipment. If several pieces of equipment are plumbed directly into the RO permeate prior to going directly into a distribution loop, the pressure of the loop should be added to the calculated pressure drops across each of the pieces of equipment to determine a value for the RO permeate pressure.

Calculating the actual membrane pressure. Once values have been determined for the various variables that can affect the membrane permeate flowrate, the actual membrane feed pressure can be calculated for the RO system. This is calculated by the following:

$$\text{feed pressure} = \frac{\text{element permeate flow} \times \text{psi}_{\text{dsgn}}{}^{*}}{\text{element design flow} \times \text{fouling factor} \times \text{temperature factor}} + \text{psig}_{\text{prm}} + \text{psid}/2 + \text{psi}_{\text{osm}}$$

where:
$psig_{prm}$ is the permeate pressure,
psid/2 is one half the hydraulic pressure differential,
psi_{osm} is the average osmotic pressure of the system,
psi_{dsgn} is the pressure at which the element is tested,
fouling factor is the fouling factor that was previously estimated, and
temperature factor is the water temperature correction factor.

Example:
Basis:
Permeate flowrate, 45 gpm
Recovery, 78%
Permeate flow (PA thin-film spiral-wound element at 225-psig, 2,000-ppm sodium chloride test conditions), 1,800 gpd
Minimum element concentrate-to-permeate ratio, 6:1
Elements per housing, 4
City water TDS, 520 ppm
City water temperature, 62 °F
City line pressure, 65 psig
SDI, 3.4
Permeate water plumbed to top of 30-foot-high storage tank

Answer:
First choose a desired driving pressure. On low-TDS brackish waters at a water temperature of 77 °F, the design permeate flowrate will be achieved by the PA thin-film membrane element if operated at 205 psig. A driving pressure of 200 psig will result in a individual-element permeate flowrate that will be slightly less than the design rate.

The SDI of 3.4 indicates that suspended solids are present in the water. This is not a particularly high SDI, but it is likely that some membrane fouling will occur. Therefore, a fouling factor of 0.85 will be chosen for this application.

The individual-element permeate flowrate can now be calculated:

$$\text{driving pressure} \times \text{fouling factor} \times \text{design permeate flowrate} \div \text{design pressure}$$

$$200 \text{ psig} \times .85 \times 1{,}800 \text{ gpd} \div (225 \text{ psig} - 20 \text{ psig})$$
$$= 1{,}490 \text{ gpd} = 1.04 \text{ gpm } (1{,}490 \div 1{,}440 \text{ minutes per day})$$

---

*The manufacturer's design pressure is based upon testing with a solution that normally includes a certain concentration of sodium chloride. To compensate for the osmotic pressure that is included in this design pressure, 20 psig should be subtracted from the design pressure if the manufacturer's test conditions included a 2,000-ppm sodium chloride concentration. If 1,000 ppm of sodium chloride was used in the manufacturer's test conditions, 10 psig should be subtracted.

Number of membrane elements =
desired permeate flow ÷ individual-element permeate rate
= 45 gpm ÷ 1.04 gpm = 43.3 elements

Number of housings = Number of elements ÷ elements per housing
= 43.3 elements ÷ 4 elements per housing
= 10.8 housings, which can be rounded up to 11

Actual number of elements = 11 housings × 4 elements/housing = 44 elements

Actual element permeate flow = 45 gpm ÷ 44 elements = 1.02 gpm

Minimum element concentrate flowrate
= element permeate flow × minimum concentrate:permeate ratio
= 1.02 gpm × 6 = 6.12 gpm

Individual housing permeate flowrate =
element permeate flow × elements/housing
= 1.02 gpm/element × 4 elements = 4.08 gpm

System concentrate flowrate
= total permeate flow ÷ recovery - total permeate flow
= 45 gpm ÷ 0.78 - 45 gpm = 12.7 gpm

Last-stage housing number
= concentrate flowrate ÷ minimum element concentrate rate
= 12.7 gpm ÷ 6.12 gpm = 2.08 housings

There is sufficient concentrate flow available to meet the minimum individual-element concentrate flowrate for two housings. In fact, there is very little excess flow that might otherwise result in an excessive pressure differential across the last stage.

The flowrates for the individual housing and for the total stage feed flowrate can now be calculated based upon using two housings in the last stage. The last-stage feed flowrate is the same as the second-to-last-stage concentrate flowrate.

Last-stage feed flow
= housing permeate × number of housings/stage + concentrate flow
= 4.08 gpm × 2 + 12.7 gpm = 20.9 gpm

Number of housings in next-to-last stage
20.9 gpm ÷ 6.12 gpm = 3.42 housings (rounded down to 3 housings)

Feed flowrate = 4.08 gpm/housing × 3 + 20.9 gpm = 33.1 gpm

Number of housings in second-to-last stage

33.1 gpm ÷ 6.12 gpm = 5.41 housings (rounded down to 5 housings)

Ten of the required eleven housings have now been allocated. Since there is only one remaining housing to be allocated, the extra housing could be added to one of the existing stages. The result would be an array with a very low system pressure differential. However, the array would either need to operate at a lower recovery than originally intended, or concentrate recycle would be required to achieve the desired recovery. The priorities of the application would normally dictate how the system should be designed and operated.

In this example, several alternatives will be evaluated. The first will be the result of adding a housing to an existing stage. Placing this housing in a stage that has the greatest excess flow after its original housing allocation will result in minimizing the volume of concentrate recycle required.

Both the second- and the third-to-last stage had an excess flow after the allocation, comparable to about 0.4 housings. In this situation, placing the housing in either of these two stages will result in the same flowrate requirement for concentrate recycle. However, placing the extra housing in the more downstream stage will result in a smaller increase in the flowrates of the other stage, and subsequently a smaller system pressure differential. Thus, the housing should be placed in the second-to-last stage, resulting in a 5-4-2 array.

Since a housing was added to the second-to-last stage, the concentrate flow of that stage will need to increase, based upon maintaining the minimum individual housing concentrate flowrate for each of the four housings. This in turn will affect the flowrates in the downstream stage (the last stage). The concentrate flow for the second-to-last stage is the same as the feed flowrate for the last stage.

6.12 gpm × 4 housings = 24.5 gpm

Last-stage concentrate flowrate = 24.5 gpm - (4.08 gpm/housing × 2) = 16.3 gpm

Concentrate recycle flowrate = 16.3 gpm - 12.7 gpm = 3.6 gpm

Because of the need to plumb in the recycle line with a control valve, recycling will add to the capital cost of the RO system. Since its use will also reduce the permeate quality, it may be desirable to instead reduce the system recovery.

It is possible to predict the effect that the concentrate recycle will have on the permeate quality. In the example, 3.6 gpm would be recycled back to the high pressure pump. As previously discussed, the concentrate TDS can be estimated from the feedwater TDS by dividing it by the value of "1 - recovery". For a 78% recovery system, this value is 4.5. The effect of the recycle on the feedwater TDS can then be estimated:

$$\text{combined TDS} = \frac{\text{makeup flow} \times \text{feed TDS} + \text{recycle flow} \times \text{recycle TDS}}{\text{makeup flow} + \text{recycle flow}}$$

where:
recycle TDS = feed TDS × 4.5

$$\text{combined TDS} = \frac{(57.6 \text{ gpm} \times 520 \text{ ppm}) + (3.6 \text{ gpm} \times 4.5 \times 520 \text{ ppm})}{(57.6 \text{ gpm} + 3.6 \text{ gpm})}$$
$$= 627 \text{ ppm}$$

The combined feed TDS can be expected to be 20% higher (627÷520 - 1) than if recycle were not used. The concentrate TDS will not be affected. Therefore, the average membrane TDS as well as the permeate TDS will be 10% higher than if recycle were not used.

Without recycle, the system recovery for this 5-4-2 array would be:

45 gpm ÷ (45 gpm + 16.3 gpm) = 73.4% recovery

The individual housing flowrates for this array would be as follows:

Stage	*Concentrate*	*Feed*
Last	16.3÷2 = 8.15	24.5 ÷ 2 = 12.2
Second-to-last	6.12	(24.5 + 4.08 × 4) ÷ 4 = 10.2
First	40.8÷5 = 8.16	(40.8 + 4.08 × 5) ÷ 5 = 12.2

It is possible to design an array with the desired recovery and without concentrate recycle. It will require an additional stage and higher individual housing flowrates, which will mean that the system will have a higher pressure differential. This will result in the need for additional membrane feed pressure.

To add an additional stage necessitates removing housings from the existing stages. This should be performed one housing at a time from the particular stage that will be least affected by the resulting higher flowrates that now have to pass through the remaining housings in the stage. This requires evaluating each stage.

Last-stage individual housing concentrate flow with one less housing:
12.7 gpm ÷ 1 housing = 12.7 gpm/housing

Second-to-last-stage individual housing concentrate flow with one less housing:
20.9 gpm ÷ 2 housings = 10.4 gpm/housing

Third-to-last-stage individual housing concentrate flow with one less housing:

33.1 gpm ÷ 4 housings = 8.27 gpm/housing

It is clear that the housing should be removed from the third-to-last stage since the resulting concentrate flow will be the lowest of the three statges. Its flowrate should then be reevaluated with another housing removed, since it will be necessary to remove another housing from one of the stages.

33.1 gpm ÷ 3 housings = 11.0 gpm/housing

This time the second-to-last stage can best handle the additional flow with a reduced number of housings. The array is now a 3-4-2-2 array. It will be necessary to remove at least one more housing to balance the array. The second-to-last stage should be re-evaluated for removing another housing. Also, the third-to-last stage will also need to have its flowrates recalculated.

The second-to-last individual housing concentrate flowrate with another housing removed would be:

20.9 gpm ÷ 1 housing = 20.9 gpm/housing

The total fccd flow to the second-to-last stage, which is the same as the concentrate flow of the third-to-last stage, will now be:

20.9 gpm + 4.08 gpm/housing × 2 housings = 29.1 gpm

If the third-to-last stage had a housing removed, its individual housing concentrate flowrate would be:

29.1 gpm ÷ 3 housings = 9.7 gpm/housing

The third-to-last stage is again the best one for removal of a housing. The array is now a 4-3-2-2. The individual housing flowrates should again be evaluated to determine if the array can be further optimized.

The third-to-last stage with another housing removed would have the following individual housing concentrate flowrate:

29.1 gpm ÷ 2 housings = 14.5 gpm/housing

The first stage with four housings would have a total concentrate flowrate of:

29.1 gpm + 4.08 gpm/housing × 3 housings = 41.3 gpm

It would have an individual housing concentrate flowrate of:

1.3 gpm ÷ 4 housings = 10.3 gpm

This flowrate is less than what the individual housing concentrate flowrate would be for the last stage (which would be the best stage from which to remove a housing) if it had a housing removed. Therefore, the 4-3-2-2 array cannot be further optimized.

The individual housing flowrates of this array would be:

Stage	*Concentrate*	*Feed*
Last	$12.7 \div 2 = 6.35$	$20.9 \div 2 = 10.4$
Second-to-last	$20.9 \div 2 = 10.4$	$29.1 \div 2 = 14.5$
Third-to-last	$29.1 \div 3 = 9.7$	$41.3 \div 3 = 13.8$
First	$41.3 \div 4 = 10.3$	$(41.3 + 4.08 \times 4) \div 4 = 14.4$

Generally, the individual housing flowrates through this array are about 20% more than the 5-4-2 array. This will result in a greater system pressure drop, particularly since there is also an additional stage to consider. A higher feed membrane pressure will be required for this array, resulting in higher operating and capital equipment costs.

The best array for this RO system will depend on the application priorities. If permeate quality and system recovery are both priorities over operating cost, the 4-3-2-2 would be the best array. If the application involves a high potential for fouling or scale formation, the higher flowrates used in the 4-3-2-2 array might be desirable to reduce the membrane fouling/scaling rate. If system recovery is not a priority, then the 5-4-2 array could be operated at lower recovery for better permeate quality. If operating costs are the priority, then the 5-4-2 array with concentrate recycle would probably be the desirable array.

For continued consideration of this example, the 4-3-2-2 array will be chosen. Other variables will now need to be considered in order to determine the required membrane pressure so that the high-pressure pump can be sized.

Calculating with temperature correction factor. In this example, the feedwater temperature is constant throughout the year at 62 °F. If the temperature varied, the system should be sized using a temperature correction factor based on the lowest temperature possible to insure that adequate permeate flow is available all year. From the Appendix, the factor is 0.81. (This means that the permeate flow at 62 °F will be 81% of what it would be at 77 °F.)

Calculating the average osmotic pressure. The arithmetic average for the TDS within the RO system would be calculated as follows:

$$\text{average TDS} = (\text{feed TDS} + \text{feed TDS} \div (1 - \text{recovery})) \div 2$$

$$= (520 \text{ ppm} + 520 \text{ ppm} \div (1 - 0.78)) \div 2 = 1{,}440 \text{ ppm}$$

Since this is a somewhat typical brackish water, we can use the approximate factor of 0.01 psig osmotic pressure/ppm to calculate the average osmotic pressure for the system.

$$1{,}440 \text{ ppm} \times 0.01 \text{ psig/ppm} = 14 \text{ psig}$$

Calculating the hydraulic differential pressure. The feed-to-concentrate pressure drop can be approximated for each of the housing stages based upon the average of the individual housing feed and concentrate flowrates. This average flow can be calculated by subtracting half the housing permeate flowrate from the individual housing feed flowrate. This flow can then be used to find the pressure differential for the stage (from the pressure differential graph in the Appendix).

Last-stage average flow and pressure drop $= 10.4 \text{ gpm} - 4.08 \text{ gpm} \div 2 = 8.36 \text{ gpm}$

The individual-element pressure drop for this flow would be about 4 psid. This would be multiplied by the four elements in the housing to come up with the pressure drop for the entire housing.

$$4 \text{ psid/element} \times 4 \text{ elements} = 16 \text{ psid}$$

Second-to-last-stage average flow and pressure drop

$$14.5 \text{ gpm} - 4.08 \text{ gpm} \div 2 = 12.5 \text{ gpm} = 8 \text{ psid/element}$$

$$8 \text{ psid/element} \times 4 \text{ elements} = 32 \text{ psid}$$

Third-to-last-stage average flow and pressure drop

$$13.8 \text{ gpm} - 4.08 \text{ gpm} \div 2 = 11.8 \text{ gpm} = 6.75 \text{ psid/element}$$

$$6.75 \text{ psid/element} \times 4 \text{ elements} = 27 \text{ psid}$$

First-stage average flow and pressure drop

$$14.4 \text{ gpm} - 4.08 \text{ gpm} \div 2 = 12.4 \text{ gpm} = 7.5 \text{ psid/element}$$

$$7.5 \text{ psid/element} \times 4 \text{ elements} = 30 \text{ psid}$$

The total system differential pressure would be the total of the stage pressure drops:

$$16 \text{ psid} + 32 \text{ psid} + 27 \text{ psid} + 30 \text{ psid} = 105 \text{ psid}$$

As predicted, this array will suffer from a fairly significant hydraulic pressure drop.

Calculating the permeate pressure. The permeate line in this example is plumbed to the top of a 30-foot-high storage tank. The storage tank will be at atmospheric pressure and so will not add any back pressure to the permeate. However, the height (also known as head) of the storage tank will provide some back pressure. The conversion of water head to pressure is 2.31 feet of head per psig of pressure. The permeate back pressure would be:

$$30 \text{ ft}/2.31 = 13 \text{ psid}$$

All of the values have now been calculated necessary to determine the actual feed membrane pressure needed to derive the desired permeate flowrate from the RO system. The pressure is calculated as:

$$\text{feed pressure} = \frac{\text{element permeate flow} \times \text{psi}_{dsgn}^{*}}{\text{element design flow} \times \text{fouling factor} \times \text{temperature factor}} + \text{psig}_{prm} + \text{psid}/2 + \text{psi}_{osm}$$

$$\{1.04 \text{ gpm} \times (225 \text{ psig} - 20 \text{ psig})\} \div (1.25 \text{ gpm} \times 0.85 \times 0.81) + 13 \text{ psig} + 105 \text{ psid}/2 + 14 \text{ psig} = 327 \text{ psig}$$

Choosing the high pressure pump. A pump needs to be chosen to provide the total flowrate (that would include the concentrate recycle flow if it were being used) of 57.6 gpm at a pressure of 327 psig. The city inlet pressure is 65 psig. If it is assumed that as much as 25 psig of this pressure may be lost as the feedwater travels through the RO pretreatment equipment, the high pressure pump inlet pressure could be as low as 40 psig (65 psig - 25 psig). This can be subtracted from the membrane feed pressure in determining the required pressure capability of the high-pressure pump.

$$327 \text{ psig} - 40 \text{ psig} = 287 \text{ psig}$$

A number of high-pressure pumps are available that could meet these flow and pressure requirements. Examples would be the Grundfos Model CR16-50N, the Tonkaflo Model 5516, or a Webtrol Model H60B20T. A 20-horsepower (hp) motor will be necessary for the pump. Pump efficiencies, materials of construction, availability, serviceability, and cost should all be considered when choosing the particular pump.

Brake horsepower. To consider energy consumption in this decision it is important to calculate the pump's brake horsepower using the following calculation:

$$\text{Brake hp} = \text{gpm} \times \text{output psig} \div 1{,}714 \times \text{pump efficiency}$$

If the pump curve fell exactly on the desired flow and pressure requirements, and had an efficiency of 60%, it would have a brake horsepower of:

$$\text{brake hp} = 57.6 \text{ gpm} \times 287 \text{ psig} \times 0.6 = 16.1 \text{ hp}$$

Predicting Permeate Quality

There are several ways to predict the permeate quality from an RO system. One method is based upon the rejection characteristics of the dissolved salts versus the rejection of the individual ions. It takes into account the effects of the other ions in the solution when estimating the salt passage; thus, it tends to provide better accuracy.

The first step in performing this calculation is to express all ion concentrations as $CaCO_3$. This is equivalent to working in concentrations that are based on the number of molecules. Conversion factors are given in the Appendix.

The next step is to estimate the average concentration in the RO membrane system for each of the ions. With membranes of high rejection (i.e., better than 92% rejection of sodium chloride), it can be approximated that nearly all ions will be rejected by the membrane. A mass balance can then be used to find a simple ratio useful in calculating each of the average ion concentrations:

$$\text{feed concentration} = (1 - \text{recovery}) \times \text{concentrate concentration}$$

or:

$$\text{concentrate concentration} = \text{feed concentration}/(1 - \text{recovery})$$

$$\text{Average concentration} = \text{feed concentration} \times (2 - \text{recovery}) \div (2 - 2 \times \text{recovery})$$

Thus the factor to use in estimating average ion concentrations is as follows:

$$(2 - \text{recovery}) \div (2 - 2 \times \text{recovery})$$

Example:

The average concentration of a salt in an RO system with a 75% recovery can be estimated by multiplying the concentration of the feedwater in the system by the following factor:

$$(2 - 0.75) \div (2 - 2 \times 0.75) = 2.5$$

The average ion concentration will be 2.5 times the feed ion concentration. The next step is to determine the concentrations for the various salts that might be formed by the cations and anions that are present in a water source. The preference of association for the salts is somewhat arbitrary, and must be derived empirically (i.e., by experiment).

Table 1-5 lists common salts in an order of assumed preference. The salt passage is given for each salt relative to that of sodium chloride. If a particular salt will permeate the membrane at 0.6 times that of sodium chloride, and a membrane is rated for 3% sodium chloride rejection (i.e., 97% rejection), 1.8% of that salt can be expected to pass the membrane. This would be equivalent to slightly better than a 98% rejection of the salt.

Once average concentrations have been calculated for the ions in a particular solution, the order of salt association given in Table 1-5 can be used to create an artificial listing of salts for the given solution. As much calcium sulfate should be listed as allowed for by the available concentrations of calcium and sulfate (as $CaCO_3$). If there is only 1 mg/L of sulfate ion, there can only be 1 mg/L of calcium sulfate, even though there may be a large concentration of calcium available. The remaining calcium would be associated with other anions, depending on the next associated salts in the listing.

Based upon the design specification for sodium chloride passage for the particular membrane, percent passage can be approximated for each of the salts. Salts made up of ions with high valence numbers (large charge values) will tend to have lower passage numbers. Factors can be used to convert the passage of each salt to the membrane's rating for sodium chloride passage.

Table 2-1
Fraction of Salt Passage Relative to Sodium Chloride
(in order of salt formation preference)

Salt	*Formula*	*Fraction Relative to NaCl*
Calcium sulfate	$CaSO_4$	0.3
Calcium bicarbonate	$Ca(HCO_3)_2$	0.6
Sodium chloride	$NaCl$	1
Sodium bicarbonate	$NaHCO_3$	0.8
Sodium sulfate	Na_2SO_4	0.3
Calcium chloride	$CaCl_2$	0.6
Magnesium bicarbonate	$Mg(HCO_3)_2$	0.6
Magnesium chloride	$MgCl_2$	0.6
Potassium chloride	KCl	1
Magnesium sulfate	$MgSO_4$	0.3
Potassium bicarbonate	$KHCO_3$	1
Potassium sulfate	K_2SO_4	0.3
Calcium nitrate	$Ca(NO_3)_2$	5
Sodium nitrate	$NaNO_3$	5
Magnesium nitrate	$Mg(NO_3)_2$	5
Potassium nitrate	KNO_3	5
Calcium fluoride	CaF_2	1
Sodium fluoride	NaF	2
Magnesium fluoride	MgF_2	1
Potassium fluoride	KF	2

(Data courtesy of Osmonics, Inc.)

Example:
Determine the permeate quality for an RO system operating at 82% recovery that uses a membrane rated for 95% sodium chloride rejection (5% passage), on a water source with the following analysis:

All concentrations given as $CaCO_3$.

Cations		*Anions*	
Calcium - Ca^{+2}	145	Bicarbonate - HCO_3^-	120
Magnesium - Mg^{+2}	38	Carbonate - CO_3^{-2}	0
Sodium - Na^+	189	Sulfate - SO_4^{-2}	89
Potassium - K^+	7	Chloride - Cl^-	162
Fluoride - F^-	0	Hydroxyl - OH^-	0
Nitrate - NO_3^-	4		

First, it should be noted that the total concentration of cations does not equal the total of the anions. The cations exceed anions by 4 mg/L. One of the anion concentrations should be increased by this amount to the analysis. (In this example, chloride will be chosen.) Next, a factor should be determined to estimate an average membrane concentration for each of the ions.

$$\text{Factor} = (2 - \text{recovery}) \div (2 - 2 \times \text{recovery})$$
$$= (2 - 0.8) \div (2 - 2 \times 0.8) = 3$$

Each ion's concentration should by multiplied by 3 to find an average ion concentration. This may not be very accurate for nitrates, which are not as well rejected by the RO membrane. Since the concentration of nitrate is relatively small in the given analysis, this will not significantly affect the results.

Average concentrations:

Cations		*Anions*	
Calcium - Ca^{+2}	435	Bicarbonate - HCO_3^-	360
Magnesium - Mg^{+2}	114	Carbonate - CO_3^{-2}	0
Sodium - Na^+	567	Sulfate - SO_4^{-2}	267
Potassium - K^+	21	Chloride - Cl^-	498
Fluoride - F^-	0	Hydroxyl - OH^-	0
Nitrate - NO_3^-	12		

The association of salts can now be performed. The results are given in Table 2-2. The permeate concentrations are expressed as $CaCO_3$. They can be converted into their concentrations as ion using Index X-1. If desired, a salt rejection can then be calculated for the given water source. Be sure that this is performed using a combined feed or average salt concentration that has been expressed as ion.

Operating and Capital Equipment Costs

The capital cost of an RO system is, of course, dependent on the size of the system. For systems of less than about 150 gpm, the cost per gallon of permeate changes dramatically with the size of the system. The frame, high-pressure pump, and labor costs tend to dominate over membrane and housing costs. In the larger systems, cost tends to be more linearly proportional to permeate flow, as

Table 2-2
Example of Permeate Quality Determined by Salt Association

Salt	*Average Concentration*	*% Passage*	*Permeate Concentration*
Calcium sulfate	267	0.015	4.0
Calcium bicarbonate	168*	0.03	5.0
Sodium chloride	498	0.05	24.9
Sodium bicarbonate	69**	0.04	2.8
Sodium sulfate	0	0.015	
Calcium chloride	0	0.03	
Magnesium bicarbonate	114***	0.03	3.4
Magnesium chloride	0	0.03	
Potassium chloride	0	0.05	
Magnesium sulfate	0	0.015	
Potassium bicarbonate	9	0.05	0.5
Potassium sulfate	0	0.015	
Calcium nitrate	0	0.25	
Sodium nitrate	0	0.25	
Magnesium nitrate	0	0.25	
Potassium nitrate	12	0.25	3
Calcium fluoride	0	0.05	
Sodium fluoride	0	0.1	
Magnesium fluoride	0	0.05	
Potassium fluoride	0	0.1	

*At this point, all of the calcium has been allocated.
**All of the sodium is now allocated.
***All of the magnesium is now allocated.
(Data and method courtesy of Osmonics, Inc.)

the membrane and housing costs tend to dominate.

Obviously, the cost of a complete RO system will also depend dramatically on other factors such as how much pretreatment equipment is used, if standby pumps are kept available, and if service contracts are included. The cost of a brackish-water spiral-wound RO system can be approximated by the following formula:

RO cost = \$850 - \$1,200/gpm permeate flow

This formula does not take into account cost increases due to inflation. Small wall-mounted systems and systems that use flexible tubing and low-pressure housings probably may cost less than what would be predicted by this formula.

Cellulose acetate membrane systems that operate at 200 psig are going to be substantially higher in cost than a 400-psig CA system or a 200-psig thin-film system, because of the additional membrane required. Although a 400-psig CA system requires a larger high-pressure pump and more expensive housings than a 200-psig thin-film system, it is usually only slightly more expensive because of the typically higher cost of thin-film membrane elements.

Operating costs for an RO system include the following:

1. Power
2. Chemical injection
3. Cartridge prefilters
4. Membrane replacement
5. Water and sewer charges
6. Cleaning chemicals
7. Maintenance labor

Power. The amount of energy to drive a pump can be calculated using the following:

$$\text{kwh/1,000 gallons permeate} = \frac{\text{pump pressure} \times 0.00728}{\text{pump efficiency} \times \text{motor efficiency} \times \text{recovery}}$$

where:
kwh is the energy required for the pump in kilowatt hours,
pump pressure is the pressure added to the system by the high-pressure pump,
pump efficiency is the efficiency of the pump at the operating flowrate, and
recovery is the RO permeate recovery as a fraction.

The cost per kwh of electrical energy can then be multiplied by the results in order to find the cost per 1,000 gallons of permeate.

Chemical injection cost is figured as follows:
chemical cost/1,000 gallons permeate
= 8.3 lb/gallons × ppm of chemical × chem cost/lb × 1,000 ÷ recovery

where:
ppm of chemical is the RO feedwater chemical concentration

Note: For most solutions, a concentration of a substance in water given in ppm is equal to one given in mg/L. One part per million of a chemical is defined as 0.000001 pounds of chemical per pound of water.

Cartridge Prefilters. Cartridge filters commonly come in sizes built around a 10-inch base size: thus, the filter industry tends to use this size as a unit of measurement for comparison purposes. A 40-inch filter would be considered equal to four 10-inch equivalents (TIE). A 9-7/8-inch or 9 1/2-inch filter would be evaluated the same as one TIE.

Based upon a design flowrate of 5 gpm per TIE, and some estimated filter life, a cost of the prefilters as a function of RO flow can be determined:

$$\text{filter cost/1,000 gallons permeate} = \frac{\text{cost} \div \text{TIE}}{5\ \text{gpm/TIE} \times \text{recovery} \times \text{days/change} \times 1.44}$$

where:
cost ÷ TIE is the relative cost of a 10-inch cartridge filter,
recovery is the RO permeate recovery as a fraction, and
days/change is the number of days the RO is operating between filter replacements.

Membrane replacement. Using a projected life for the RO membrane elements, a cost per 1,000 gallons of permeate flow can be determined.

$$\text{Membrane cost/1,000 gallons permeate} = \frac{\text{cost/element}}{\text{permeate gpm/element} \times \text{days/change} \times 1.44}$$

where:
cost/element is the cost of a single membrane element,
permeate gpm/element is the individual-element permeate flowrate of the RO system in gpm, and
days/change is the number of days the RO is operating between membrane element replacements.

Water and sewer charges. Municipalities typically bill for both water and sewer use on the basis of the incoming water gallonage. This rate is usually on a 100-cubic foot (ft^3) basis, which is equal to 748-gallon increments. If this is the case, the cost for the RO usage is calculated by the following:

$$\text{water and sewer cost/1,000 gallons permeate} = \frac{\text{cost/ft}^3}{\text{recovery} \times 0.748}$$

where:
cost/ft^3 includes the cost of both the water and the sewer charge per ft^3 of water consumed, and
recovery is the RO permeate recovery as a fraction.

Cleaning Chemicals. For most systems the frequency of RO cleaning is low enough that the cost of the chemicals is negligible when compared to the entire volume of permeate produced between the cleanings. However, the cleaning cost can be significant for heavily fouling applications.

$$\text{Cleaning chemical cost/1,000 gallons permeate} = \frac{\text{chemical cost/cleaning}}{\text{permeate gpm} \times \text{days/cleaning} \times 1.44}$$

where:
chemical cost/cleaning is the cost of cleaning chemicals to perform a comprehensive cleaning/sanitization of the RO system,
permeate gpm is the RO permeate flowrate in gpm, and
days/cleaning is the number of days between cleanings.

Maintenance labor. For smaller RO systems, the labor needed to maintain the RO is a significant cost of operating an RO system. The time required to record operating data and clean an RO is not appreciably less for a smaller system than it is for a large system. Typically, the time to record data for one RO array (independent of its size) is less than 30 minutes per day. The labor required to clean and rinse one RO array is usually around 4 man-hours.

$$\text{Labor cost/1,000 gallons permeate} = \frac{\text{cleaning time} \times \text{cost/man-hour}}{\text{permeate gpm} \times \text{days/cleaning}} + \frac{\text{monitoring hours/day} \times \text{cost/man-hour}}{\text{1.44 Permeate gpm} \times 1.44}$$

where:
cleaning time is the hours required to clean the RO,
cost/man-hour is the total cost per hour of labor,
monitoring hours/day is the number of hours required each day to monitor the RO

system,
permeate gpm is the RO permeate flowrate in gpm, and
days/cleaning is the number of days between system cleanings.

Example:
Basis:
Permeate, 100 gpm flowrate, RO operates 70% of the time
Recovery, 75%, with no recycle flow
$0.08/kwh
245 psig, 60%-efficiency pump with 94%-efficiency motor
Scale inhibitor, 7 ppm at $750/500-lb drum
Prefilter, $2/TIE, changed monthly
4 gpm/element at $1,200/element
Water and sewer charge, $0.65/100 ft^3
Cleaning solution, $200/6 months
Cleaning, 4 man-hours at $20/man-hour

$$\text{kwh/1,000 gallons permeate} = \frac{\text{pump pressure} \times 0.00728}{\text{pump efficiency} \times \text{motor efficiency} \times \text{recovery}}$$
$$= 245 \text{ psig} \times 0.00728 \div 0.6 \div 0.94 \div 0.75 = 4.2 \text{ kwh}$$
Power \$/1,000 gallons permeate = 4.2 kwh × \$0.08/kwh = \$0.34

Chemical cost/1,000 gallons permeate
$$= 8.3 \text{ lb/gal} \times \text{ppm of chemical} \times \text{chemical cost/lb} \times 1{,}000 \div \text{recovery}$$
$$= 8.3 \text{ lb/gal} \times 0.000007 \times \$750/500 \text{ lb} \times 1{,}000 \div 0.75 = \$0.12$$

$$\text{Filter cost/1,000 gallons permeate} = \frac{\text{cost} \div \text{TIE}}{5 \text{ gpm/TIE} \times \text{recovery} \times \text{days/change} \times 1.44}$$
$$= \$2/\text{TIE} \div 5 \div 0.75 \div 30 \text{ days/change} \div 0.7 \div 1.44 = \$0.02$$

$$\text{Membrane cost/1,000 gallons permeate} = \frac{\text{cost/element}}{\text{permeate gpm/element} \times \text{days/change} \times 1.44}$$

$$= \$1{,}200/\text{element} \div 4 \text{ gpm/element} \div 1{,}095 \text{ days/change} \div 0.7 \div 1.44 = \$0.27$$

$$\text{water and sewer cost/1,000 gallons permeate} = \frac{\text{cost/ft}^3}{\text{recovery} \times 0.748}$$

$$= \$0.65/100 \text{ ft}^3 \div (0.75 \times 0.748 \text{ kgal}/100 \text{ ft}^{3)} = \$1.16$$

Labor cost/1,000 gallons permeate =

$$\frac{\text{cleaning time} \times \text{cost/man-hour}}{\text{permeate gpm} \times \text{days/cleaning}} + \frac{\text{monitoring hours/day} \times \text{cost/man-hour}}{1.44 \text{ Permeate gpm} \times 1.44}$$

$$= \frac{4 \text{ man-hours} \times \$20\text{/man-hour}}{100 \text{ gpm} \times 182 \text{ days/cleaning}} + \frac{0.5 \text{ hour} \times \$20\text{/man-hour}}{1.44\ 100 \text{ gpm} \times 1.44} = \$0.07$$

Power	\$0.34
Chemical injection	\$0.12
Cartridge prefilters	\$0.02
Membrane replacement	\$0.27
Water and sewer charges	\$1.16
Cleaning chemicals	\$0.01
Maintenance labor	\$0.07
Total	\$1.99/1,000 gallons permeate

In this example, raw water and water disposal costs dominate the cost of operating the RO system. In some areas these costs may be far less significant and power costs may be higher. It is important that the particular variables of the specific appliaction be evaluated in estimating the RO operating costs.

This concludes the chapter on RO design. The next chapter will cover design and maintenance considerations of the RO pretreatment equipment.■

CHAPTER 3

RO PRETREATMENT

The purpose of RO pretreatment is to optimize the performance and life of the RO membrane elements. Pretreatment equipment is engineered to accomplish this task by meeting the following three objectives:

1 Insure the compatibility of the feed stream with the RO membrane.

2 Reduce or eliminate the potential for scale formation on the RO membrane or in the element flow channels.

3 Reduce the RO cleaning frequency by reducing the fouling potential of the feed stream.

Membrane Compatibility

The RO membrane has to be able to tolerate the solutes or solvents it will be processing if an economical life is to be expected from the membrane. If a certain component of the feed stream is capable of significantly attacking the RO membrane or any material that goes into the construction of the membrane element, either that component will need to be removed for most applications to be a success, or a different membrane element will need to be used. The most common example of this compatibility issue is the need to remove chlorine or chloramines from the feed stream of a polyamide membrane RO system. If this feed stream component is not removed, these membrane elements will have a significantly reduced life expectancy.

For special process applications, it is sometimes necessary to compromise on membrane life. Sometimes it is not feasible to remove an incompatible contaminant in the feed stream, yet the economics of the application can justify a reduced membrane life.

Cellulose acetate membrane compatibility. Unlike PA membranes, CA membranes (discussed in Chapter 2) are tolerant of normal concentrations of oxidizing agents that will be present in most municipal water sources. Most CA membrane systems can safely handle up to a 1.0-ppm continuous feed concentration of free chlorine.

Whereas PA membranes are fairly tolerant of pH extremes, CA membranes are not. Cellulose acetate is somewhat unstable in water, continually losing acetyl groups. The rate of this breakdown, also known as hydrolysis, depends upon the pH and temperature of the RO feedwater. (Refer back to Figure 1-8.) To maximize CA membrane life, pH adjustment is usually required.

pH control for reduction of hydrolysis. The pH of most municipal water sources is greater than 7. From Figure 2-2, it can be deduced that the optimum pH to reduce the rate of membrane hydrolysis is about 4.8. It is likely that a noticeable reduction in membrane life will occur if a CA membrane system is operated at a pH greater than 7 for an extended period of time.

For most water sources, the RO concentrate will have a higher pH than the feedwater due to the higher concentration of bicarbonate ion in the RO concentrate stream. The membrane elements in the tail end of the RO system (i.e., the elements that see the water last as the feedwater passes through the system) will be exposed to this pH that is higher than the feedwater pH. In a 75%-recovery RO system, it is common for the RO concentrate pH to be more than 0.5 higher than the feedwater pH. The subsequent loss in membrane performance due to pH-related hydrolysis will be most evident in the tail end membrane elements.

The desired operating pH for a CA membrane system is usually not based solely upon minimizing the rate of CA membrane hydrolysis. To achieve the optimum pH of 4.8 requires that nearly all of the alkalinity in the feedwater be converted to carbon dioxide. Depending on the concentration of alkalinity in the feedwater, this would require a substantial concentration of acid. With the absence of the alkalinity, any excess acid would dramatically affect the water pH; therefore for most applications, attempting to maintain a feedwater pH of 4.8 for minimal membrane hydrolysis would be impractical and difficult to control.

Most industrial CA membrane RO systems operate with a feedwater pH between 5.5 and 6.5. In fact, the particular pH is usually chosen based upon the pH required to maintain the solubility of calcium carbonate in the RO concentrate stream, rather than on the rate of CA membrane hydrolysis.

With small CA RO systems with low permeate recovery rates, it may be economically feasible to operate with a feedwater pH as high as 8. This might be the case if there are few membrane elements to replace, and the feedwater pH never exceeds 8.0. The alternative of purchasing and maintaining a pH control system would cost more than it would to replace the membrane elements on a more frequent basis. (The system and membrane manufacturers can assist in making these determinations.)

Sulfuric acid (H_2SO_4) is commonly used for pH control because of its relatively low cost, and because it is less aggressive than hydrochloric acid toward metal parts in the system. Its dissociation is close to complete, contributing two hydrogen ions for every molecule of acid. Also, the acid's anion component, sulfate ($SO_4^=$), which is divalent, is relatively well rejected by the RO membrane; therefore, sulfuric acid injection does not significantly reduce the quality of the RO permeate water.

Sulfuric acid is dangerous to handle, as it is very aggressive toward organic materials (such as skin and most clothes). Many areas now require double containment of acids. This means that any acid drums must be kept in an area capable of retaining the volume of chemical contained in the drums. (Consult with local authorities.)

With some water sources, the additional sulfate contribution from sulfuric acid injection can increase the potential for sulfate precipitation. Calcium, barium, and strontium sulfate all have limited solubilities, which are usually better at low pH; therefore, acid injection is beneficial at reducing the potential for scale formation, but not necessarily if sulfuric acid is used.

Hydrochloric acid (HCl) is sometimes used in place of sulfuric acid. It usually is more expensive, and can be more aggressive with stainless steel parts in the water system. However, it will not increase the potential for sulfate scale formation. The anion component, chloride (Cl^-), is not as well rejected as the sulfate ion. If an RO system were to switch from sulfuric acid to hydrochloric acid injection for pH control, there would be a noticeable drop in the overall salt rejection for the system.

It is possible to use other acids for pH control of the RO. All other acids will likely be significantly more expensive than sulfuric or hydrochloric acid.

Compound X. A problem first encountered in the southwestern United States that results in the degradation of the lead end CA membrane elements (i.e., the first elements to come into contact with the RO feed stream) has been dubbed "Compound X." It typically occurs with well waters that are chlorinated. The degradation is fairly rapid, with the destruction of the lead end membrane elements occurring within a few days.

The Compound X phenomenon, first encountered during the early years of reverse osmosis, has been surrounded with mystery and misinformation. Frequently, membrane degradation that was really caused by other problems, such as pH or temperature excursions, was attributed to Compound X. This resulted in the compilation of a lot of inaccurate information about the Compound X phenomenon.

Some conjectured that Compound X was caused by bacteria that thrived in the presence of low concentrations of chlorine. Others proposed it was caused by a chemical species created in the presence of chlorine. In any case, the solution was to remove the chlorine by injecting a reducing agent, or by filtering the water through activated carbon. Once the chlorine was removed, it was frequently necessary to use some other biocide to prevent biological fouling and/or membrane attack caused by other types of bacteria.

In recent years there have been fewer reported cases of Compound X. This may be due to the improved resistance of the newer CA membranes, or to the increased attention now given by the membrane and system manufacturers when designing a new RO system for operation in the southwestern United States.

Study at the Yuma Desalting Facility. Investigative work into the Compound X phenomenon has been performed at the Bureau of Reclamation, Yuma Desalting Facility, located in Yuma, Arizona. The Yuma facility is a multimillion-gallons-per-day RO facility designed to remove salt from the Colorado River prior to its entrance into Mexico. This salt removal is dictated by a treaty with Mexico requiring that the river water be returned to its original quality.

A membrane element was pulled from the lead end of the RO system. It had demonstrated symptoms of membrane degradation similar to those associated with Compound X. A piece of its membrane was analyzed using scanning electron microscopy (SEM) and energy-dispersive X-ray (EDX) analyses. (See "Analytical Methods for RO Troubleshooting" in Chapter 6.) Figures 3-1 and 3-2 show the presence in the membrane skin layer of holes that are the apparent cause of the poor salt rejection and increased permeate flow from the element.

An EDX analysis was performed on the center of one of the holes, and also, as a control, on a piece of membrane from the tail end of the same housing. This

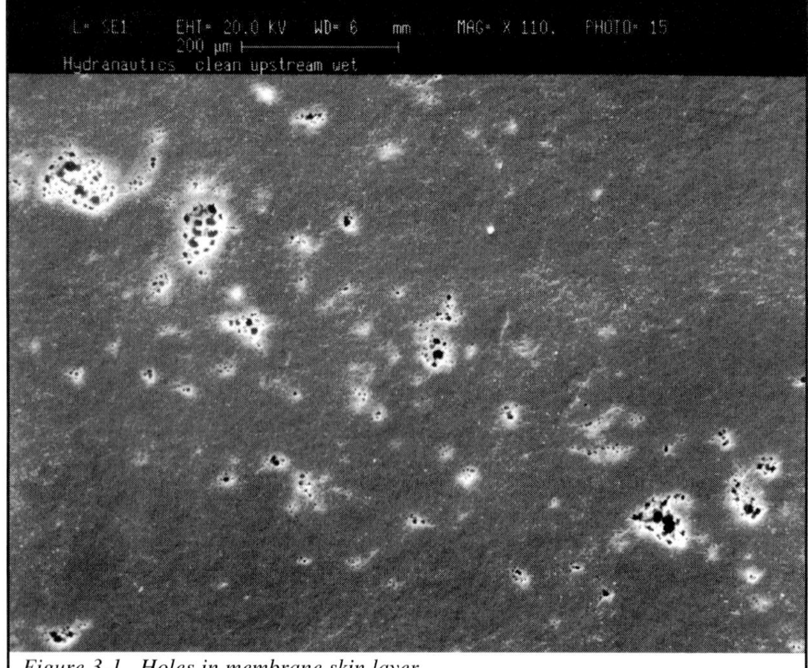

Figure 3-1. Holes in membrane skin layer.
(Courtesy of Hydranautics.)

control element had not experienced significantly lower rejection losses or increases in permeate flux.

The purpose of performing EDX on the tail end membrane was to obtain a basis for comparison. This membrane should have present many of the foulants that would also be present on the lead end membrane. Differences in the X-ray spectrum between the two pieces of membrane could indicate a greater concentration of a contaminant that might be related to the membrane degradation.

The two spectrums are shown in Figures 3-3 and 3-4. The lead end membrane is identified as "Hydranautics upstream," whereas the intact membrane is identified as "Hydranautics downstream." The lead end membrane X-ray identifies iron and manganese. Although some iron is present on the control membrane, it is not as significant in concentration.

The source for the iron and manganese was the raw water, which had concentrations approaching 0.1 mg/L. It was suspected that the iron and manganese were falling out of solution in the lead end elements. On the membrane surface, the iron and manganese would act as catalysts to increase the rate of membrane oxidation caused by the 0.2 to 0.3 mg/L of free chlorine that was present in the feedwater. It is suspected that this resulted in the local oxidation of the membrane around the iron and manganese particles (see Figure

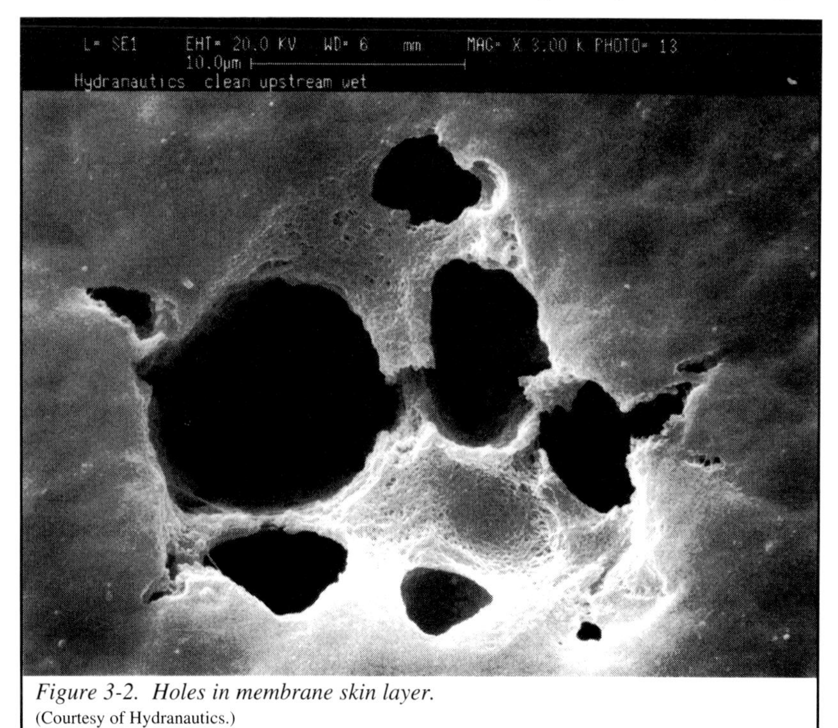

Figure 3-2. Holes in membrane skin layer.
(Courtesy of Hydranautics.)

3-5) (Mark Wilf, July 1993).

The ability of transition elements to increase the oxidation potential of oxidizers is documented for PA membranes. Certain elements, such as iron, manganese, or copper, can change their valence states in an oxidation/reduction reaction. This ability allows the metals to act as a catalyst to increase the rate of oxidation caused by an oxidizing agent. (W. Byrne and M. Roesner, 1991).

The metals in the RO feedwater would be in a highly oxidized valence state due to the presence of the chlorine. Since iron and manganese in these valence states are mostly insoluble, they would tend to fall out of solution on the surface of the lead end membrane elements. These metals can be chemically reduced (i.e., the metals can take one or more electrons into the outer valence shell of their atom). When free chlorine is present around this environment, it can more aggressively steal electrons from the membrane (i.e., oxidize the membrane).

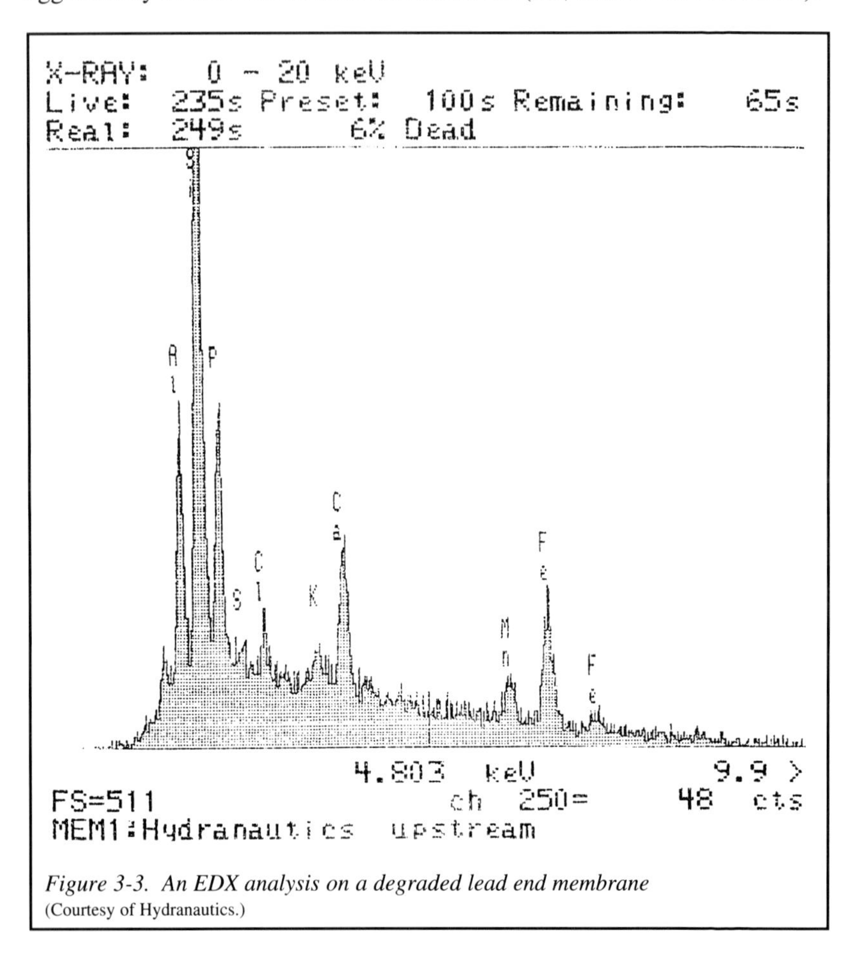

Figure 3-3. An EDX analysis on a degraded lead end membrane
(Courtesy of Hydranautics.)

With a PA membrane, the presence of transition elements in the RO feedwater can make it possible for trace concentrations of chlorine to attack and oxidize the membrane. Although CA membranes are much more resistant to oxidation, it is apparent that the combined presence of free chlorine and transition metals can also result in CA membrane deterioration.

The solution at the Yuma Desalting Facility was to inject ammonia into the RO feedwater. The ammonia reacted with the free chlorine to form chloramines, which as oxidizing agents are not nearly as aggressive as free chlorine. Yet chloramines are still biocides; thus, the membrane will still have some amount of protection against biological activity.

Before the system modification at Yuma, the lead end elements were lasting from 6 months to 1 year prior to their required replacement due to poor performance. Since the modification, no symptoms of membrane deterioration

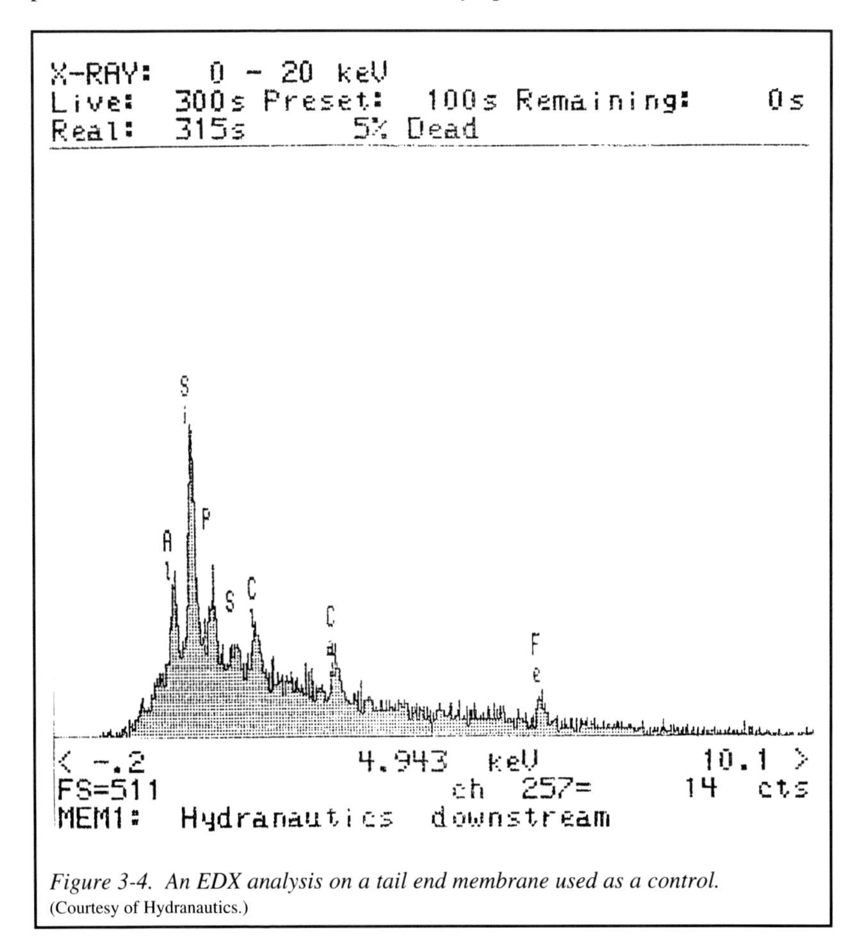

Figure 3-4. An EDX analysis on a tail end membrane used as a control.
(Courtesy of Hydranautics.)

have been apparent. Based on the membrane performance, it is anticipated that the membrane elements will last at least 3 years, if not more.

The findings at Yuma are consistent with information obtained from other facilities and membrane manufacturers. It would seem that the mystery of Compound X has been solved. The lessons learned about the ability of transition metals and oxidizing agents to interact are important and should be considered when designing an RO pretreatment system.

Chlorine removal with PA membrane. Most polyamide RO membranes can tolerate little or no free chlorine. Depending on other constituents of the feed stream, even trace concentrations of chlorine can break down the polyamide. As discussed in the previous section, the presence of iron, manganese, copper, or other transition metals can act as catalysts to speed the aggressiveness of the chlorine's oxidative effects. This also holds true for other oxidizing agents such as iodine, bromine, and ozone.

Some membrane manufacturers quantify the concentration of chlorine that can be tolerated by their PA membrane element. This number may be expressed in ppm-hours and is calculated by the concentration of chlorine in the feedwater multiplied by the hours the membrane is in contact with the concentration; thus, a PA membrane with a reported tolerance of 1,000 ppm-hours could supposedly be capable of handling a feedwater concentration of 0.1 ppm of free chlorine for 10,000 hours prior to experiencing significant membrane deterioration. In actuality, this tolerance is heavily dependent on the makeup of the feedwater. Most membrane manufacturers would prefer to have the feedwater chlorine concentration reduced to undetectable concentrations.

Many municipal water systems now use chloramines as biocides in the city water. Chloramines are less reactive than free chlorine; thus, they are more stable. Their advantage over free chlorine is that they do not react with organics in the water to form trihalomethanes (THMs), which are believed to be carcinogenic.

Chloramines in low concentrations are compatible with many PA thin-film

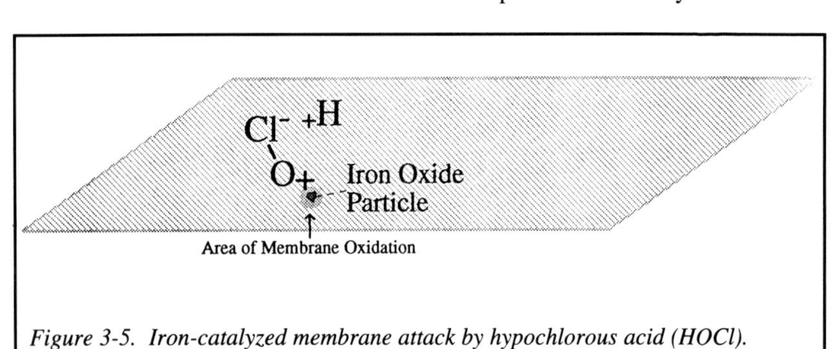

Figure 3-5. Iron-catalyzed membrane attack by hypochlorous acid (HOCl).

membranes, yet their presence in a water source can still represent the potential for PA membrane deterioration. Municipal water treatment systems will typically react ammonia with free chlorine to create the chloramines. The occasional presence of free chlorine along with the chloramines can lead to the demise of the PA membrane; therefore, it is nearly always required by the PA membrane manufacturers that the pretreatment system include some means of removing the chlorine or chloramines if either are present in the municipal water (Filteau, April 1993).

pH >5:

$$HOCl + NH_3 \longrightarrow NH_2Cl + H_2O$$

(Bauer and Snoeyink, 1973).

The removal of oxidizing agents can be achieved by filtering the water through activated carbon, or by injecting the proper concentration of chemical reducing agent. Typical reducing agents used for this purpose include sodium sulfite (Na_2SO_3), sodium bisulfite ($NaHSO_3$), sodium metabisulfite ($Na_2S_2O_5$), and sodium thiosulfate ($Na_2S_2O_3$). They should be injected at a sufficient concentration to equal the molar quantity of the oxidizing agent, plus leave a residual of one part per million. The reactions with hypochlorous acid are as follows:

Sodium sulfite:

$$HOCl + Na_2SO_3 \longrightarrow HCl + Na_2SO_4$$

Sodium bisulfite:

$$HOCl + NaHSO_3 \longrightarrow NaCl + H_2SO_4$$

Sodium metabisulfite:

$$2HOCl + Na_2S_2O_5 + H_2O \longrightarrow 2NaCl + 2H_2SO_4$$

Sodium thiosulfate can react with chlorine in a number of different ways which are dependent on pH.

Sodium bisulfite and sodium metabisulfite are probably the two most common reducing agents used in RO pretreatment systems. Sodium bisulfite is available in an aqueous solution whereas sodium metabisulfite is provided as a dry granular material. When mixing sodium metabisulfite with water, the fumes can be very irritating (Shurtleff, 1993).

Because of the critical nature of removing oxidizing agents upstream of PA RO systems, it is suggested that two different injection pumps and day tank systems be used to inject the reducing agent. Thus, if one tank is allowed to run dry or if the injection pump loses its prime, the other pump is still on line to protect the PA membrane.

Another safety option when relying on the injection of a reducing agent is to monitor the water with an Oxidation/Reduction Potential (ORP) meter. This meter measures whether the feedwater of the RO is under an oxidative or a reductive atmosphere. It should be tied into an alarm/shutdown control system with the RO. To insure the elimination of oxidizing agents in an RO feedwater, the ORP (redox potential) reading should be less than 175 millivolts (mV).

Reductive water environments can result in the proliferation of anaerobic bacteria. One type of anaerobic bacteria that reduces sulfate (SO_4^{-2}) to hydrogen sulfide (H_2S) functions best in water with a redox potential in the range of -100 to -200 mV. Maintaining a slightly positive redox potential can assist in controlling this type of biological activity (F. Bernardin, 1983).

Example:
Note: It may be useful for some readers to review the procedure for determining weight concentration ratios when working with chemical reactions. This example will cover how to convert weight concentrations into molar concentrations, and how to use the stoichiometric chemical reaction formulas.

What concentration of sodium bisulfite should be injected into an RO feed stream to eliminate a free chlorine concentration of 0.3 mg/L?

Figure 3-6. Barium sulfate scale on a membrane.

Answer:
As discussed in Chapter 1, free chlorine (which exists as a dissolved gas) is in equilibrium with hypochlorous acid as follows:

$$Cl_{2(g)} + H_2O \text{ <====> } HOCl + HCl$$

According to this reaction, one mole of free chlorine will dissociate in water to form one mole of hypochlorous acid. According to the stoichiometric reaction for the chemical reduction by sodium bisulfite, it takes one mole of sodium bisulfite to neutralize one mole of hypochlorous acid; therefore, it will take one mole of bisulfite to react with one mole of chlorine expressed as Cl_2.

Moles are an expression of quantity based upon a certain number of molecules. The quantity is based around the weight of one proton atom, which is the same as that of one neutron atom. The mole was arbitrarily defined as the number of protons (or neutrons) whose weight would total one gram. This quantity was later determined to be $6.02 \text{ X } 10^{23}$ (Avogadro's number). The weight of a mole of a particular element or compound depends on the number of protons and neutrons in its nucleus, and is called its molecular weight (expressed in grams per mole, g/mole).

The hydrogen molecule has only one proton and no neutrons, which gives it a molecular weight of 1.0. This means that a mole of hydrogen molecules will have a combined weight of one gram. A certain molar quantity of hydrogen would be converted into a weight-based concentration by multiplying the molar quantity by its molecular weight (in this case, by 1.0).

The molecular weight of sodium bisulfite can be determined by adding the molecular weight contribution from each atom in the sodium bisulfite molecule. The molecular weights for individual ions can be obtained from the periodic table in the Appendix.

Atom	*Molecular Weight (g/mole)*
Na	23
H	1
S	32.1
3O	$16 \times 3 = 48$
$NaHSO_3$	104.1

The molecular weight of sodium bisulfite is 104.1 grams per mole (g/mole). This is equivalent to 104,100 milligrams per mole (mg/mole).

The chlorine concentration (expressed as free chlorine) can also be converted to a molar concentration using the molecular weight of free chlorine (Cl_2).

$$35.5 \times 2 = 71 \text{ g/mole} = 71{,}000 \text{ mg/mole}$$

The chlorine concentration of 0.3 mg/L is converted into a molar concentration by dividing by 71,000 mg/mole. The result is a molar concentration of 0.0000042 moles/L. Given the one-to-one relationship between the sodium bisulfite and chlorine in their chemical reaction, this concentration would also be the molar concentration of sodium bisulfite required to react with the chlorine.

The bisulfite concentration can then be converted into a weight-based concentration by multiplying the 0.0000042 moles/L by its molecular weight, 104,000 mg/mole, with a result of 0.44 mg/L. This is approximately 1.5 times the weight concentration of chlorine. To insure the complete reaction of chlorine in the water, and to increase the rate of reaction, it is common to inject an excess concentration of the reducing agent. If an excess of 1 mg/L of sodium bisulfite were injected in addition to that required for a complete reaction, the desired injection concentration would be 1.45 mg/L of sodium bisulfite.

The weight concentration ratios for a complete reaction with free chlorine for each the four reducing agents mentioned previously would be as follows:

Reducing Agent	*Molecular Weight*	*Weight Concentration Ratio*
Sodium sulfite	126.1	1.8
Sodium bisulfite	104.1	1.5
Sodium metabisulfite	190.2	1.35
Sodium thiosulfate	158.2	dependent on pH

Prevention of Scale Formation

As pure water permeates through an RO membrane, it leaves behind a higher concentration of whatever does not pass through the membrane; thus, the concentration of salts in the water on the feed side of each membrane increases as the water passes across each element on its way out the RO concentrate. As this occurs, certain dissolved salts may exceed their solubility. If salts fall out of solution on the membrane surface, or in the element flow channels, the membrane performance will be affected. Figure 3-6 shows an example of barium sulfate scale on a membrane.

Excessive scale formation can result in the demise of the membrane elements. Either prevention of the scale formation, or the proper cleaning of scale if it has occurred, are priorities for the success of the RO system.

The importance of a complete water analysis was discussed in Chapter 1. Certain slightly soluble salts can potentially form scale in an RO membrane system. A complete analysis will give the concentrations of the ionic makeup of those salts. With this information, the scaling potential of each of the slightly soluble salts can be determined.

Calculating the potential for scale formation. Salts dissolve in water by separating into their cation and anion components. If the concentration of a salt's cation and associated anion concentration exceeds the solubility of that salt, crystal formation followed by precipitation (i.e., the falling out of solution by salt particles) may occur. This can easily happen within an RO system as the salts are concentrated within the system.

Each salt has its own characteristics that will affect the seriousness of its potential for crystal formation. Some will fall out of solution faster than others. Some precipitated salts will be easier to clean than others. Generally, it is desirable to prevent the formation of any salt within the membrane system, as this will eliminate the need to clean scale from the membrane.

The solubility of a salt will depend on the concentration of its components, the water pH, the temperature, and the concentration of other salts in the solution. Different methods are available to predict the solubility of each particular salt. The relative significance of the water's pH, temperature, and the makeup of the water source on the particular salt's solubility will vary.

When evaluating a water source for scaling potential, it is suggested that all of the slightly soluble salts that will be discussed in this chapter be evaluated for their solubility in the RO concentrate stream. If any are close to exceeding their solubility, a means for controlling that scale formation should be considered. This will likely be the case for calcium carbonate, and may be required for other salts as well. The slightly soluble salts that are most likely to be found in a naturally occurring water source would include the following:

Slightly Soluble Salts in Naturally Occurring Water Sources

Calcium carbonate	$CaCO_3$
Calcium sulfate	$CaSO_4$
Strontium sulfate	$SrSO_4$
Barium sulfate	$BaSO_4$
Calcium fluoride	CaF_2
Silica	$Si(OH)_4$
Calcium silicate*	$CaSiO_3$
Magnesium silicate*	$MgSiO_3$
Ferrous silicate*	$FeSiO_3$

*It is necessary to consider silicate solubilities only if the pH of the RO concentrate stream will be greater than 8.

Metals such as ferric iron (Fe^{+3}), manganic manganese (Mn^{+3}), or aluminum (Al^{+3}) are not included in the previous listing because, for all practical purposes, their salts are insoluble. If present in an RO feedwater, filtration of their salts will be required to prevent their settling within the RO membrane elements.

Estimating salt concentrations in the RO concentrate. The greatest potential for scale formation is in the RO concentrate where the salt concentrations are their highest. For most systems with a high-rejecting membrane (95% sodium chloride rejection or better), the RO concentrate total dissolved solids (TDS), as well as any individual ion concentrations, can be approximated for a particular RO permeate recovery by dividing the concentration in the feedwater by the following:

$$1 - \text{recovery (as a fraction)}$$

where:

$$\text{recovery} = \text{permeate flowrate/feed flowrate}$$

The reciprocal of this expression can be expressed as a fraction. It can then be multiplied by the feedwater concentration of any well-rejected ion to estimate the concentration of that ion in the RO concentrate.

$$\text{Concentration factor} = 1 \div (1 - \text{recovery})$$

For example, an RO system with a 75-gpm permeate flowrate and a 25-gpm concentrate flowrate would have a recovery of 75%. With a standard high-rejecting membrane, the factor that can be multiplied by the feedwater TDS to determine the concentrate TDS would be as follows:

$$1 \div (1 - 0.75) = 4$$

In actuality, the concentration of salts at the membrane surface of the previous example may be more than four times the feedwater concentration. As pure water permeates the RO membrane, a higher concentration of salts will be left behind directly at the membrane surface. How well these salts diffuse back into the bulk stream will determine how much higher in concentration the salts at the

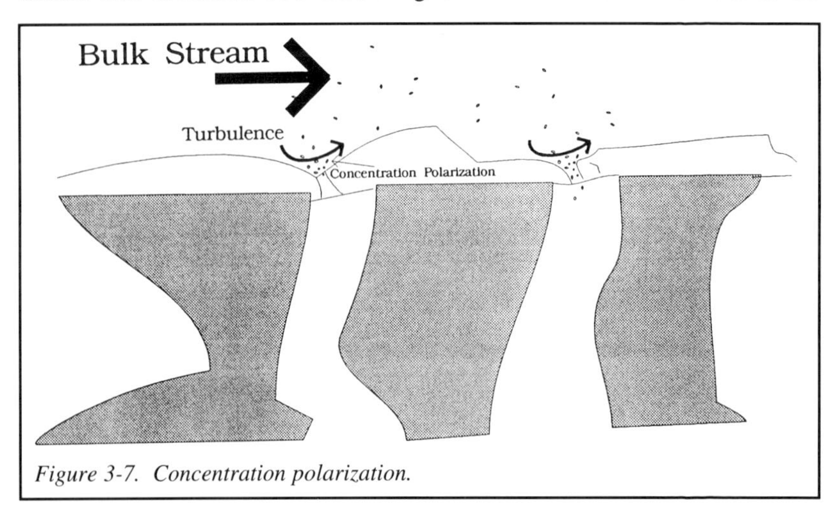

Figure 3-7. Concentration polarization.

membrane surface will be over that in the bulk stream. This phenomenon has been called *concentration polarization.* The term simply refers to the concentration gradient at the membrane surface created by the less-than-immediate re-dilution of salts left behind as water permeates through the membrane (Figure 3-7).

The extent of the concentration polarization at the membrane surface will depend on the turbulence of the bulk stream, and the ability of that turbulence to place shear forces on the salts. It has been estimated for a standard (clean) spiral-wound membrane element that the concentration of salts at the membrane surface is 13% to 20% greater than in the bulk stream. This polarization should be considered when calculating scale formation potential. (Fluid Systems predicts a 13% greater concentration at the membrane surface, whereas Hydranautics predicts a 20% concentration polarization factor in their membrane performance literature.)

The concentration factor can be written as follows:

$$\text{concentration factor} = \text{polarization} \div (1 - \text{recovery})$$

where polarization is the ratio of the membrane surface concentration to the concentration in the bulk stream.

Calcium Carbonate. As discussed in Chapter 1, bicarbonate (HCO_3^-) is present in most naturally occurring water sources. Because of the ability of water to dissolve calcium carbonate from the ground (thus forming bicarbonate ion), or carbon dioxide (CO_2) from the air (which can convert to bicarbonate in water), natural water sources will typically be near saturation in dissolved bicarbonate. As these waters are concentrated in an RO system, calcium carbonate is going to be one of the first salts to precipitate.

Langelier saturation index. The potential for calcium carbonate scale can be estimated using the Langelier saturation index (LSI; also Langelier stability index):

$$LSI = pH - pH_s$$

where:
pH_s is the pH at which the water is saturated in calcium carbonate, and is calculated by the following:

$$pH_s = (9.3 + A + B) - (C + D)$$

where:
$A = (\text{Log}_{10}[\text{TDS}] - 1)/10$
$B = -13.12 \times \text{Log}_{10}(^{\circ}\text{C} + 273) + 34.55$
$C = \text{Log}_{10}[\text{Ca}^{+2} \text{ as } CaCO_3] - 0.4$
$D = \text{Log}_{10}[\text{alkalinity as } CaCO_3]$

and values in brackets [] indicate molar concentrations (moles/L), except for TDS, which is expressed as milligrams per liter (mg/L).

Note: The values for A, B, C, and D can also be derived from Figure 3-8.

At an index of less than 0, a water source will have a very limited scaling potential. At increasing positive index values, the scaling potential increases. Using the Langelier index, an estimate can be made for the concentration of salts possible in the RO concentrate before there is a potential for scale formation. The Langelier can also be used to determine if there is a need for the injection of a scale inhibitor.

The Langelier index can also be used to predict the corrosivity of the water (i.e., how aggressive the water may be towards various materials of construction being employed in the water distribution system). As the Langelier value becomes increasingly negative, the corrosivity of the water will increase.

The RO concentrate pH is a function of the alkalinity and the dissolved carbon dioxide (CO_2) concentration in the RO concentrate water. To estimate the pH in the RO concentrate requires first calculating the CO_2 concentration in the feedwater. Since dissolved gases (such as CO_2) will readily permeate the RO membrane, the concentrate gas concentration will typically be about the same as that in the feed. The concentrate pH can then be approximated using the CO_2 concentration and the estimated alkalinity in the concentrate.

The relationship between pH and the ratio of alkalinity to CO_2 is shown in Figure 3-9, and can also be expressed by the following equation:

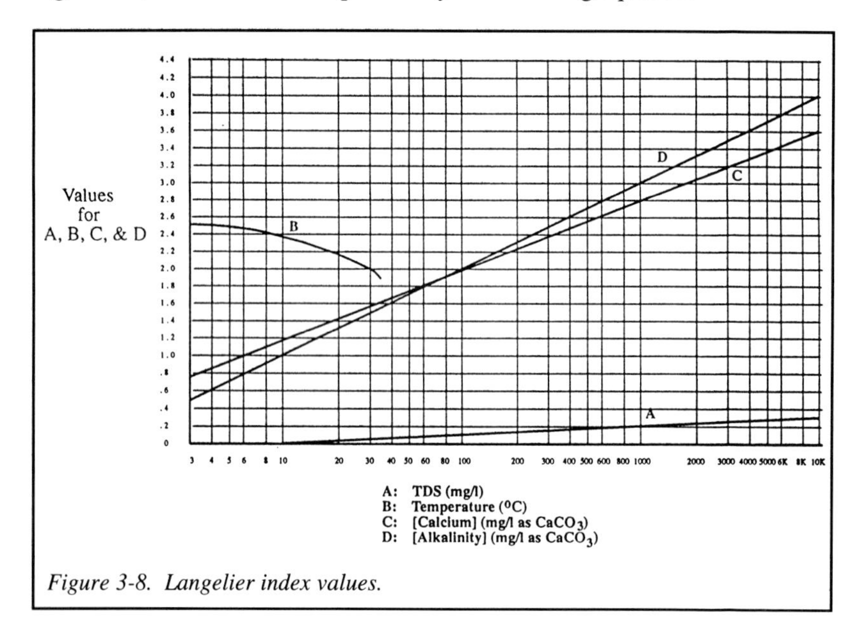

Figure 3-8. Langelier index values.

$$pH = Log_{10}([\text{alkalinity as } CaCO_3]/[CO_2]) + 6.3$$

To determine the CO_2 concentration in the feed water, the previous equation can also be expressed as follows:

$$10^{(pH-6.3)} = [\text{alkalinity as } CaCO_3] \div [CO_2] \text{ or:}$$

$$[CO_2] = [\text{alkalinity as } CaCO_3] \div 10^{(pH-6.3)}$$

At values for pH greater than 8.2, the dissolved CO_2 concentration becomes negligible. At higher pH values, carbonate ion (CO_3^{-2}) forms from the bicarbonate (HCO_3^{-}). The ratio between the bicarbonate concentration (methyl orange alkalinity) and the carbonate concentration (phenolphthalein alkalinity) as a function of pH is given in Figure 3-10.*

Example:
Determine if there is potential for calcium carbonate formation in an RO operating at 60% recovery using a high-rejecting spiral-wound membrane, on a water source with the following characteristics:

pH	8.2
Ca^{+2} hardness	35 mg/L as $CaCO_3$
Bicarbonate alkalinity	134 mg/L as $CaCO_3$
TDS	380 mg/L
Temperature	62°F

Note: For this example, a concentration polarization ratio of 1.13 will be used.

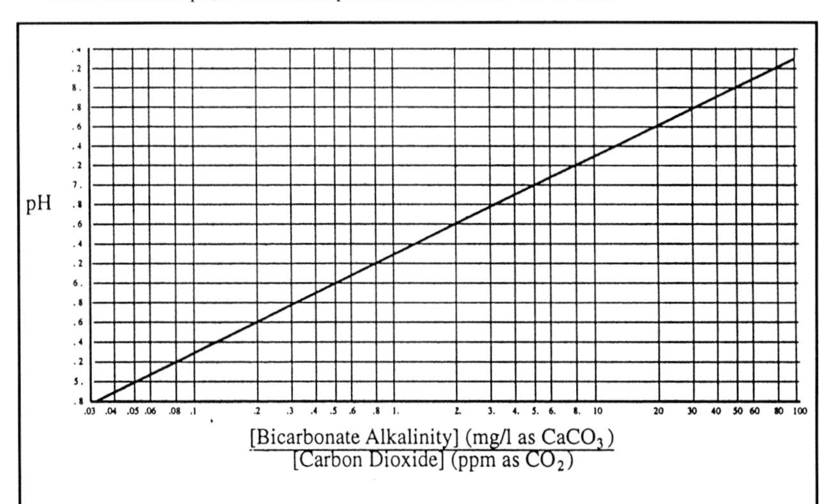

Figure 3-9. Relationship between bicarbonate alkalinity, carbon dioxide, and pH. (Courtesy of Rohm & Haas Co.)

Concentration factor = polarization ÷ (1 - recovery) = 1.13 ÷ (1 - 0.6) = 2.825

Concentrate membrane surface TDS: 380 mg/L × 2.825 = 1073

$$A = (\mathrm{Log}_{10}[\mathrm{TDS}] - 1) \div 10 = 0.203$$

Converting temperature from Fahrenheit to Centigrade:

$$(62^\circ\mathrm{F} - 32) \times 5/9 = 16.7^\circ\mathrm{C}$$

$$B = -13.12 \times \mathrm{Log}_{10}(^\circ\mathrm{C} + 273) + 34.55 = 2.25$$

Concentrate membrane surface [Ca^{+2}]:

$$35\ \mathrm{mg/L} \times 2.825 = 98.9$$

$$C = \mathrm{Log}_{10}[Ca^{+2}\ \mathrm{as}\ CaCO_3] - 0.4 = 1.60$$

Concentrate membrane surface alkalinity [HCO_3^-]:

$$134 \times 2.825 = 379$$

$$D = \mathrm{Log}_{10}[\mathrm{alkalinity\ as}\ CaCO_3] = 2.58$$

pH of calcium carbonate saturation:

$$pH_s = (9.3 + A + B) - (C + D)$$

$$= (9.3 + 0.203 + 2.25) - (1.60 + 2.58) = 7.57$$

As discussed, determining the pH of the concentrate water requires knowing the CO_2 concentration in the feedwater, which will also be about the same concentration in the RO concentrate water:

$$[CO_2] = [\mathrm{alkalinity\ as}\ CaCO_3] \div 10^{(pH - 6.3)} = 134\ \mathrm{mg/L} \div 10^{(8.2 - 6.3)} = 1.7\ \mathrm{ppm}\ CO_2$$

The concentrate pH (at the membrane surface) can then be calculated:

$$pH = \mathrm{Log}_{10}([\mathrm{alkalinity\ as}\ CaCO_3] \div [CO_2]) + 6.3 = \mathrm{Log}(379/1.7) + 6.3 = 8.65$$

$$LSI = pH - pH_s = 8.65 - 7.57 = +\ 1.08$$

The positive value for the Langelier index indicates a potential for calcium carbonate formation in the RO concentrate. Since pH is a logarithmic scale, a value of greater than +1 indicates a strong potential for the formation of calcium carbonate. Unless something is done to control or prevent this situation, it is likely that scale will form in the tail end RO membrane elements (i.e., within the elements through which the feedwater stream passes prior to exiting the RO as the concentrate stream).

Stiff and Davis stability index. The Langelier index loses its accuracy in predicting the solubility of calcium carbonate for water with salt concentrations

*"Engineering Manual for the Amberlite^R Ion Exchange Resins", Rohm and Haas Company.

exceeding 4,000 mg/L. Stiff and Davis modified Langelier's equation for high-TDS water sources using empirically derived values. The Stiff and Davis stability index is calculated as follows:

$$\text{Stiff \& Davis stability index} = pH - pCa - pAlk - K$$

where:
pCa is the negative log of the calcium concentration (expressed as molarity);
pAlk is the negative log of the total alkalinity concentration (expresses as molarity); and

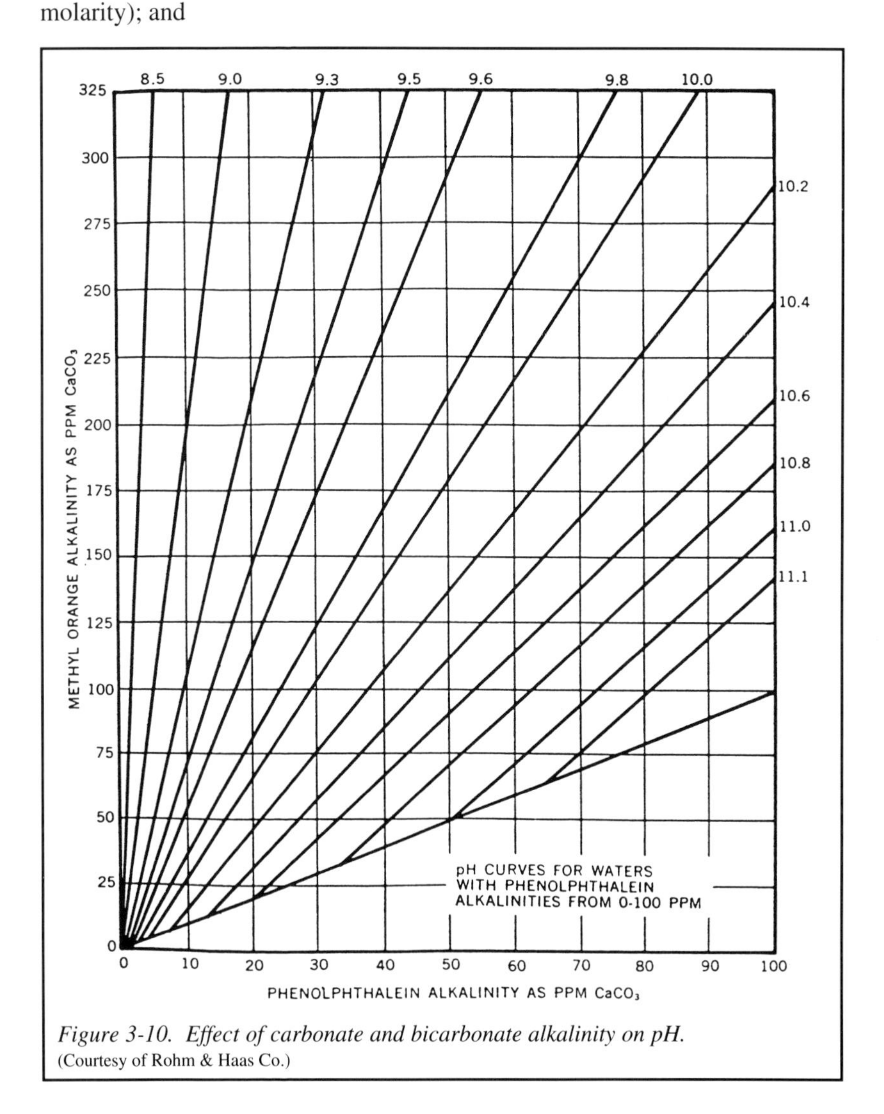

Figure 3-10. Effect of carbonate and bicarbonate alkalinity on pH.
(Courtesy of Rohm & Haas Co.)

K is a constant whose value depends on water temperature and ionic strength (ionic strength is related to the total salt concentration).

These values can be obtained using Figures 3-11 and 3-12. Like the Langelier index, a positive Stiff and Davis index indicates a potential for calcium carbonate

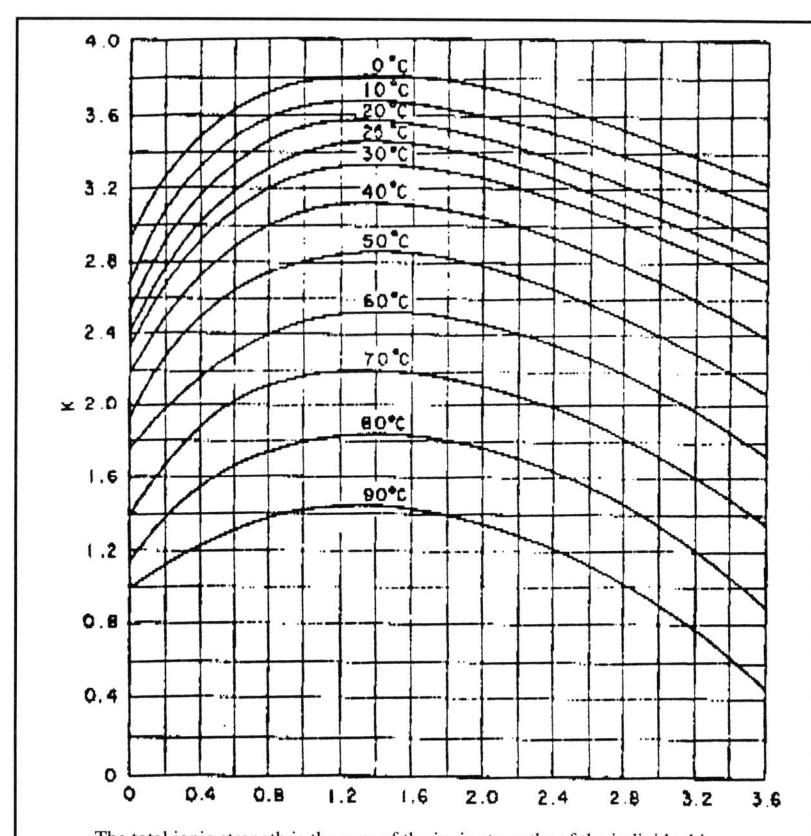

The total ionic strength is the sum of the ionic strengths of the individual ions

Ion	*Factor, ppm*
Na	2.2×10^{-5}
Ca	5.0×10^{-5}
Mg	8.2×10^{-5}
Cl	1.4×10^{-5}
HCO_3	0.8×10^{-5}
SO_4	2.1×10^{-5}

Figure 3-11. K values at various ionic strengths and factors for converting water analyses to ionic strength.
(Stiff and Davis, 1952)

precipitation, and a negative value indicates the potential for corrosion.

Preventing calcium carbonate precipitation. Preventing the formation of calcium carbonate is desirable and usually required for most RO systems. For small RO systems, this can be accomplished by simply operating at low permeate flow recovery to minimize the concentration of bicarbonate in the concentrate. In larger industrial systems, higher water recoveries must be used for the RO system to be economical. The ways calcium carbonate scale prevention can be accomplished on higher-recovery RO systems are as follows:

1. Inject acid in the RO feed stream to convert much of the bicarbonate to carbon dioxide, as well as to increase the solubility of calcium carbonate due to the lower pH.
2. Inject a scale inhibitor in the RO feed stream to slow the formation of scale.
3. Remove the hardness in the RO feedwater by exchanging it for sodium on the resin in a water softener.

These methods will be discussed later in this chapter. Some of these methods may also be effective at preventing the precipitation of slightly soluble salts other than carbonates, should this also be required.

Solubility product. The previous discussion of the Langelier index calculation was one example of how to determine the potential for the formation of a particular salt. The salt in this case was calcium carbonate. Other salts can also fall out of solution if their solubilities are exceeded. They will typically not fall out of solution as quickly as calcium carbonate; however, these other salts will tend to be more resistant to cleaning if they should happen to precipitate. They may thus pose a greater long-term threat to the RO system performance.

The solubility for most salts is related to the concentration of the salt's cation and anion components. At the point of saturation, the solubility product constant (K_{sp}) is related to the salt's makeup ion concentrations according to the following relationship:

$$K_{sp} = [\text{cation}]^{\#} \times [\text{anion}]^{\#}$$

where:
K_{sp} is a value for the solubility product,
[cation] is the cation concentration,
[anion] is the anion concentration, and
is the quantity of the particular ion present within the salt molecule.

The value for the solubility product constant will also be a function of pH, temperature, and the characteristics of other salts in the solution. As a means of roughly predicting the solubility of a salt, however, the effects of these variables can be neglected. The ionic concentrations of a water source can be interjected into the previous equation. The resulting value can be compared to the solubility product constant to determine if a salt is supersaturated, and can thus precipitate.

For a particular salt, if $[cation]^{\#} \times [anion]^{\#} > K_{sp}$ then the solubility of the salt may have been exceeded. The risk of the salt's precipitation will depend on how far its solubility has been exceeded.

If $[cation]^{\#} \times [anion]^{\#} < K_{sp}$ then the ionic concentrations have not exceeded the solubility of the salt.

For many salts, the solubility product constant is not a strong function of pH, temperature, or the characteristics of the other salts in the solution. It is also true that non-carbonate salts tend to precipitate slowly, which makes it difficult to determine the exact concentrations when precipitation begins to occur. For these reasons, estimating the concentrations at which precipitation is going to occur in an RO system is not exact.

Since an RO system is continually displacing its water, there is little concern about the precipitation of a salt on the edge of its solubility limit. It will likely be washed out of the RO long before it forms a crystal large enough to fall out of suspension. The solubility product constant can thus be used as a tool to determine if many salts are in a concentration range where solubility is a concern. (See Table 3-1)

Example:
Determine if the concentration of calcium fluoride will exceed its solubility in an 80%-recovery RO system using a high-rejecting RO membrane. The raw water

Table 3-1
Solubility Product of Common Slightly Soluble Salts
(expressed in molar concentrations, or in mg/L as ion)

Salt	*Molar Solubility Product*	*Mg/L - Solubility Product*
$CaSO_4$	2.5×10^{-5}	96,300
$SrSO_4$	6.3×10^{-7}	5,300
$BaSO_4$	2.0×10^{-10}	2.64
CaF_2*	5.0×10^{-11}	723
$Si(OH)_4$**	2.0×10^{-3}	96 mg/L as SiO_2

*The fluoride concentration should be squared (i.e., raised to the power of 2) when calculating the solubility product of calcium fluoride. There are two fluoride ions in a calcium fluoride salt molecule
**Silica can change its chemical structure as a function of pH. Its solubility will depend on its structure and is also temperature-dependent. If silica will be present in the RO concentrate stream at a concentration exceeding 20 mg/L, its potential for scale formation should be scrutinized. This will be discussed in the next section. (Stumm and Morgan, 1981).

calcium concentration is 230 mg/L and the fluoride concentration is 0.6 mg/L.

Answer:
The factor to determine the surface concentration in the RO concentrate would be:

concentration factor = polarization*/(1 - recovery) = 1.13 ÷ (1 - 0.8) = 5.65

The concentration of calcium in the RO concentrate would be approximately

$$230 \text{ mg/L} \times 5.65 = 1{,}300 \text{ mg/L}$$

The approximate concentration of fluoride in the RO concentrate would be

$$0.6 \text{ mg/L} \times 5.65 = 3.39 \text{ mg/L}$$

The solubility product constant (K_{sp}) of calcium fluoride (CaF_2) at saturation would be determined by the concentration of calcium (Ca^{+2}), multiplied by the concentration of fluoride (F^-) raised to the power of 2.

$$K_{sp} = [Ca^{+2}] \times [F^-]^2 = 723$$
(concentrations expressed in mg/L as ion)

The actual concentration of calcium can be multiplied by the actual concentration of fluoride squared and can then be compared to the solubility product constant for calcium fluoride salt.

$$[Ca^{+2}] \times [F^-]^2 = 1{,}300 \times (3.39)^2 = 14{,}900$$

Since this product is far greater than the solubility product constant, the potential for calcium fluoride precipitation exists. Some means of controlling or preventing that precipitation should be considered.

Silica solubility. At pH values below 8, dissolved silica is in the silicic acid form ($Si(OH)_4$, also expressed as H_4SiO_4). If the solubility of silicic acid is exceeded, silica will precipitate as shown in the following:

$$Si(OH)_4 \longrightarrow SiO_2 + 2H_2O$$

The solubility product constant for silica at saturation is represented by the following:

$$K_{sp} = [Si(OH)_4]$$

where:
$[Si(OH)_4]$ is the silicic acid concentration expressed in mg/L as SiO_2.

---

*It is probably not essential to consider the concentration polarization at the membrane surface in the solubility calculations due to the slow rate of precipitation of calcium fluoride. However, including the factor will provide an extra margin of safety against salt formation.

At 25°C and a neutral pH, the solubility product constant for silica is about 96 mg/L (as SiO_2). At silicic acid concentrations greater than 96 mg/L, it is possible for silica to begin to precipitate; however, silica is slow to crystallize. Many RO systems can operate safely with silicic acid concentrations as high as 140 mg/L in the RO concentrate without experiencing scale formation.

Silica solubility is a strong function of temperature. At temperatures approaching freezing (32°F), silica becomes increasing insoluble. A linear relationship can be drawn for the effect of temperature on silica solubility (DuPont Bulletin 502, Page 15). Figure 3-13 can be used to adjust the solubility product for temperature. The dashed line indicates the ability of RO systems to operate with slightly greater concentrations of silica than the theoretical limit.

The solubility of silica increases at low and at high pH. Acid injection can therefore be used as a means to increase silica solubility. Refer to Figure 3-14 for solubility as a function of pH (DuPont Bulletin 502, Page 16).

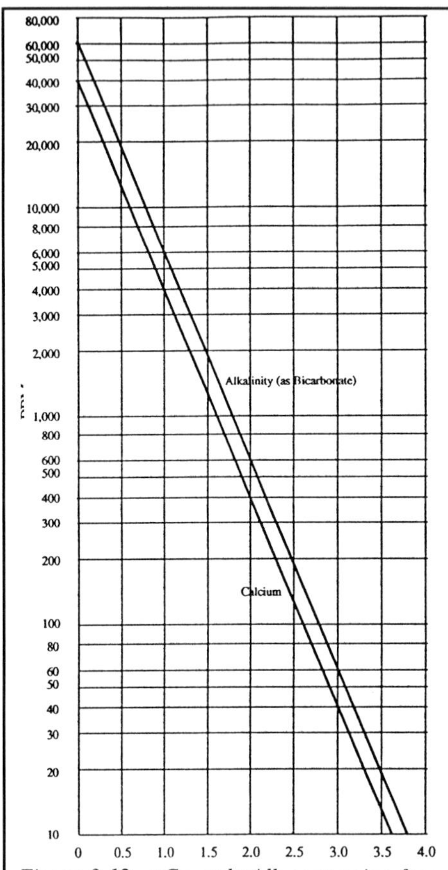

Figure 3-12. pCa and pAlk conversion from concentration in ppm.

As the pH of water increases beyond a value of 8, silicic acid (H_4SiO_4) will dissociate into the silicate anion (SiO_3^{-2}). This will increase the solubility of the silica unless certain multivalent cations are available in significant concentration. These cations can form insoluble silicate salts above certain pH values.

At a pH greater than 8, the presence of iron or aluminum with silicates will likely cause the precipitation of insoluble silicates. Other multivalent cations will also tend to precipitate with the silicates at a particular pH greater than 8. This precipitation has been noted to occur at a pH value 1 to 2 points below the point at which hydroxide solubility would be exceeded. (Stumm et al., 1981.)

Hydroxide solubilities of multivalent cations have been well documented. Hydroxide precipitation is commonly used in industrial waste treatment as a means of removing metals from water. These solubility curves have been modified in Figure 3-15a by decreasing the pH by 2. The shaded area indicates a cation concentration at a pH where precipitation may begin to occur. It will likely occur if the pH is higher than that given in the shaded area. The curves can thus be used as a conservative estimate of silicate solubilities for RO systems operating at high pH conditions. Since the RO concentrate water may be as much as 1 pH unit higher than the RO feedwater pH, silicate solubility is a concern for RO systems operating with a feed pH greater than 7.5.

Silicate particles, such as those present in most clays, may coagulate at low pH in the presence of highly charged cations such as aluminum or iron. The aluminum or iron may be residuals leftover from municipal clarification systems.

Acid injection. The most effective way to prevent calcium carbonate scale formation is to lower the feedwater pH via the injection of an acid. The acid will convert bicarbonate ion into carbon dioxide by the following reaction:

$$2NaHCO_3 + H_2SO_4 \Longleftrightarrow CO_{2(g)} + 2H_2O + Na_2SO_4$$

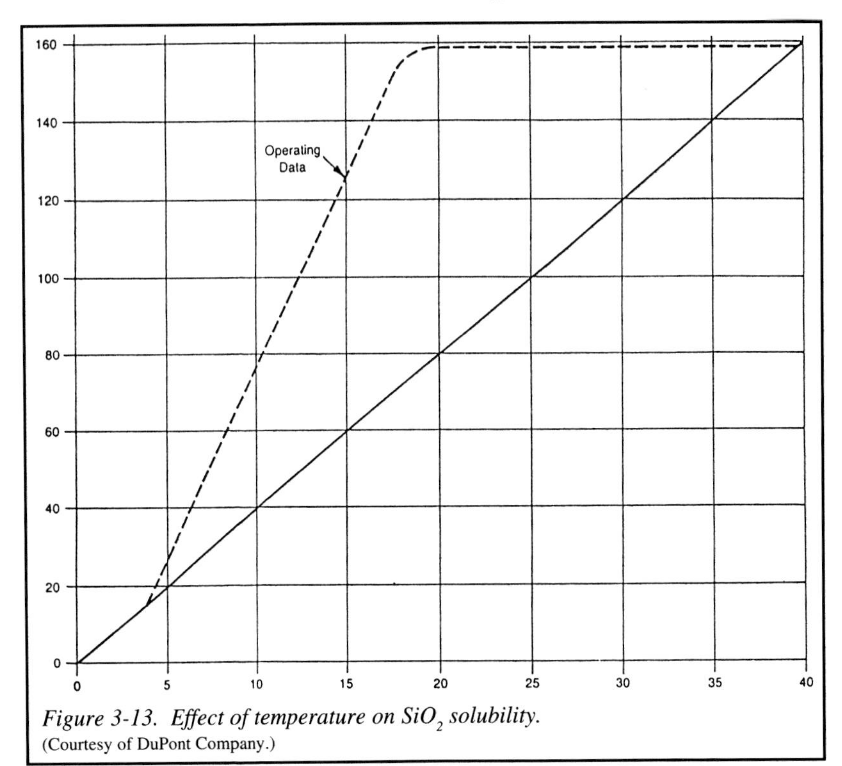

Figure 3-13. Effect of temperature on SiO_2 solubility.
(Courtesy of DuPont Company.)

Bicarbonate provides the carbonate ion in forming the calcium carbonate salt; therefore, the solubility of calcium carbonate depends greatly on the bicarbonate ion concentration. It also depends on the pH of the water. Sufficient concentrations of acid injected into the RO feedwater will eliminate calcium carbonate formation by reducing the bicarbonate concentration; and by increasing the calcium carbonate solubility by lowering the water pH.

An additional advantage of using acid injection as a means to prevent the formation of calcium carbonate is that a sufficient acid concentration will redissolve calcium carbonate that has already fallen out of solution. If the injection system should somehow fail, resulting in the formation of calcium carbonate scale, once the correct concentration of acid is again in the system, the scale will likely go back into solution. However, this will not be the case if the carbonate scale is in the RO system long enough to catalyze the precipitation of sulfate or silica scale (Luss, 1993).

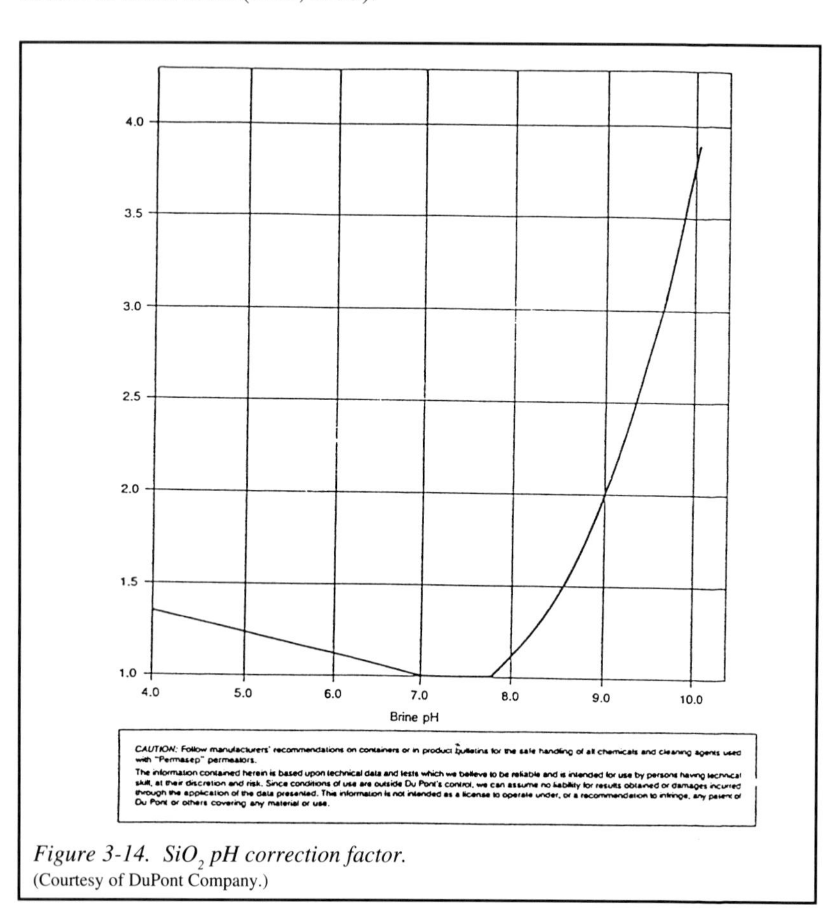

Figure 3-14. SiO_2 pH correction factor.
(Courtesy of DuPont Company.)

As discussed in a previous section, acid injection is necessary with most cellulose acetate membrane systems to reduce the hydrolysis rate of the membrane. It is therefore the likely choice as the primary means to prevent calcium carbonate formation in CA membrane systems.

When an RO system that utilizes acid injection shuts down, the supersaturated carbon dioxide will tend to separate from the water. As this occurs, the water pH will rise. In the tail end membrane elements where the concentrations of salts are at their highest, the increase in pH may be sufficient to allow the precipitation of

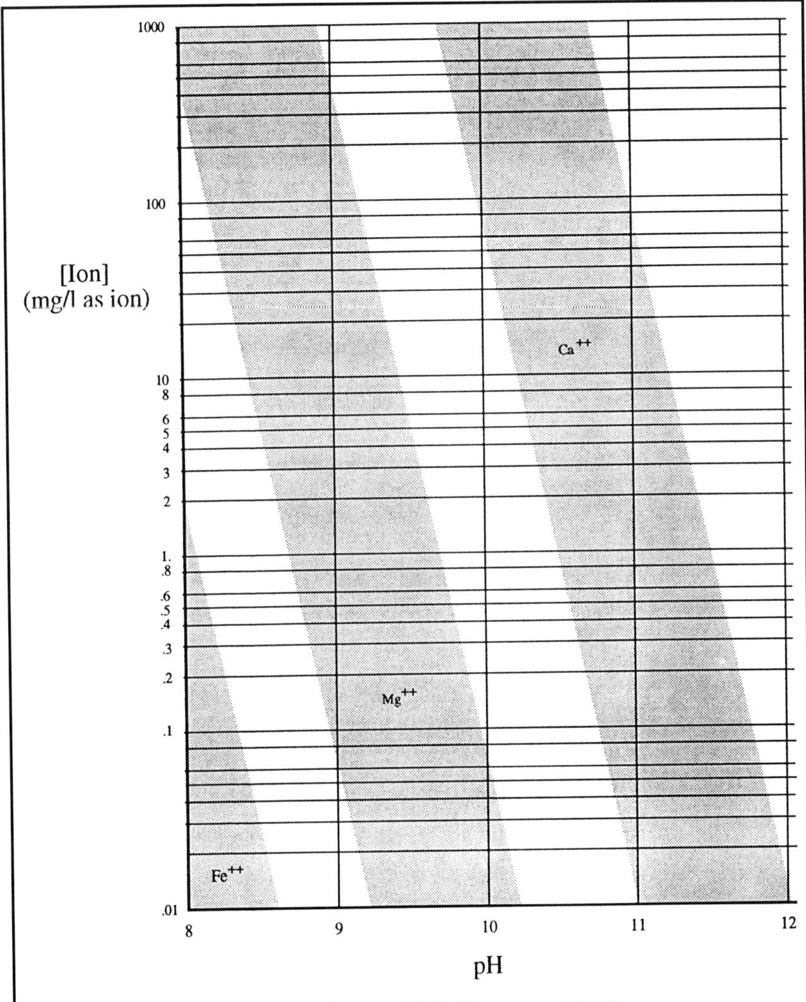

Figure 3-15a. Metal concentration at which silicate precipitation can occur. (Strumm and Morgan, 1981, courtesy John Wiley & Sons.)

calcium carbonate. This calcium carbonate will go back into solution as soon as the RO is again on-line and the pH decreases. With extended shutdowns, however, the presence of calcium carbonate on the membrane surface can cause an increased rate of membrane hydrolysis with CA membrane elements. This may affect the life of the tail end membrane elements.

For CA membrane systems that rely solely on acid injection (to prevent calcium carbonate precipitation), it is suggested that excess acid be injected in the feedwater. In other words, the feedwater pH should be lowered to some value less than what is required to marginally meet a zero value for the Langelier index in the RO concentrate stream. An alternative safety measure would be to inject a small concentration of scale inhibitor in addition to acidifying the feedwater, which will slow the precipitation process.

As discussed in Chapter 1, the injection of acid can cause the coagulation of suspended solids, and their subsequent falling out of suspension. It is therefore suggested that the point of injection for acid be well upstream of media filters in waters of high suspended solids concentration. The materials of construction of the media filter and piping must be compatible with the acid.

The primary disadvantage in using acid injection is the formation of carbon dioxide. The supersaturated CO_2 will readily permeate the membrane, is an undesirable water contaminant in many industries, and increases the loading on downstream anion exchange resins. Removing this CO_2 using degasification will be discussed in Chapter 4.

Estimating acid concentrations for pH adjustments. A rule of thumb in using acid injection to prevent calcium carbonate formation is that lowering the feedwater pH to 6.0 will reduce the bicarbonate concentration by about 80%. For most water sources and typical RO permeate recoveries, an 80% reduction of bicarbonate will be sufficient to prevent calcium carbonate. This may not be the case for some water sources with either high hardness or high alkalinity, or for RO systems operating with higher than 75% permeate recovery.

The Langelier index is useful in determining if a particular feedwater pH will be sufficient to prevent carbonate formation. The calculations discussed in the previous section can be used to verify that sufficient acid will be injected.

Estimating how much acid is required to drop the RO feedwater pH to a desired level requires calculating how much bicarbonate ion is converted into carbon dioxide to achieve the lower pH. The concentration of carbon dioxide in the feedwater prior to the acid injection must first be determined, using the ratio of bicarbonate to carbon dioxide given as a function of pH in Figure 3-9. This ratio can also be calculated as follows:

$$[CO_2] = [HCO_3^-] \div 10^{(pH - 6.3)}$$

where:
$[HCO_3^-]$ is expressed in mg/L as $CaCO_3$, and
$[CO_2]$ is expressed in mg/L as ion.

The carbon dioxide concentration in the feedwater can be converted into a concentration expressed as $CaCO_3$ by multiplying it by its conversion factor, 1.14. (See Appendix.) It can then be added to the incoming alkalinity to find a combined "carbon oxides" concentration. This total will be maintained even after acid injection:

$$\text{total carbon oxides} = [HCO_3^-] + [CO_2] \times 1.14$$

where:
total carbon oxides is a relative value for the total of the alkalinity and carbon dioxide concentrations expressed as $CaCO_3$.

The equation for CO_2 can be substituted in the previous equation to obtain

$$\text{total carbon oxides} = [HCO_3^-] + [HCO_3^-] \div 10^{(pH - 6.3)} \times 1.14$$

or

$$[HCO_3^-] = \text{total carbon oxides} \div (1 + 1.14 \div 10^{(pH - 6.3)})$$

The difference in the bicarbonate concentration from its previous concentration (expressed as $CaCO_3$) will equal the concentration of acid to be injected (expressed as $CaCO_3$). All the information needed to calculate the Langelier index (to determine if calcium carbonate will be completely soluble at the desired pH) would now be available.

Example:
If sulfuric acid is injected to achieve a pH of 6 in the feedwater of an 80%-recovery RO system, will calcium carbonate scale be prevented? The feedwater has the following characteristics:

Temperature	70°F
pH	7.8
Calcium	130 mg/L as $CaCO_3$
Alkalinity	153 mg/L as $CaCO_3$
TDS	300 mg/L

Answer:
concentration factor = polarization ÷ (1 - recovery) = 1.13 ÷ (1 - 0.8) = 5.65

concentrate [TDS]: 300 mg/L × 5.65 = 1695

$$A = (\text{Log}_{10}[TDS] - 1) \div 10 = 0.223$$

Converting temperature from Fahrenheit to Centigrade

$$(70\ ^{\circ}F - 32) \times 5/9 = 21.1\ ^{\circ}C$$

$$B = -13.12 \times Log_{10}(^{\circ}C + 273) + 34.55 = 2.16$$

concentrate membrane surface $[Ca_{+2}]$: 130 mg/L × 5.65 = 734.5

$$C = Log_{10}[Ca^{+2} \text{ as } CaCO_3] - 0.4 = 2.47$$

The carbon dioxide concentration in the feedwater prior to the acid injection is

$$[CO_2] = [HCO_3^-] \div 10^{(pH - 6.3)} = 153 \div 10^{(7.8 - 6.3)} = 4.8 \text{ mg/L}$$

Total concentration of carbon oxides will be

$$1.14 \times (4.8 \text{ mg/L } CO_2) + 153 \text{ mg/L } HCO_3^- = 158.5 \text{ mg/L}$$

After acid injection, the alkalinity in the RO feedwater will be

$$[HCO_3^-] = \text{total carbon oxides}/(1 + 1.14/10^{(pH - 6.3)})$$
$$= 158.5 \text{ mg/L} \div (1 + 1.14 \div 10^{(6 - 6.3)}) = 48.4 \text{ mg/L}$$

and the carbon dioxide concentration would be

$$(158.5 \text{ mg/L} - 48.4 \text{ mg/L})^2 \div\ = 96.6 \text{ mg/L}$$

concentrate membrane surface alkalinity $[HCO_3^-]$: 48.4 × 5.65 = 273.5

$$D = Log_{10}[\text{alkalinity as } CaCO_3] = 2.44$$

$$pH_s = (9.3 + A + B) - (C + D) = (9.3 + 0.223 + 2.16) - (2.47 + 2.44) = 6.77$$

The pH in the RO concentrate can be determined using the ratio of bicarbonate to carbon dioxide:

$$pH = log_{10}([HCO_3^-] \div [CO_2]) + 6.3 = log(273.5 \div 96.6) + 6.3 = 6.75$$

$$LSI = pH - pH_s = 6.75 - 6.77 = -0.02$$

There would be very minimal potential for formation of calcium carbonate in the RO tail end membrane elements. If the membrane is cellulose acetate, it would be suggested that a larger safety margin be used for the solubility of carbonates. Any carbonate scale on a CA membrane can cause an accelerated rate of membrane hydrolysis.

Many RO systems will use the injection of scale inhibitor in addition to acid injection to increase the margin of safety against carbonate scale formation. If other salts exceed their solubilities in the RO concentrate, the injection of scale inhibitor may be required to prevent the precipitation of the other salts. In such a case, the scale inhibitor injection will also assist in preventing the precipitation of calcium carbonate.

In addition to calcium, other multivalent cations can also form slightly soluble salts with carbonates. All the methods of controlling the formation of calcium carbonate scale will also be effective against the formation of other carbonate scales. Unless the concentration of the multivalent cation is greater than that of calcium (which would be unlikely in a naturally occurring water), it is sufficient to rely on the solubility of calcium carbonate as an indicator of the possibility for any carbonate scale formation. If calcium carbonate solubility is exceeded, there becomes a greater risk that other carbonate solubilities will also be exceeded.

Scale inhibition and dispersion. If the concentration of a salt exceeds its solubility, precipitation may begin to occur. Crystals will start to form that will catalyze the formation of more of the salt crystal on the crystal surface. These salt crystals will eventually reach such a size and density that they fall out of suspension. The scale precipitation process will continue until the ions left in solution are at their solubility limit.

Scale inhibitors are very effective at preventing the fouling of RO membrane elements due to the formation of scale. They slow the precipitation process by inhibiting salt crystal growth. They absorb onto the plane of the forming salt crystals, slowing the expansion of the salt crystal by preventing the attraction of more supersaturated salt to the crystal surfaces. The crystals never grow to a size or concentration sufficient for them to fall out of suspension.

Most scale inhibitors have some dispersive qualities, the extent depending on the inhibitor. Dispersion involves surrounding particles of suspended salt, iron, or organic solids with the anionically charged scale inhibitor that will tend to repulse other similarly charge-surrounded particles. This prevents the agglomeration of the particles into larger particles that might otherwise fall out of suspension.

Scale inhibitors are effective only at slowing the scale formation process or slowing the process of scale particle agglomeration. In an RO system, they have to perform this function only long enough for the salts to flow out of the RO system through the RO concentrate.

Loss of RO membrane performance due to scale formation is not as common as fouling by suspended solids or biological activity. This is because scale formation is relatively easy to control or eliminate with proper pretreatment. If the precipitation of scale occurs, it may be a great concern or it may be a small concern, depending on the membrane type, the particular scale, and its quantity.

Many systems can be cleaned of calcium carbonate simply by increasing the rate of acid injection upstream, or by recirculating an acidic cleaning solution. Some polyamide thin-film spiral-wound polyamide RO facilities rely strictly on scale inhibitor injection as a means of preventing the formation of calcium carbonate.

Operating at minimal levels of scaling prevention is not suggested with CA membrane. Should scale formation occur, it can create areas of localized high pH at the membrane surface that will increase the hydrolysis rate of the cellulose acetate. These systems have been known to lose most of their salt rejection capabilities after a few days of being allowed to operate with significant calcium carbonate scale present on the CA membrane.

Location of injection point. When both sulfuric acid and scale inhibitor injection are in use, and the sulfate salt solubility is being exceeded, the location of the scale inhibitor injection point must be upstream of the point of acid injection. Otherwise, the sulfate precipitation process may begin before the scale inhibitor has an opportunity to slow it down. The high concentrations of sulfate ions localized at the point of the acid injection will promote the nucleation of sulfate salt crystals unless scale inhibitor is present to slow the process.

Over-injection of inhibitor. If a scale inhibitor is injected at too high a concentration, it is possible to exceed the solubility of the cation/inhibitor complex. If this occurs, the cation/inhibitor complex will tend to fall out of solution in the membrane system, thus defeating its own purpose. The potential for this occurrence can be minimized by using a blended scale inhibitor that uses more than one type of inhibitor. The inhibition effects are additive (and sometimes synergistic), yet there is less chance that the inhibitor itself will fall out of solution. In fact, one inhibitor will assist in keeping another inhibitor in solution.

Comparison of scale inhibitors. Many common scale inhibitors consist of molecules that contain carboxylic acid (-COOH) or phosphate (PO_4^{-3}) functional groups. Lower-molecular-weight polyacrylate molecules (molecular weight distribution from 1,000 to 5,000) contain multiple carboxylic acid functional groups and are commonly used in many inhibitors. They are best at inhibiting carbonate and sulfate formation, but have limited dispersive qualities.

Sodium Hexametaphosphate (SHMP, also called sodium hex') was once the inhibitor of choice because it offered good inhibition at a low cost. Its big disadvantage is that it is unstable and difficult to mix. In fact, if it is not remixed every 3 days, the hexametaphosphate can hydrolyze into phosphates that can precipitate with calcium to form calcium phosphate at neutral or alkaline pH. This salt can then end up fouling the membrane system. The popularity of sodium hex' as a scale inhibitor for RO systems has declined because of its heavy dependence on proper usage.

Organophosphonates are an improvement over sodium hexametaphosphate in that they are more stable. They offer inhibition and dispersive characteristics similar to those of sodium hexametaphosphate. Unlike sodium hex', however, their functional groups will tend to stay attached.

Higher-molecular-weight polyacrylates (molecular weight distribution in the range of 6,000 to 25,000) are best at dispersion. However, they are not as effective at scale inhibition as the lower-molecular-weight acrylates.

Blend Inhibitors. As previously discussed, blend inhibitors offer advantages over monochemical inhibitors. With a monochemical inhibitor, there is a greater possibility that over-injection of the inhibitor can result in the inhibitor itself falling out of solution with a multivalent cation. With blend products, the other inhibitor present can help prevent the precipitation of the first inhibitor. Also, there will be a smaller concentration present of each of the individual components of the blend product.

Some blend inhibitors offer both low- and high-molecular-weight polyacrylates for good dispersive and inhibitor qualities. Others offer a blend of lower-molecular-weight acrylates and organophosphonates, again for good dispersive as well as inhibitor characteristics.

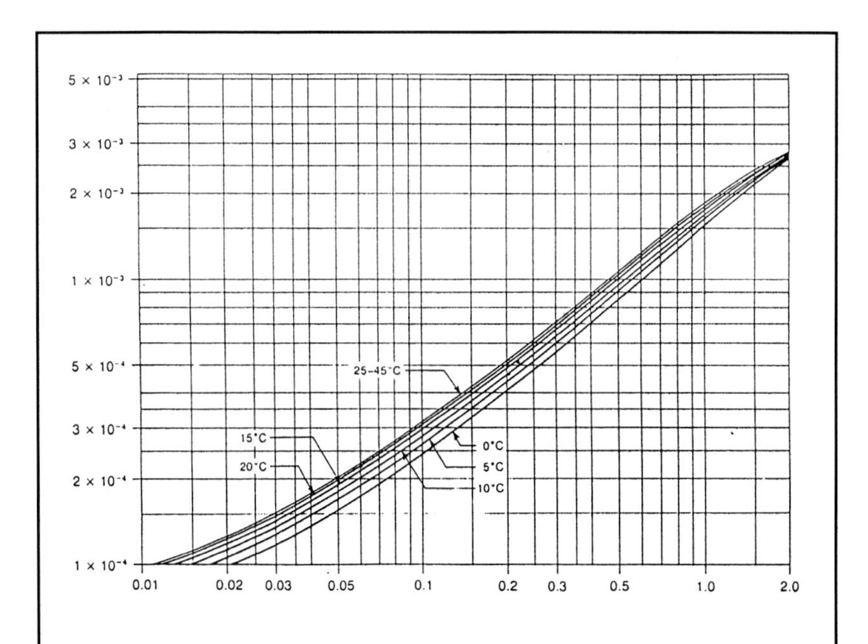

Figure 3-15b. K_{sp} for $CaSO_4$ versus ionic strength.
(Courtesy of E.I. Dupont).

Day tank maintenance. Biological activity in the scale inhibitor day tank can sometimes cause problems. The biogrowth can plug the inlet of the injection pump suction line and stop the flow of the scale inhibitor.

Many scale inhibitors/dispersants include a biogrowth retardant in their formulas to prevent growth in the day tank or in the concentrated chemical. The retardant will be effective only if there is a sufficient concentration. If it is necessary to dilute the scale inhibitor in the day tank, the supplier should be contacted to determine the maximum dilution at which the retardant is effective.

Biogrowth can also be prevented by either dropping the pH to 4 by adding acid, or by adding a minimum of 200 ppm of sodium bisulfite. It is critical that the biogrowth inhibitor is compatible with the RO membrane material. Check with the chemical supplier or the membrane manufacturer. (Luss, 1993).

Depending on the advice of the manufacturers of the scale inhibitor and membrane, it is possible to exceed the solubility of calcium carbonate (and certain other salts) with the correct addition of inhibitor. Some are willing to push the LSI to a value as high as +2 in the RO concentrate. However, some of the membrane manufacturers limit the Langelier index to +1.5 with the addition of scale inhibitor, and some to +0.5 when using sodium hexametaphosphate.

Certain membrane manufacturers also limit the extent of supersaturation for sulfate salts that are acceptable in the RO concentrate stream with the injection of scale inhibitors. An example of the limits are as follows:

Calcium sulfate ($CaSO_4$)	230% saturation
Strontium sulfate ($SrSO_4$)	800% saturation
Barium sulfate ($BaSO_4$)	6,000% saturation

(Courtesy of Hydranautics, Inc.)

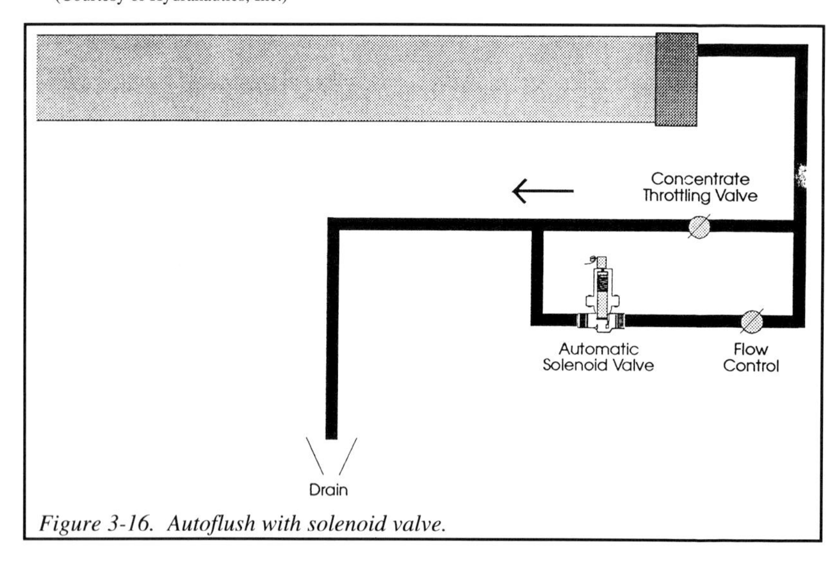

Figure 3-16. Autoflush with solenoid valve.

Many RO systems may operate on water sources that are near the saturation point of calcium sulfate. This means that if they operate at a typical permeate recovery of 75%, the supersaturation of calcium sulfate in the RO concentrate stream will approach 400%. This will well exceed the ability of scale inhibitors to prevent scale formation. To more accurately determine the solubility of calcium sulfate so as to better define the limits of supersaturation, Figure 3-15b can be used to find the solubility product constant at saturation. It takes into account water temperature and ionic strength. Ionic strength is calculated by using one half the sum of the molar concentrations of all the ions in the water source, each multiplied by its valence number squared: $I = 1/2\, \Sigma m_i z_i^2$. (Note: The bottom of Figure 3-11 can also be used to calculate the ionic strength).

Because scale inhibitors simply slow the precipitation process and do not actually stop it, it is important that the RO does not sit stagnant for extended periods of time while containing supersaturated salts. It may be necessary to install an automatic solenoid valve in parallel with the RO concentrate valve, as shown in Figure 3-16. This solenoid valve opens just prior to the RO shutting down. In this manner, supersaturated salts can flush out of the system. This is particularly critical for CA membrane that can hydrolyze when exposed to calcium carbonate scale.

For small RO systems that have a limited number of chemical injection systems, a frequent desire is to use one system to inject two chemicals. This practice is possible with certain chemicals, although generally not recommended. The supplier should always be consulted prior to mixing two chemicals in the same day tank. When using a mixture of two chemicals, some flexibility will be lost in not being able to temporarily increase or decrease the dosage of one chemical without affecting the injected concentration of the other chemical.

Water softening using ion-exchange resins can prevent scale formation within reverse osmosis systems by removing the hardness (multivalent cations) in the

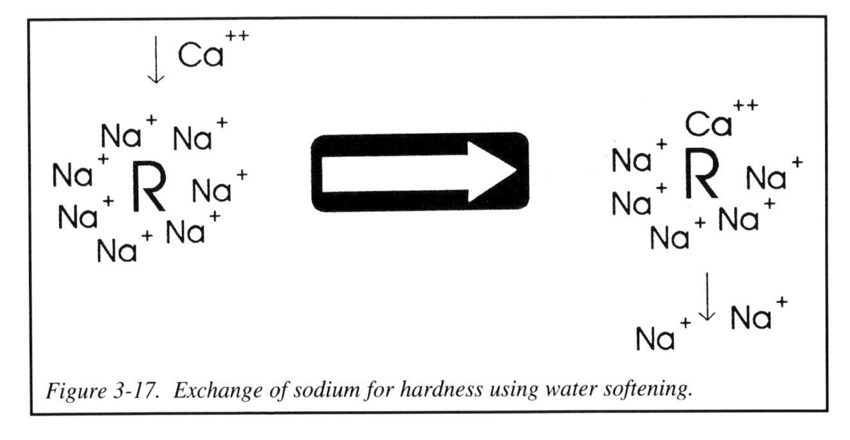

Figure 3-17. Exchange of sodium for hardness using water softening.

RO feedwater. The most common of these ions include calcium (Ca^{+2}), magnesium (Mg^{+2}), and iron (ferrous [Fe^{+2}], which tends to be soluble; or ferric [Fe^{+3}], which is usually insoluble).

A water softener uses a resin media that has functional groups with strongly negative charge characteristics. This gives the resin the ability to attract various positively charged cations. It is regenerated with a sodium chloride (common table salt, NaCl) solution. When multivalent hardness or iron passes through the resin, it is preferred to the monovalent sodium and makes an exchange. For divalent cations (i.e., with a valence of +2, such as calcium or magnesium), two atoms of sodium will be displaced for every atom of hardness removed. (See Figure 3-17.)

Regeneration of a water softener typically involves four steps:

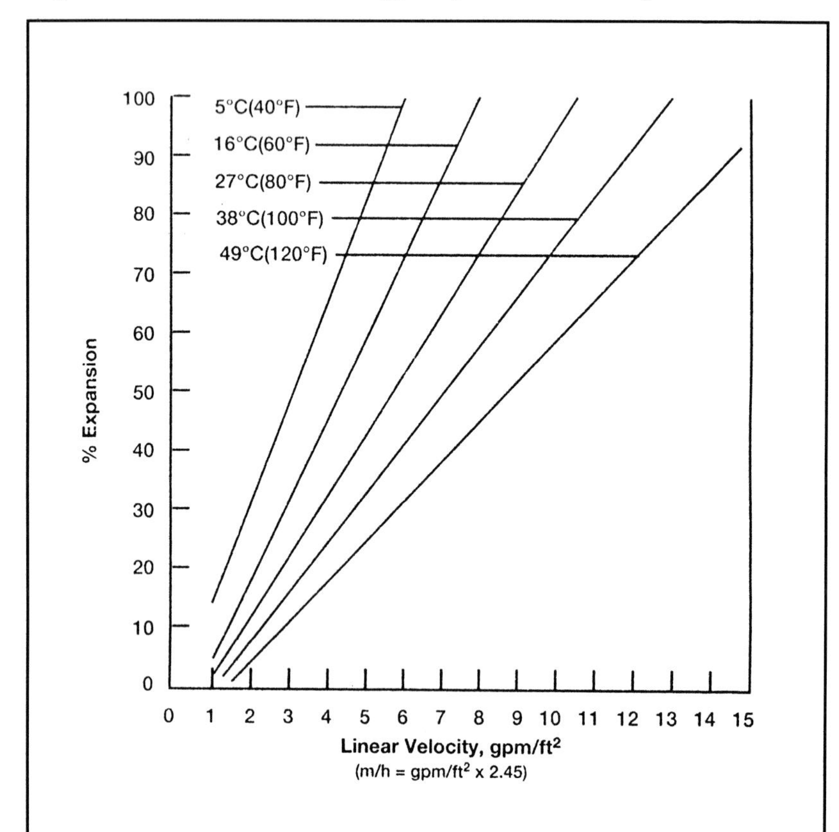

Figure 3-18. The bed expansion for various backflow rates as a function of temperature for a typical cation exchanger (Amberjet IR-1200(H)). (Courtesy of Rohm & Haas Co.)

Backwash. To remove the suspended solids and resin fines that have been accumulated by the resin, it is necessary to backwash the softener. This requires putting a flow upward through the resin at a flowrate sufficient to lift the resin within its vessel. The desired expansion for a softener resin is 50% of the bed depth. The time required for backwashing is usually around 10 minutes, or until the backwash effluent is clear. (Betz, 1982).

The bed expansion is a strong function of water temperature. Figure 3-18 gives the bed expansion for various backflow rates as a function of temperature. At 40°F, a backflow rate of about 4.5 gpm/ft^2 is sufficient to expand the bed by more than 50%, whereas this rate increases to better than 6 gpm/ft^2 if the water temperature is 70°F. As water temperature drops during winter months, care should be taken to not backwash out resin through the top of the bed.

Brining. The exchange of cations on a resin bead is an equilibrium process. The resin will preferentially attract hardness ions over sodium ions because of their greater valence charge characteristics. However, this preference can be overcome by increasing the ratio of sodium ions to hardness ions. A softener will remove little if any hardness from a feedwater (such as seawater) containing an extremely high concentration of sodium.

Regenerating a softener requires passing a high concentration of sodium chloride (usually about a 10% solution) in the normal flow direction through the resin bed. This shifts the equilibrium of the resin toward giving up the hardness in exchange for the sodium. The flowrate during this brining (Figure 3-19) should be about 1.0 gpm/ft^3 of resin.

The capacity for hardness that can be removed by a softening resin will depend on the sodium chloride (NaCl) concentration that is used in the regeneration. (See Table 3-2.) This capacity will affect the required regeneration frequency, as well as the percentage of influent hardness removed by the bed. Since it is

Table 3-2
Strong Acid Cation Resin Capacity as a Function of Regenerant Salt Dosage

Regeneration Concentration (lb NaCl/ft^3)	*Capacity Grains/ft^3 (as $CaCO_3$)*
5	18,000
10	25,000
15	29,000
25	34,000

(Data courtesy of Rohm & Haas)

desirable to remove the vast majority of hardness from an RO feedwater, a high concentration of sodium chloride, such as 15 pounds of salt/ft^3 of resin, is typically used for the regeneration of softeners used for RO pretreatment. If it is desirable to further reduce the effluent hardness concentration, the softener can be regenerated with an even higher concentration of salt. This is not a particularly optimal usage of salt, but the RO performance may benefit by it.

Slow rinse. A slow rinse of 1 gpm/ft^3 of resin is used downward through the resin bed to displace all of the highly concentrated salt solution. This slow rinse allows the resin at the bed bottom to receive adequate contact time with the salt solution. After the concentrated salt solution is displaced, the system is ready for fast rinse.

Fast rinse. A fast rinse of about 1.5 gpm/ft^3 of resin is used to rinse residual hardness and excess sodium chloride from the bed. The hardness will tend to be rinsed out first. For some applications, the unit can be put back into service after the effluent hardness concentration returns to normal; however, a longer backwash period will be required for the effluent sodium chloride concentration to equal that of the influent water.

Softener maintenance. When used to prevent the precipitation of calcium carbonate in an RO system, the proper operation of the water softener is critical for the optimal performance of the RO system. Its effluent hardness should be checked and recorded daily using a test kit that has a sensitivity to calcium carbonate of 1 mg/L. The automatic valves for on-site regenerable units should be serviced annually. If the cation exchange resin is more than 5 years old and its effluent hardness concentration is increasing, it should be replaced.

If the resin fouls with iron or aluminum, it may require treatment with a cleaner normally available from suppliers as a proprietary formulation. The high

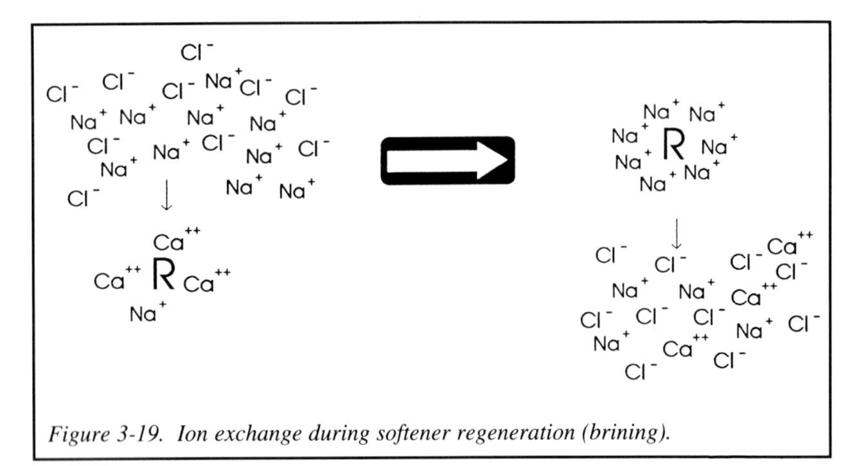

Figure 3-19. Ion exchange during softener regeneration (brining).

charge characteristics of iron or aluminum make these metals difficult to regenerate from the cation resin.

Occasionally, softeners will allow undesirable concentrations of calcium or magnesium to pass through. This can occur for a number of reasons:

1 The resin can become fouled with iron. (The resin has a stronger affinity to hold onto iron than to calcium or magnesium.)

2 The resin can become organically or biologically fouled.

3 The resin can age.

4 The regeneration system can malfunction.

5 The brine tank may not have been properly filled with the salt (sodium chloride) required for the regeneration.

The malfunctioning of the softener can be an immediate problem for an RO system, as precipitation is likely to occur. A backup method of preventing calcium carbonate scale formation, such as the injection of acid or a scale inhibitor, or a backup softener, is recommended.

An inexpensive backup to a softening unit would be to use a portable softening column. If the primary softener is functioning properly, the backup softening column will require infrequent regeneration (performed off-site by a water service company).

Water softening offers certain advantages over acid injection, if used as the sole means of preventing the formation of calcium carbonate. By removing the cations that cause hardness, softening of a feedwater will also eliminate the possibility of precipitating other insoluble salts, such as sulfates or phosphates. Softening will also remove iron and manganese that might fall out of solution if

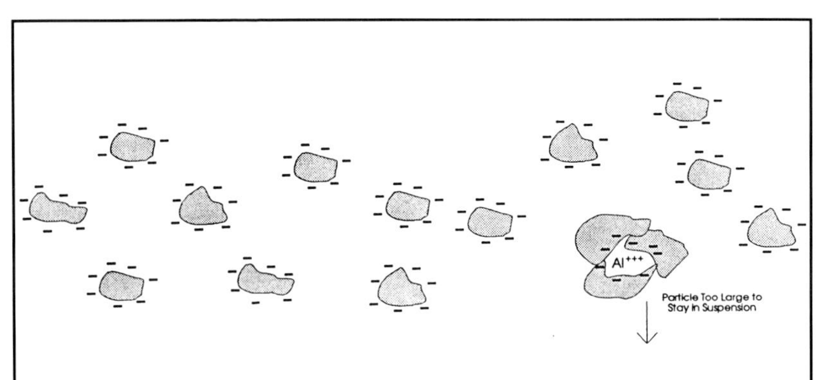

Figure 3-20. Left: suspension via repulsion; right: neutralization of colloid repulsion caused by presence of trace metal.

present in their oxidized states.

Colloid stabilization. A softening column is effective at stabilizing colloids that might be present in the RO feedwater. Colloids will tend to stay dispersed in water by the repulsive forces of their various negatively charged surface groups. If a highly charged cation is present, its positive charge will neutralize the charge present on the surface of the colloid particles, allowing them to group with the cation and fall out of suspension. Replacement of the highly charged cations with monovalent sodium ions enables greater concentration of the colloidal material in the RO system before it will fall out of suspension (Figure 3-20).

Removal of residual cationic coagulant. If a cationic coagulant is being injected upstream of a media filter to promote the coagulation and filtration of colloids, softening is a way to prevent the possibility of the coagulant breaking though the pretreatment system and fouling the RO membrane. The coagulant will be attracted to the strongly negative charge on a standard strongly acidic type of cation-exchange resin used in softeners. Removal of the coagulant from the resin may require backwashing the resin with a high chlorine concentration (Iverson, 1993).

Acid-regenerated softening. The softener discussed above uses what is called a strongly acidic cation resin. It has a strongly negative charge that enables it to better hold onto sodium during its regeneration.

If the resin used in the softener is a weakly acidic cation resin, it will not be as capable as the strongly acidic resin of holding onto sodium. Such a bed has been employed upstream of reverse osmosis to provide additional protection against scale formation. The weakly acidic resin can be regenerated with a strong acid such as hydrochloric acid or sulfuric acid. The acid will displace any hardness that would have been removed during its service cycle.

While in service, the bed will exchange hydrogen ions (H^+) for any hardness that is present in the raw water. This reduces the potential for scale formation. Depending on the makeup of the water source, the resulting pH will typically be between 5.0 and 6.5. This lower pH will also reduce the potential for calcium carbonate precipitation by converting much of the bicarbonate ion into carbon dioxide. In addition, the lower pH will increase the solubility of the carbonate and sulfate scale. This is all accomplished without adding additional anions to the water, as would be the case with an acid injection system; and without adding sodium to the water, as would be the case with a standard softening system.

Not adding sodium into the RO feedwater can be advantageous in applications where trace sodium is a concern. This tends to be the case in semiconductor and beverage applications.

One major disadvantage of this type of pretreatment system is that carbon dioxide is created in the RO feedwater because of the conversion of bicarbonate ion into carbon dioxide. This carbon dioxide will permeate the RO membrane and may require downstream removal. Another disadvantage is that this type of system has a higher cost of operation because of the higher cost of regenerating with acid versus salt. Acid-regenerated softening should be considered however, as a pretreatment option for water sources with a high potential for scale formation.

A summary of the commonly used methods for prevention of scale formation is given in Table 3-3 as a function of the salts for which each method is effective.

Fouling Control

Any membrane system, such as an RO system, can foul if exposed to the

Table 3-3
Common Methods for Preventing Scale Formation

Scale	*Method of Prevention*	*Manner of Operation*
Carbonate	Acid injection	Converts HCO_3^- ion into CO_2; increases carbonate solubility
	Water softening	Removes cation hardness
	Scale inhibitor/dispersants	Binds up cation hardness, slows agglomeration of scale particulates
	Lower RO permeate recovery	Lowers the concentration of hardness & carbonates
Sulfate	Water softening	Removes cation hardness
	Scale inhibitor/dispersants	Binds up cation hardness, slows agglomeration of scale particulates
	Acid injection	Lower pH increases sulfate solubility
	Lower RO permeate recovery	Lowers the concentration of hardness & sulfates
Silica	Lower RO permeate recovery	Lowers the concentration of hardness & silica
	Increase water temperature	Increases solubility
	Dispersant injection	Slows agglomeration of scale particulates
Iron or manganese*	Oxidation w/ filtration	Manganese greensand or oxidizing agent injection oxidizes it to insoluble state that can be filtered
	Water softening	Removes iron/manganese

*To be discussed in next section.

suspended solids that are present in many raw water sources. These suspended solids can consist of silt, humic and fulvic acids (which come from the natural breakdown of organic matter), and other organic or inorganic materials. They are most commonly found in surface waters (i.e., rivers, lakes, or reservoirs). As they are concentrated within an RO system, the suspended particles are brought into close proximity that can allow them to overcome their similar surface charge that normally keeps them apart. They can then group into particles large enough to fall out of suspension on the membrane surface. Once this occurs, surface attraction forces will tend to keep the foulant adhered to the membrane. This will affect the membrane performance, particularly its permeate flowrate.

If corrected in time, most foulants can be removed from a membrane surface by cleaning the system with a cleaning agent designed to redissolve and disperse the foulants. Therefore with proper maintenance, fouling is not necessarily a problem for most membrane systems.

Because cleaning a system costs time and money, it is desirable to minimize the frequency at which the system requires cleaning. This can be achieved through a variety of means, from coagulation and flocculating agents, to media filters and cartridge filters. Again, the primary objective of removing the suspended solids is to reduce the RO operating costs.

Media filtration. Media filters (i.e., sand filters) are an excellent means of reducing the concentration of suspended solids in an RO feedwater. Media filters

Table 3-4
U.S. Standard Screen Equivalents

Mesh/ Sieve No.	*Opening in Millimeters*	*Opening in Inches*
12	1.68	0.0661
14	1.41	0.0555
16	1.19	0.0469
18	1.00	0.0394
20	0.84	0.0331
25	0.71	0.028
30	0.59	0.0232
35	0.50	0.0197
40	0.42	0.0165
45	0.35	0.0138
50	0.297	0.0117
60	0.250	0.0098
70	0.210	0.0083

(Rohm and Haas Company, *Engineering Manual for the Amberlite[R] Ion Exchange Resins*)

use a filtration bed consisting of one or more layers of media granules. These granules may be sand, garnet, and/or anthracite granules, and may use more than one size range of granule. The granules have the ability to remove suspended particulates via a number of different filtration mechanisms.

The original media filters, still used commonly in city water treatment systems, are called gravity filters. Water is percolated through a bed of sand under the force of gravity. Two filtration mechanisms are the predominant factors in gravity filtration. One is the physical screening of particles too large to fit through the spaces between the media granules. Most of this screening takes place in the top layer of media. The other is the force of gravity, whereby particles settle onto the upper surfaces of the individual media granules.

Gravity filters use a fine sand media to achieve excellent effluent water clarity. The velocity is so low through such a filter, however, as to require an extremely large filter. Although possibly worth considering for high fouling applications, such a low-velocity filter for most RO applications is not economically practical.

The effectiveness of gravity filters is dependent on achieving low water velocities through the media bed. At higher velocities, the shear forces of the water flow against the suspended particles will tend to reduce the effectiveness of gravity filtration. It will also result in high pressure drops. The media will tend to trap particles in the top few inches of media. As the holes between the media granules plug, the pressure drop across the media and foulants will increase. The pressure drop will dictate when to backwash the filter.

Pressure filters operate by pushing water through a filtration media using the pressure supplied by a pump. A common type of pressure filter, called a multimedia filter, uses different sizes and densities of media granules. This is in an effort to achieve a high solids removal capacity at higher flow velocities.

A multimedia filter is designed to make better use of the bed depth in the removal of a greater volume of suspended solids. This is achieved by loading larger media granules of lower density at the top of the bed, with at least one layer of smaller media granules of higher density below it. The larger granules remove the larger suspended particles, leaving the smaller particles to be filtered by the finer media. These beds will be operated at flow velocities as high as 20 gpm/ft^2. However, their best efficiency in the removal of fine particles occurs when they are operated in the range of 3 to 5 gpm/ft^2 (McPherson, 1981).

Smaller granules of a given media density will be lifted higher in the filter vessel during backwash than will larger granules of that same media density. At the completion of the backwashing, the media will settle back down, with the larger granules settling first. Since it is desirable to keep the larger granules at the top of the filter media, different media densities must be used in a multimedia filter.

Anthracite, which is processed from a hard coal, serves this purpose because of its lower density than sand, and larger granule size. In this manner, it will lift higher than sand during backwashing and will settle on the top of the media bed. Anthracite has irregular surface characteristics that create flow eddies, assisting it in filtering suspended solids. However, it also has a nonuniform particle size that tends to create a larger pressure drop than that created by a media whose particle size is more uniform.

Aluminum silicate granules (also known as Filter-AG from Clack Corporation) are sometimes used as a substitute for anthracite because the irregularities of their surface and size are similar to anthracite. The granules are even lighter than anthracite, however, for an even greater separation in density from the denser media layers. For single-media pressure filters, the lighter density of the aluminum silicate granules enables reduced backwash rates.

Silica sand is sometimes used as the next layer after anthracite in a multimedia filter because of its smaller granule size with higher density. However, the difference in density between anthracite and silica sand is not enough to prevent the sand from intermixing with the anthracite. When this occurs, the intergranular space is reduced, and the advantage of using two different medias is subsequently lost. Most of the filtration will occur on the surface of the bed, and it will tend to suffer from excessive pressure drops. (Michaud, 1993).

Crushed garnet rock offers similar granule sizing to that of silica sand, but with higher density. Its density is different enough from that of anthracite to create a fairly sharp definition between media layers. This enables the bed to obtain better depth filtration.

A common makeup of a multimedia bed, going from the top of the bed to the bottom, might include the following:

Media	*Granule Size*	*Density*	*Bed Depth*
Anthracite	0.8-0.9 mm	54 lb/ft^3	14 - 16"
Garnet	30 × 40 mesh	130 lb/ft^3	8 - 10"
Garnet	8 × 12 mesh	144 lb/ft^3	6"

(Michaud, Chub, Personal Communication, March, 1993)

The 8 X 12 mesh garnet is present to act more as support for the finer 30 X 40 mesh garnet than as a filtration layer. It will be located below the 30 X 40 mesh garnet, but is larger in granule size.

The mesh rating for media corresponds to the number of holes per linear inch used in a screen to separate the media. In a 30 X 40 mesh media, 90% of the media granules are larger than the holes in a 40-mesh screen, but smaller than the holes in a 30-mesh screen. If the rating just gives a single number mesh rating, it refers to the upper limit on granule size, with 90% of the media being smaller in size than the screen mesh rating. Screen sizing is presented in Table 3-4.

The particular grading used in a multimedia filter can be altered, depending on the size of foulants anticipated in the RO feedwater. Some water sources may contain a high concentration of fine silt, removal of which may require a comparatively fine media. In such cases, anthracite of a granule size of 0.6 to 0.8 mm may be substituted for the larger anthracite granules. An additional 4-inch layer of 60-mesh garnet may be placed on top of the 30 X 40 mesh garnet.

The pressure drop through this finer media filter will be about 6 psid at a flowrate of 8 gpm/ft^2. This is compared to a typical pressure drop of about 3 psid for the standard media at this same flow velocity (Michaud, 1993).

The pressure drop across finer media will tend to increase more quickly as the media removes suspended solids, and it will thus require more frequent backwashing. To reduce the backwash frequency, the filter vessel can be sized with a larger diameter for a lower flow velocity. This will also help the removal efficiency.

Media filters should be backwashed if their pressure drop increases by 10 psid. (They may have to be backwashed at lower pressure drops if there is insufficient pressure for the high-pressure RO pump inlet.) At pressure drop increases of more than 15 psid, the suspended solids can compact to form a cake of higher density. Such a foulant cake can be difficult to break up and remove during backwashing.

Media filters are sometimes designed with crossbars to break up foulant cakes during backwashing. Some are designed with compressed air intakes to blow in air during backwashing as a means of breaking up the filter cakes. (Note: A pressure vessel that uses compressed air will for reasons of safety require special design features that may add dramatically to its cost.)

To prevent the blinding off by the finer filter media of the lower distribution lateral in a media filter, gravel support layers are typically used to cover the lateral. This gravel has very little to do with filtration, except to support the

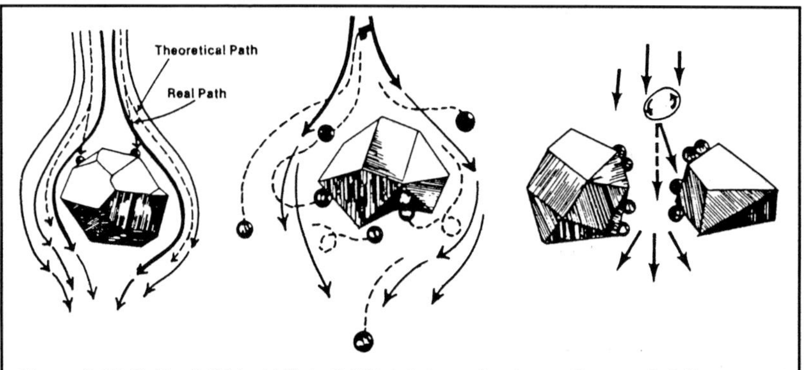

Figure 3-21 (left), 3-22 (middle), 3-23(right), mechanisms of internal, diffusive, and hydrodynamic forces to bring particles into contact with media granules.
Coccagna, 1984 (Courtesy of Culligan).

filtration media. It does not lift during backwashing of the media. A common gravel grading might include the following:

Media	*Granule Size*	*Density*	*Bed Depth*
Fine gravel	1/8" × 1/4"	100 lb/ft^3	4"
Medium gravel	1/4" × 1/2"	100 lb/ft^3	4"
Coarse gravel	1/2" × 3/4"	100 lb/ft^3	4-6"

Media filters should be backwashed at a rate between 14 and 16 gpm/ft^2. At this rate, the media filter will need up to 50% freeboard. Freeboard is the empty area above the media within the filter vessel to accommodate the media expansion. At the suggested backwash rate, most of the media will lift, but there will be little lifting of the support gravel. A 50% freeboard in a filter with 3.5 feet of media depth would translate to 1.75 feet of space above the media. The sideshell of the filter vessel would therefore be at least 5.25 feet.

Filtration mechanisms other than just the physical screening of particles become more important in true depth filtration. Multimedia filters are capable of achieving much finer filtration than what might be indicated by the spacing between media granules. This ability is achieved via transport mechanisms that bring the smaller particles into close proximity with the media granules, and attractive forces that hold the particles onto the media.

Inertial forces that increase with higher flowrates can cause the momentum of a particle to bring it into contact with the media granules, rather than stay in the bulk water flow (Figure 3-21). Diffusive forces of submicron particles in water will create a random motion that can bring the fine particles into contact with the

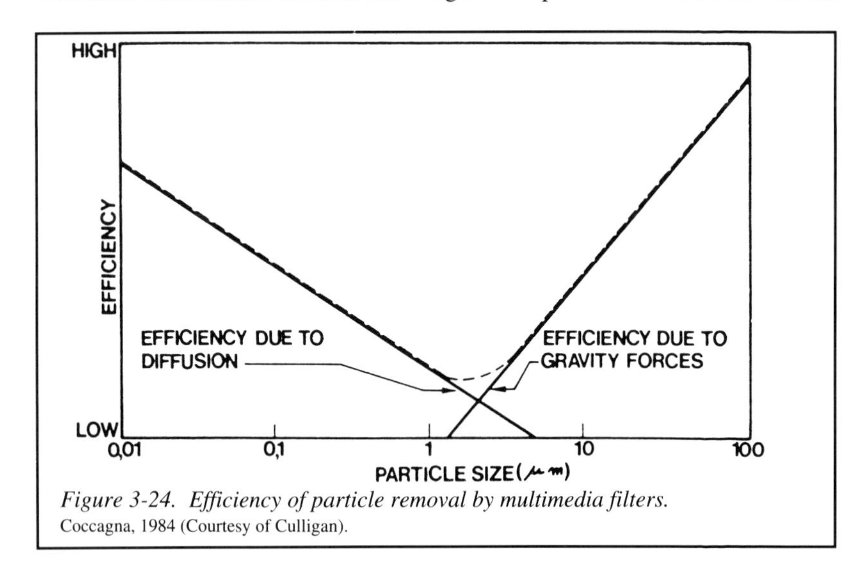

Figure 3-24. Efficiency of particle removal by multimedia filters.
Coccagna, 1984 (Courtesy of Culligan).

granules (Figure 3-22). Hydrodynamic forces caused by the flow of water through the spacing between media granules will tend to give particles a rotational movement within the space. The centrifugal force from the rotation will drive particles to the walls of the granules (Figure 3-23).

Once particles are at the granule walls, electrostatic attractive forces, known as van der Waals forces, will tend to hold onto the particles. For the particles to stay attached, these forces must be sufficient to overcome the shear forces of the water flowing by the particles. The smaller particles will be better able to withstand the shear forces, because there is less surface area to drag against the flowing water (Coccagna, 1984).

Multimedia filters are thus able to remove many submicron particles. As discussed, they also can remove larger particles via screening and inertial forces. The overall performance is such that the worst removal is of particles that are approximately 1 μm in size. (See Figure 3-24.)

The media in a multimedia filter will require periodic replacement. With time, the surface irregularities on the granules will be lost due to abrasion against other granules. In addition, the media can become coated with organic material that is not removed by backwashing. The media will also over time collect dense particles that will not backwash out.

Depending on the quality of the water source, the media will probably need to be replaced every 5 to 10 years. If the after-backwash pressure drop of the filter has increased substantially since the media was new, it is time to replace the media.

It is suggested that the gravel underbedding be replaced when the rest of the media is replaced. This will remove all fines and dense debris that may have settled into the gravel.

When water is acidified (as may be required for the RO system), the positively charged hydrogen ions (acidic ions) can dispel some of the anionic charge of the suspended solids. This can allow the solids to coagulate and fall out of solution.

For this reason, it is a good practice to locate acid injection systems upstream of the media filters used in reducing the concentration of suspended solids. The materials of construction of the media filters must be compatible with the acidified water. With an acidified feedwater, standard (unlined) carbon steel, brass, or copper should not be used for the filter tank, laterals, or the face-piping.

Manganese greensand or Birm™ filters. An unusual glauconite sand, mined in New Jersey and called *greensand*, can be chemically treated such that it is covered with manganese oxide groups. Now called manganese greensand, it has the ability for the manganese oxide groups to change valence states. It is supplied in an oxidized form that will oxidize dissolved iron, manganese, and hydrogen sulfide in water. Once contaminants are oxidized to their insoluble states, the greensand acts as an excellent filter to hold onto them. Media filters that include greensand are able to remove iron, manganese, or hydrogen sulfide down to concentrations as low as 0.01 mg/L (Aronovitch, 1993).

The oxidation of these species by the manganese oxide occurs as follows:

$$2(Z{:}MnO_2) + 2Fe(OH)_2 \longrightarrow Z{:}Mn_2O_3 + Fe_2O_3 + 2H_2O$$

$$2(Z{:}MnO_2) + 2Mn(OH)_2 \longrightarrow Z{:}Mn_2O_3 + Mn_2O_3 + 2H_2O$$

$$2(Z{:}MnO_2) + H_2S \longrightarrow Z{:}Mn_2O_3 + S + H_2O$$

The oxidation potential of the greensand can be depleted. Its capacity is approximately as follows:

	lb/ft³	*ppm-gallons/ft³*	*grains/ft³*
Iron	0.1	12,000	700
Manganese	0.06	7,000	400
Hydrogen sulfide	0.03	4,000	230

If the media becomes exhausted, it can be regenerated by feeding the bed with a potassium permanganate ($KMnO_4$) solution. The potassium permanganate should be diluted so that the media is exposed to between 0.5 and 2 ounces of potassium permanganate per cubic foot of greensand. This will restore the media oxidation potential.

$$2(Z{:}Mn_2O_3) + 2KMnO_4 \longrightarrow 4(Z{:}MnO_2) + K_2MnO_4^{*} + MnO_2$$

In some localities, sanitary discharge requirements limit the use of potassium permanganate. It is possible to reduce or eliminate the need to regenerate the greensand by injecting sodium hypochlorite (chlorine) upstream of the bed. The chlorine will oxidize any iron, manganese, or hydrogen sulfide in the water, thus reducing the oxidation demand placed upon the greensand.

In water sources containing hydrogen sulfide, it is critical that the oxidation

*The K_2MnO_4 can chemically reduce further in state, offering more oxidation potential in the process.

Table 3-5
Chlorine Injection Requirements for Upstream Oxidation
(of 1 mg/L of contaminant)

Oxidation	*Requirement (as Cl_2)*	*Minimum Recommended (as Cl_2)*
Fe^{+2} (as ion)	1.3 mg/L	1.3 mg/L
Mn^{+2} (as ion)	2.6 mg/L	2.6 mg/L
H_2S (as sulfur, S)	4.4 mg/L	5 mg/L

characteristics of the manganese greensand not be allowed to be depleted. The reduction potential of hydrogen sulfide is so great that it will strip the manganese oxide coating off the greensand if the hydrogen sulfide is not oxidized upstream. It is strongly recommended when removing hydrogen sulfide, or high concentrations of iron or manganese, to continuously inject sodium hypochlorite or potassium permanganate upstream. The injection concentration should be sufficient to better than match the concentration of the iron, manganese, and hydrogen sulfide when their concentration is expressed as $CaCO_3$, or as moles/liter.

The recommended concentrations for oxidation of the contaminants are given in Table 3-5. It should be noted that for hydrogen sulfide removal, the concentration of chlorine required to oxidize the hydrogen sulfide will be 5 mg/L or more, per mg/L of hydrogen sulfide (expressed as sulfur).

An alternative to manganese greensand for iron and manganese removal is a product called Birm (a registered trademark of Clack Corporation). Birm utilizes a manganese dioxide coating on a light silica core. It catalyzes the reaction between iron or manganese and dissolved oxygen in order to precipitate the metal. The Birm itself does not perform the oxidation.

Birm requires that a certain concentration of dissolved oxygen be present in the feedwater, and that the water have an alkaline pH. Media filters using Birm may need some sort of air eduction system to increase the oxygen concentration in the feedwater. Hydrogen sulfide or oil should not be present in the feedwater, as they can foul the Birm.

There are several advantages to using Birm instead of greensand for the removal of iron or manganese. One advantage is that no chemicals are required. Also, the Birm tends to be less expensive than greensand.

Coagulation aids. Positively-charged (cationic) polymers can be injected into a water source to neutralize the negatively charged suspended solids. These polymers thus assist in the coagulation of the solids. They also act as flocculating agents, holding together groups of coagulated solids, which further promotes the falling out of solution of the solids. A cationic coagulant would typically be injected upstream of media filters designed for removal of the solids.

Zeta potential can be used as an indirect monitor that the correct concentration of coagulants is being injected. Theoretically, for the best removal of suspended solids, just enough coagulant would be injected to obtain a neutral zeta potential. If coagulant is over-injected to the point of obtaining a positive zeta potential, the solubility of the suspended solids can actually increase. (See "Zeta Potential" in Chapter 1.)

Injection of cationic polymeric coagulants was a method used to reduce suspended solids concentrations prior to the arrival of RO technology. Unfortunately, using such a method as part of the RO pretreatment is frequently not

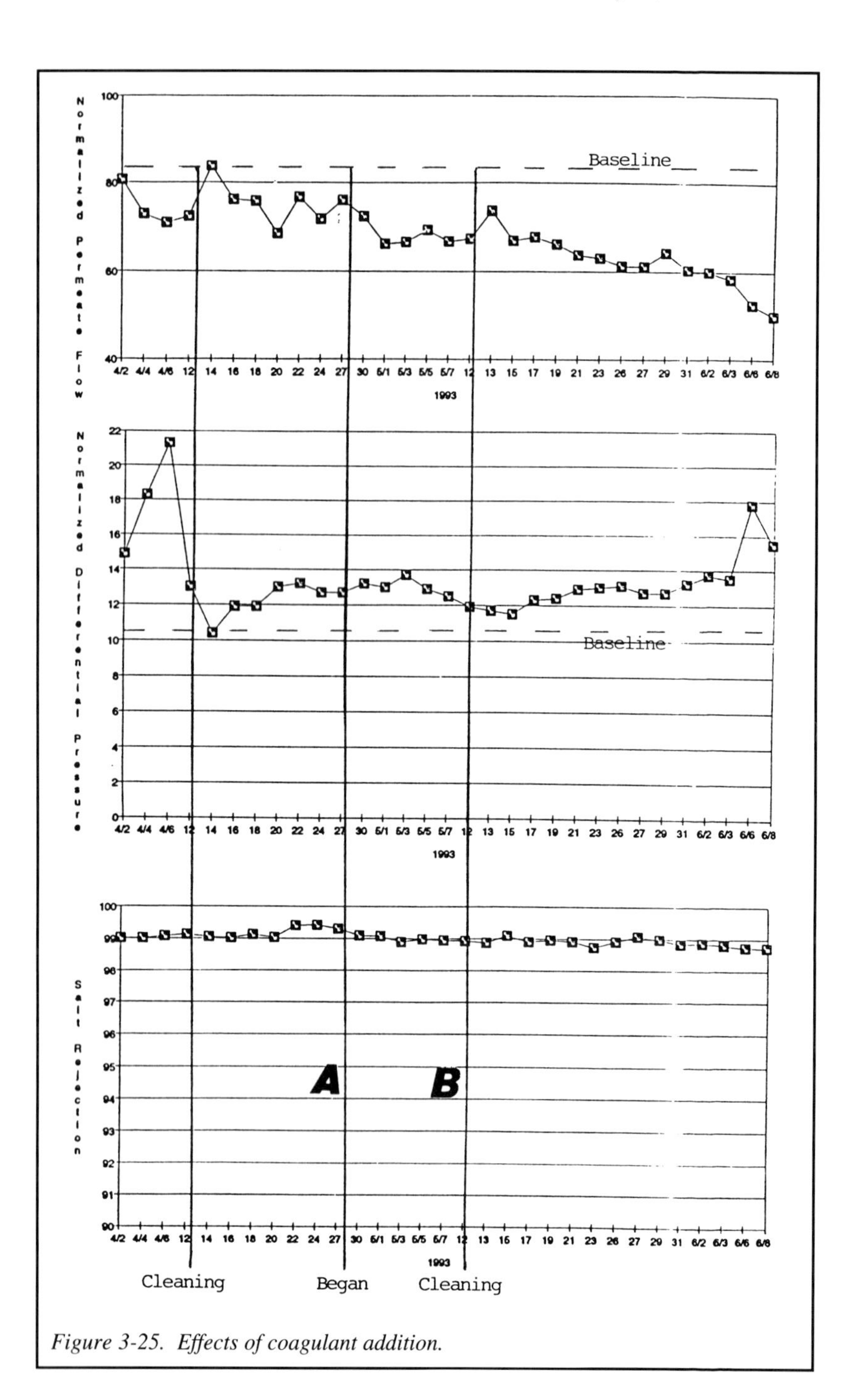

Figure 3-25. Effects of coagulant addition.

effective as a means of reducing the fouling rate of an RO system. Although polymeric coagulant injections may be very effective at reducing the suspended solids concentration in an RO feedwater, they can lead to a dramatic increase in the rate of RO membrane fouling.

Media filters are not 100% efficient at the removal of coagulants. If any coagulant should break through the media filter, it will tend to be attracted to the RO membrane surface. This is particularly true with polyamide membranes that are negatively charged (which is the case for most PA thin-film membranes). The positively charged coagulant will attach itself to the membrane, and may be impossible to remove.

Cationic coagulants will also tend to react with scale inhibitors that are employed to reduce the potential for scale formation. Since most scale inhibitors are negatively charged (anionic), they will react with the positively charged coagulant. This reaction forms an insoluble complex that will readily foul a membrane system. Depending on the nature of the coagulant and the scale inhibitor, the foulant may be extremely difficult to clean (McPherson et. al., 1982).

Figure 3-25 demonstrates the performance of a PA-membrane RO system that had a polymeric coagulant injected upstream of its media filters. This cationic coagulant contained quaternary ammonium functional groups. The normalized RO permeate flowrate indicates a direct correlation between when the injection began, and the beginning of a increase in the RO fouling rate (Point A). Cleaning was not nearly as effective at restoring original membrane performance after the coagulant addition (Point B).

Even after the coagulant addition was taken off-line, the RO continued to experience a high fouling rate. Attempts to clean the foulant were made using aggressive cleaning solutions. However, it was not possible to restore the original membrane performance. The effects of the polymer were so stable that it seemed the membrane had a new performance baseline. This baseline was about 15% to 20% lower in normalized permeate flowrate, and 30% to 40% higher in the system pressure differential (difference between the membrane feed and the concentrate pressure).

It might have been possible for the enduser to accept the altered performance baseline if the rate of membrane fouling had not increased so dramatically. After cleaning, the performance would quickly fall off the baseline. It was as if the polymer would attract the suspended solids to the membrane surface. Most of these suspended solids would have likely flushed through to the RO concentrate if the polymer were not present on the membrane. The end result was the requirement for frequent cleaning in order to maintain minimal permeate flowrates.

In most cases of fouling related to a cationic polymer, the complete removal of the coagulant from the membrane elements is nearly impossible. Cleaning solutions with extreme pH (exceeding normal guidelines of the membrane manufacturer) may remove some of the polymer. Solutions of methanol have been known to remove some of the polymer. (Note: Methanol can attack some of the materials used in the manufacturing of membrane elements.)

A number of facilities have reported similar problems related to using coagulant injection systems. With one such system, their requirement for cleaning increased to the point that the system was being cleaned every 4 days.

Typically, the chemical companies that promote the use of cationic polymers are not familiar with the subtleties of reverse osmosis. They do not perform the preliminary testing to insure that the correct coagulant dosage is used, and that the coagulant cannot end up in the membrane system.

Contact time should be designed into a coagulant injection system. The coagulant will be the most effective if it has time to neutralize the charge of the suspended solids, and still has time to group with the suspended solids. The coagulant/suspended solids complexes will then be of sufficient size to be removed by a media filter.

The desire to reduce the concentration of suspended solids in the RO feedwater is usually what drives endusers to use coagulant addition. Their water may contain a relatively high silt density index. Because of the SDI requirements that some RO membrane manufacturers have arbitrarily chosen for their elements, the misconception is given that the RO system will not operate properly with the high SDI water.

In most cases, however, these systems would have operated normally with the high SDI water, except for possibly requiring a higher cleaning frequency. This cleaning frequency would have likely been far less than the frequency required after the membrane fouled with coagulant.

If the injection of a cationic polymeric coagulant is used as part of an RO pretreatment system, then a water softener is strongly recommended downstream of the media filter. The cation-exchange resin in the softener will attract any residual coagulant that might break through the media filter. If the softener is being used only for the purpose of removing residual coagulant, it may not be necessary to regenerate it with salt (sodium chloride). Depending on the nature of the coagulant, an occasional 1% chlorine treatment may be sufficient to remove some coagulants. (This solution should not be allowed to come into contact with the RO membrane. Iverson, 1993).

Activated carbon filtration is used in the removal of dissolved and suspended organics, and in the removal of chlorine and chloramines. It can be of great assistance in reducing contaminant concentrations, and yet also be a large contributor of contaminants. Some industries rely on activated carbon filtration heavily, whereas other industries avoid its use.

Carbonaceous materials are activated using steam and high temperatures (and can also be activated chemically). In the process, non-carbonaceous material is removed, resulting in highly porous carbon granules. The carbon has the ability to adsorb organics, attracting carbon-based organics to the activated carbon surface. The attraction is associated with van der Waals forces.

Suspended solids removal. Activated carbon in a 12 X 40 mesh rating offers better filtration than many multimedia filters. It has been reported to be effective at removing suspended particles as small as 8 to 10 μm in size (Michaud, 1993). Using activated carbon to remove high concentrations of suspended solids will necessitate frequent backwashing of the media, based on pressure drop. Depending on the carbon hardness, backwashing can create carbon fines that can be shed into the effluent water.

The ability of carbon to remove any particular organic depends on the nature of the organic. Carbon tends to be best in the removal of higher-molecular-weight organics, such as humic and fulvic acids. Carbon has a limited ability to remove certain highly polar organic molecules, such as trihalomethanes.

Trihalomethane removal by activated carbon. The presence of THM (e.g., chloroform) in a municipal water source is typically the result of a chemical reaction between free chlorine and trace organics. Trihalomethanes are considered potentially carcinogenic, so are a concern when found in drinking water. As a trace contaminant, they pose a threat to many industrial processes as well.

Trace trihalomethanes are not well removed by reverse osmosis, ion exchange, or ultrafiltration. If present in a feedwater, they can be a major concern in the semiconductor industry. Their presence in a high-purity water system can be noted directly after a UV-light sanitizer as a decrease in water resistivity. This lower resistivity is the result of chloride ions being broken off the trihalomethane molecule by the UV light (Fulford, 1993).

Activated carbon media that has been in service past its capacity for THMs has been known to shed higher concentrations of THMs when starting up after an extended shutdown. These THMs can affect water quality throughout an entire water treatment system.

Activated carbon is used frequently for chlorine and chloramines removal upstream of RO systems. This is necessary to protect polyamide membranes, as well as being required to meet water quality specifications for many applications. Activated carbon is excellent at adsorbing and breaking down chlorine, but is not quite as efficient at breaking down chloramines.

Residual chlorine and chloramines concentrations are large concerns in dialysis water treatment. Should they be present in significant concentrations in the dialysis water used to remove contaminants from a patient's blood, they can contaminate the blood and cause hemolysis of the blood cells.

The U.S. Department of Health and Human Services (DHHS) has made recommendations regarding the volume of carbon to be used in the removal of chlorine or chloramines in dialysis applications. They base their recommendations on an empty-bed contact time (EBCT). This is the time it takes for water to flow through the area within a carbon filter where the carbon would be located. This time will be somewhat proportional to the time the water will be in contact with the activated carbon. Following the DHHS guidelines should insure that enough activated carbon is present for complete removal.

The recommended EBCT for dialysis water treatment is 6 minutes for free-chlorine removal and 10 minutes for chloramines removal. In other applications, an EBCT of 3 minutes is usually sufficient for chloramines removal (Michaud, 1993). With the given EBCT, the required volume of carbon can be calculated as follows:

$$V = (Q \times EBCT)\ 9\ 7.48$$

where:
V = volume of activated carbon required (ft^3)
Q = water flowrate (gpm)
EBCT = empty-bed contact time

Example:
Estimate the volume of activated carbon required for chloramines removal in a 10-gpm dialysis application.

Answer:

$$V = (10 \times 10) \div 7.48 = 13.4\ ft^3$$

With a typical bed depth of 4 feet, the cross-sectional area required for the vessel to hold the carbon would be $13.4\ ft^3 \div 4\ ft = 3.35\ ft^2$. The cross-sectional area of a filter is related to its diameter by the following:

$$\text{Cross-sectional area} = 0.785 \times D^2$$

where D is the filter diameter. Therefore the required inside diameter of the filter vessel would be

$$D^2 = 3.35\ ft^2 \div 0.785$$
$$D = 2.07\ \text{feet}$$

In this case, it would be reasonable to use a 2-foot-diameter vessel with slightly more than a 4-foot bed depth to meet the carbon volume requirement. To use a larger-diameter vessel, such as a 2.5-foot vessel, would result in a flow velocity through the media of 10 gpm $\div$ $\{0.785 \times (2.5\ ft)^2\}$, or 2 gpm/ft^2. This low velocity could result in channeling within the media, which would affect the performance of the filter.

Adsorption capacity. An important parameter for specifying activated carbon is its relative adsorption capacity. This is frequently expressed as an iodine number or a molasses number. The U.S. Department of Health and Human Services recommends using an activated carbon with an iodine number of 1,000 for the removal of chloramines. The California Department of Health Services and Office of Statewide Health Planning and Development requires a minimum iodine number of 900 for chloramines removal in dialysis water treatment.

Reaction between chlorine or chloramines and activated carbon. In water, an equilibrium concentration of hypochlorous acid (HOCl) is formed by the dissociation of chlorine.

$$Cl_{2(g)} + H_2O <\longrightarrow> HOCl + HCl$$

Activated carbon removes free chlorine by reducing it to the chloride ion as follows:

$$C^* + HOCl \longrightarrow> C^*O + HCl$$

The carbon surface thus becomes oxygenated with time. With high chlorine concentrations, the carbon surface can become oxidized to carbon monoxide (CO) or carbon dioxide (CO_2) (DeJohn, 1976). In either case, the carbon is limited in its chlorine-removal capacity.

It has been estimated that 1 gram of carbon has the capacity to react with 3 to 5 grams of free chlorine as Cl_2 (Snoeyink and Suidan, 1975). For an activated carbon filter designed to operate at 10 gpm/ft^3 with a 4-foot bed depth and an incoming chlorine concentration of 1 mg/L, the carbon would have a minimum life expectancy of over 10 years. In spite of this, it is suggested that the carbon be replaced every 6 months to a year to reduce biological activity, and to prevent the shedding of organic contaminants. (Some applications may require more frequent replacement.)

The reaction between chloramines and carbon takes place as follows:

$$C^* + NH_2Cl + H_2O \longrightarrow> C^*O + NH_{3(g)} + HCl$$

Activated carbon is very efficient at removing free chlorine from water. It is so good, in fact, that the chlorine is eliminated within the first several inches of the bed depth. This leaves the rest of the bed free of any biocide. As the carbon has also served to removed organics from the feedwater, its pores, complete with organic nutrients, are now an excellent breeding area for bacteria. This bacteria will tend to be shed into the carbon filter effluent, which then can contaminate a downstream RO system. Biological fouling is a common problem for polyamide RO systems using activated carbon filtration as part of their pretreatment.

It is possible to reduce the biological activity of an activated carbon filter chemically. Caustic (high pH) can break up biological foulants, making it possible to remove the bulk of the foulants during backwash. Oxidizing biocides can be somewhat effective when used in the backwash water. This should not be attempted with a high hydrogen peroxide concentration, however, as the evolving gases can be violent.

Backwashing of an activated carbon filter. Backwashing an activated carbon filter will remove some of the suspended solids and large debris that may have been filtered by the top layer of the carbon. It will also lift the media if compaction has occurred. A backwash flowrate of 12 gpm/ft^2 is suggested.

One drawback of backwashing carbon is that the bed will re-stratify, which may result in the shedding of organics. It can also create carbon fines. After a backwash, the bed should be forward rinsed to drain for a minimum of 10 minutes, or until an SDI measurement results in the SDI filter being free from carbon particles.

It is possible to regenerate and disinfect the carbon using steam. This requires a stainless steel tank, valves, and piping that may triple its cost. Stainless steel carbon filters are becoming more popular in biologically critical applications.

Specifications for activated carbon. The mesh designation for the carbon indicates the hole spacing (number of holes per inch) of the screens used to isolate the carbon granules. For example, a 12 × 40 mesh activated carbon has mostly

Table 3-6
Ash Analysis of Activated Carbon (in %)

	Coal	*Coconut Shell*
Total ash	13.7	2.6
N	0.8	Neg.
S	2.0	Neg.
SiO_2	2.8	0.4
CaO	1.0	0.3
FeO_3	4.6	0.7
Cl	Neg.	Neg.
Na_2CO_3	0.4	0.2
K_2CO_3	0.3	1.3
Al_2O_3	1.2	Neg.
PO_4	1.3	0.15

Na and K concentrations are expressed as carbonate.
Neg. implies a negligible concentration below the test sensitivity (typically less than 0.02%).
(Courtesy of Cameron Yakima).

granules that are larger than a 40-mesh screen (40 holes per inch) and smaller than a 12-mesh screen (12 holes per inch). It can be generalized that the smaller the granules, the better the surface area contact between the carbon and the water flow. This usually results in improved adsorption. The granules should not be too fine, however, because of the difficulty in preventing the shedding of fine granules into the RO feedwater.

The initial fines content of the carbon can be reduced by additional screening of the activated carbon. The carbon is sifted through the finer screen several times to achieve better removal of the finer granules.

It is important that a very hard activated carbon, with a hardness rating of 95 or better, be used if the filter is directly upstream of the RO. Soft wood-based and coal-based carbons can generate carbon fines during handling or backwashing. These fines have been known to physically plug RO membrane elements located downstream. With respect to carbon shedding, coconut-shell-based carbon tends to perform well when used for RO pretreatment (Walsh, 1993).

It is also desirable to use a carbon with a low ash content (0.5% water-soluble ash or less). The ash is the salt content of thc carbon. These salts can dissolve into the RO feedwater. If the carbon has a high ash content, it can increase the concentration of certain dissolved metals in the RO feed and permeate water. This is particularly a concern with coal-based carbon. In dialysis water applications, coal-based activated carbon has been known to increase the concentration of aluminum, which can be a dangerous contaminant for dialysis patients, in the RO feedwater (Beseman, 1993).

A total ash analysis of a typical coal-based and a coconut shell activated carbon is shown in Table 3-6. The water-soluble salt content (i.e., the salt that would dissolve into the RO feedwater) would be a smaller percentage of the contaminant concentrations.

In critical applications, such as for dialysis water, activated carbon should be acid-washed to reduce the carbon ash content. Both coconut shell and coal-based carbons are available acid-washed from specialty carbon suppliers.

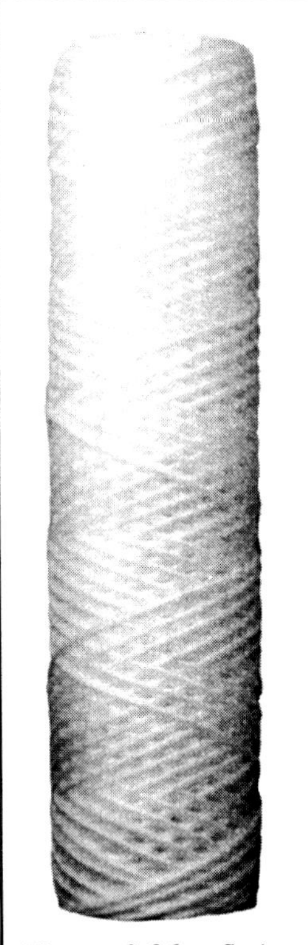

Figure 3-26. String-wound depth filter.
(Courtesy of FilterCor)

Cartridge prefilters are recommended directly upstream of the RO high-pressure pump to protect the pump and lead end RO membrane elements from debris. These filters can also serve to remove suspended solids that might otherwise foul the RO membrane. The vast majority of RO systems use some sort of cartridge filter upstream.

Cartridge filters have minimal surface areas. It is therefore questionable whether relying on cartridge filters to remove suspended solids is economical. If the filters are plugging quickly, it is probably more economical to install a media filter. In some cases, it may be less expensive to use a prefilter with a larger pore rating for longer life, and plan on cleaning the RO more frequently (McPherson, 1981).

The RO membrane manufacturer will usually require a prefilter with a pore size rating sufficient to protect the lead end membrane elements from large debris, usually from a 5- to a 25-µm nominal rating. Such debris could lead to excessive pressure differentials across the RO membrane element, and subsequently to telescoping of the element.

The filters most commonly used in such an application are called depth filters because they offer a thick media in which they can trap a relatively large volume of contaminants. The media may be graded in density toward the filter core so as to trap finer particulates as the water passes through. This allows them to achieve greater solids removal prior to replacement based upon increasing pressure drop across the filters. Depth filters tend to be less susceptible to *blinding off,* whereby a limited volume of foulant plugs the outside of the filter, necessitating its replacement.

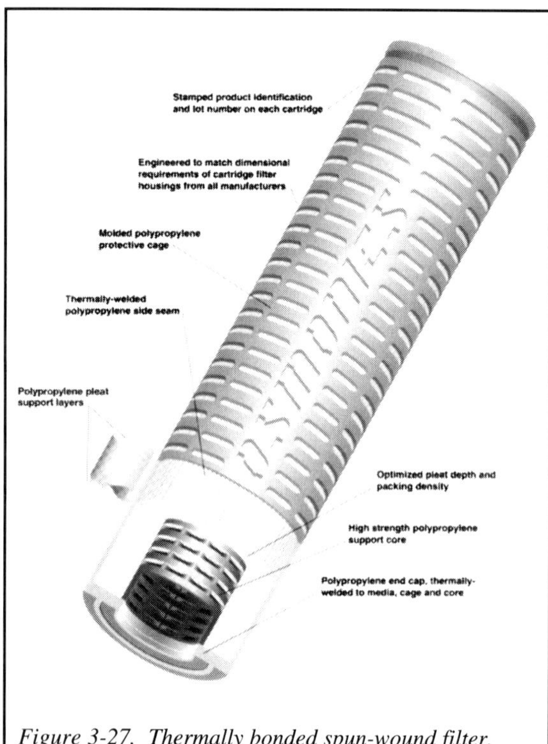

Figure 3-27. Thermally bonded spun-wound filter.
(Courtesy of Osmonics, Inc.)

Pleated membrane filters rely more on membrane surface filtration for their mechanism of filtration. They consist of a flat sheet of membrane that has been pleated like an accordion. In this manner, the filter can contain a fairly large surface area of membrane. Pleated filters are typically more expensive than depth-type cartridge filters; however, they can be useful if the filtration of fine particulates is required.

Depth filters do not offer as tight a pore sizing as pleated membrane filters. String-wound filters (see Figure 3-26; they look like their name) are typically 60 to 80% efficient at removing particles the size of the manufacturer's pore rating. Thermally-bonded spun-wound filters (Figure 3-27) are between 80% and 90% efficient. Pleated membrane filters and some other types of prefilters are greater than 98% efficient at the manufacturer's pore rating. (Hillestad, 1994).

The method most commonly used to seal a depth-type cartridge filters with its housing is called a knife-edge seal. The housing has a sharp protruding edge at each end of the housing that cuts into the filter end. Some type of post that fits the hollow inside of the filter is usually used by the housing to guide the filter into alignment with the circular edge. (See Figure 3-28.)

It is difficult to achieve an absolute sealing method between the filter and housing using a knife-edge seal. Some amount of leaking around the seal is common. Since depth filters are not intended to remove 100% of the incoming contaminants, this is not a problem unless the bypassing is significant. This bypass can frequently be observed on the end of filters that have been in use. A

Figure 3-28. Indent created by knife-edge seal.
(Courtesy of Osmonics, Inc.)

stained area will be present from the outside of the filter all the way to the inside edge.

With a prefilter rated at 1 μm or less, it becomes more critical to use a better means of sealing the cartridge against the housing. Single or double O-ring end configurations offer a seal that is more likely to be absolute.

It is critical that RO prefilters be constructed of materials that are compatible with the pH extremes, oxidizing/reducing agents, and cleaning chemicals to which the filters might be exposed. Polypropylene filters are recommended. Cotton prefilters are not. Cotton filters have been known to come apart, shedding their fibers downstream into the RO system. This will increase the RO system pressure differential and can cause an increased rate of fouling.

Pressure drop versus flow curves for the prefilters should be considered when sizing the filters and housing. This is particularly critical if the incoming water pressure is minimal. It may be necessary to oversize the prefilter housing to reduce the pressure drop across the filters. For most prefilters, it is common to size the housing such that each 10-inch equivalent (TIE) filter does not encounter a flowrate of more than 5 gpm. At this flowrate, the clean pressure drop for the filter should not be more than 3 psid.

For example, if the RO inlet flowrate is 22 gpm, a minimum of five TIE filters would be required to keep the flowrate per TIE filter at less than 5 gpm. With five equivalent filters, the actual flowrate per TIE is 4.4 gpm. It should be possible

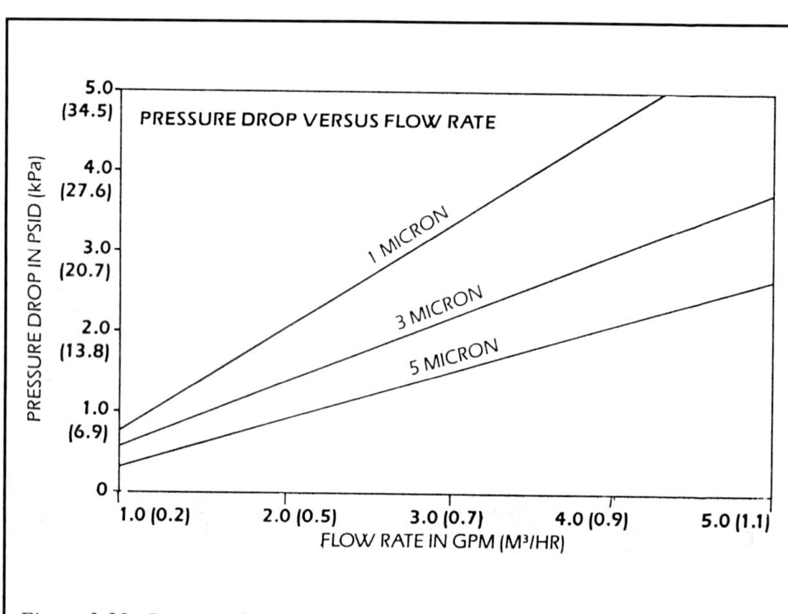

Figure 3-29. Pressure drop versus flowrate for a common TIE filter.
(Courtesy of Osmonics, Inc.)

to get the pressure drop for the filter at this flowrate from the filter manufacturer's literature. Since all the filters in the housing will be in a parallel configuration, the pressure drop of an individual filter should be the same as that for the entire housing (Figure 3-29).

To house exactly five TIE filters, the only option is called a five-round, single-high housing. *Five round* means that the housing holds five actual filters in parallel. *Single high* means that the filter length is the same as a single 10-inch filter.

It is possible that a lower pressure drop through the filters might be desirable, or that less expensive housings are available than the five round. (The larger diameter housings tend to be more expensive than longer, smaller-diameter housings.) In this case, a three-round, two-high housing might be a better alternative. The flow per TIE filter would now be 3.7 gpm (22 gpm divided by six TIE filters).

Filter housings can come with diameters large enough to hold more than two dozen parallel cartridges and can be long enough to house 50-inch-long filters. With these tall housings, however, it can be difficult to get the housing casing up and over the filters. In some facilities, it has been necessary to use an overhead lift mechanism in order to change cartridge filters. In these longer housings, replacing filters can also be difficult if the filters cannot be easily lined up with the housing connections.

It is common with the smallest RO systems to use single-filter plastic housings manufactured by Ametek, Inc. The 10-inch version of their single filter housing is actually best fit by a filter that is 9 $^{7}/_{8}$ inches in length, but 20-inch filters work well in their 20-inch housing.

Prefilters should be replaced at least every 2 months, or when the pressure differential across the filters exceeds 15 psid. Foulant layers can begin to compact or collapse at larger pressure differentials. The 2-month minimum is chosen to prevent the prefilters from becoming a source of bacteria contamination for the downstream RO system.

Ultrafiltration. Since UF is considered an economical means of removing suspended solids from water, it is a common misconception that it would be a good choice for RO pretreatment. In some cases it is. For the majority of cases, however, it is not.

The biggest advantage of relying on UF to remove foulants prior to RO is that the more aggressive cleaning solutions required to dissolve and remove certain stable foulants can be used with the UF membrane.

If fouling is a problem with an RO membrane, however, it will likely be a greater problem with a UF membrane. In side-by-side testing, UF membranes will typically foul at least ten times faster than RO membranes; thus their required cleaning frequency will be more than ten times greater.

As discussed in Chapter 1, the mechanism utilized by RO membranes makes better use of the membrane area than does ultrafiltration. With UF, water has to pass through a limited physical area made up by the combined UF pore cross-sectional area. This results in a much greater localized velocity perpendicular to the membrane surface. With greater water velocity, it is difficult for foulants to diffuse back into the UF bulk stream (Figure 1-3).

Also, there is greater penetration of the foulant into the membrane pore structure of UF membranes, making it much more difficult to remove the foulants. In applications involving heavy fouling, it is common to see a significant and permanent loss of permeate flow from a UF membrane during its initial operation. The amount of permanent flow loss between cleanings will decline unless the UF is operated at greater pressures, thus driving more contaminants into the pores.

Given that most UF membranes begin with a relatively high flux, these initial permanent losses are not a concern. The fouling rate between cleanings after the initial permanent fouling has occurred is more critical.

It is not a good idea to install a UF system upstream of an RO if the sole intention is to minimize the maintenance requirements of the RO. The requirements will just be transferred to the UF system, and possibly magnified. The exception would be if the foulant must be removed by a special cleaning agent that is not compatible with the RO membrane.

When designing a UF system for a high-fouling application, it is important to dramatically overdesign the system. At constant pressure, the membrane manufacturer's design permeate flowrate may decline by 50% to 75%. Much of this loss may be irretrievable because of pore fouling within the UF membrane structure. The extent of permanent loss will vary depending on the nature of the foulants, and on the permeate flux. Systems with higher permeate flux (because of a higher membrane driving pressure) will tend to suffer greater pore fouling, and thus have greater permanent losses in normalized permeate flowrate.

Control of biological fouling. Reducing the extent of biological activity present in a membrane system is critical to its success. If little or no chlorine is present in cellulose acetate membrane systems, some types of bacteria can create biofilms with low pH characteristics that can increase the hydrolysis rate of the CA membrane (Luss, 1993). This is not as common with newer CA blend membranes that are more resistant to low pH. But even these newer membranes are prone to biological fouling that can also lead to the demise of the membrane.

If activated carbon filters are used upstream of CA RO systems as a means of reducing the extent of organic fouling of the membrane, it may be necessary to inject sodium hypochlorite downstream of the carbon. Since the carbon will break down any chlorine or chloramines that might have been present in the raw water, the injection of hypochlorite will replace the biocide.

If carbon has been used upstream to remove chloramines, and sodium hypochlorite (chlorine) is injected downstream of the carbon, some chlorine will tend to react with the residual ammonia (NH_3) that will have been created when the chloramines reacted with the carbon. A free chlorine residual will not be measurable until all the ammonia has reacted to re-form chloramines, which are not nearly as effective as free chlorine as a biocide.

The presence of free chlorine downstream of the RO system may be a concern for certain applications. Ion-exchange resins will remove the free chlorine, but may be degraded in the process. Continuous exposure to as little as 0.3 mg/L of chlorine has been known to cause resin damage if used over the life of the resin (usually about 5 years). The chlorine can oxidize cation resins whose by-products can subsequently foul anion resins. Oxidized resins have been known to shed total organic carbon (TOC) into ion-exchange effluent water, which is a contamination concern for many high-purity water applications (Mukhopadhyay, 1993).

If activated carbon is used downstream of the RO to remove the chlorine, the carbon filter will usually grow bacteria. It can then shed the bacteria into its effluent stream, possibly contaminating downstream equipment. Using carbon in this manner somewhat defeats the purpose of RO's ability to remove particulates and bacteria.

Injecting a reducing agent (such as sodium bisulfite) in the RO permeate water will chemically reduce the chlorine to chloride ion. This will add to the downstream ionic loading of the permeate water; thus, it will increase the required regeneration frequency of resins beds located downstream.

Some chlorine will degas when passing through a forced-draft or induced-draft degasifier. A properly designed vacuum degasifier will remove the vast majority of the chlorine.

The problem of biological control in PA membrane RO systems has been one of the most significant reasons that these membranes have not completely dominated the RO membrane market. Given the intolerance of PA membranes to biocidal concentrations of free chlorine, their application to surface waters has been somewhat limited.

As discussed earlier in this chapter, activated carbon filters tend to become a source of biological activity. Reducing agents are commonly used as an alternative to activated carbon for chlorine or chloramines removal upstream of PA membrane systems.

Any time the atmosphere of an RO feedwater is altered, such as by pH adjustment or by changing the oxidative/reductive atmosphere of the water, the alteration has a biostatic action on bacteria in the water. It may not kill a significant number of bacteria, but it will slow their growth. Therefore, one advantage of using the injection of reducing agent to remove the chlorine or chloramines is that there is this biostatic effect. This advantage is limited,

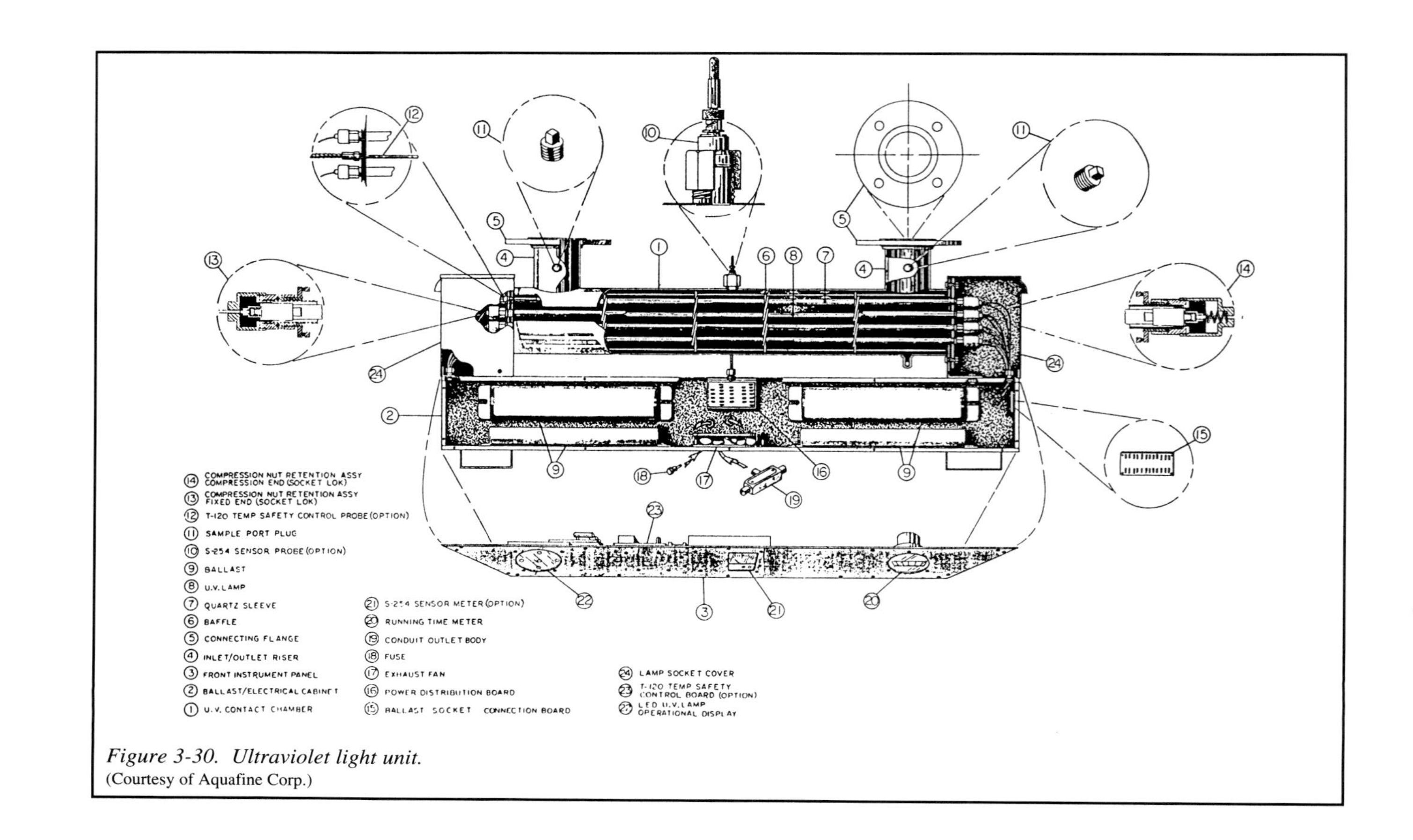

Figure 3-30. Ultraviolet light unit.
(Courtesy of Aquafine Corp.)

however as far as significantly controlling biological growth in a PA membrane system.

Iron bacteria. Certain types of bacteria thrive in water under a reductive atmosphere. Although all bacteria need oxygen, anaerobic bacteria obtain their oxygen from anions dissolved in the water, such as the sulfate (SO_4^{-2}) or nitrate (NO_3^-) ions. One commonly found bacteria of this type is known as iron bacteria.

Iron bacteria is so called because it oxidizes iron in its metabolic process. This iron is excreted as insoluble ferric oxide (Fe_2O_3) (Cubicciotti and Licina, 1990). Its presence is characterized by a heavy rusty slime that can rapidly foul RO prefilters as well as the RO membrane elements.

With iron bacteria fouling problems, it is often wrongly thought that the water source provides the source of iron to support the iron bacteria. This assumption is usually made because the rusty slime is visually present in pretreatment filters upstream of the RO. However, an analysis of the water source frequently will indicate a very low concentration of iron (less than 0.05 mg/L).

Actually, the source of iron is usually the result of corrosion of steel or stainless steel caused by the iron bacteria itself. The bacteria can form a pocket on the inside wall of stainless steel vessels or on piping that is protected by a coating of iron oxide. Within the pocket, the bacteria excrete acids that can dissolve the iron in the stainless steel. The bacteria are then able to oxidize the iron. Over time, the corrosion of the stainless steel can be significant (Borenstein, 1988).

Prefilter housings, high-pressure manifolding, pump casings, and membrane housings can thus provide the iron to sustain the bacteria. As the iron bacteria proliferates, some will be shed into the water and will tend to contaminate downstream equipment.

Due to the protective shell around the bacteria, iron bacteria can be extremely difficult to completely clean out of an RO system. Sodium ethylene-diaminetetraacetic acid (EDTA) in an alkaline pH solution can be effective at chelating the iron while breaking up the organic mass, if the volume of iron bacteria is not too severe.

For heavy iron bacteria fouling problems, a more detailed cleaning procedure is suggested. A formulation that uses a strong reducing agent with a surfactant can be used as the first step of a three-step cleaning. Such a solution will remove the protective shell around the bacteria. A highly alkaline solution can then be used to effectively break up and dissolve the organic bacteria slime. Finally, to slow the recolonization of the system, a peracetic acid solution can be used to disinfect the system.

With CA membrane systems, iron bacteria can be prevented by injecting sodium hypochlorite (chlorine) upstream of the pretreatment equipment, or directly upstream of the RO. Since PA membrane systems cannot tolerate strong oxidants, the chlorine will have to be removed prior to a PA RO. Batch disinfection using peracetic acid or some other compatible biocide will therefore be necessary with a PA membrane RO system.

Sufficient chlorine injection at the front of the pretreatment system should be effective at eliminating the source of the iron bacteria. With periodic batch disinfection downstream, it may be possible to significantly delay the recolonization of the RO system.

It may be possible to prevent the growth of iron bacteria simply by maintaining a neutral oxidation-reduction (redox) potential. If a reducing agent is being injected to eliminate the presence of chlorine or chloramines in the water, just enough reducing agent should be used to achieve a neutral or slightly positive redox potential. This has been known to prevent the growth of anaerobic bacteria (Bernardin,1976).

Ultraviolet light sanitization. Ultraviolet light units can be effectively used upstream of a PA RO system to reduce the numbers of bacteria entering the RO system. This can slow the recolonization of the RO system, and thus reduce the maintenance requirements for the RO system.

Ultraviolet kills bacteria using a 254-nanometer (nm) wavelength light bulb (Figure 3-30). The UV disables the bacteria's chromosomes, which kills them. At a minimum intensity of 30,000 microwatts per second per square centimeter, UV is reported to be 99.9% effective at killing bacteria (Aquafine[R] Owner's

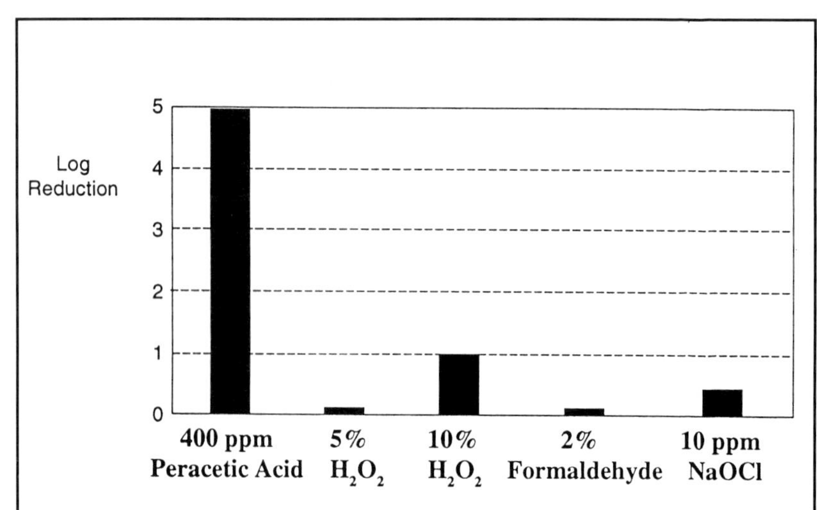

Figure 3-31. Biocide efficacy showing log reduction of B. subtillis var niger (ATCC 9372) spores after 30 minutes exposure to various biocides.
Maltais and Stern, 1989 (Courtesy of Minntech Corporation).

Manual, 1992).

The UV unit output intensity is usually rated for when the bulbs are a year old. The bulbs will typically lose about half of their intensity due to the effects of the UV light on the glass in the bulb. The bulbs should be replaced after a year (more frequently if the UV unit turns on and off with the RO).

The UV should not be allowed to operate for an extended period of time without water flow through the unit. Otherwise, the water can heat up, possibly causing damage to the UV unit, the piping, or the RO membrane. However, frequent UV shutdowns and start-ups will shorten the bulb life.

The UV is not 100% effective at killing all of the feedwater bacteria. This means that some bacteria will survive the UV to enter the RO. Once in the RO, the bacteria are free to grow. However, a UV light unit will tend to decrease the required frequency of RO system sanitization. Since the unit does not require a great deal of power, it is not a significant operating expense.

Peracetic acid.. Certain biocidal concentrations of peracetic acid solutions (also called peroxyacetic acid) have been found to be compatible with most polyamide thin-film RO membranes. Although the peracetic acid solutions have oxidation potential, in their recommended concentrations it is insufficient to attack the membrane.

Peracetic acid is one of the most effective chemical biocides available, offering faster killing ability than other commonly used sanitizing agents. Its attack very specifically deactivates the bacteria's enzymes. This allows the peracetic acid to be biocidal while having a minimal oxidation potential (Byrne and Roesner, 1991).

The effectiveness of peracetic acid is compared to other biocides in Figure 3-31. It should be noted that many of the biocides presented in that figure are incompatible with PA membranes.

Peracetic acid is an unstable molecule that is created by the equilibrium reaction of acetic acid (CH_3COOH) with hydrogen peroxide (H_2O_2):

$$CH_3COOH + H_2O_2 <====> CH_3COOOH + H_2O$$

For the water treatment industry, peracetic acid solutions are typically available in formulations containing roughly 4% to 5% peracetic acid with about 20% hydrogen peroxide. The recommended concentration for batch sanitization of membrane and piping systems is 400 mg/L of peracetic acid (which would be present with 2,000 mg/L of hydrogen peroxide). This is approximately a 1:100 dilution of the concentrated solution. The 400-ppm peracetic acid concentration should be considered the maximum concentration acceptable for thin-film membranes. (Consult with the membrane manufacturer.)

The use of peracetic acid should be avoided with new thin-film membranes.

As with some other biocides such as formaldehyde or glutaraldehyde, the peracetic acid solution can react with residual amines that may still be present within the membrane structure. Several days of operation should be allowed prior to sanitization.

Transition metals such as iron or manganese can increase the oxidation potential of the peracetic acid/hydrogen peroxide, and thus should be removed using an appropriate cleaning solution prior to the sanitization. Heavy organic fouling should be cleaned out with an appropriate alkaline cleaning solution.

Contact time for the peracetic acid will depend on the severity of the biological fouling. Usually 1 hour is sufficient for most RO systems. Monitoring of bacteria counts is a way to judge the effectiveness of the sanitization. An effective sanitization will mean that the RO system may operate longer before bacteria can recolonize the RO system.

Continuous injection of peracetic acid. Development work is currently being performed on the continuous injection of peracetic acid into the feed stream of PA thin-film RO systems. In several pilot studies, it has been demonstrated that continuous exposure to low concentrations of peracetic acid can reduce the biological fouling of equipment located downstream. In one particular pilot study, it was shown that slightly higher concentrations (1 mg/L of peracetic acid), when injected continuously, were capable of dramatic reductions in bacteria counts within an RO system.

A thin-film RO system located in San Diego, California, was losing nearly 30% of its normalized permeate flowrate over a period of 1 week. The RO was located downstream of activated carbon filters that were shedding high numbers of bacteria into the RO feed stream. The numbers were typically in excess of several thousand bacteria colonies per milliliter. A temporary restoration of flow occurred each week when the RO feedwater pH was dropped to a pH of 3 for a period of 30 minutes.

A 0.4-mg/L peracetic acid concentration (present with 2 mg/L of hydrogen peroxide) was injected into the RO feedwater while the RO was in operation, which was roughly 60% of the time. The permeate flow decline was immediately arrested, although there was no noticeable decline in bacteria counts. Over the next 2 months, the permeate flow was stable at the higher flowrate. When the peracetic acid injection was taken off-line, the permeate flow started dropping as before. No effect on the membrane salt rejection was noted over the testing period (Byrne, 1990).

In the Phoenix, Arizona area, a water softener located downstream of activated carbon was experiencing heavy biofouling that necessitated its premature regeneration. A concentration of between 0.6 and 1.0 ppm of peracetic acid was injected upstream of the softener. This eliminated problems with high pressure drops across the softener, thus extending its service cycle. The softener backwash water no longer showed the visible presence of biofilm. No oxidative effects were noticed over the 8-month period of this test (Connors, 1993).

The water treatment system of a kidney dialysis center located outside of St. Cloud, Minnesota, was experiencing ongoing bacteria growth problems in their two parallel RO systems, their distribution storage tank, and their piping system. It was severe enough to require weekly batch sanitization of the water treatment system.

The pretreatment to the RO systems consisted of softening, followed by independent activated carbon filtration, followed by cartridge filtration. The distribution system included a mixed-bed ion-exchange unit, followed by a UV light, followed by submicron filtration (Figure 3-32).

The pilot study was begun after a standard system sanitization. Peracetic acid was injected into the RO feed stream at a concentration of 1.0 ppm. The particular RO chosen (of the two) was located downstream of a heavily biologically contaminated activated carbon filter. This carbon filter was shedding bacteria consistently in the too-numerous-to-count (TNTC) range.

The peracetic acid solution was able to permeate the RO membrane. Rejection was only 15% at the standard RO operating pressure. The passage of peracetic acid into the RO permeate was beneficial in that it prevented the growth of bacteria in the low-velocity permeate carrier material located on the inside of a spiral-wound membrane envelope.

However, the residual peracetic acid, hydrogen peroxide, and acetic acid components were not desirable in the water to be used for dialysis, or for other purposes in the facility. (They can be readily removed by the anion-exchange resin in a mixed bed.) This dialysis facility was chosen for the test because its water treatment system included a mixed-bed ion-exchange system.

It is not known what the long-term oxidative effects were on the anion resin. The peracetic acid/hydrogen peroxide did not affect the performance of the

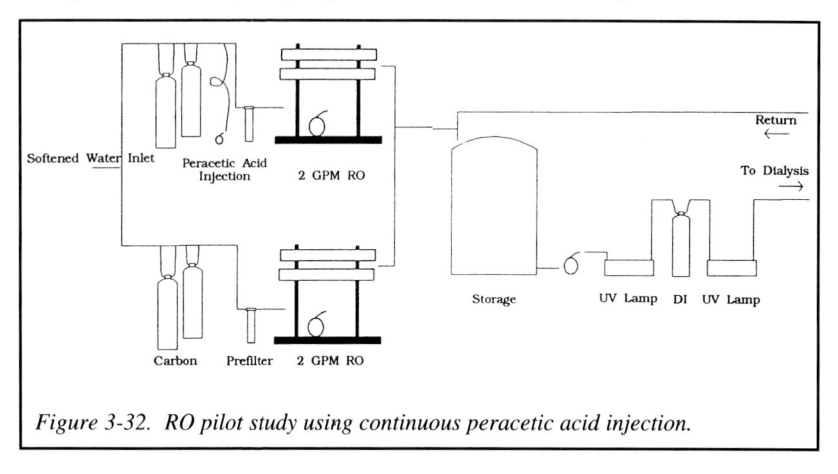

Figure 3-32. RO pilot study using continuous peracetic acid injection.

mixed bed over the period of the test.

The heterotrophic bacteria plate count dropped with the peracetic acid injection from TNTC in the RO feedwater to 0 (per 100 mL), or near 0, in the RO concentrate. The permeate bacteria count consistently was 0. This occurred over a period of a month and a half. After this time, the RO started experiencing rising bacteria counts in the concentrate, coinciding with decreased operating time. Low feed pressure to the facility was causing the RO to automatically shut down. It is likely that this lack of operation was the reason for the increasing biological activity in spite of the presence of peracetic acid.

When peracetic acid was being injected, it was discovered that the off-gassing of the hydrogen peroxide from the concentrated solution would frequently cause the solenoid (diaphragm) injection pump to lose its prime. This would occur even when the pump was mounted such that it had a flooded suction.

Peristaltic pumps have been found capable of pumping peracetic acid/ hydrogen peroxide solutions. These pumps are slightly more expensive than solenoid pumps (Cohen, 1993).

Peracetic acid/hydrogen peroxide will react with chemical reducing agents if present in the RO feedwater. When reducing agents are injected upstream to remove chlorine or chloramines in the RO feedwater, a greater injected concentration of peracetic acid/hydrogen peroxide will be required.■

CHAPTER 4

RELATED WATER TREATMENT EQUIPMENT

The operation of a reverse osmosis system is dependent on a number of ancillary unit operations and treatment methods. This chapter discusses various treatment equipment operations and their design requirements for an efficient operation of the RO unit.

Pressure Filter Design

Multimedia filters, activated carbon filters, and water softeners all use some sort of pressure vessel. The particular device will dictate the requirements for the internal design of its pressure vessel. Generally, water softener vessel design is more critical than that for a multimedia filter because of the need to achieve low effluent hardness concentrations. The performance of the softener is sensitive to the distribution of flow through the vessel, and to the presence of contaminants introduced by the vessel itself. Although multimedia and activated carbon filters benefit from balanced distribution of flow through the media, this balance is not as critical to their operation as it is to a softener.

Lateral design. One distribution lateral is located in the top and one in the bottom of the softener.

The laterals evenly disperse the influent flow and collect the effluent flow in order to achieve nearly uniform flow velocity through the cross-sectional area of the media. The laterals should be designed so that the flow into or out of the lateral is met with some resistance. This resistance increases as the flowrate increases. In this manner, the flow dispersion or collection will tend to balance itself uniformly along the lateral.

If the laterals are incorrectly designed, or if the laterals have become blocked or damaged, higher flow velocities can occur in certain areas of the vessel. With time, media can actually drift away from an area with higher localized flow velocity. The problem subsequently becomes worse, since there is now even less resistance to flow in that area of the vessel. The flow essentially short-circuits itself through that area of the vessel, and this will affect the performance of the media.

In small-diameter vessels (8 inches or less), lateral design is not as critical as with larger vessels, as it is difficult for the flow to become too far out of balance over such a small cross-sectional area. Small-diameter vessels may have fairly simplistic laterals , if they have any at all (Figure 4-1).

With multimedia and activated carbon filters, manufacturers have been known to compromise on the upper lateral design if the bottom lateral is well designed. They may use a simple standpipe setup, or will use a flat plate located directly above a standpipe to offer slightly better distribution of flow than just the standpipe arrangement.

Much better distribution is achieved using a hub radial or a header lateral design. The hub radial design (Figure 4-2) uses lateral spokes radiating off from a central point. With vessels larger than 36 inches in diameter, the hub radial lateral will tend to have reduced flow velocities near the vessel wall. Also, with larger vessels, the spacing of the notches or holes in a hub radial lateral must vary along the length of the radial in order to take into account the greater flow requirement near the vessel wall.

A header lateral (Figure 4-3) is a better design for larger-diameter vessels. The header lateral disperses flow to a number of parallel arms across the cross-sectional area of the vessel. In this manner, the holes or notches can be equally distributed across the cross-sectional area. It is important with larger-diameter vessels to properly support the header lateral to prevent damage that could be caused by the movement of the media. This is particularly important if the laterals are plastic.

Some laterals may have holes drilled along the length of the arm to collect or distribute the water. These holes are covered with a fine screen to prevent the media from entering or plugging the holes. More commonly, finely notched arms (as in Figures 4-2 and 4-3) are used with the lateral. Electrodeposition machining (EDM) makes it possible to manufacture stainless steel lateral arms with sufficiently fine slots to prevent media particle intrusion.

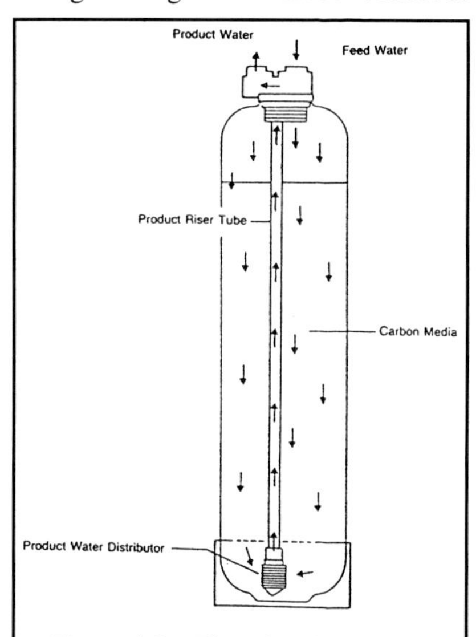

Figure 4-1. Flow distribution in a small-diameter exchange carbon filter.

Flowrates during backwashing of the media are usually greater than the service flowrates. The laterals must be capable of handling these increased rates. Usually there is more working pressure available during the backwash cycle since the effluent out of the top of the vessel is near atmospheric pressure. It is common to use some sort of orifice or flow control device on the backwash effluent line to obtain the desired backwashing flowrate.

The laterals should be constructed of stainless steel or plastic. Carbon steel pipe is not suggested for the laterals or the face-piping on the tank. Carbon steel will result in the introduction of additional iron into the RO feedwater.

Steel pressure vessels: *Tank linings.* The tank lining of carbon steel pressure vessels is critical when used for RO pretreatment. A poor lining will result in oxidation of the tank wall. The extent of this oxidation will depend on the aggressiveness of the RO feedwater. If chemicals, such as acids, are being injected upstream of the pressure vessel, the water will more quickly attack the tank wall. This can result in the premature failure of the tank, and in the shedding of iron into the RO feedwater.

Lining small-diameter steel vessels can be extremely difficult; thus, fiberglass vessels are more commonly used for the smaller tanks. The fiberglass may not have the longevity of a properly-lined carbon steel vessel, but its cost will be substantially less. If a fiberglass tank is used for a water treatment system installed outdoors, the tank should either be painted with a ultraviolet-resistant paint, or covered. Sunlight can dramatically decrease thc lifc of a fiberglass vessel.

Some commonly used means of lining a carbon steel tank include galvanizing the tank; lining it with rubber; or coating it with an epoxy, enamel, vinyl ester, or a polyvinyl chloride compound. Zinc galvanizing has a poor reputation because chunks of the lining can break off the tank wall. Epoxy and PVC linings can be difficult to apply. Rubber and vinyl ester linings can be relatively expensive.

Generally, how well the lining process is performed has a lot to do with its

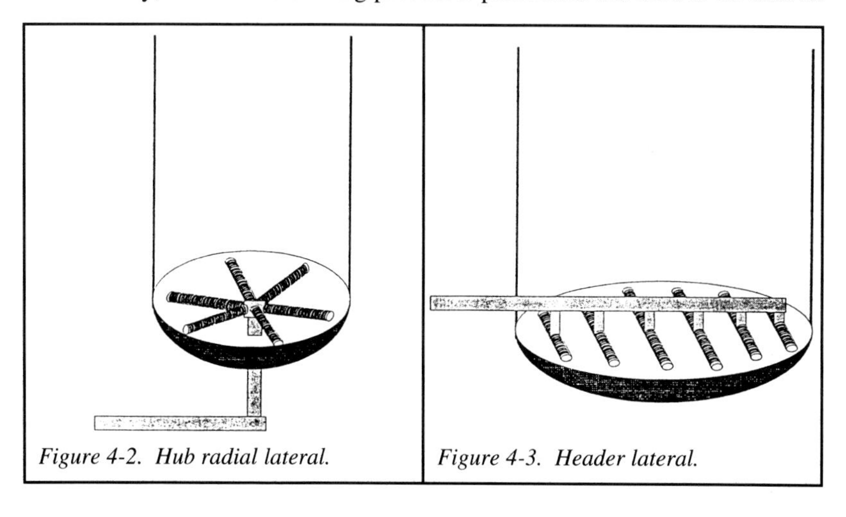

Figure 4-2. Hub radial lateral.

Figure 4-3. Header lateral.

long-term success. The vessel should first be sandblasted and cleaned. A rust-inhibiting primer should be used that is compatible with the lining material. The vessel manufacturer should have experience in using the lining material, or should subcontract the lining to a company that specializes in lining pressure vessels.

Coding of Vessels. Pressure vessels are manufactured that are considered

1 Non-code,

2 ASME code-stamped, or

3 Built to ASME code but not stamped.

The pressure rating for a vessel built to code is less than the rating for one that is not. For instance, a pressure vessel that is built to code may be rated for operation at pressures up to 100 psig. A non-code vessel with the same wall thickness may be rated by the manufacturer for operation at pressures up to 125 psig. The ASME code-stamped vessel will have been checked by an authorized inspector.

Which vessel to use will depend on the margin of safety desired, the degree of confidence in the vessel manufacturer, and the liabilities involved. If the manufacturer is reputable and if there are minimal liabilities involved in the application (i.e., if the vessel were to fail, it would not result in a major financial loss for the enduser), then non-code vessels will usually perform quite adequately. Pressure vessels used in the pretreatment of RO systems are typically not in a position where their failure will result in a catastrophic failure of the entire water treatment system. Independent engineering firms may want to specify code-stamped vessels, however, in order to reduce their liability just in case a failure should occur.

Skid mounting. Installation expense can be reduced if the pressure vessels are provided skid-mounted. It is generally less expensive to do as much plumbing as possible at the factory, rather than to plumb on-site.

As with the RO skid (see "RO structural frame" in Chapter 2), the supporting frame and pressure vessels should be sandblasted, cleaned, and painted with a rust-inhibiting primer. They should then be powder-coated, or painted with two coats of a high-quality epoxy or enamel paint.

Temperature Control/Heat Exchangers

During the winter months, water temperatures can drop dramatically for surface water sources. This will result in a substantial change in the permeate flowrate from an RO system. In order to keep up with flow requirements, RO feedwater is sometimes heated. (Temperature correction factors are given in the Appendix). It may also be heated because a particular process requires a certain process water temperature.

In many cases, it is not necessary to heat the water to meet the water requirements of the process, especially if the RO was overdesigned originally. In these cases it is suggested that the heat exchanger (or whatever the heating mechanism is) not be used.

There are several reasons not to heat the water if the higher water temperature is not required. Higher water temperatures will frequently result in greater biological activity within the membrane system, particularly if a biocide is not present within the system. Bacteria multiply faster in higher temperatures. Also, if the RO operates on water demand, increasing the RO permeate flowrate will increase its downtime. Bacteria will grow more rapidly in a stagnant RO system.

The potential for carbonate and sulfate scale formation is greater at higher water temperatures. Also, the higher permeate flux at higher temperatures will increase the concentration polarization effects at the membrane surface. This also can increase the potential for scale formation, and can increase the fouling rate by suspended solids.

Generally, operation at lower temperatures should result in lower maintenance requirements for the RO system. The trade-off for the lower permeate flux is that the RO will need to operate more, thus consuming more electrical power. However, this energy consumption is typically less than what the heat exchanger would use in heating the water.

Unless the source of heat for the water is a waste heat, it is more economical to operate the RO longer, or at higher pressure, than it is to heat the water. For example, it takes approximately 220,000 British thermal units (Btu) over an hour to raise the temperature of water from 55 to 77 °F for an RO system with a 20-gpm feedwater rate. The RO permeate flowrate would be about 30% less if the RO were operated at 55 °F instead of 77 °F. For a typical RO system that operates at 260 psig, the extra energy consumed to operate longer in making up for the reduced permeate flow (an additional 26 minutes) is about 18,000 Btu.

One alternative to heating an RO feedwater during colder seasons is to design the RO system with a variable-speed drive and an oversized high-pressure pump and motor. Variable-speed drives alter the rotational speed of the high-pressure (centrifugal) pump. When the inlet water is colder, higher pressures can be created by ramping up the high-pressure pump and motor, thus maintaining the desired permeate flowrate. With warmer water temperatures, the driver can scale back the pump output. This is accomplished without the loss of energy efficiency that would occur if the high-pressure pump outlet were throttled with a valve.

The RO system should not be operated at pressures that might cause a dramatic increase in membrane compaction. The upper limit for cellulose acetate is about 550 psig. During colder months, this pressure should not be exceeded as a means of maintaining the permeate flowrate. With polyamide thin-film membrane, however, membrane compaction is not a concern over most pressure ranges. (Consult the membrane manufacturer.)

Proper control of the heating mechanism (frequently a heat exchanger) is extremely important upstream of an RO system. It is difficult to protect the RO membrane from high-temperature surges that may occur, particularly after the RO and the heat exchanger have been shut down for a period of time. When RO feedwater sits stagnant inside a heat exchanger, it can become the same temperature as the heat source. When the RO goes back into operation, it will be slugged with this high-temperature water, possibly causing instantaneous membrane or element damage. This is particularly likely if the source of heat for the exchanger is steam. It is safer to use warm water as the heat source because of its lower heat content.

Some smaller RO systems have directly mixed the building's hot water supply with their cold water supply to achieve the desired feedwater temperature. This can sometimes cause additional RO fouling if the hot water system contains scale or other debris. Such debris is more likely in a hot water piping system. The hot water is more aggressive in dissolving metals from the piping materials. Also, carbonate and sulfate salts are more likely to precipitate in the hot water. The presence of these scale particles in the RO feedwater may catalyze the precipitation of supersaturated salts within the membrane system.

An industrial hot water source should not be mixed directly with the cold water RO supply if the hot water contains corrosion inhibitors. These chemicals may not be compatible with the RO membrane and system. In fact, these chemicals can sometimes cause problems with RO membrane systems even without direct mixing if a leak exists in the upstream heat exchanger.

Some systems will attempt to avoid the control problems associated with a heat exchanger by putting an automatic dump valve on the RO feedwater line. Before the RO membrane is exposed to it, the feedwater is dumped for a set time, allowing the temperature and the injected chemical concentrations to stabilize prior to the RO coming on-line.

The dump line should be throttled or flow-regulated to match the flowrate that would normally feed the RO unit if it were on-line. The system should be timed to allow low-pressure flow through the RO after the dumping period and prior to the engaging of the high-pressure pump. This will fill the RO membrane housings and ensure that the system is flooded prior to the engaging of the pump. There will thus be less possibility of damage to the membrane elements and to the pump.

When using a heat exchanger with steam or extremely hot water as its heat source, an emergency dump valve is suggested downstream of the heat exchanger. It would be opened if high water temperature is sensed upstream of the RO. This should reduce the possibility of the hot water coming into contact with the RO membrane. (McPherson, Hobson 1981).

It is also suggested for heat exchangers using steam that a portion of the exchanger effluent water be recirculated back to the feed of the heat exchanger. (See Figure 4-4.) This will insure that the heat exchanger's temperature sensor is exposed to the true temperature of the water inside the heat exchanger. This recirculation will also assist in buffering temperature variation while the RO is operating.

The heat exchanger's effluent temperature reading should be tied into the RO alarm control system. The RO should be prevented from starting if the water temperature in the heat exchanger is too high. Unless the water temperature drops, the RO inlet valve should not open.

Chemical Injection Pumps

Chemical injection pumps that are used with RO systems less than 250 gpm in size are usually solenoid-driven. An electric solenoid pushes a diaphragm in and out. Check valves on the diaphragm cavity inlet and outlet direct fluid up into the cavity and out into its effluent line. (See Figure 4-5.)

The pump has a stroke setting that determines to what percentage of maximum extension the solenoid will push the diaphragm into the cavity. The pump also has an internal solenoid speed setting, or can have its speed controlled externally by an instrument/controller. The speed setting determines the frequency of the pump stroke. Settings are based on the percentage of maximum pump output. For a particular stroke and speed setting, the pump output flow is calculated as follows:

$$\text{Daily pump output} = \text{stroke } \% \times \text{speed } \% \times \text{maximum pump output}$$

The stroke and speed setting can be adjusted along with the dilution of the injected chemical to achieve the desired chemical concentration in the RO feed stream. A high-speed setting will assist in the mixing of the chemical as it is injected into the bulk stream. A low setting should be avoided unless the system

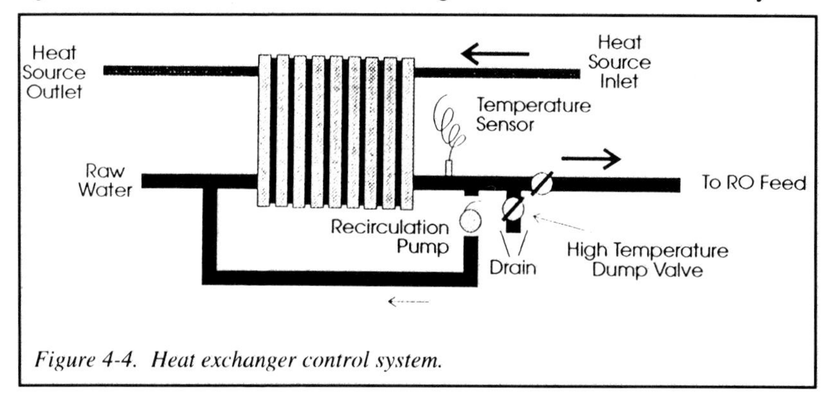

Figure 4-4. Heat exchanger control system.

is designed for in-line mixing downstream.

Most solenoid-driven pumps will have a minimum stroke setting to keep the pump from losing its prime. Too high a stroke rate may result in a reduced life expectancy for the diaphragm.

Priming an injection pump requires getting fluid all the way through the inlet tubing/piping, and through the diaphragm cavity. If an air pocket exists in the inlet line or in the diaphragm, it will tend to compress under the pressure of the pump stroke, rather than be pumped into the effluent line; thus, the air pocket will prevent the pump from pumping fluid.

Priming the pump with strong chemicals can be dangerous, as the chemical may be sprayed on personnel. If possible, the pump should be primed using water, and the chemical added later. It is also suggested that pumps used for strong chemicals be purchased with an automatic priming valve of some sort.

Protective clothing, head gear, and gloves should be worn if it is necessary to directly prime an injection pump with a chemical (or when filling the day tank). Company and government safety regulations should be considered.

When priming with water, the pump's effluent tubing can be removed from the point of injection in order to remove the back pressure on the line. (Note: A check valve should be installed at the point of injection to prevent any leaking of the process fluid.) The reduced back pressure will make it easier for the pump to push out any residual air in the lines. The pump suction line can then be disconnected and manually filled with wa-

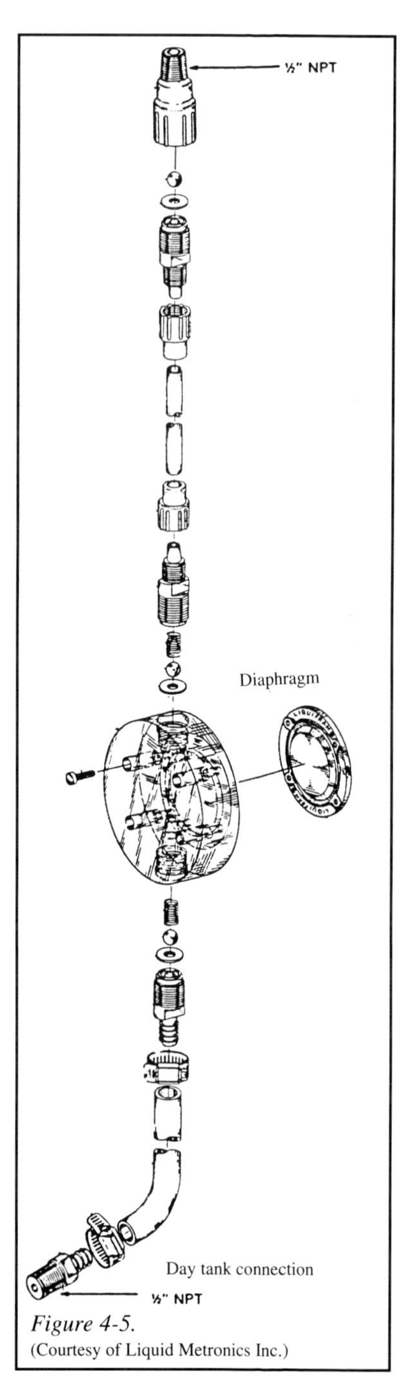

Figure 4-5.
(Courtesy of Liquid Metronics Inc.)

ter. It should be quickly reconnected to avoid any draining of the tubing. Maximum stroke and speed settings should be used with the pump until it is primed and air is out of the tubing.

It will be easier to keep the pump primed if the pump and its inlet tubing are located below the fluid level in the chemical day tank. In this manner, the inlet tubing and pump are always flooded. This is particularly helpful when pumping a chemical that can degas, such as sodium hypochlorite (chlorine) or hydrogen peroxide solutions. There are day tanks designed to have the injection pump mounted in the base of the tank, with the inlet line of the pump plumbed directly into a fitting at the tank base (Figure 4-6).

The only tubing to use with the injection pump should be that supplied or specified by the pump manufacturer. Pump fittings are designed for a tubing with a specific diameter and wall thickness. Slight variations in the tubing can result in leaks or catastrophic failure. For most compression fittings used by many of the pump manufacturers, the tubing should be pushed all the way into the fitting before the nut is tightened.

For pumping harsh/hazardous chemicals, it is suggested that a clear tubing of larger diameter (that is compatible with the chemical) be placed around the outside of the pump tubing and fittings. This will act as secondary containment in case a leak occurs in the pump tubing. Using clear tubing will allow the leak to be visible so that it can be repaired.

Secondary containment may be required in certain localities. This may be true for the day tank as well. Check with local authorities.

A spring-loaded anti-siphon check valve should be installed in the tubing at the RO feed stream injection point. This serves to prevent flow from the main line from backing up into the injection tubing. The spring in the valve prevents the siphoning of chemical through the tubing if the main line experiences a pressure vacuum. Such a vacuum frequently occurs in RO systems, sometimes caused by a concentrate or permeate line that is plumbed with its outlet well below the chemi-

Figure 4-6. Solution tank with recess for mounting pump.

cal day tank level. The spring in the valve can handle a vacuum of 3 to 5 psig. This translates to a capability of preventing a vacuum caused by a line opening located between 7 and 12 feet below the injection point.

Some amount of back pressure from the line is necessary for the proper operation of the injection pump. The minimum back pressure suggested by one manufacturer is 25 psig. (Clouthier 1993).

Options are usually available as to the materials of construction of the pump head, diaphragm, and check-valve balls. The difference in price is usually not significant between the various options. For the greatest flexibility in using the pump for injecting various chemicals, extremely inert materials are suggested. For most applications, it would be suggested to use a diaphragm of Teflon® (registered trademark of DuPont), a PVDF head, and ceramic check-valve balls. This combination will be capable of handling oxidizing agents, reducing agents, strong acids, caustic, and scale inhibitors.

Even though solenoid injection pumps are considered positive displacement pumps, their output rate can be affected by the line pressure into which the fluid is being pumped, and by the viscosity of the solution being pumped. A pump with a pressure output capability of at least 15 psig more than the line pressure is recommended. This should be sufficient to overcome the 3- to 5-psig drop that occurs through the anti-siphon check valve (which should be installed at the injection point).

If a long injection pump effluent line is used between the pump and the injection point, additional pump pressure may be required. After all, a solenoid injection pump needs to move its fluid only during the fraction of a second that the solenoid is driving the diaphragm. The fluid velocity at that moment can be fast enough to create quite a significant back pressure in the pump line.

Injection pumps can effectively pump into lines containing higher pressure than the rated pump pressure. One manufacturer claims the ability to handle a back pressure up to 130% of the pump's pressure rating. The pump manufacturer will usually supply a graph to determine the effect of this higher pressure on the pump's flowrate.

The viscosity or density of concentrated chemicals may make pumping difficult. This should be considered when sizing an injection pump. Standard pumps are limited to pumping a fluid of a maximum viscosity of about 400 centipoise (cP) (Cloutier, 1993). Dilution of the injection chemical can be used to lower the viscosity of the injection fluid, as limited by the maximum flowrate of the pump.

It should be noted that it is common to inject a chemical upstream of an RO's prefilters. The prefilters aid in mixing the chemical. The prefilter housing should be constructed of materials compatible with the particular injection chemicals.

An in-line static mixer can be beneficial in promoting the mixing of a chemical upstream of an RO system. A static mixer is simply a piece of pipe that is filled with baffles designed to create mixing as fluid passes through it.

The day tank chemical dilution will dictate how often the tank will have to be filled. It will also impact the injection pump size and settings. If the chemical is to be pumped from a 50-gallon day tank, and the operator wants to remix the chemical once per week, an injection pump should be chosen that injects about 6 gallons per day (gpd) as an average rate. This would amount to 42 gallons over the week. The remaining gallonage in the day tank after the week should be sufficient to keep the pump's inlet tubing submersed.

It is suggested that an attempt be made to use pump settings that are in the middle of their range. This will offer greatest flexibility if adjustments are later needed. If an injection pump has a maximum output of 24 gpd, a normal pump setting might be a 50% stroke setting and a 50% speed setting. This pump should inject about 6 gpd at these settings under continuous RO operation. This flowrate would thus be consistent with the 42-gallons-per-week consumption rate if the RO were operated continuously. Noncontinuous operation would result in a lower weekly usage of diluted chemical.

Once the injection pump flowrate is determined, the dilution concentrations for the chemical day tank can be determined. This can be calculated using the relationship in the following equation:

$$\frac{\text{Chemical volume} \times \text{chemical concentration} \times \text{pump rate}}{\text{day tank volume}} = \text{desired concentration} \times \text{RO feed gpd}$$

where:
chemical volume is the amount of concentrated chemical in the day tank;
chemical concentration is the concentration of chemical as provided by the supplier;
pump rate is the set flowrate of the pump in gpd;
day tank volume is the total volume in the day tank;
desired concentration is the chemical concentration in the RO feedwater, and
RO feed is the RO feed flowrate in gpd.

Example:
If an RO has a feed flow of 24 gpm (34,560 gpd) and it is desired to inject 0.5 mg/L of sodium hypochlorite (chlorine), determine the amount of bleach (5.25% sodium hypochlorite) to dilute in a 30-gallon day tank when using a 24-gpd injection pump set at a 50% stroke and a 65% speed setting.

Answer:

$$\text{Pump flow} = \text{stroke \%} \times \text{speed \%} \times \text{maximum pump output}$$
$$= 0.5 \times 0.65 \times 24 \text{ gpd} = 7.8 \text{ gpd}$$

This value can now be used to find the volume of chemical to dilute in the 30-

gallon day tank.

$$\text{Chemical amount} \times (0.0525) \times (7.8 \text{ gpd}) \div 30 \text{ gallons}$$
$$= 0.0000005 \times 34{,}560 \text{ gpd flow}$$
$$\text{Chemical amount} = 1.25 \text{ gallons}$$

Therefore at the chosen injection pump settings, 1.25 gallons of bleach should be diluted in 30 gallons of water to achieve the desired chlorine concentration.

For these calculations, it is common to assume that the density of the chemicals is close to that of water. It is therefore not usually necessary to take the chemical density into account. This will be true for most chemicals, though it may not be the case when injecting strong acids. However, the rate at which the injection pump delivers the chemical is usually the greatest inaccuracy in the equations. Therefore, the calculations are just to get the approximate concentrations required, and any fine tuning can be performed with the injection pump settings based upon the actual feedwater concentrations as measured on-site.

If the chemical is such that it is not possible to verify the concentration in the RO feedwater, such as might be the case with a reducing agent or a scale inhibitor, the injection flowrate of the pump should be verified. This can be accomplished by placing the pump inlet tubing into a graduated cylinder filled with the desired chemical at its day tank concentration. (Use proper safety equipment if the chemical is hazardous.) The rate at which the fluid is drawn out of the cylinder and pumped into the RO inlet line can be timed to calibrate the flow settings for the injection pump.

After the initial dilution of the 30-gallon day tank, it is unlikely and undesirable that the tank be completely depleted prior to remixing the chemical. With some chemicals, such as sodium hexametaphosphate, the remaining chemical should be dumped out and a new batch of solution mixed fresh. With other chemicals, it may be acceptable to simply top off the day tank using the correct day tank concentration. This dilution will use the same ratio of concentrated chemical to dilution water. The volume of both the chemical and the dilution water will be proportionally less.

In the example, if only two-thirds of the day tank has been injected when remixing the day tank chemical, only two-thirds of the 1.25 gallons of bleach would be needed to top off the day tank. The remainder of the tank volume would be filled with water. If it is suspected that some of the chlorine in the day tank has dissipated, it probably would be more accurate to dump the remainder of the tank and start fresh. (Note: Local municipal discharge requirements must be met if the chemical is to be dumped to a sanitary sewer.)

Degasifiers

Acid injection is required as pretreatment for some RO systems to reduce the potential for scale formation, or to reduce the rate of CA membrane hydrolysis. When acid is injected into water sources containing the carbonate ion, carbon dioxide is released into the water. The chemical reaction of the bicarbonate ion with sulfuric acid is as follows:

$$2HCO_3^- + H_2SO_4 \longrightarrow 2CO_{2(g)} + SO_4^{-2} + 2H_2O$$

If 80% of the alkalinity is converted into carbon dioxide by the acid injection, the carbon dioxide concentration will frequently be in excess of 100 ppm. This carbon dioxide will readily permeate an RO membrane, usually at a much higher concentration than the other anions in the RO permeate.

Carbon dioxide has been known to cause problems in water systems. Certain processes are sensitive to its presence. It can add to the ionic loading on ion-exchange systems (if used downstream). Carbon dioxide will be removed by anion resin, but can dramatically increase the required regeneration frequency of the ion-exchange equipment. It is common in this situation for the carbon dioxide to more than triple the regeneration requirements of the ion-exchange system.

When water is supersaturated in carbon dioxide gas, the carbon dioxide will degas naturally if the water is exposed to the atmosphere. The equilibrium concentration of carbon dioxide for water in contact with air is usually less than 10 ppm. If the water is allowed to sit in a storage tank for an extended period of time, the carbon dioxide will degas. However, most water systems do not have storage tanks large enough to give the carbon dioxide the time needed to degas.

Degasifiers are really very simple devices designed to facilitate the natural degassing of water containing supersaturated gases. Under atmospheric pressure, water is sprayed over plastic packing in a tall column; while air is blown or sucked upward, countercurrent to the water. The packing is designed to provide optimum surface area contact between the water and the air. The water falls from the packing to a collection tank at the base of the column. (See Figure 4-7.)

For just the removal of carbon dioxide, forced-draft or induced-draft degasifiers are employed. Forced-draft degasifiers blow air upward against the falling water; induced-draft degasifiers suck air up through the column. Although induced-draft degasifiers are slightly more effective because of the lower pressure that is present in the column, they do require a column that is structurally stronger to prevent its collapse under the slight vacuum conditions. Both types of degasifiers are effective at dropping the carbon dioxide in the water to less than 10 ppm.

For high-purity water applications, both forced- and induced-draft degasifiers require the inlet air to be filtered to prevent the introduction of airborne contaminants. This is typically performed using a high-efficiency particulate (HEPA) filter that is capable of removing better than 99.97% of all particles larger than 0.3 µm.

A concern with forced- and induced-draft degasifiers in high-purity water applications is the possible introduction of contaminants when a degasifier is used downstream of the RO system. Even with the HEPA filter, contaminants can be introduced in a size range that the RO would have removed. Airborne hydrocarbons can also be introduced into the water that may be difficult to remove by downstream treatment and equipment.

To utilize the RO for removing some of the contaminants introduced by the dagasifier, the degasifier is sometimes located upstream of the RO system. As the carbon dioxide degasses, the pH of the water will increase to as high as 7. This can lead to increased hydrolysis with CA membrane systems. If a second acid injection system is used downstream of the degasifier, it can be difficult to control, as most of the bicarbonate buffering would have already been removed

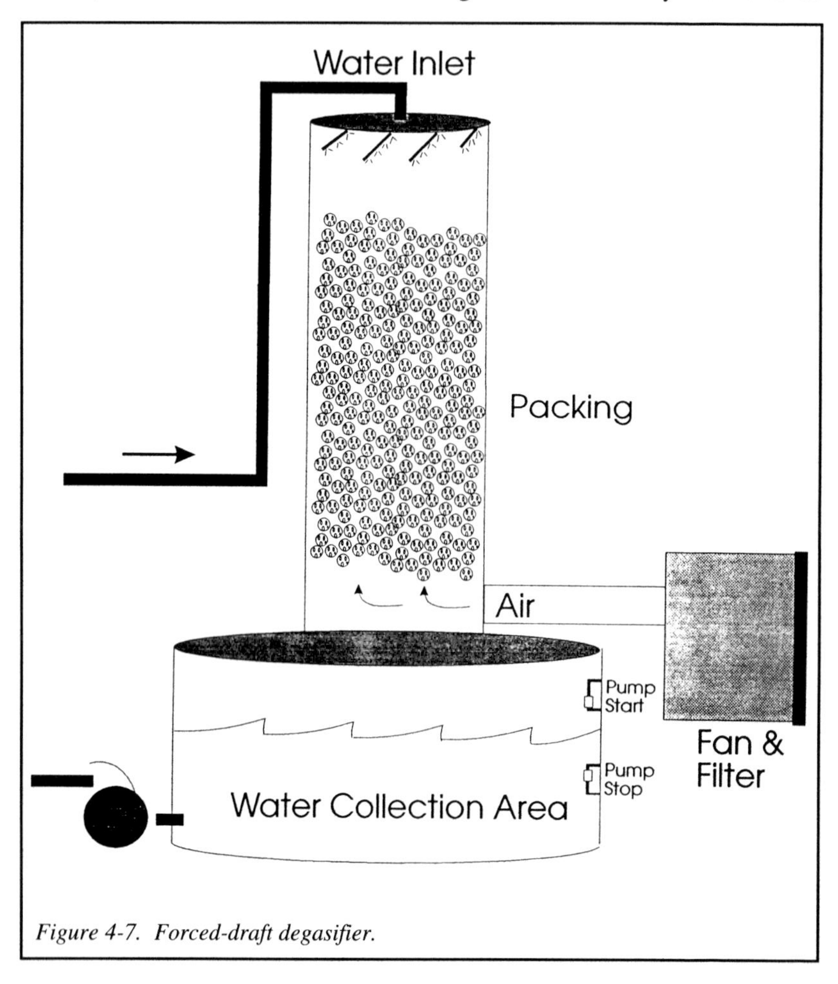

Figure 4-7. Forced-draft degasifier.

by the first injection system. Such a system is prone to wide pH fluctuation. Also, any remaining bicarbonate that is converted to carbon dioxide by the second acid injection system will tend to pass through into the RO permeate, thus placing a greater loading on any ion-exchange resins located downstream.

As newer CA membranes can better tolerate higher pH conditions than the older CA membranes, this second injection system may not be necessary. It may make more sense to reduce the pH further with the first injection system, such that the degasifier effluent pH is low enough for the water to go directly into the RO system without re-acidification. (Dropping the pH to 4.8 will theoretically convert all the bicarbonate to carbon dioxide, although it would be difficult to control the pH at values in this range.)

Vacuum degasifiers are sometimes used in high-purity water applications. Their benefit is their ability to remove both carbon dioxide and dissolved oxygen down to very low concentrations. In power industry applications, this offers the advantage of making the water less corrosive for stainless steel. In all industries using deionized water, the reduced oxygen content offers the advantage of making the water extremely bacteriostatic. Without the presence of oxygen in the water as dissolved oxygen, or in the form of anions such as sulfate (SO_4^{-2}) or nitrate (NO_3^-), the bacteria cannot grow. Significant biological activity in such systems is unlikely (Fulford, March 1993).

Vacuum degasifiers use a much taller column (as high as 28 feet or more) than do forced- or induced-draft degasifiers. A strong vacuum is placed on the column as water is dispersed over the top of the packing. Under the vacuum conditions, dissolved carbon dioxide and oxygen will degas to the parts-per-billion (ppb) level. The column height is necessary to provide the time for the degassing to occur.

The tall column is also required to provide enough water height (head) to allow a transfer pump to suck water out of the storage area in the base of the column. Even with the water head, there will normally be some amount of vacuum still present in the transfer pump suction line. The pump should be chosen so that it has a low inlet pressure requirement. This requirement is called the net positive suction head (NPSH), and is a measurement of the minimum head pressure necessary to prevent cavitation. The stored water height within the column (relative to the height of the transfer pump) must always be greater than the NPSH value at the pump's flowrate. The NPSH values are usually given in pump curves.

The vacuum pumps used in vacuum degasification tend to be relatively expensive. Also, the degasifier column has to be constructed of materials that can withstand the vacuum. Vacuum degasifiers are therefore expensive, and typically used only when the water treatment application demands it.

Storage Tanks

Although the output permeate flowrate of an RO system is fairly constant, the demand for water in many applications is not. In applications where the control of biological activity is critical, the answer is to oversize the RO and recirculate the unused permeate water back to the feed stream of the RO. The RO operates continuously, using feedwater that consists of the combination of pretreated makeup water and recirculated RO permeate water. With continuous operation, there is reduced biological activity in the RO, and there will be no stagnant water storage areas.

There are two disadvantages, both economic, of the continuously operating RO design:

1. The oversizing of an RO means that the capital cost of the system will be much higher, up to 50% to 200% higher than a system sized for the average water usage.

2. Energy usage, water, sewer, and pretreatment costs will be higher due to the continuous operation of the high-pressure pump and the continuous loss of RO concentrate.

The additional operating costs given in Point 2 above may be offset by the operational savings incurred by not having to fight biological warfare with the water treatment system. Biological problems can tie up manpower, lead to increased membrane replacement, and sometimes lead to additional system shutdowns.

The higher operating costs are such that continuously operating RO systems are typically used only in smaller, biologically critical applications. Water

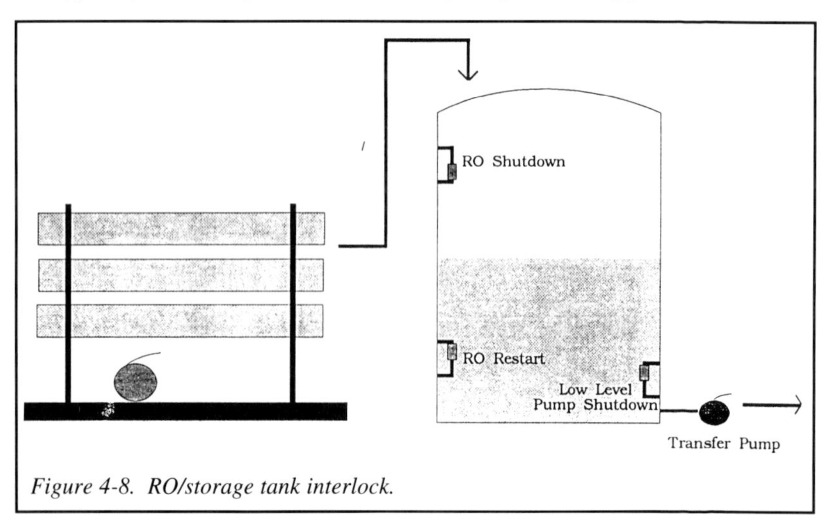

Figure 4-8. RO/storage tank interlock.

systems used in dialysis centers are frequently designed in this manner. These systems tend to have permeate flowrates that are less than 6 gpm, and the additional costs of a continuous RO are not as significant as they might be for a larger system.

The economics of most larger systems require using a storage tank to meet the peak demands of the system. The RO can then be sized to meet the peak average flowrate, versus sizing to the instantaneous peak demand. If the peak usage at a facility is 90 gpm, but the highest average usage over the day is only 30 gpm, the RO can be sized for the 30-gpm permeate flow. Water from the storage tank can be used to make up for the additional water requirements.

When the storage tank is full, RO operating expense is reduced by shutting down the RO system. The most common means of controlling the RO system to meet the specific permeate demands of the application is to interlock the RO system operation with a storage tank located downstream. The RO control circuitry is wired into the level controls in the storage tank. When the storage tank reaches a high level the RO automatically shuts off. When the tank empties to a low level, the RO automatically starts up (Figure 4-8).

In some applications it is desirable to better control biological activity in the RO system (more common with PA-membrane RO systems), and yet not desirable to size the RO for the entire piping distribution flowrate. For these, the RO control system can still be designed so that the RO operates continuously by interlocking it with the storage tank, so as to recirculate storage tank water to feed the RO when the storage tank is full (Figure 4-9).

In such recirculating systems, when the storage tank level control hits its high level, a valve on the city water line closes and a line from the storage tank opens to feed the RO system. This design requires higher operating costs due to the

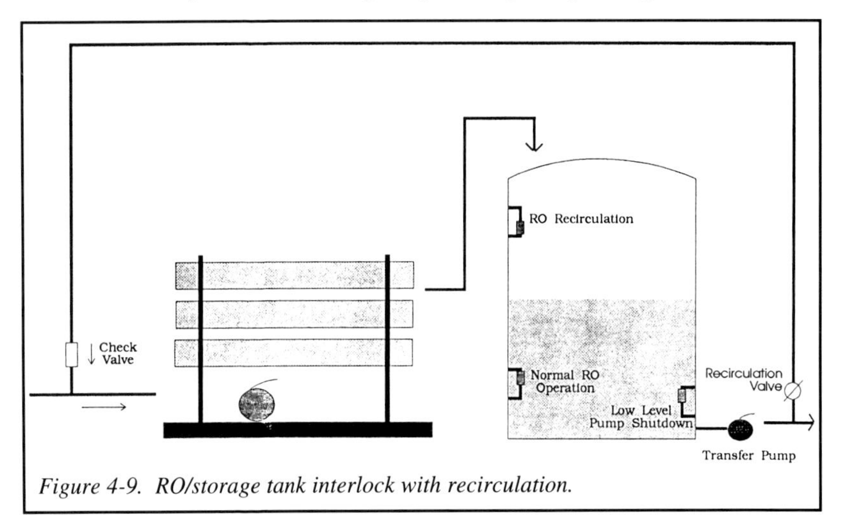

Figure 4-9. RO/storage tank interlock with recirculation.

additional RO operation and wasted concentrate water. This additional concentrate flow can be minimized if a dual RO concentrate flow throttling system is used such that the RO operates at higher permeate recoveries during the recirculation mode of operation. One control valve throttles the concentrate stream for the desired flowrate during the recirculation mode. During normal operation, a solenoid valve opens to allow additional concentrate flow through a second throttling valve.

Storage tank sizing. The storage tank should be sized to handle periods of peak flow demand. This will require knowing the maximum possible usage on an hourly basis. If these values cannot accurately be estimated, then flow totalizers should be used to obtain this data.

To determine an adequate buffer capacity, evaluate those periods of time in which the rate of usage exceeds the rate of RO makeup. Usage should also be evaluated over various time periods of peak flow usage. This will insure that the tank is sized to handle the demand over an extended usage period. The following example illustrates this point.

Example:

Basis: 60-gpm RO permeate flow (3,600 gallons per hour)

Hourly Consumption on a Peak Day

Time	*Reading (gph)*	*Time*	*Reading (gph)*
7:00 a.m.	1,500	7:00 p.m.	1,000
8:00 a.m.	14,000*	8:00 p.m.	2,000
9:00 a.m.	10,000*	9:00 p.m.	1,000
10:00 a.m.	12,000*	10:00 p.m.	1,000
11:00 a.m.	1,500	11:00 p.m.	1,000
12:00 a.m.	1,000	12:00 p.m.	1,000
1:00 p.m.	1,000	1:00 a.m.	1,000
2:00 p.m.	5,000*	2:00 a.m.	1,000
3:00 p.m.	1,000	3:00 a.m.	1,000
4:00 p.m.	6,500*	4:00 a.m.	1,000
5:00 p.m.	2,000	5:00 a.m.	1,000
6:00 p.m.	1,500	6:00 a.m.	1,000

*Hour during which water usage exceeds RO makeup capacity.

Usage for the day adds up to 70,000 gallons (gal). This is well within the daily RO capacity, 86,400 gallons.

The majority of consumption appears to occur during the first shift (46,000 gallons are consumed during the hours between 6:00 a.m. and 2:00 p.m.). If the storage capacity were based on the shift usage, a storage tank would need to have a minimum capacity of:

$$46{,}000 \text{ gal} - (3{,}600 \text{ gph} \times 8 \text{ hours}) = 17{,}200 \text{ gal}$$

Actually, consumption exceeds the RO capacity by a greater margin over the period between 7:00 a.m. and 4:00 p.m. If our tank were based upon this usage, its size would need to be

$$51{,}500 \text{ gal} - (3{,}600 \text{ gph} \times 9 \text{ hours}) = 19{,}100 \text{ gal}$$

Since most of this usage occurs in the morning, however, the rate of usage needs to be evaluated over a shorter period of time. If just the peak 3 hours of the morning are considered, the minimum storage tank sizing needed would be

$$36{,}000 \text{ gal} - (3{,}600 \text{ gph} \times 3 \text{ hours}) = 25{,}200 \text{ gal}$$

If the shift usage had been used to size the storage tank, the user would be running out of water during the peak usage mornings; therefore, it is critical that all time periods of excessive usage be evaluated independently as well as cumulatively.

One method of evaluating all peak periods is to begin by evaluating all periods of consecutive hours where consumption exceeds makeup. In the previous example, this would be the following time periods:

Time Period	*Min. Storage*
7:00 a.m. to 10:00 a.m.	25,200 gal
1:00 p.m. to 2:00 p.m.	1,400 gal
3:00 p.m. to 4:00 p.m.	1,900 gal

Then the time periods between and over these times should be evaluated. Be sure to include the period from the last peak usage period back to the first.

Time Period	*Min. Storage*
7:00 a.m. to 2:00 p.m.	19,100 gal
1:00 p.m. to 4:00 p.m.	1,700 gal
3:00 p.m. to 10:00 a.m.	-7,900 gal

Again, the time periods between and over the previous periods should be evaluated.

Time Period	*Min. Storage*
7:00 a.m. to 4:00 p.m.	19,300 gal
1:00 p.m. to 10:00 a.m.	-9,100 gal
3:00 p.m. to 2:00 p.m.	-13,800 gal

All of the peak periods have now been evaluated. It can safely be derived (until usage increases) that the tank should be sized to at least have a capacity of 25,200 gallons.

Piping Distribution Systems

It is common in many industries for the RO permeate water to be plumbed directly into a water distribution system. How this piping system is designed will directly affect the RO permeate pressure. Predicting the pressure losses in a directly fed piping system will make it possible to predict the RO permeate pressure and its affect on the RO permeate flowrate.

The pressure drop through straight piping is related to the velocity of water within the pipe. In most water piping systems, the flow will be turbulent, with numerous eddies continually being created within the bulk stream as it passes through the pipe. In such systems, the pressure drop through the pipe will be proportional to the flowrate squared (i.e., to the power of two).

Piping elbows, tees, valves, and constrictions will add to the pressure drop through piping. Sharp elbows will create a greater pressure loss than long sweep elbows. A few sharp restrictions will result in a larger pressure drop than numerous sweep elbows. When plastic piping is glued or welded together, it is common for a lip to be created at the joint by the excess glue (Figure 4-10) or by melted plastic from the welding process. This lip creates a flow constriction that will affect the pressure drop.

The inside diameter (i.d.) of piping will differ depending on the schedule classification. Schedule 80 PVC and polypropylene pipe i.d. will be less than the nominal pipe size (Tables 4-1 and 4-2). Schedule 40 PVC pipe i.d. will be greater

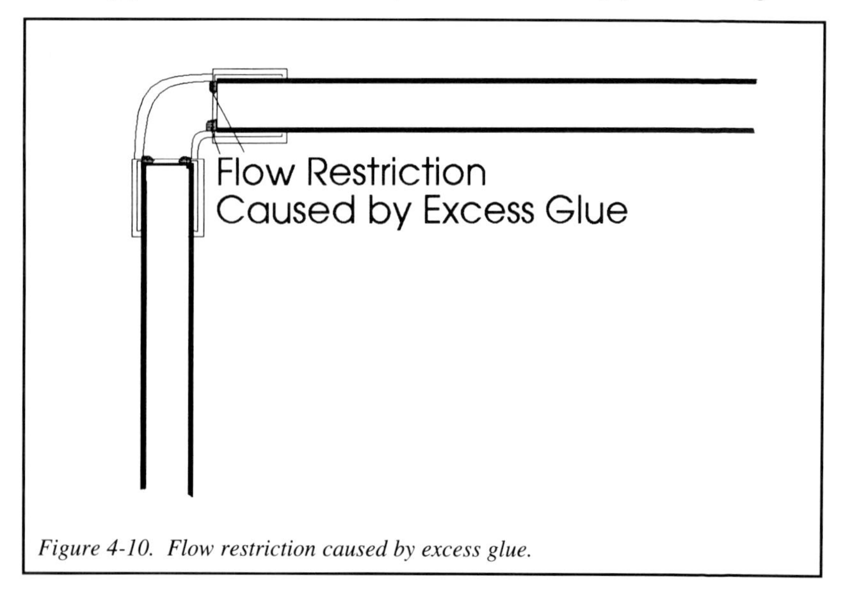

Figure 4-10. Flow restriction caused by excess glue.

than the nominal pipe size (Table 4-3).

The shear of the inside pipe wall as flow goes through is a major factor in pressure drop. This shear will be greater with piping whose inside wall is rough, which is generally the case with PVC piping that has been in service for more than a year. Purified water tends to leach the plasticizers out of piping, leaving the inside wall brittle and rough. This effect will be more dramatic in smaller pipe sizes.

Piping should not be oversized to eliminate pressure drop concerns, however, as it has been shown that most water piping systems suffer from the growth of bacteria on inside walls (Patterson et al., 1991). If this growth becomes excessive, it will tend to shed bacteria into the water. These bacteria are a concern in many industries either directly, or because they are particulates. Biofilm growth will tend to be more excessive in piping systems with low water velocities (Figure 4-11).

The friction of the inside wall against the water going by is such that the flowrate directly at the pipe wall is zero, even with high bulk water velocities. The local water velocity increases with distance away from the pipe wall and toward the center of the pipe (Figure 4-12). How quickly this local velocity increases is a function of the flowrate through the pipe (Greenkorn and Kessler, 1972).

High flowrates in piping create turbulence that will assist in minimizing the biofouling layer on the pipe wall. For water that no longer contains a biocide, it is suggested that the bulk velocity in the piping not fall below 3 ft/sec for water. In fact, it is desirable as a means of reducing the biofoulant layer to maintain a

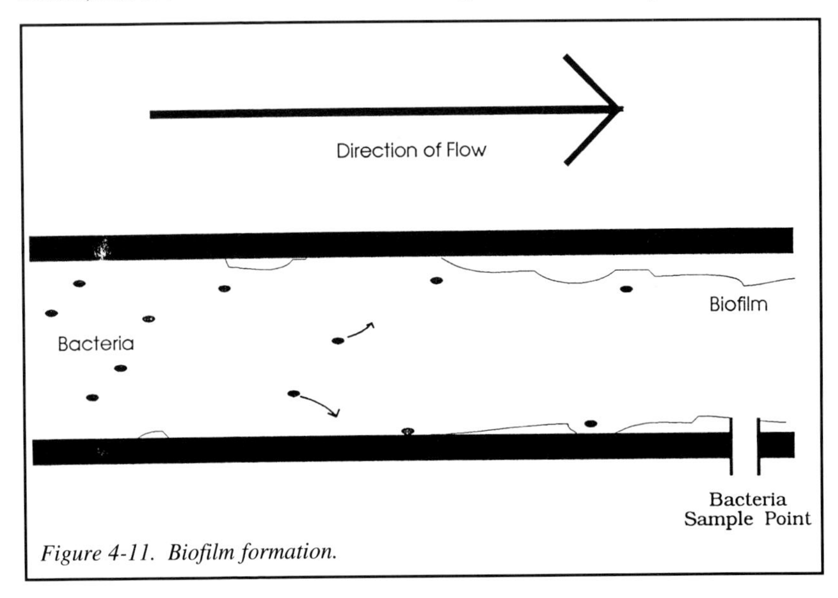

Figure 4-11. Biofilm formation.

minimum velocity of 5 ft/sec. Some high purity piping systems are designed with a minimum velocity of 7 ft/sec.

To obtain a continuous flow through the piping, the distribution piping is looped back to the water treatment equipment, either to a storage tank or to the feed of the RO system. To maintain a minimum pressure in the piping system, either a valve is used to throttle the flow going back into the system, or a pressure regulator is used. The pressure regulator adjusts flow to maintain the desired pressure upstream of the regulator. For direct RO feed systems, this back pressure adds to the required permeate pressure of the RO system.

Distribution systems will have usage points where water is removed from the system. This reduces the flowrate of the remaining water. How much water is removed and where these removal points are located in the piping system have a great deal to do with the overall distribution-loop pressure drop.

The following calculation (Greenkorn and Kessler, 1972) can be used to approximate the pressure losses through a straight section of pipe:

$$\text{psid} = 0.323 \times f \times L \times v^2/D$$

where:
f is a friction factor that depends on the roughness of the pipe inside wall and the pipe i.d.,
L is the length in feet of straight pipe,
v is the bulk water velocity in feet per second, and
D is the pipe inside diameter in inches.

The friction factor, f, can be determined from Figure 4-13 if a number is estimated for the relative roughness, K. For older PVC piping systems, the value

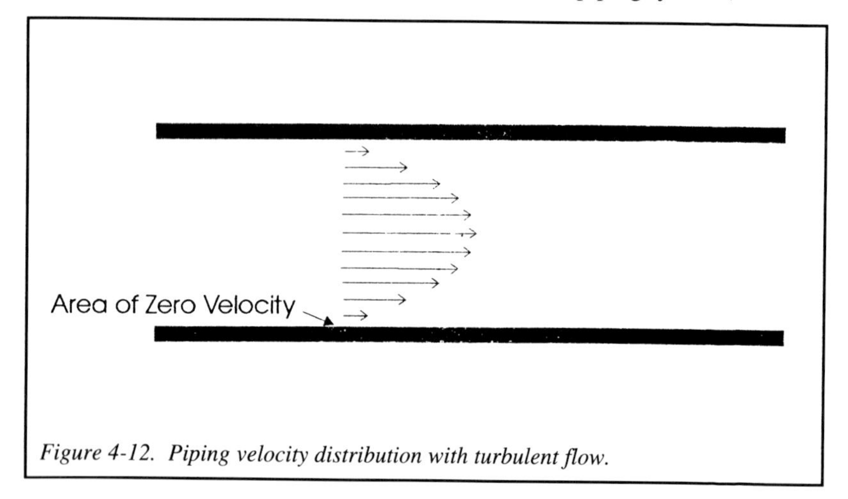

Figure 4-12. Piping velocity distribution with turbulent flow.

for K may be as high as 0.01. For smoother piping materials, this factor will be lower. For stainless steel piping, it will be about 0.002. Piping of PVDF will often be in this range also, depending on how well its fittings are connected.

Ridges between fittings will result in a substantially greater relative roughness factor. Since this is difficult to predict, a conservatively low relative roughness factor should be used to estimate pressure losses in plastic piping systems.

In order to calculate the pressure drop through a piping system, equivalent lengths of straight pipe can be substituted for elbows, tees, valves, and known constrictions in the system. These can be determined using Figure 4-14 and then totaled.

In calculating the pressure drop through an entire distribution system, the pressure drop through each segment of the system must be determined individually. A segment would consist of a section of piping of various pieces of straight pipe and fittings, all with the same inside diameter and flowrate. After a point of use where the flowrate has changed, the next segment of pipe will need its pressure drop calculated separately. All of the pressure drops through the various segments of pipe in a distribution loop are then added together to find the drop across the entire piping loop.

A piping distribution system that uses a single loop that is plumbed to all usage points before returning to its source is called a serpentine, or "snake" system (Figure 4-15, top). A distribution system that uses numerous parallel subloops that feed back into a common return line is called a "ladder" system (Figure 4-15, bottom).

The advantage of a snake system over a ladder system is that there is no concern about balancing the flow between the various subloops. If a usage point on one of the subloops in a ladder system draws water excessively, it is possible to get zero flow downstream of the usage point, or even reverse flow where flow feeds back from the return line to the usage point.

For a snake system to operate correctly, it must be designed with a high enough flowrate so that the total peak usage is a minor part of the distribution flowrate. In this manner, the velocity through the piping system does not vary significantly between periods of peak usage and periods of minimal usage. This means that the pressure at the various usage points will not vary dramatically.

It is suggested for a snake system that the peak usage be no more than 30% of the bulk flow. The pipe should then be sized for this high bulk flowrate, keeping in mind the desired minimum bulk water velocity. If this minimum velocity is more than 5 ft/sec (as suggested), the pressure drop through the loop will probably be substantial. A relatively large distribution pump may be required to meet the high flow and pressure requirements.

A long snake distribution system will require a large distribution pump, and will tend to generate a significant amount of heat in the recirculating water. If the water usage is minimal, there will be minimal water being made up to the system. Under these conditions, the water temperature will increase in the loop. In some applications where the control of water temperature is critical, such as in semiconductor manufacturing, heat exchangers may actually be used in the distribution loop to remove this excess heat.

Ladder piping systems are most commonly used when numerous usage points are required over a substantial distance. The advantage of the ladder distribution system in this type of situation is that the pressure drop across the system will be significantly less than in the snake system.

Parallel subloops will have equal pressure drops. If the individual flowrates through each subloop are not known, but the total flowrate is known, the subloop flowrates can be determined by finding the particular flows that will result in equal pressure drops in the subloops. In ladder-type piping systems where there are numerous subloops that feed back into a common return line, calculating all the flowrates required to balance the various subloops is a complicated, iterative process that is best performed by a computer.

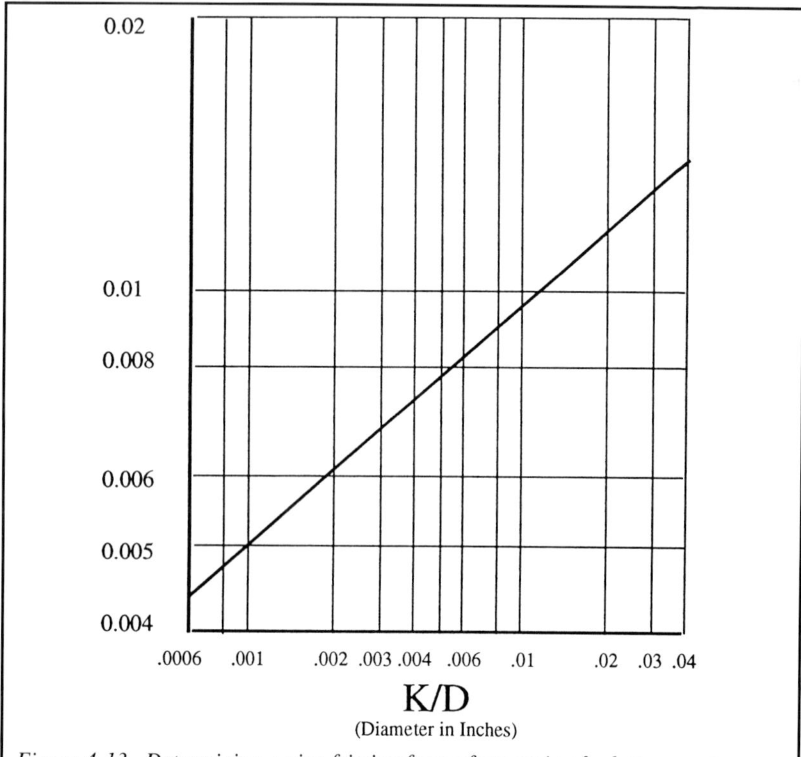

Figure 4-13. Determining a pipe friction factor from ratio of relative roughness to diameter.
Greenkorn and Kessler, 1972.

A source of pressure loss that can be significant in larger water distribution systems that use a storage tank is the difference in water height between the water level in the storage tank and the point where the return water re-enters the tank. In tall storage tanks, water may return to the top of the tank, which may be a number of feet above the tank water level. The difference in water "head" in the tank can account for several psi of pressure loss. The factor to use in determining the pressure loss resulting from a difference in water height is 0.43 psi per foot of water.

Pressure Drop Calculation

Example:

A purified water loop is being designed such that an RO will directly feed a distribution loop that returns to the RO feed stream. The piping loop will consist of 1-inch nominal schedule 80 PVC with three usage points, each preceded by four standard 90° elbows and 15 feet of straight pipe. After the last usage point, 10 feet of straight pipe is followed by a pressure regulator, just prior to entering the RO feed stream. The regulator is set at 40 psig (which is greater than the RO feedwater pressure). Each usage point draws 3 gpm. It is desired to maintain a bulk velocity in the piping of 5 ft/sec. In order to size the RO system and its high-pressure pump, it is necessary to know what the flowrate and pressure need to be at the beginning of the loop.

Answer:

From Table 4-1, the inside diameter for 1-inch schedule 80 pipe can be found to be 0.957 inch.

The minimum flowrate in the piping as it returns to the RO feed stream can be calculated using the i.d. of the pipe:

$$\text{Flow} = \text{velocity} \times \text{pipe cross-sectional area}$$

where:
the cross-sectional area = $\text{i.d.}^2 \times 0.785$
i.d. is the inside diameter of the pipe

$$\text{cross-sectional area} = (0.957")^2 \times 0.785 = 0.719 \text{ square inch (in}^2\text{)}$$

$$\text{flow} = 5 \text{ ft/sec} \times 0.719 \text{ in}^2 \times (7.5 \text{ gal/ft}^3 \times 60 \text{ sec/min} \times \text{ft}^2/144 \text{ in}^2) = 11.2 \text{ gpm}$$

This flowrate becomes 14.2 gpm upstream of the last usage point, 17.2 gpm upstream of the next-to-last usage point, and 20.2 gpm before the first point.

As shown in Figure 4-14, each elbow can be converted to its equivalent length of straight pipe, 2.7 feet. The total equivalent feet of straight pipe now becomes 25.8 feet (15 ft + 4 × 2.7 ft) between each usage point. The pressure drops

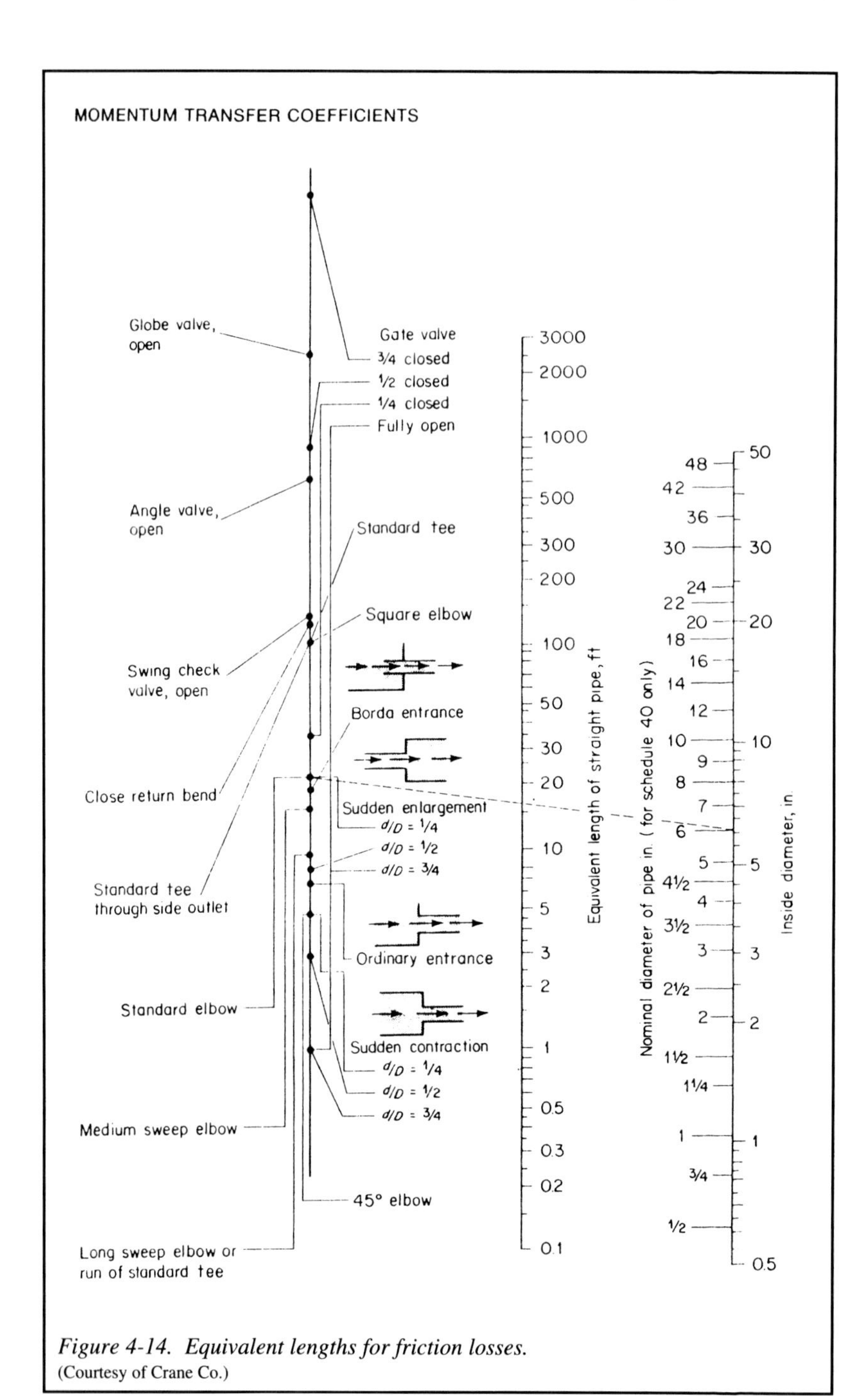

Figure 4-14. Equivalent lengths for friction losses.
(Courtesy of Crane Co.)

between each point can then be calculated using the particular flowrate at each segment.

Using a relative roughness factor of 0.01, a friction factor can be found from Figure 4-13. For 1-inch piping, this factor would be about 0.0095.

$$\text{psid} = 0.323 \times f \times L \times v^2 \div D$$
$$= 0.323 \times 0.0095 \times 10 \text{ ft} \times (5 \text{ ft/sec})^2 \div 0.957 \text{ inch} = 0.8$$

Between this node and the previous one, the flow and its velocity in the piping changes. Therefore, the velocity in the piping is calculated by the following:

$$\text{velocity (ft/sec)} = 0.41 \times \text{flowrate (gpm)/i.d.}^2$$

where:
i.d. is the inside diameter of the pipe in inches, and

$$\text{velocity} = 0.41 \times 14.2/0.957^2 = 6.36 \text{ ft/sec}$$

$$\text{psid} = 0.323 \times 0.0095 \times 25.8 \times 6.36^2 \div 0.957 = 3.35$$

Again, the flow and velocity will change for the upstream node:

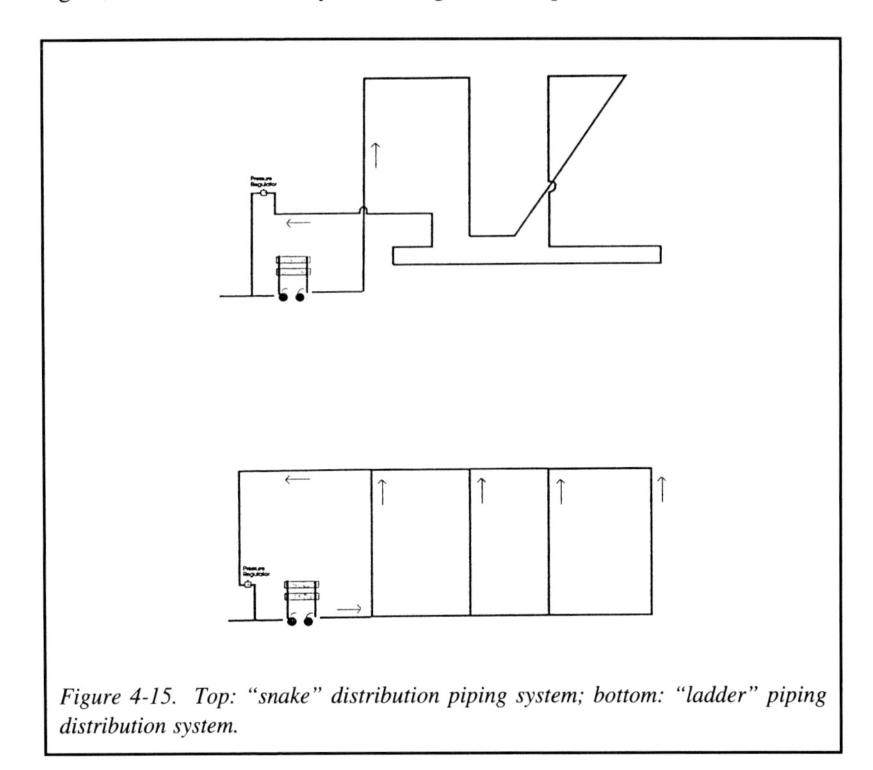

Figure 4-15. Top: "snake" distribution piping system; bottom: "ladder" piping distribution system.

$$\text{velocity} = 0.41 \times 17.2 \div 0.957^2 = 7.70 \text{ ft/sec}$$
$$\text{psid} = 0.323 \times 0.0095 \times 25.8 \times 7.7^2 \div 0.957 = 4.9$$

And prior to the first usage point:

$$\text{velocity} = 0.41 \times 20.2 \div 0.957^2 = 9.04 \text{ ft/sec}$$
$$\text{psid} = 0.323 \times 0.0095 \times 25.8 \times 9.04^2 \div 0.957 = 6.8 \text{ psig}$$

The pressure drops can now be all totaled and added to the setting of the pressure regulator to estimate the pressure at the beginning of the piping system:

$$40 + 0.8 + 3.35 + 4.9 + 6.8 = 55.85 \text{ psig}$$

In sizing the RO system and the high-pressure pump, a permeate flowrate of 20.2 gpm will be required at the minimum water temperature for the system. The permeate back pressure on the system will be 55.85 psig. This will result in an additional pressure requirement for the high-pressure pump (over that required for the desired permeate flowrate) of 55.85 psig.■

Table 4-1
PVC Schedule 80 Pipe Dimensions

Nominal Size (Inches)	*Outside Diameter (Inches)*	*Minimum Wall Thickness (Inches)*
1/8	0.405	0.095
1/4	0.540	0.119
3/8	0.675	0.126
1/2	0.840	0.147
3/4	1.050	0.154
1	1.315	0.179
1-1/4	1.660	0.191
1-1/2	1.900	0.200
2	2.375	0.218
2-1/2	2.875	0.276
3	3.500	0.300
4	4.500	0.337
5	5.563	0.375
6	6.625	0.432
8	8.625	0.500
10	10.750	0.593
12	12.750	0.687
14	14.000	0.750
16	16.000	0.843

Table 4-2
Polypropylene Schedule 80 Pipe Dimensions

Nominal Size Inches	*Outside Diameter Inches*	*Inside Diameter Inches*	*Minimum Wall Thickness (Inches)*
1/2	0.840	0.546	0.147
3/4	1.050	0.742	0.154
1	1.315	0.957	0.179
1-1/4	1.660	1.278	0.191
1-1/2	1.900	1.500	0.200
2	2.375	1.939	0.218
3	3.500	2.900	0.300
4	4.500	3.826	0.337
6	6.625	5.761	0.432

Table 4-3
PVC Schedule 40 Pipe Dimensions

Nominal Size (Inches)	*Outside Diameter (Inches)*	*Minimum Wall Thickness (Inches)*
1/8	0.405	0.068
1/4	0.540	0.088
3/8	0.675	0.091
1/2	0.840	0.109
3/4	1.050	0.113
1	1.315	0.133
1-1/4	1.660	0.140
1-1/2	1.900	0.145
2	2.375	0.154
2-1/2	2.875	0.203
3	3.500	0.216
4	4.500	0.237
5	5.563	0.258
6	6.625	0.280
8	8.625	0.322
10	10.750	0.365
12	12.750	0.406
14	14.000	0.438
16	16.000	0.500

CHAPTER 5

RO MAINTENANCE

Instrumentation and Monitoring

The basic operation of a reverse osmosis system is simple. A high-pressure pump provides pressure that drives pure water to diffuse through a semipermeable membrane. The basic operation is easily monitored once the proper parameters and their relative effects are considered.

An RO system cannot foul, scale, or suffer from some sort of membrane deterioration problem without affecting the system monitoring instrumentation. Proper system monitoring will keep the operator informed of fouling and scaling trends, and will reveal the signs indicative of membrane deterioration.

The concern is sometimes raised that the effects of membrane deterioration are being covered up by membrane fouling. It is unlikely, however, that the two problems are occurring simultaneously and exactly canceling out the effects of each other. In the worst case, the extent of a problem may appear less significant than if it were not overshadowed by the effects of a concurrent problem. For example, when calcium carbonate scale formation occurs on cellulose acetate membrane, this results in both membrane hydrolysis which causes an increase in the normalized permeate flowrate and a reduction in normalized permeate flow (due to the scale fouling of the membrane).

If there is a concern that some sort of membrane deterioration might be occurring, cleaning the system should reveal the full extent of the deterioration problem. It is common for membrane fouling to improve the rejection characteristics of the membrane. After cleaning, the rejection may drop, frequently leaving the mistaken impression that the cleaning solution attacked the membrane and caused the loss in rejection.

Daily monitoring of key parameters. In order to see performance changes that might be indicative of fouling, scaling, or a membrane deterioration problem, the right instrumentation must be used and its readouts properly recorded and evaluated. A sample RO data collection sheet is shown in Figure 5-1. This particular sheet also includes some of the key operating parameters for some common RO pretreatment equipment. A data collection sheet of this type should be used to record the following equipment performance data on a regular basis (daily for most systems):

Pretreatment Equipment

Media filter inlet/outlet pressure. This measures the pressure of the inlet and outlet water streams of the sand, multimedia, or activated carbon filter.

Media filter pressure drop. The difference between the inlet and outlet pressures is an indicator of the extent of media fouling, and can be used to determine when to backwash the filter.

Effluent chlorine is an on-site measurement of the free or total chlorine concentration of the activated carbon filter effluent water, or of the RO feedwater.

WEEK _______

	Monday	Tuesday	Wednes.	Thursday	Friday	Saturday	Sunday
Sand Filter Inlet/Outlet PSI	\	\	\	\	\	\	\
Pressure Drop							
Effluent Chlorine							
Carbon Inlet/Outlet PSI	\	\	\	\	\	\	\
Pressure Drop							
Effluent Chlorine							
Softener Inlet/Outlet PSI	\	\	\	\	\	\	\
Pressure Drop							
Gallons Since Regen							
Outlet Hardness							
Prefilter Inlet/Outlet PSI	\	\	\	\	\	\	\
Pressure Drop							
Silt Density Index (SDI)							
Water Temperature							
Chlorine Concentration							
RO Hour Meter							
Feed pH							
Membrane Psi							
Concentrate Psi							
Pressure Drop							
Permeate Flow							
Concentrate Flow							
Recycle Flow							
Feed TDS/Cond.							
Permeate TDS/Cond							
% Salt Rejection							
Misc./Comments							

Technician:

Figure 5-1. Reverse osmosis data collection sheet.

For CA membrane systems, there should be sufficient free or total chlorine (total chlorine includes chloramines) to prevent biological activity. For PA membrane systems, there should be no free or total chlorine present. The reading can be used to determine if an activated carbon filter is performing adequately, or if a sufficient concentration of reducing agent is being injected into the RO feedwater.

Softener regeneration. The gallonage processed by the softener since it was last regenerated should assist in tracking the regeneration frequency of the softener, and in determining if its resin still has adequate hardness removal capacity.

Softener effluent hardness. When relying on a water softener to prevent the precipitation of calcium carbonate scale on the RO membrane, it is essential that the effluent hardness concentration be kept at minimal levels. If hardness is appreciably present, the softener may need to be regenerated more frequently, or with a higher salt dosage.

Prefilter inlet/outlet pressure shows the pressure of the feedwater as it enters and exits the RO system prefilter. The outlet pressure will also confirm that there is sufficient pressure feeding the high-pressure pump to prevent cavitation.

Prefilter pressure drop. The difference between the prefilter inlet and outlet pressures is an indicator of how badly the prefilter is fouled.

Silt density index is a measurement of the suspended solids concentration in the RO feedwater. (See "Investigation of Design Conditions - Silt Density Index" in Chapter 1.) It may be desirable to perform an SDI measurement after each media filter in the RO pretreatment system to monitor individual performances. Turbidity measurements may be substituted for the SDI.

Water temperature is most accurately measured in the RO concentrate or permeate water (so that the effects of any increase in temperature caused by the high-pressure pump are included).

Water pH. The pH of the RO inlet water after pH adjustment is critical for systems using pH correction as a means of controlling calcium carbonate scale formation. It should be noted that the RO permeate and concentrate pH values are usually not the same as that of the RO inlet water.

Reverse osmosis system

RO hour meter. A measurement of the time the RO has been actively operating is helpful for determining if the RO is operating most of the time, or just sitting stagnant. It also can be used as an indicator of how much water is being used by the facility.

RO membrane pressure. The pressure at the entrance of the membrane housings is also called the feed or primary pressure. If the high-pressure pump is being throttled by a valve, this pressure should be measured on the effluent side of the valve. (The pressure gauge should not be located directly downstream of a throttling valve, as the higher velocities can cause an aspiration effect that will cause the pressure gauge to read lower than the actual downstream pressure.)

RO concentrate pressure. The pressure at the exit of the tail end housing of the membrane array is also called the brine or final pressure. This is the pressure prior to the concentrate throttling valve.

RO pressure drop. The difference between the RO membrane and concentrate pressures is an indicator of the resistance to flow as the RO feedwater passes through the membrane elements. This parameter is also called the differential pressure or the "delta-p."

RO permeate flow is the flowrate of the RO permeate water, which is often referred to in water purification systems as the product water.

RO concentrate flow is the flowrate of the RO concentrate water, which is also known in water purification applications as the brine.

RO permeate recovery is the ratio of the RO permeate flowrate to the feedwater flowrate. This parameter will impact the concentration of salts as the feedwater passes through the RO system, and thus is related to the potential for scale formation within the membrane elements.

RO recycle flow is the flowrate of the concentrate recycle stream, if one is used.

RO normalized permeate is the permeate flowrate after it has been normalized for temperature and for pressure variations. The calculations will be discussed later in this chapter.

RO feed TDS is the solute concentration as the water enters the RO system. For pure water systems, the solute is considered to be the total dissolved solids (TDS), or the salt concentration as measured by conductivity.

RO permeate TDS is the concentration of the solute that has permeated the membrane, usually measured as TDS or conductivity.

RO percent salt rejection is the percentage of solute that is rejected by the RO membrane. This calculation will be discussed later in this chapter.

Weekly/monthly monitoring of key parameters. To provide even more information about the performance of an RO system, it is suggested that the following parameters be monitored on a less frequent basis:

Concentrate concentration. The TDS/conductivity of the RO concentrate can be used along with the feedwater concentration to calculate an average membrane concentration, which can then be used to calculate an average membrane salt rejection. It can also be used in a mass balance calculation to check the accuracy of the flow and concentration readings. A mass balance is simply based on the fact that in a stable and continuously flowing system, whatever matter comes into the system has to go out of the system.

$$\text{Permeate flow} \times \text{permeate TDS}^* + \text{concentrate flow} \times \text{concentrate TDS}^* = (\text{feed flow}) \times \text{feed TDS}^*$$

where:
feed flow = permeate flow + concentrate flow

Individual housing permeate concentration. The TDS/conductivity of the permeate water from each individual housing is very helpful when trying to isolate a performance trend within the system. This will be discussed further in Chapter 6, "RO Troubleshooting." Some means of identifying which housing corresponds to which concentration reading should be employed, such as the method used in Figures 5-2 and 5-3.

Individual housing permeate flowrate is the permeate flow from each individual housing. As in the individual housing permeate concentration, this value is useful when isolating performance trends.

Interstage pressures. The pressures between the RO banking (stages) are useful in determining the hydraulic pressure differential across each stage as a means of isolating an increase in pressure differential across the entire system array.

It would be time-consuming to perform the above four readings on a daily basis. However, a full set of these readings, if recorded at the start-up of a new set of membrane elements, would establish a baseline for future comparison. They should be recorded on a regular basis thereafter. If a problem were ever suspected, all the readings could again be recorded to provide valuable information that might help in solving the problem.

Calibration. Regular calibration of the RO system instrumentation is essential to insure that the data obtained is reliable and accurate. A great deal of time and frustration can be saved by implementing a regular routine of instrument calibration.

pH Meters are notorious for drifting readings. Typically, this is a result of the pH transducer's reference probe losing some of its potassium chloride solution.

It is suggested that the water pH be verified weekly using a laboratory

*Conductivity must be converted to a TDS value before the equation will be accurate. Even so, its accuracy will be limited by the estimation being used to convert the conductivity reading into a TDS value. (Note: Even TDS meters are only measuring the conductivity, then using an internal conversion factor to obtain a TDS reading.) The actual relationship between the TDS and what its conductivity will read in water depends on the specific conductivities of the salts making up the TDS. A way to achieve a more accurate mass balance would be to run the calculation using specific ion concentrations (such as the chloride ion concentration) instead of the TDS.

instrument. On a monthly schedule, the pH probe should be buffered according to the manufacturer's recommendations. This involves placing the probe into buffer solutions with particular pH values and calibrating the instrument to those values. Should it be found that the values are drifting every time the meter is checked, it is likely that the reference probe needs to be replaced, or have its potassium chloride solution replenished.

Flowmeters are best calibrated by diverting the flow downstream of the meter into a large calibrated container and checking its filling rate with a stopwatch. With large flowrates this can be impractical. Sometimes flow readings can be verified by temporarily installing another flowmeter in-line, or by replacing the meter with another comparable meter to check its reading. With paddlewheel-type flowmeters, devices are available that emulate a transducer signal in order to calibrate the meter independently of the paddlewheel. However, this calibration will only be as accurate as the paddlewheel's output signal, which typically is the most likely to go out of calibration.

Conductivity meters probes should be cleaned on a monthly basis. If a conductivity meter is not reading accurately according to a standardized solution, this is usually not critical as long as the readings do not drift. The relative readings over time are the ones most critical in monitoring the RO performance.

Pressure gauges can be calibrated and checked using a special pressure device designed for that purpose. Another option is to purchase a specially calibrated gauge that can be used to verify the reading of each pressure gauge in the system.

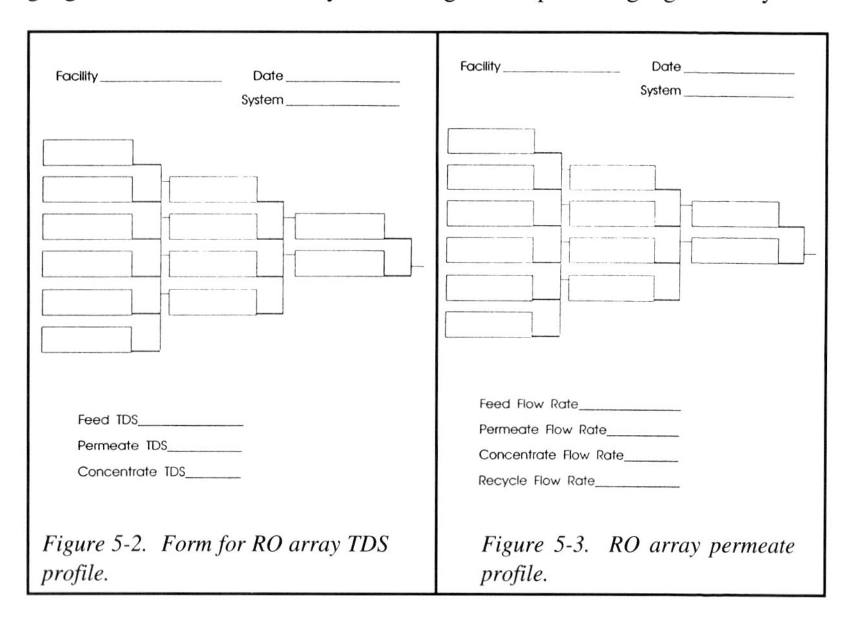

Figure 5-2. Form for RO array TDS profile.

Figure 5-3. RO array permeate profile.

Quick-connect pressure taps can be installed in the system at all the array interstage locations (Figure 5-4). In this manner, the same calibrated gauge can be used to easily check the pressure at all the critical locations. This insures that the differential pressure readings will be accurate.

Data normalization is a process whereby the RO system operating data is converted into a form that most accurately reflects the system performance. Compensation is made for operational variables so that changes in system performance are directly evident. The performance changes that can be monitored in this manner would include the following:

- Membrane fouling and/or scaling — the presence of foulants and/or scale directly on the membrane surface.
- Hydraulic plugging — the presence of material (possibly foulants or scale) that has ended up in the flow channel spacing between the membrane leaves of spiral-wound elements, or between the fibers in hollow-fiber elements.
- Membrane degradation — loss of the ability to remove dissolved salts due to a chemical change in the membrane's structure.
- Mechanical failure — the direct passage of salt into the permeate through mechanical lesions in the membrane, or via a broken O-ring or element glue line (the sealing on the edge of a spiral-wound membrane envelope).

These concerns can not occur without affecting the readings of the RO system

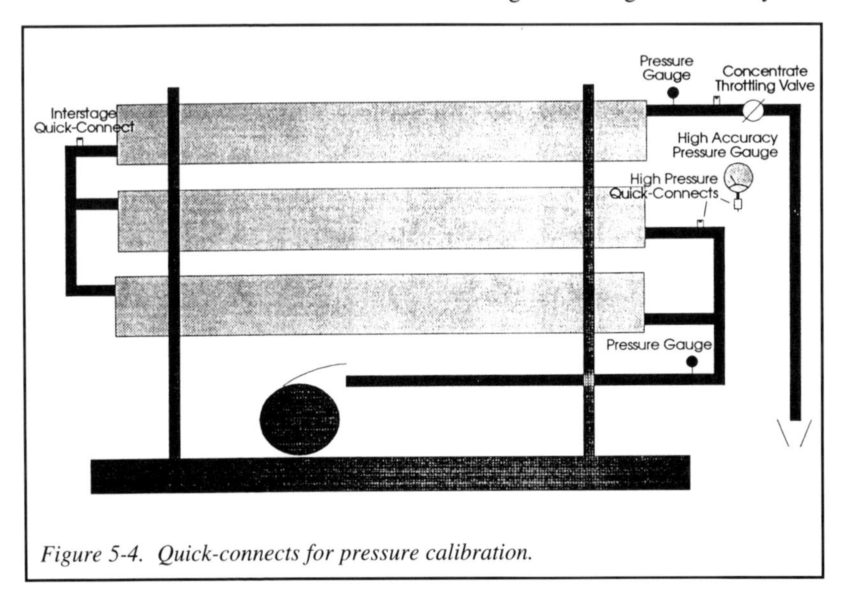

Figure 5-4. Quick-connects for pressure calibration.

instrumentation. Because many of the RO performance parameters are interdependent, however, it can be difficult to realize the full extent of changes in the performance of an RO system.

For example, if an RO membrane fouls, the permeate flowrate will decrease (unless the operator compensates by increasing the membrane feed pressure). Since less flow will pass through the centrifugal high-pressure pump and its throttling valve (if one exists), the pump will operate higher on its curve, thus putting out more pressure. This higher pressure will result in a higher permeate flowrate, although not as high as the permeate flowrate prior to the fouling.

If a throttling valve is in use downstream of the high-pressure pump, its pressure drop will be less with the lower flowrate. The result is that the membrane feed pressure increases as the permeate flowrate declines. Again, this tends to mask the full extent of the membrane fouling, since the higher operating pressure results in a higher permeate flowrate.

Normalizing data standardizes the various interdependent parameters to constant values, so that variation of the remaining variable reflects only changes in a primary parameter. The RO permeate flowrate can be modified for changes in pressures, salt concentrations, and water temperature so that it reflects only the state of the RO membrane and the effects of any material adhered to the membrane surface (i.e., inorganic scale or organic foulants).

The key RO performance parameters can be fully represented by three variables that are calculated from the RO operating data. Thus, the effects of four key concerns listed previously — membrane fouling and/or scaling, hydraulic plugging, membrane degradation, and mechanical failure — can be directly monitored. The three variables are the following:

- Salt rejection
- Normalized differential pressure
- Normalized permeate flowrate

Salt rejection, since it reflects RO permeate quality, is the most widely known method of monitoring the performance of an RO system. Unfortunately, many problems that eventually show up as a loss in salt rejection could have been noticed and corrected sooner by monitoring other parameters. It is often possible that a problem could have been resolved prior to a system ever suffering a significant loss in salt rejection.

However, salt rejection is still a good method of monitoring whether or not an RO is doing its basic job of removing salts. Salt rejection is calculated as follows:

$$\%\ \text{salt rejection} = (\text{membrane TDS} - \text{permeate TDS}) \div \text{membrane TDS} \times 100$$

where:

membrane TDS is some measurement of the salt concentration on the pressurized side of the RO membrane, and
permeate TDS is some measurement of the salt concentration of the permeate stream.

The membrane TDS will change as the feedwater salts are concentrated within an RO system. There are two different methods commonly used to calculate salt rejection, and they reflect different ways to interpret a value for the membrane TDS.

One method utilizes the RO feedwater concentration for the membrane TDS. This practice neglects the effects of the specific RO permeate recovery. A salt rejection calculated in this manner will be lower than the actual individual element salt rejection. The extent of the variation will depend on the recovery of the RO system (i.e., how concentrated the salts become in the RO concentrate stream).

The other commonly used method uses a mathematical average of the feed and concentrate TDS to approximate the average TDS within the RO system. Values obtained using this method will more closely match the individual element salt rejection values. This method will also normalize for changes in salt rejection that occur due to changes in the RO permeate recovery. With the previous method, an increase in permeate recovery would likely result in a decline in the calculated salt rejection, even though the membrane performance had not really changed.

Thus, an average feed TDS is suggested as a more accurate way to calculate salt rejection:

$$\text{average feed TDS} = (\text{membrane feed TDS} + \text{concentrate TDS})/2$$

And the salt rejection calculation becomes:

$$\%\ \text{salt rejection} = (\text{average feed TDS} - \text{permeate TDS}) \div \text{average feed TDS} \times 100$$

If the concentrate TDS has not been measured, it can be estimated using the permeate recovery of the system where recovery is expressed as the permeate flow fraction of the feedwater flowrate:

$$\text{concentrate TDS} = \text{membrane feed TDS} \times [1/(1 - \text{recovery fraction})]$$

This equation can be substituted in the previous equation with the following result:

$$\text{average feed TDS} = \text{membrane feed TDS} \times [1 + 1 \div (1 - \text{recovery})] \div 2$$

Rejection of individual ions. Since the rate of rejection varies for each of the particular salts in the feedwater, a variation in the makeup of an RO feedwater will result in a change in the overall percent rejection of the TDS. For better accuracy, individual ion concentrations can be used in the equation instead of the TDS. At comparable operating conditions (i.e., similar pressures and flowrates), the individual ion rejection will remain constant unless something has occurred to affect the membrane performance.

When a system is started up with a new membrane, it is a good idea to record an individual ion rejection in order to have a basis for future performance comparison, should a concern arise. A rejection based upon a monovalent ion such as chloride or sodium will offer a stable (except for membrane performance changes) yet sensitive basis for future membrane performance comparison.

A rejection calculated using a divalent ion such as calcium can be used to distinguish between a mechanical leak in the system and membrane deterioration. A mechanical leak in a membrane, spiral-wound glue line, hollow fiber, or O-ring will result in a comparable drop in rejection for both monovalent and divalent ions, whereas the rejection decline will be more severe for monovalent ions in the case of membrane deterioration.

Conductivity readings can be substituted for TDS values in the salt rejection equation. This will usually result in a rejection that is lower than if TDS values were used in the calculation. This is due to the changing ratio between TDS and conductivity over the range of values between the feed and permeate concentrations.

Hydraulic pressure differential has also been called the system pressure drop, the "delta p," and sometimes the differential pressure. It is the difference between the pressure (called the feed pressure, primary pressure, or membrane pressure) of the RO feedwater as it enters the first membrane elements, and the pressure of the water leaving the final membrane elements (called the concentrate or brine pressure). The pressure differential is a measure of the pressure lost as the water passes through the flow channels of all the elements in the system. It should not be confused with the pressure difference between the feed pressure and the permeate pressure.

The pressure differential is a way of monitoring the resistance to flow through the RO system. At constant flowrates, an increase in the pressure differential indicates that something is blocking the flow. This might be physical debris that has broken through the system prefilters, pump shavings from the high-pressure pump, or scale or biofilm particulates. The telescoping (downstream unraveling) of spiral-wound elements can also cause an increase in the hydraulic pressure differential.

The system pressure differential is a function of the permeate and concentrate flowrates. Since these rates may vary from day to day for most RO systems due to variation in water temperature or some other changing parameter, it may be difficult to directly compare the system pressure differentials.

For many spiral-wound RO arrays, the following formula can be used to normalize the system pressure differential for variation in flowrates:

$$\text{normalized differential pressure} = \text{psid}_{actual} \times \frac{(2 \times \text{concentrate flow}_{dsgn} + \text{permeate flow}_{dsgn})^{1.5}}{(2 \times \text{concentrate flow} + \text{permeate flow})^{1.5}}$$

where:

psid_{actual} is the difference between the feed pressure and the concentrate pressure;
concentrate flow_{dsgn} is the RO concentrate flowrate at start-up, or is simply a value typical of the operation of the particular system; and
permeate flow_{dsgn} is the RO permeate flowrate at start-up, or is simply a value typical of the operation of the particular system (possibly the sum of the design individual permeate flowrates of all the elements).

For systems with varying flowrates, the equation is useful for enabling the direct comparison of system differential pressure under different conditions. A percent change in normalized differential pressure since system start-up can be calculated, and can be used as a basis for judging when to clean an RO system that is suffering from blockages in the element flow channels.

Normalized permeate flowrate is probably the most important monitoring parameter for an RO system, as it best reflects changes in the RO membrane performance. If the membrane degrades, the normalized permeate flowrate will typically increase. If the membrane fouls, the normalized flowrate will decrease.

From the initial discussion of reverse osmosis theory, we know that the permeate flow through a particular RO membrane is proportional to the membrane pressure minus the osmotic and permeate pressures. This is at constant temperature. Normalizing the permeate flowrate to a set of standard conditions is a matter of taking into account the effects of the various variables on the combined RO permeate flowrate.

Normalizing for the effects of pressure, temperature, and solute concentration (osmotic pressure) on permeate flowrate will cause the resulting flow value to reflect only changes that are directly due to characteristics of the RO membrane, the membrane surface, or the integrity of the membrane elements. Thus the normalized permeate flowrate can be used to monitor the following:

1. The extent of fouling/scale formation on the membrane surface;
2. Membrane compaction;
3. The integrity of the membrane system (i.e., mechanical leaks in the system such as in O-rings); and
4 The extent of membrane deterioration (should it occur).

Regular monitoring of the normalized permeate flowrate will allow the direct comparison of operating data from different days and under different operating conditions. In this manner, the full extent of fouling, scaling, or membrane compaction can be judged. If some sort of membrane deterioration or mechanical bypass is occurring, it will be evident as an increasing normalized permeate flowrate.

In Chapter 1, the following relationship between the RO permeate flowrate and other variables was discussed:

$$\text{permeate flux} = K_T \times (\text{membrane psid - osmotic psid})$$

where:
permeate flux is the flowrate through the membrane per unit of membrane area,
K_T is some constant determined by the membrane and is a function of temperature,
membrane psid is the difference between the pressures on each side of the membrane (i.e., the head pressure), and
osmotic psid is the difference between the osmotic pressures of the solute on each side of the membrane.

Membrane pressure. The previous equation can be used to relate the permeate flux to the difference between the membrane pressure and the permeate pressure at some point in a membrane system. In a full-scale RO system, the membrane pressure decreases because of hydraulic pressure losses as the water flows through the membrane elements. The lower membrane pressures in the tail end elements result in lower permeate flowrates from the tail end elements. An arithmetic average pressure can be used to compensate for the effect on the permeate flow of this declining membrane pressure:

$$\text{average membrane psi} = (\text{membrane feed psi - concentrate psi}) \div 2$$

Permeate pressure. The RO permeate pressure will be negligible if the RO permeate stream is plumbed to an open tank. However, if the permeate is plumbed directly into a water distribution system or to another treatment process, its pressure may not be negligible.

For example, if the RO permeate is plumbed directly into a deionization column that gives the water a 50-psig back pressure, the 50 psig should be subtracted from the membrane pressure to determine a true membrane driving pressure.

Osmotic pressure. In most reverse osmosis applications, the osmotic pressure of the permeate will be negligible when compared to the osmotic pressure on the feedwater side of the membrane. This will be true if the overall salt rejection is better than 90%. In such cases, the permeate water osmotic pressure can be neglected, leaving just the average osmotic pressure on the feedwater side of the membrane to be considered.

As pure water is permeating through the membrane, the remaining solution gets more and more concentrated as the feedwater passes through the RO system; thus, the osmotic pressure of the solution will increase as the water goes through the system. Arithmetic averaging can be used to approximate an average osmotic pressure for the entire RO system.

Such an approximation is accurate enough for systems with low osmotic pressures (concentrate osmotic pressures of less than 100 psig). However, systems with a high recovery and/or a high feed concentration should have the osmotic pressures of each stage (bank) of the system evaluated independently for greater accuracy.

Each solute in the feedwater will have a different osmotic pressure. Each of these individual osmotic pressures has an additive effect in creating the solution osmotic pressure. For example, sodium chloride has an osmotic pressure of about 0.011 psig/ppm; while sodium sulfate has about half that osmotic pressure, around 0.0057 psig/ppm. Generally, the ions with lower valence charges will have greater osmotic pressures. For organic molecules, the smaller the molecule, the greater the osmotic pressure it will generally contribute to the solution.

In applications where a solute is not well rejected by the RO membrane, the osmotic pressure of the permeate water must be considered in the permeate flow equation. Based on the concentration of the solute in the permeate, the osmotic pressure should be calculated, and then subtracted from the osmotic pressure on the membrane side in order to determine the osmotic pressure differential across the membrane.

For most municipal water sources, the osmotic pressure in the feedwater can be estimated by multiplying the feed TDS concentration by 0.01 psig/ppm. An average osmotic pressure for the system would be calculated by first calculating the average TDS of the system:

$$\text{average TDS} = (\text{feed TDS} + \text{concentrate TDS}) \div 2$$

The average osmotic pressure is then as follows:

$$\text{average osmotic pressure} = \text{average TDS} \times 0.01 \text{ psig/ppm}$$

For many municipal waters with a feedwater TDS of less than 400 ppm, the average osmotic pressure is small enough (less than 10 psig) that it can be neglected in the calculations.

Temperature correction factor. In the previous equation, the permeate flowrate is proportional to the membrane driving pressure at constant temperature. Since water temperature will vary in many treatment systems, its effect must be taken into account in a normalized permeate flow calculation.

The membrane manufacturers have different means of compensating for temperature variation. Some use data tables and others use equations to derive a temperature correction factor. This factor is multiplied by the permeate flowrate to estimate a permeate flow that the RO would have produced if the water temperature were 77 °F (25 °C). This number will be greater than the actual permeate flowrate if the actual temperature was less than 77 °F, and less than the actual rate if the temperature was greater than 77 °F.

With a temperature correction factor, an average membrane pressure, an average osmotic pressure differential, and a permeate pressure for an RO system, the equation for the normalized permeate flow of the system becomes as follows:

$$\text{normalized permeate flowrate} = K \times F_{77} \times (\text{average membrane psig - average osmotic psid - permeate psig})$$

where:
K is some constant determined by the membrane,
F_{77} is a factor to compensate for temperature,
average membrane psig is an arithmetic average of the feed and concentrate membrane pressures, and
average osmotic psid is the difference between the osmotic pressure on the feed side of the membrane and that of the permeate.

The constant, K, can be eliminated from the expression by using a ratio between the actual pressures and some fixed pressure values. These values can be based on the original system start-up values, or on the system design values. For example, if at start-up an RO system was producing a 100-gpm permeate flowrate at 70 °F (temperature correction factor of 0.9), with a feed pressure of 250 psig, a pressure differential of 50 psig, and negligible osmotic and permeate pressure, the normalized flow equation then becomes as follows:

$$\text{normalized permeate} = \frac{F_{77} \times (\text{average membrane psig - average osmotic psid - permeate psig})}{F_{77o} \times (\text{average membrane psig}_o \text{ - average osmotic psid}_o \text{ - permeate psig}_o)} \times \text{actual permeate flowrate}$$

where:
F_{77} is a factor to determine the approximate permeate flow for operation at 77 °F
average membrane psig is a feed/concentrate membrane pressure average;
average osmotic psid is the difference between the osmotic pressure on the feed side of the membrane and the osmotic pressure of the permeate;
permeate psig is the permeate pressure;

F_{77o} is a factor to determine the approximate permeate flow for start-up operation at 77 °F;
feed psig$_o$ is the membrane feed pressure at start-up
average membrane psig$_o$ is the start-up average system membrane pressure;
average osmotic psid$_o$ is the difference between the osmotic pressure on the feed side of the membrane and the osmotic pressure of the permeate at start-up; and
permeate psig$_o$ is the permeate pressure at start-up

$$= F_{77} \times (\text{average membrane psig}) \div (0.9 \times 250 \text{ psig} - 50 \text{ psid} \div 2)$$
$$= F_{77} \times (\text{average membrane psig}) \div 202.5 \text{ psig}$$

This equation can now be used to calculate a normalized permeate flowrate that can be directly compared to the permeate flowrate at start-up. Any difference from the start-up flowrate would be due to compaction, fouling and/or scaling, membrane deterioration, or a mechanical leak of some sort (in the membrane elements or in the system). (Note: New membranes should be operated for at least 30 minutes before recording this start-up data.)

It may be desirable to compare the actual permeate flowrate of a membrane element to the design permeate flowrate for that element. In this case, typical design values for a single brackish-water thin-film (spiral-wound) element would be 225 psig of operating pressure and 20 psig of osmotic pressure. Typical values for a cellulose acetate membrane element would be 420 psig of operating pressure and 20 psig of osmotic pressure. Most manufacturers test their elements with a 2,000-ppm sodium chloride solution that has an osmotic pressure of roughly 20 psig. Design temperature would be 77 °F (temperature correction factor of 1.0). The permeate pressure would be zero for a design element, and the pressure differential would be negligible.

For a typical CA membrane element, the preceding equation (taking into account the possibility of a significant osmotic pressure) now becomes as follows:

$$\text{normalized permeate} = \frac{F_{77} \times \text{actual permeate flowrate} \times (\text{average membrane psig} - \text{average osmotic psid} - \text{permeate psig})}{(\text{feed psig}_D - \text{osmotic psid}_D)}$$

where:
F_{77} is a factor to determine the approximate permeate flow for operation at 77 °F
average membrane psig is the average membrane pressure;
average osmotic psid is the difference between the osmotic pressure on the feed side of the membrane and the osmotic pressure of the permeate;
permeate psig is the permeate pressure;
feed psig$_D$ is the membrane design pressure; and
osmotic psid$_D$ is the osmotic pressure of the design test solution.

$= F_{77} \times$ (average membrane psig - average osmotic psig - permeate psig) $\times$ actual permeate flow $\div$ (225 psig - 20 psig)

$= F_{77} \times$ (average membrane psig - average osmotic psig - permeate psig) $\times$ actual permeate flow $\div$ (205 psig)

Example:
In this example, it is desired to determine how an RO system performs as compared to its design permeate flowrate. The design permeate flow will be based upon the manufacturer's rated design performance for the individual elements.

Basis:

Design pressure of 420 psig with a 2,000-ppm NaCl feed
Design permeate flow of 5 gpm/element, 97.5% rejection
Twenty 8-inch-diameter spiral-wound elements in the system
Start-up concentrate flow of 33 gpm
Start-up permeate flow of 100 gpm @ 75 °F
Start-up membrane feed pressure of 410 psig
Start-up concentrate pressure of 340 psig
Start-up permeate TDS of 16 ppm
Current permeate flowrate of 85 gpm
Current concentrate flowrate of 28.3 gpm
Current membrane feed pressure of 450 psig
Current concentrate pressure of 380 psig
Municipal water with a TDS of consistently 350 ppm
RO permeate TDS of 15 ppm
Negligible permeate pressure (permeate goes to open tank)
Water temperature of 70 °F

Answer:
Salt rejection:

% rejection = (average feed TDS - permeate TDS) $\times$ 100/average feed TDS

where:

average feed TDS = (membrane feed TDS + concentrate TDS) $\div$ 2

The concentrate TDS is not given. Although it would be more accurate to have the actual measurement, the concentrate TDS can be estimated based upon the permeate recovery (since the membrane rejection rate is high enough to result in a negligible salt concentration in the permeate stream as compared to the feed and concentrate streams).

$$\text{Recovery} = \text{permeate flow/feed flow} = 85 \text{ gpm} \div (85 \text{ gpm} + 28.3 \text{ gpm}) = 0.75$$

$$\text{Average feed TDS} = 350 \times (1 + 1 \div (1\text{-}0.75)) \div 2 = 875 \text{ ppm}$$

$$\text{Salt rejection} = (875 - 15) \times 100 \div 875 = 98.3\%$$

This calculation can also be performed to obtain the start-up salt rejection:

$$\text{start-up salt rejection} = (875 - 16) \times 100/875 = 98.2\%$$

Normalized Differential Pressure

$$\text{differential pressure} = \text{membrane feed psi} - \text{concentrate psi}$$
$$= 450 \text{ psig} - 380 \text{ psig} = 70 \text{ psig}$$

$$\text{Normalized differential pressure} = \frac{\text{differential pressure} \times (2 \times \text{start-up concentrate flow} + \text{start-up permeate flow})^{1.5}}{(2 \times \text{Concentrate flow} + \text{Permeate flow})^{1.5}}$$

$$= 70 \text{ psig} \times (2 \times 33 \text{ gpm} + 100 \text{ gpm})^{1.5} \div (2 \times 28.3 \text{ gpm} + 85 \text{ gpm})^{1.5} = 89 \text{ psig}$$

Normalized permeate flowrate

Comparison with element design flowrates:

$$\text{Normalized permeate} = \frac{F_{77} \times \text{actual permeate flowrate} \times (\text{average membrane psig} - \text{average osmotic psig} - \text{permeate psig})}{(\text{feed psig}_D - \text{osmotic psid}_D)}$$

$$\text{Temperature correction factor} = 1.11 \text{ (from Appendix)}$$
$$\text{Average membrane pressure} = (450 \text{ psig} + 380 \text{ psig}) \div 2 = 415 \text{ psig}$$

Although the average osmotic pressure will probably be so small it could be neglected, for the purposes of this example its effect will be included. It can be calculated using the previously determined average feed TDS. Because the RO membrane has high salt rejection characteristics, the osmotic pressure of the permeate stream will be less than 1 psig and can be neglected.

$$\text{Average osmotic pressure} = 0.01 \text{ psig/ppm} \times \text{average TDS}$$
$$= 0.01 \text{ psig/ppm} \times 875 \text{ ppm} = 8.75 \text{ psig}$$

The permeate flowrate normalized to design element conditions can now be calculated:

$$\text{normalized permeate} = \frac{1.11 \times (415 \text{ psig} - 8.75 \text{ psig}) \times 85 \text{ gpm}}{} = 95.8 \text{ gpm}$$

(420 psig - 20 psig)

If this flow is divided by the number of elements, it can be directly compared with the individual element permeate flowrate according to the membrane manufacturer.

$$\text{Normalized element permeate flow} = 96.8 \text{ gpm} \div 20 \text{ elements} = 4.84 \text{ gpm/element}$$

Comparison with start-up permeate flow:

$$\text{Normalized permeate} = \frac{F_{77} \times \text{actual permeate flowrate} \times (\text{average membrane psig - average osmotic psid - permeate psig})}{F_{77o} \times (\text{average membrane psig}_o \text{ - average osmotic psid}_o \text{ - permeate psig}_o)}$$

The temperature correction factor for the 75 °F water temperature at start-up is 1.03.

The start-up average membrane pressure is (410 psig + 340 psig) ÷ 2 = 375 psig.

The start-up osmotic pressure will be the same as the actual osmotic pressure since the feedwater TDS and the permeate recovery have not changed. The permeate flowrate normalized to start-up conditions can now be calculated:

$$\text{normalized permeate} = \frac{1.11 \times (375 \text{ psig} - 8.75 \text{ psig}) \times 85 \text{ gpm}}{[1.03 \times (420 \text{ psig} - 70 \text{ psig}/2 - 8.75 \text{ psig}]} = 98.9 \text{ gpm}$$

	Current	*Start-up*	*Design*
Salt rejection	98.3%	98.2%	97.5%
Normalized differential psid	89 psid	70 psid	
Normalized permeate flow	98.9 gpm	100 gpm	
Normalized element permeate	4.84 gpm		5.0 gpm

Now the performance parameters can be directly compared to see if any changes have occurred in the system performance. The normalized permeate flowrate can be compared against either the normalized permeate flowrate at start-up or against the individual element design specifications. As most membrane systems do not perform exactly according to their design specifications, it is usually more valuable to track the system performance by comparing the data to the start-up data.

The salt rejection in the example is slightly higher than both the design rejection and the start-up salt rejection. The slight increase since start-up is not significant. Such increases in rejection are not unusual for new RO systems.

As compared to the 100-gpm start-up permeate flow, the normalized permeate flow is 1.1% low [(100 gpm - 98.9 gpm) ÷ 100 gpm]. As compared to the element design permeate flow, the flow is low by 3.2%. This difference in flow is not significant.

The normalized differential pressure has increased by 27%. This is significant and indicates that something is probably blocking the flow channels within the membrane elements. The cause of this increase should be further investigated, and cleaning might be required.

Graphing. Due to inaccuracies in instrumentation as well as in how the instruments are read, there will be some fluctuation in the normalized readings. It is critical to discern whether these fluctuations are truly due to inaccuracy in data collection, or whether they represent a trend in system performance. This can best be accomplished by graphing the normalized data against time.

If operating data is being collected on a daily basis, the normalized permeate flowrate, normalized differential pressure, and salt rejection can be graphed on the *y* axis against the respective day on the *x* axis. Curves or lines can then be drawn over the data to show performance trends.

A decreasing trend in normalized permeate flowrate, or an increasing trend in the normalized pressure differential, can be graphically followed to determine when cleaning will be necessary. An increasing trend in the normalized permeate flow will be graphically visible if a membrane deterioration problem is occurring. If the salt rejection is declining, it is an indicator either of a fouling problem, or of a membrane deterioration problem. Graphing all three parameters will make it easy to correlate the parameter trends. If the normalized permeate flow is increasing at the same time the salt rejection is declining, a membrane deterioration problem or some sort of mechanical bypass is likely occurring.

The raw operating data from one particular RO system is graphed in Figures 5-5a to 5-5g. The salt rejection, normalized permeate flowrate, and system pressure differential are shown in Figures 5-6a to 5-6c. The RO system performance trends are much easier to track in the second set of graphs. In these, the extent of fouling can be clearly determined. It is also clear that the cleaning was successful at restoring the baseline performance (assuming that the initial data on the graph indicates the clean-membrane baseline performance).

The membrane manufacturers will provide guidelines for the extent of performance change tolerable before it is necessary to clean a system. It is suggested that these guidelines be followed closely, as the probability of a successful cleaning is much greater if the system is cleaned according to these guidelines.

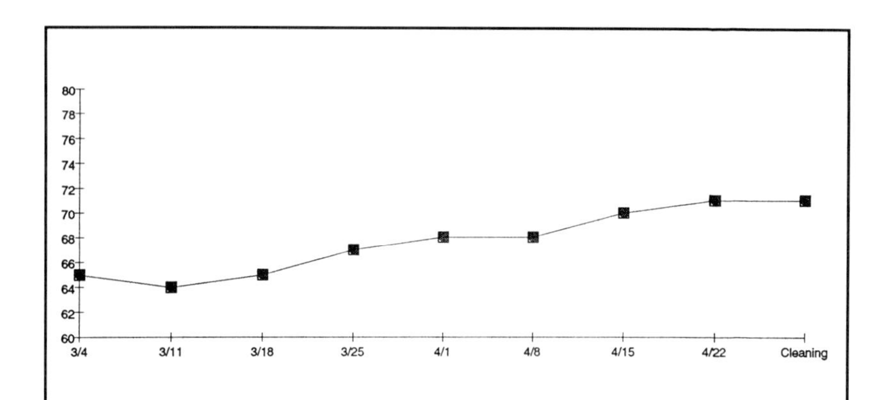

Figure 5-5. Raw operating data from one RO system. a. Temperature.

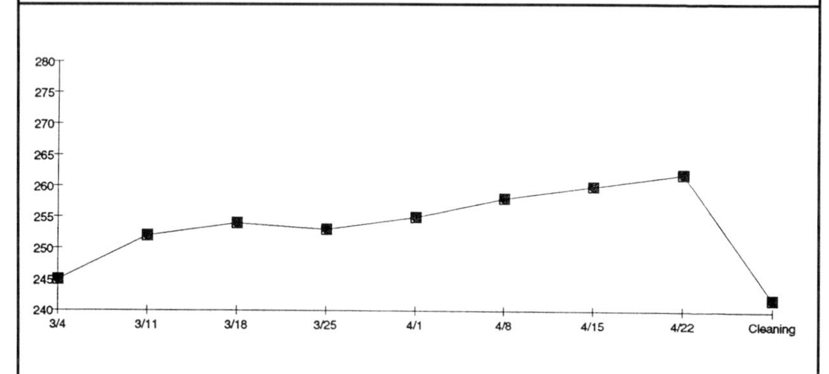

b. Membrane feed pressure.

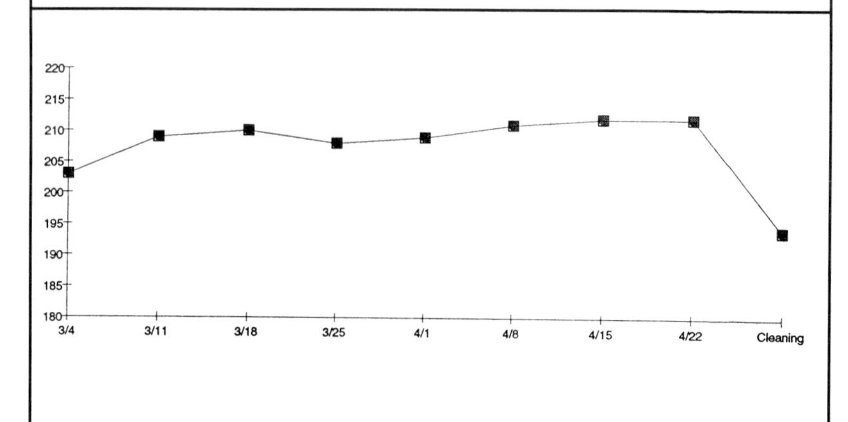

c. Concentrate pressure.

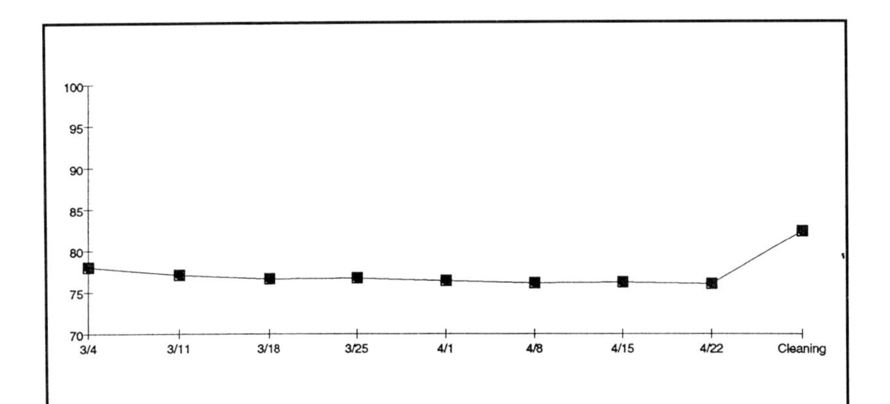

Figure 5-5. Raw operating data from one RO system. d. Permeate flowrate.

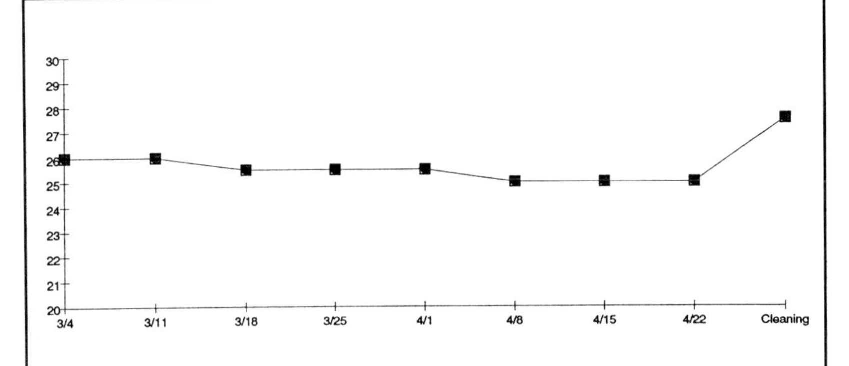

e. Concentrate flowrate.

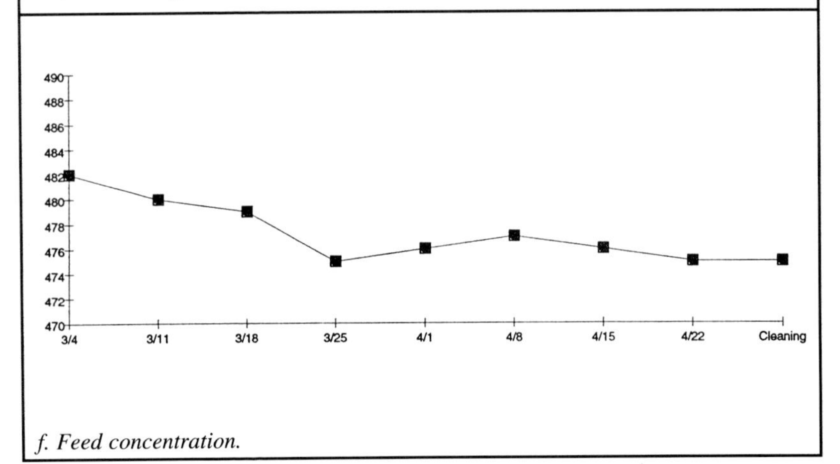

f. Feed concentration.

Cleaning

Cleaning frequency. Fouling is a normal phenomenon of most RO systems. With the correct cleaning frequency, most foulants can be removed from the membrane; thus, it makes no difference what the fouling rate is, as long as original performance is restored after the fouling/cleaning cycle. Even RO systems that experience a high rate of fouling can achieve an extended membrane life expectancy if the system is cleaned when required.

A foulant layer on the membrane surface or in the membrane element flow channels may affect the ability of salts to be re-diluted in the bulk stream. In other words, foulant layers can increase the concentration polarization of salts at the membrane surface. If this is occurring, one of the symptoms will be a reduction in salt rejection (i.e., an increase in the RO permeate salt concentration).

When foulant layers increase the concentration polarization, there is an increased likelihood that scale formation will occur. For RO systems operating at recoveries whereby the solubilities of sulfate or silica salts have been exceeded, a potentially dangerous symptom is a decrease in salt rejection coinciding with a drop in normalized permeate flow or with an increase in pressure differential. This is a sign that an increase in the concentration of these salts is occurring on the membrane surface. If sulfates or silica should happen to precipitate, they can be extremely difficult to clean. Cleaning with the appropriate solution should be performed as soon as sulfate or silica scale is suspected.

Many foulants, particularly clay-type soils, can compact with time as the foulant layer increases in depth. As the foulant compacts, it will become more difficult to get back into solution during cleaning.

In cases of a potentially compacting foulant, or in cases of RO systems

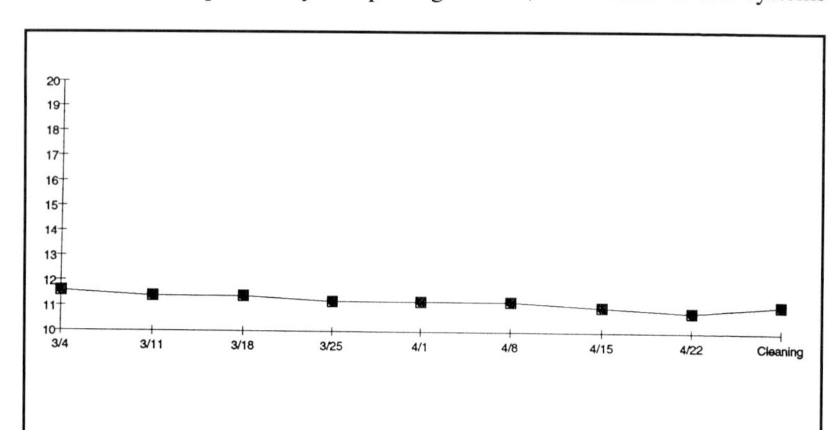

Figure 5-5. Raw operating data from one RO system. g. Permeate concentration.

operating with supersaturated sulfates or silica, cleaning prior to a significant change in membrane performance is critical. The membrane manufacturer should be consulted for specific recommendations for these types of conditions.

Calcium carbonate scale can readily go back into solution at lower pH. This can create on the membrane surface a local pH condition of about 10.5. This high pH can cause a fairly rapid increase in the rate of hydrolysis of CA membrane. If the presence of calcium carbonate scale is suspected, the membrane should be cleaned as soon as possible. Fortunately, it is relatively easy to clean calcium carbonate from a membrane surface with an acidic cleaning solution.

If an increasing pressure differential is allowed to proceed without cleaning, the eventual result can be a blockage of flow channels between the membrane leaves of a spiral-wound element, or around the fibers of a hollow-fiber element. If the channels are blocked, the ability to get cleaning agents into the fouled areas is reduced. Most of the cleaning agent will divert around the severely fouled area, and it may not be possible to ever get the membrane element clean using standard methods.

Membrane manufacturers will provide guidelines for the extent of performance change tolerable before the need to clean a system. If manufacturer's guidelines are not available, the following can be used as the maximum amount of performance change to allow before cleaning for standard foulants:

1. Loss of 10 to 15% in normalized permeate flowrate
2. Increase of 10 to 15% in normalized differential pressure
3. Decrease of 1 to 2% in salt rejection

Most foulants will affect the normalized permeate flow or differential pressure more significantly than they will affect the salt rejection. For most systems, the salt rejection decline guideline is offered more as a safety measure to insure that calcium carbonate formation is not occurring with CA membrane, or that scale formation is not proceeding beyond the ability to clean it.

The percentages given above are relative to the start-up values after an initial membrane break-in period. With CA membrane systems, there will be some loss of flow due to compaction. The majority of this compaction occurs during the first 200 hours of operation. Some loss of flow may occur for both CA and PA membrane due to foulant plugging of some lower-velocity areas of the membrane element. Typically, from 10% to 15% of the normalized permeate flow can be permanently lost during the first few days of operation. After this period of breaking in, baseline membrane performance will tend to be more stable (McPherson, March 1981).

Compaction can occur more quickly any time a CA RO membrane is operated at greater than design pressures. The membrane's performance will be permanently impaired if operated at pressures greater than 550 psig for an extended

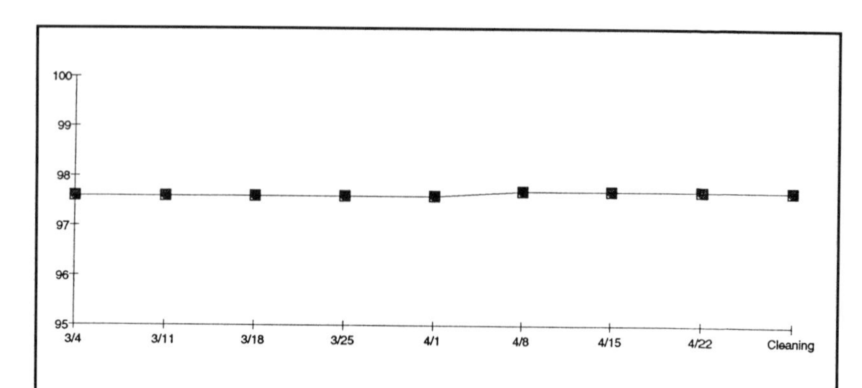

Figure 5-6. Salt rejection, normalized permeate flowrate, and system pressure differential. a. Percentage of salt rejection.

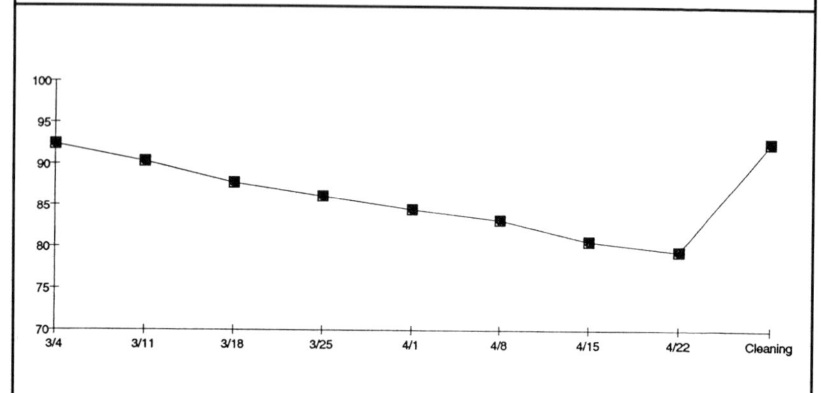

b. Normalized permeate flowrate.

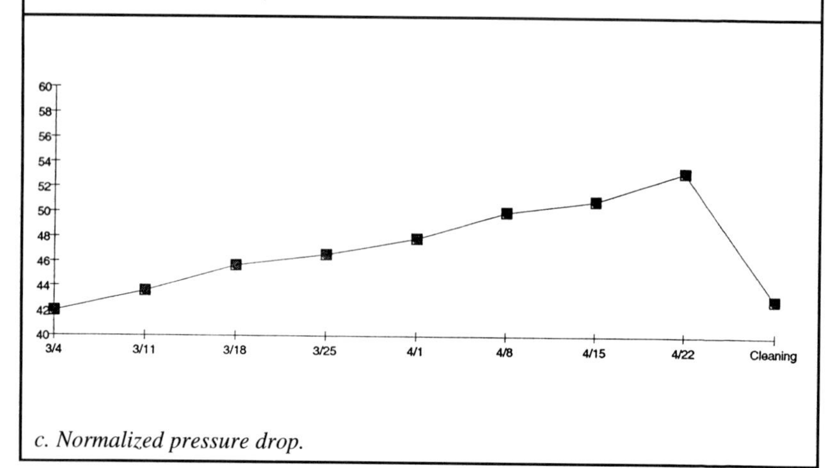

c. Normalized pressure drop.

period of time. In contrast, a PA membrane tends to have greater structural strength, and can be operated at higher than normal pressures with little concern about compaction. However, it is important that the permeate flux (i.e., permeate flow per unit of area) be kept at or below the manufacturer's guidelines. Otherwise, a rapid increase in membrane fouling rate can occur.

Different cleaners for different foulants. The performance variable most significantly affected when an RO fouls is an important clue to the type of foulant/scale that caused the change in performance. Each foulant/scale component will be best removed by a cleaning solution or cleaning regimen that best fits the particular solubility characteristics of that component.

Frequently, foulant/scale formation on the membrane surface will be in varying layers. These layers are best removed if the appropriate solution is used first for cleaning the top layer. Once this is removed, it will be easier to remove the next layer using *its* appropriate solution. For difficult foulant/scale layers, it may be necessary to use a low pH (acidic) cleaning solution, followed by a high pH (alkaline) solution, followed by another low pH solution (with rinsing between cleaning steps). Or it may be necessary to start with an alkaline solution before using an acidic solution, all depending on the nature of the foulant/scale layers.

With smaller RO systems, the determination of the optimum cleaning regimen is sometimes accomplished by trial and error. If the incoming water analysis indicates high iron or manganese, a cleaning solution formulated for cleaning metal precipitants is used. If the water is high in suspended solids, or if high bacteria counts indicate possible fouling due to biological activity, an alkaline cleaning solution might be used. If a cleaning solution is ineffective, some other approach may be taken. If a solution is only partially effective, a two-step cleaning procedure may be in order.

For larger RO systems, trial and error is not a practical process for finding a successful cleaning regimen. It is more cost effective to analyze a fouled element to investigate the nature of the foulant layers. A cleaning procedure can then be devised with a greater chance for success. A supplier of cleaning chemicals may be able to provide this service.

Inorganic Scale

Salts may crystalize and fall out of solution in an RO membrane element because their concentrations have exceeded their solubilities. Certain metals, as well as hydrogen sulfide, will fall out of solution if they become oxidized. The metals may also come out of solution if the pH of the water increases. It is usually possible with proper pretreatment to control or prevent such scale formation or fouling from occurring.

If fouling by inorganic scale should occur, it will affect the RO membrane performance. The difficulty in cleaning the scale from the membrane will depend on the nature of the scale.

Carbonate scale. Calcium carbonate scale formation tends to occur on the membrane surface where the salts are most concentrated. It shows up as a loss in normalized permeate flow unless the membrane is CA and hydrolysis is simultaneously occurring. Frequently, it may also cause a slight decline in salt rejection. The increased passage of salt is the result of the ability of calcium carbonate to form an equilibrium reaction with the water. Some of the calcium carbonate is continually going into and out of solution. This affects the membrane as if a large concentration of calcium and carbonate ions were present at the membrane surface; thus, the membrane tends to pass a greater concentration of the ions through the membrane.

Calcium carbonate scale will readily go back into solution at lower pH. Just about any acidic cleaning solution will be successful at redissolving calcium carbonate scale. However, it is common for the calcium carbonate scale to catalyze the precipitation of calcium sulfate scale because of the locally high concentrations of calcium that occur around the scale formations. Cleaning a carbonate salt complex may require a more aggressive acidic cleaning formulation.

Calcium carbonate scale can quickly hydrolyze a CA membrane. If its presence is suspected in a CA system, the RO should be cleaned as soon as possible. An acidic cleaning solution should be recirculated through the elements. The presence of a surfactant in the cleaning solution formulation, by reducing the surface tension of the foulant layer, will aid in wetting out the scale layer with the acidic solution.

Sulfate scale. Calcium, barium, or strontium sulfate can be dispersed into solution by an acidic cleaner if the structure of the scale is not extremely pure. If the crystalline structure is interrupted by other scale, organics, or scale inhibitor, there will be a greater likelihood of removing it. The other materials will more easily go back into solution, thus breaking up the scale/foulant matrix. This will make it possible to disperse the sulfate scale particles into the cleaning solution.

Pure formations of sulfate scale may not affect the RO salt rejection or the normalized permeate flowrate. Depending on where the scale formation occurs within the membrane element, sulfate scale may behave like an inert blockage to flow. In this case, the effect on the RO performance data will be an increase in the system pressure differential. If the sulfate scale is sufficient to affect the flow patterns at the membrane surface, effects will be apparent in the salt rejection and/or the normalized permeate flowrate.

Severe calcium sulfate scale can sometimes be coaxed back into solution using a special cleaning solution that contains high carbonate concentrations. This solution also contains ethylenediaminetetraacetic acid (EDTA). It works based on the following salt transformation on the membrane surface:

$$CaSO_{4(s)} + Na^{+} + HCO_3^{-} <====> CaCO_{3(s)} + Na^{+} + H^{+} + SO_4^{-2}$$

$$CaCO_{3(s)} + Na^{+} + H^{+} + SO_4^{-2} <====> Ca^{+2} + HCO_3^{-} + Na^{+} + SO_4^{-2}$$

The dissolved carbonate salt will be in the form of bicarbonate ions. The bicarbonate can displace the sulfate in the calcium sulfate scale, allowing the sulfate to go into solution. The chelating properties of the EDTA can then be effective at getting the calcium carbonate into solution. This cleaning solution is also effective with light strontium sulfate scale.

There is nothing currently available that is effective at redissolving heavy barium sulfate scale and that is also compatible with an RO membrane element. If such scale formation occurs, the replacement of the membrane elements will probably be required. It is therefore critical that barium sulfate precipitation be prevented, which (fortunately) most scale inhibitors are very effective at doing.

Iron can be present in water in its ferrous (Fe^{+2}) or ferric(Fe^{+3}) state. Ferric iron is relatively insoluble in water, whereas ferrous iron is fairly soluble except at high pH. The iron will be oxidized to the ferric state if chlorine or some other oxidizing agent is present in the water. If this oxidation occurs upstream of media filters, the larger iron particles can be filtered.

If small iron particles happen to break through the upstream filters and settle in the RO system, they will tend to show up in the RO operating data as an increase in the system pressure differential. Most of this increase will be localized in the lead end membrane elements, unless the iron is suspended with other contaminants. In this case the suspended complex will tend to come out of solution as it is concentrated within the RO system.

Light iron fouling can be readily cleaned using a standard acidic or EDTA-based cleaning solution. The mechanisms of iron removal with these solutions are chelation and dispersion.

With heavy iron precipitation, a standard acidic or EDTA-based solution may not be successful at removing all of the iron. Iron is particularly difficult to remove if the precipitation occurs directly on the membrane surface, forming a crystalline structure on the membrane. This may show up as a decrease in the normalized permeate flowrate. In this case, the best chance of cleaning the iron will be to reduce it from the ferric to the ferrous state, using a strong reducing agent such as sodium bisulfite. Once reduced, the iron will readily go back into solution at lower pH. The optimum pH is about 3.5. (Luss 1993).

Silica can be present in an RO feedwater as particulate silica, colloidal silica (also called unreactive or amorphous silica), or dissolved silica (also called molybdate-reactive silica). Particulate silica may be present in clays and soils that have been introduced into a surface water source by runoff.

Particulate silica can foul an RO system by acting as a physical blockage within the element flow channels. Its presence will result in an increase in the system pressure differential. Cleaning it out of the system can be difficult. A standard alkaline cleaning solution may be able to remove enough of the clay and soil binding up the silica to allow the silica to be dispersed into the cleaning solution.

A 0.4% ammonium bifluoride solution has been found effective at dissolving some heavy silica scale formations. Ammonium bifluoride is a hazardous chemical that requires special handling and disposal considerations. Contact the membrane manufacturer and local authorities prior to its use.

Colloidal silica tends to be present in acidic water conditions. It is well rejected by an RO membrane and will therefore tend to be concentrated by RO. The higher concentrations may tend to fall out of solution on the membrane surface. The effects of colloidal silica on RO performance will be similar to those of particulate silica, and may require an ammonium bifluoride cleaning solution.

As discussed in other chapters, the solubility of dissolved silica will depend on pH and temperature. Once the solubility limit is exceeded, silica scale is slow to crystallize. Should it precipitate, it can be extremely difficult to clean and will also require an ammonium bifluoride cleaning solution.

At a pH greater than 8, dissolved silica dissociates from silicic acid (H_2SiO_3) to the silicate (SiO_3^{+}) anion. Insoluble silicate salts can form with multivalent cations such as calcium, magnesium, iron, or aluminum. If these salts are light in concentration, a standard acidic cleaning solution can be effective at redissolving the silicates, particularly if other forms of scale are present in the scale matrix. However, in heavier silicate scale formations, acidic cleaning solution can cause the silicates to combine to form crystal silica (SiO_2). If a standard acidic cleaning solution is not effective, the ammonium bifluoride solution will be necessary.

Organic foulants. Surface waters that originate in rivers, lakes, or reservoirs will contain humic acids that are created by the natural decay of vegetation. Some amount of these humic acids and other silt will tend to bleed through city water treatment systems and RO pretreatment systems, and can subsequently foul an RO membrane, causing a decrease in the normalized permeate flowrate. In most areas of the United States, this fouling will tend to be worse during the spring of the year when lakes thaw and snow runoff occurs.

This sort of fouling is best cleaned using an alkaline cleaning solution containing surfactants. The surfactants will wet out the foulant layer, allowing it to be dispersed into the cleaning solution. With cellulose acetate membranes, the suggested maximum pH would be 7.5 to be sure that minimal hydrolysis of the membrane occurs during cleaning. With thin-film membrane it is possible to clean at a pH exceeding 11, which allows for better solubility of organic oils and fats.

Biological foulants. In water sources containing minimal concentrations of any biocide (such as chlorine or chloramines), bacteria growth can readily foul a membrane system. This tends to be more of a problem with surface water sources containing a lot of suspended solids that act as food for the bacteria.

Biogrowths can reduce the permeate flowrate if they form a layer of slime on the membrane surface. They can also act as a physical blockage to flow in the channels between the membrane leaves (Figure 5-7) or across the hollow fibers.

The same alkaline cleaning agents used for removing organic suspended solids also tend to be good cleaners for biological growths. The surfactants wet out the bacteria bodies and reduce the cell surface tension to the point that the cell bodies will actually explode. The cell components can then be solubilized by the cleaning solution. This is best performed at a ph of 11 or greater, which is not possible with CA membrane without damaging the membrane.

Biological fouling tends to be more of a problem with thin-film membrane because of its intolerance of oxidizing chemicals. Since the best biocides tend to be oxidizing agents, their removal tends to result in biological fouling. With CA membrane, biological fouling can typically be controlled by injecting additional free chlorine in the feedwater.

Figure 5-7. Biogrowths block the flow channels between membrane leaves.

Iron bacteria. Bacteria that thrive in the presence of iron pose a special cleaning challenge. The organic biological mass tends to protect the iron that is present within the bacteria. The iron tends to hold the biological mass together.

Sodium ethylenediaminetetraacetic acid solutions at high pH tend to be somewhat effective. The EDTA will chelate the iron, and the high pH will break up the organic mass. Several cleaning passes may be required.

Cleaning first with a good iron-removing chemical, rinsing, and following this with an alkaline cleaning solution has also been effective. Frequently, one cleaning pass with each chemical may be sufficient.

Cleaning solutions: *CA membrane acidic cleaning solution.* An acidic cleaning solution that is commonly used in the RO industry for cleaning light inorganic scale from CA membrane consists of the following:

	Supplier	Item	Dilution
Thin-Film Low pH Cleaner (Inorganics)	Argo Scientific	IPA 403	1 lb/5 gallons
		Bioclean 103A	1 lb/5 gal
	King Lee	KL-1000	1 lb/10 gallons
		Diamite LPH	1 gal/40 gal
	American Fluid	Filtrapure™ Acid Cleaner	1 lb/15 gal
Cellulose Acetate Low pH Cleaner (Inorganics)	Argo Scientific	HPC 303	1 lb/4 gallons
		Bioclean 103A	1 lb/5 gal
	King Lee	KL-3030	1 lb/4 gallons
		Diamite LPH	1 gal/40 gal
	American Fluid	Filtrapure™ Acid Cleaner	1 lb/15 gal
Thin-Film Alkaline Cleaner (Organics)	Argo Scientific	IPA 411	1 lb/5 gallons
		Bioclean 511	1 lb/6 gal
	King Lee	KL-2000	1 lb/12 gallons
		Diamite AFT	1 gal/40 gal
	American Fluid	Filtrapure™ TF	1 lb/10 gal
Cellulose Acetate Alkaline Cleaner (Organics)	Argo Scientific	HPC 307	1 lb/4 gallons
		Bioclean 107A	1 lb/5 gal
	King Lee	KL-7330	1 lb/4 gallons
		Diamite ACA	1 gal/40 gal
	American Fluid	Filtrapure™ CA	1 lb/10 gal
Thin-Film/CA Iron Cleaning	Argo Scientific	N/A	
	King Lee	KL-3000	1 lb/12 gallons
	American Fluid	Filtrapure™ Iron Remover	1 lb/10 gal
Thin-Film Sanitizing	Argo Scientific	Bioclean 882	1 gal/9 gallons
	King Lee	Microtreat-TF	1 gal/1000 gal
	American Fluid	Flocide® 375 Peracetic Acid	1 gal/400 gal

Filtrapure is a trademark of H.B. Fuller Company. Flocide is a trademark of FMC.

Figure 5-8. Partial listing of proprietary cleaning solutions.

Citric acid 2%
Triton-X100 0.1%
Adjust pH to 2.5 with ammonium hydroxide (NH_4OH).

Note: The Triton X-100 is a nonionic surfactant that is not compatible with PA thin-film membranes. The PA membranes have anionic charge characteristics; therefore, the Triton X-100 should not be used to clean thin-film composite membrane elements. (Courtesy of Fluid Systems)

PA membrane acidic cleaning solution. The following solution can be used for cleaning carbonates and light iron from PA membrane:

Citric acid 2%
Adjust pH to 4.0 with sodium hydroxide (NaOH).

(Courtesy of Hydranautics)

CA membrane alkaline cleaning solution. A commonly used generic cleaning solution used for cleaning organic foulants from CA membrane elements includes the following:

Trisodium phosphate (TSP)	2%
Sodium tripolyphosphate (STPP) may be substituted for the trisodium phosphate.	
Sodium ethylenediaminetetraacetic acid(EDTA)	0.8% (powder) or 2% (39% liquid)
Triton X-100	0.1%
Adjusted to a pH of 7.0 using sulfuric acid (H_2SO_4) or hydrochloric acid (HCl).	

(Courtesy of Fluid Systems)
This solution is not compatible with PA thin-film membrane.

PA membrane alkaline cleaning solution. Commonly referred to as the 1-1-1 solution, the following cleaning solution is effective for removing organic foulants from PA membrane systems:

Sodium tripolyphosphate (STPP)	1%
Trisodium phosphate (TSP)	1%
Sodium ethylenediaminetetraacetic acid (EDTA)	1%

Cleaning frequency. A common concern is what frequent cleaning does to membrane life expectancy. Because of the extreme pH and aggressiveness of many cleaning solutions, with long-term exposure they can have an effect on the performance of the membrane. The extent of this will depend on the frequency of cleaning and the aggressiveness of the cleaning solution.

Inadequate cleaning has resulted in far more loss of membrane life expectancy than is caused by the effects of frequent cleanings. Most of the cleaning solutions approved by the membrane manufacturers can be used frequently, and their cumulative effects will be negligible over the normal life of the membrane (McPherson, May 1981).

Proprietary cleaning solutions. Cleaning solutions are commercially available that are specific for particular foulant and/or scale concerns. These solutions are already buffered for the optimum pH. There is no possibility of causing membrane damage if too much cleaning chemical is used. Some of the proprietary surfactants used in the formulations are extremely effective at wetting out and dispersing the foulants of concern. Proprietary cleaning solutions tend to be more effective than generic formulations for harder-to-clean foulants. These solutions should be approved by the membrane manufacturer prior to use (Figure 5-8).

How a cleaning solution is used is just as important as the cleaning solution itself. Higher cleaning temperatures and flowrates will increase the effectiveness of any cleaning solution. Given that membrane cleaning can be labor-intensive, it is important to take advantage of everything possible to increase the cleaning effectiveness.

Examples of membrane fouling. If an RO system is designed to rely upon chemical scale inhibitors or ion-exchange resin softening as a means of preventing calcium carbonate scale formation, there will be a greater chance of carbonate precipitation occurring than if acid injection is being employed. If an acid injection system fails to operate for some amount of time, any carbonate formed will go back into solution once the pH is dropped to below the saturation level for calcium carbonate (as long as the calcium carbonate scale is not too heavy).

If standard RO systems without acid injection experience a decline in normalized permeate flow, possibly coinciding with a decline in salt rejection, calcium carbonate scale formation should immediately be suspected. In the case of CA membrane, just in case the problem is calcium carbonate, it is critical that the system be cleaned immediately to minimize any damage due to membrane hydrolysis.

An acidic solution should be used to clean the membrane, and RO operating data should be collected after the system rinses. If the performance was restored by the cleaning, then it is confirmed that inorganic scale was what caused the loss in RO performance. With a CA membrane, if the cleaning results in an improvement in normalized permeate flowrate but a decrease in the salt rejection, then the scale was probably a carbonate. Unfortunately in this case, the loss in rejection would have been caused by hydrolysis of the CA membrane due to the presence of the calcium carbonate.

If iron is present in the RO feedwater in a concentration exceeding 0.05 ppm, and the RO is losing normalized permeate flow and/or increasing in system pressure differential, iron fouling should be suspected. If an oxidizing agent (such as chlorine) is present in the water or was present upstream, the iron will be in the ferric state and will be insoluble. It is extremely likely that the iron is falling out in the membrane elements.

If no biocide is present in the RO system, and the system suffers from a declining normalized permeate flowrate and/or increase in system pressure differential, biological fouling should be suspected. High bacteria counts from water samples taken from the RO concentrate will add more evidence to the likelihood of biological fouling. (It is common in these cases to find bacteria in quantities that are too numerous to count.) An alkaline cleaning solution should be used to clean the membrane. This should be performed after the acidic cleaning solution if the system also happened to meet the criteria for inorganic scale formation.

If the alkaline cleaning solution is successful at restoring original membrane performance, a sanitizing solution should be considered as a way of extending the time before biological activity again fouls the system. Recirculation of a sanitizing solution through the system kills any remaining bacteria that would have recolonized the system.

If iron is present in the RO feedwater little or no biocide is present such that high bacteria counts are present in the RO concentrate, iron bacteria should be suspected. Sometimes the problem will be severe enough to be visible as an orange- to brown-colored slime mass. This can frequently be found in the prefilter housing and/or in the membrane element housings.

It may not be known whether there really are iron bacteria present, or whether there is simultaneous fouling occurring by both iron and a more standard biological activity. (See Table 5-1 for suggestions on dealing with multiple foulants.) A cleaning approach that covers both would be to clean first with an iron-removing chemical, rinse, and follow with an alkaline cleaning solution. Operating data should be recorded between cleanings to determine if the iron cleaning is removing a significant amount of iron. (It is important to make this judgment on the basis of the operating data and not on the basis of how dirty the cleaning solution becomes. A significant concentration of iron may cause only a light yellowing of the iron-removing cleaning solution.)

If the source for an RO is a surface water, and the system is experiencing a loss in normalized permeate flow, fouling by organic suspended solids should be suspected. An alkaline cleaning solution should be used. If iron is also suspected, the system should be cleaned first with the iron-removing solution.

Polyamide thin-film membrane performance can be temporarily affected by cleaning solutions with extreme pH. Alkaline cleaning solutions with a pH greater than 11 can cause the membrane to "open up." The permeate flow will be greater than the original one coinciding with a lower salt rejection. This condition is only temporary and performance should return to normal within a couple days.

An acidic cleaning solution with a pH less than 3 can cause the polyamide membrane to "tighten up." The normalized permeate may be slightly less than original, with better salt rejection. As with the reaction to the alkaline cleaning solution, the original performance should return within a couple days of operation.

Table 5-1
Suggested Types and Order of Use for Cleaning Solutions When Encountering Multiple Foulants

Foulants	*Iron*	*Carbonate*	*Sulfates*	*Silicates*	*Organics*	*Biological*
Iron	Iron	Iron	Acid	Acid	Iron Alkaline	Iron Alkaline Sanitizer
Carbonates	Iron	Acid	Acid	Acid	Acid Alkaline	Acid Alkaline Sanitizer
Sulfates	Acid	Acid	Acid	Acid	Alkaline Acid*	Alkaline Acid* Sanitizer
Silicates	Acid**	Acid**	Acid**	Acid**	Alkaline Acid**	Alkaline Acid** Sanitizer
Organics	Acid	Acid	Alkaline Acid**	Alkaline Acid*	Alkaline	Alkaline Sanitizer
Biologic	Iron Alkaline Sanitizer	Acid Alkaline	Alkaline Sanitizer Sanitizer	Alkaline Sanitizer	Alkaline Sanitizer	Alkaline Sanitizer

where:
Iron = iron cleaner.
Acid = acidic cleaning solution for inorganics.
Alkaline = alkaline cleaning solution for organics.
Sanitizer = membrane sanitization chemical.

*Additional cleaning cycles between alkaline and acidic cleaning solutions may be required to restore original performance.
**If this is unsuccessful at restoring original performance, a cleaning solution specific for silica removal may be required.

Cleaning procedure: ***RO cleaning skid design.*** A suggested cleaning system is shown in Figure 5-9. With smaller RO systems, however, it may not make economical sense to buy or build something this extravagant. For instance, the RO prefilters may be relied upon to remove any debris that might be introduced by the cleaning system.

Any cleaning solution is more effective at higher temperatures. It is recommended to raise the temperature of the solution to 105 °F, if possible, when dealing with a difficult foulant or scale. Raising the temperature of the solution can take considerable time with most standard electric heaters. Recirculating the cleaning solution back to the tank with the transfer pump will help to heat as well as mix the solution.

The solution should be recirculated through the RO with the RO concentrate valve wide open. This will help to achieve a high rate of flow across the membrane surface while also reducing the pressure in the system. It is desirable to clean with as low a pressure as possible. This reduces the permeate flow to negligible levels, which reduces the amount of force holding the foulant and/or scale onto the membrane surface. If possible, the pressure should be less than 60 psig. At this pressure the osmotic pressure of most cleaning solutions is such that there will be no permeate flow (other than from small mechanical leaks in the system).

Recommended cleaning flowrate and solution volume. For difficult foulants, the flow velocity across the membrane is critical to cleaning the membrane. Flow blockages slow the velocity in the affected areas, thus limiting the ability of the

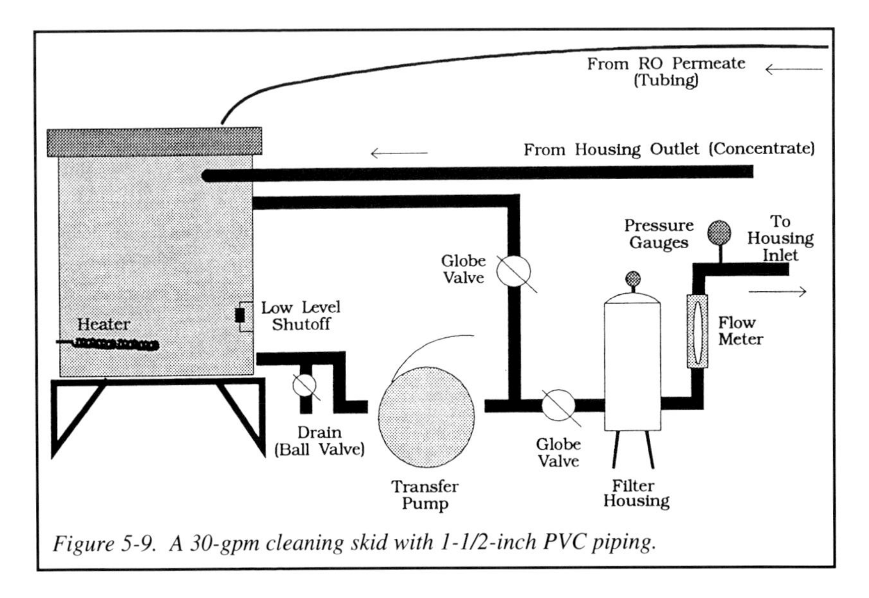

Figure 5-9. A 30-gpm cleaning skid with 1-1/2-inch PVC piping.

cleaning solution to dissolve the foulants and keep them in solution. This requires that sufficient velocities be used to create the turbulence to overcome the resistance caused by these blockages.

To protect against telescoping of the element, however, there are limits to how high the flowrate should be. Membrane manufacturers provide recommendations for the maximum flow through their membrane elements. If such a recommendation is not available, the conservative guidelines given in Table 5-2 can be used for spiral-wound elements.

The volume of cleaning solution should be sufficient to keep the volume of foulants or dilution water in the RO system from being a factor in the cleaning effectiveness. Guidelines are listed in Table 5-2.

Even with a conservatively large solution volume, an effort should be made to prevent the dilution of the cleaning solution by water within the RO. The RO should be drained prior to recirculating the cleaning solution. The return water from the RO back to the cleaning skid tank should be dumped until the presence of the cleaning solution is evident in the RO concentrate stream.

If all stages are cleaned at the same time, the cleaning flowrate will be limited by the number of housings in the last membrane stage. This means that the flowrate in the lead end stages will be dramatically less than the optimum cleaning flowrate. Given that certain foulants are prone to foul the lead end stage of membrane systems, the result can be an inadequate cleaning.

Cleaning effectiveness can be improved by isolating membrane stages during cleaning. Cleaning each stage separately makes it possible to achieve the maximum flow in every housing. For example, cleaning the leading stage of a 2-1 array (the first two housings) and then cleaning the last housing in the array will make it possible to achieve maximum flow in all housings.

Table 5-2
Recommended Cleaning Flow and Volume

Element Diameter	*Maximum Flow/ Housing*	*Solution Volume/ 40" Element*
2"	2.5 gpm	1.5 gallons
2.5"	4 gpm	2 gallons
4"	10 gpm	5 gallons
6"	20 gpm	10 gallons
8"	40 gpm	20 gallons
12"	80 gpm	40 gallons

RO cleaning directions.

1. Fill the cleaning tank with RO permeate water.
2. Drain as much water from the RO system as possible.
3. Mix the cleaning solution according to the supplier's instructions. Use a mixer, or recirculate the solution around the cleaning tank with the transfer pump, to insure that the cleaner is completely dissolved.
4. If possible, heat the solution to improve cleaning effectiveness. Do not exceed 105 °F or the membrane manufacturer's temperature guidelines.
5. Completely open the RO concentrate throttling valve to minimize operating pressure during cleaning. Gradually fill the RO system with cleaning solution. Dump any concentrate or permeate water until the presence of cleaning solution in the RO concentrate stream is evident.
6. Recirculate the solution in the normal direction of flow through the RO system, with the RO concentrate flow being returned to the cleaning tank. Throttle the transfer pump to achieve the maximum individual housing flowrate without exceeding the membrane manufacturer's guidelines. If the permeate flow is significant during the cleaning, it also should be recirculated back to the cleaning tank.
7. Recirculate for a minimum of 30 minutes. Additional recirculation and soaking can be used to improve cleaning effectiveness.
8. Discharge the cleaning solution to drain. Rinse the remaining solution out of the system at low pressure using the transfer pump or the normal system feedwater. (Note: Neutralization of spent solutions may be required at your facility. Check with local authorities.)
9. Prior to restarting the RO system, return the concentrate throttling valve to its normal position. Divert the RO permeate water to drain until conductivity returns to normal.

RO system sanitization. Reverse osmosis systems that do not use a continuous biocide in their feedwater will likely suffer from biological activity. In some cases, the extreme pH of the membrane cleaning solutions is sufficient to sanitize the system during the cleaning process. In other cases, the biological activity is heavy enough to require a greater reduction in bacteria than what can be achieved with normal cleaning formulations. If the permeate carrier material in a spiral-wound membrane element has been contaminated with bacteria, it is unlikely that the cleaning solution will be capable of sanitizing the permeate carrier.

In water, sanitization is frequently defined as a 3-logarithm (log) reduction in the number of bacteria. This is the same as a 1,000-fold reduction. Disinfection is defined as a 6-log reduction in the number of bacteria, a 1,000,000-fold reduction. (U.S.-EPA 1976).

If bacteria numbers are not reduced to near zero concentrations in the RO system, the bacteria colonies still present will begin to propagate as soon as the sanitizing solution is rinsed out. This can result in the rapid biological refouling of the system.

A true disinfection of the system has been known to extend the time prior to the recolonization of the system with bacteria. In a biologically clean element, bacteria from the feedwater will have some difficulty finding a place to attach themselves. This attachment must occur before they can begin to propagate and spread. Unfortunately, in systems that are constantly being fed by a high concentration of bacteria, batch disinfections may do little to slow the recolonization of the system.

The biocide used for the batch disinfection must be compatible with the membrane, element, and system materials of construction. A 10-ppm sodium hypochlorite solution (free chlorine) is frequently used for disinfecting CA membrane. A 400-ppm peracetic acid solution (also containing 2,000 ppm of hydrogen peroxide) can be used to disinfect PA thin-film RO membranes. New RO polyamide membrane elements must have been in service for several days prior to using peracetic acid solutions to avoid reaction with residual amines that are sometimes present in new membrane.

A 0.5% formaldehyde solution can be used to disinfect CA or PA membrane. The membrane elements should be cleaned prior to using formaldehyde (or glutaraldehyde). Formaldehyde is considered a fixative and can make it more difficult to clean the biological foulants from the system. As with peracetic acid, residual amines must be rinsed from PA membrane prior to its use. (Note: Formaldehyde is considered a carcinogen, and its use may not be allowed.)

The solution should be recirculated through the RO system for a sufficient time to achieve disinfection. This is typically a minimum of 30 minutes in the case of peracetic acid disinfection. Formaldehyde and chlorine may require more contact time.

Chlorine, peracetic acid, or formaldehyde solutions can be recirculated at either high or low pressure, since either will readily permeate RO membranes over the entire pressure range of the membrane. Other sanitizing solutions may require that they be used in a low-pressure recirculation, if the chemical is rejected by the membrane at its normal operating pressure.

The pretreatment system should also be disinfected if it does not normally see a continuous concentration of biocide. It is best if this is performed with a biocide that is compatible with the downstream RO membrane. If not, special care should be taken to completely rinse the biocide from the pretreatment equipment. This should be verified with a highly sensitive test method prior to putting the RO back into service.

Extended RO shutdowns. Occasionally, the need arises for shutting down an RO system or even an entire facility. More frequently, a water treatment system will have redundant RO units, and not need the capacity of the additional RO. An RO sitting stagnant for more than a day is an invitation to biological activity. Heavy biological activity can result in the demise of the membrane elements.

If fouling or scale formation is suspected in an RO system (suspected usually because of changes in the normalized performance data), the system should be cleaned prior to shutting it down. This should be performed through the means discussed above.

A stable biostatic or biocidal solution should be used to "pickle" an RO being shut down. This solution can be a strong reducing agent such as sodium bisulfite. It can be a fixative, such as formaldehyde or glutaraldehyde, where those are acceptable for use. Sometimes a high surfactant concentration can be used in conjunction with the biocide as a means of preventing the agglomeration of bacteria colonies.

As previously discussed, certain membrane storage agents are not compatible with the residual amines that are sometimes present in new PA membrane. These incompatible agents include formaldehyde, glutaraldehyde, peracetic acid, and hydrogen peroxide solutions. Their usage should be avoided until the system has been operated for several days, and the membrane manufacturer is confident that the residual amines have rinsed from the system.

Certain storage agents may not be compatible with the RO system components. With long term exposure a strong reducing agent solution can cause pitting of stainless steel welds. This can occur in welds of poor quality.

The membrane storage solution should be recirculated through the system for 15 minutes, or until it is certain that it is well mixed throughout the RO system. The recirculation should then be stopped and the system closed up. All lines should be valved off to prevent the system from draining, and to prevent the introduction of airborne bacteria. If an isolation valve does not already exist on the RO permeate line, a temporary valve may have to be added to the line. When the system is restarted, the solution will need to be rinsed out. (The membrane manufacturer should always be consulted for specific recommendations.)

Membrane rejuvenation. It is possible to improve the rejection characteristics of certain damaged RO membrane elements by placing a temporary surface coating on the membrane. The term coined for this procedure is "sizing" the membrane. Whether a membrane will respond to such a treatment will depend greatly on the type of deterioration, the extent of damage, and the cleanliness of the membrane. The treatment will have little effect if the membrane damage is mechanical, as where membrane lesions or glue line failures have occurred. It is most likely to be successful in cases of membrane degradation if the membrane salt rejection has not declined to less than 75% (Amjad et al.,).

For the treatment chemical to adhere to the membrane surface, the membrane must be free of any foulants or scale; therefore, the system should be cleaned just prior to treating it. The effectiveness of the cleaning may dictate whether the membrane treatment is successful.

Colloid 189 (from Colloids of America Co.) is a polyvinyl acetate copolymer, usually available as a 7.5% to 12% solution with ammonium present for a pH of between 8 and 9. It has been found effective particularly with CA membrane at restoring rejection to more than 95%, and has had some success with PA thin-film membrane. It is effective only with systems that operate with acidic conditions throughout the membrane system.

If the Colloid 189 "sticks," the RO normalized permeate flow will typically decline by 10% to 15%. There may also be some increase in the system differential pressure.

The colloid is soluble in alkaline pH (greater than 7.0), but will still be attracted to the membrane surface at slightly alkaline pH conditions. It is introduced into the RO system by injecting it while the RO is operating with the pH adjustment disengaged. When the pH of the feedwater is dropped to normal acidic operating conditions, the colloid will tend to become insoluble on the membrane surface. Many different ways have been used to introduce the chemical into the RO feed stream. A solenoid injection pump is probably the most common, but just dumping some into an upstream filter housing has been effective in some cases. (Following the provided directions is recommended, however.) The membrane manufacturer should be consulted prior to using any membrane rejuvenation chemical.

If the treatment is successful, the length of its effectiveness will depend on the particular RO operating conditions. Whenever the RO is operated at alkaline conditions, the colloid will tend to come off the membrane. Cleaning, particularly alkaline cleaning, will also tend to dissolve the colloid. It is common for the improved rejection to last between 1 and 6 months.

Polyvinyl methyl ether (PVME) from BASF and tannic acid have both been found effective at restoring the rejection characteristics of DuPont's B-9 and B-10 PA hollow-fiber permeators. There will be some loss in permeate flow, or increase in differential pressure, but typically not as much as with Colloid 189 treatments.

Like Colloid 189, PVME and tannic acid treatments will also be removed from the membrane with time, or during cleaning. In fact, it is recommended that B-10 permeators be treated with tannic acid after any cleaning operation.

The PVME is effective at improving the membrane rejection characteristics by plugging mechanical imperfections in the fibers. The tannic acid works by absorbing onto the membrane surface and restricting the diffusion of salt through the membrane fibers.

Colloid 189 membrane surface treatment procedure.

1. Thoroughly clean each stage of the CA membrane system with an acidic cleaning solution. Rinse the system and proceed with a thorough alkaline cleaning solution. Again rinse the system.
2. Record the standard operating data for the system, as well as the individual housing permeate concentration values.
3. Readjust the acid injection system for an RO feedwater pH of between 7 and 7.5. Disengage any other chemical injection systems in the RO pretreatment system.
4. The Colloid 189 should be diluted in a day tank so that it can be injected into the RO feedwater at the rate of 15 to 25 ppm (as Colloid 189). (See "Chemical Injection Pumps" in Chapter 4.) If the pH of the day tank solution is less than 8.0, adjust it up with ammonium hydroxide. Inject the colloid until the permeate concentration stabilizes (usually less than an hour).
5. Disengage the colloid injection and continue operating the RO at a pH between 7 and 7.5 for up to 15 minutes. Return the RO to normal operating conditions.
6. Record the RO operating data and the individual housing permeate concentrations. Note any improvement in performance.
7. Thoroughly clean out the Colloid 189 day tank and injection pump with an alkaline solution prior to using them for any other purpose.

PVME and tannic acid treatment procedure. DuPont recommends either batch treating the permeators, or directly injecting the treatment agent into the RO feedwater. PVME is called PT-A and tannic acid is called PT-B in the following procedures (Figure 5-10).

Off-site element cleaning and rejuvenation. A number of membrane service companies offer an off-site service for cleaning and rejuvenating damaged or fouled membrane elements. This service can be particularly beneficial for small RO systems that are not set up well for cleaning. Where endusers may have avoided cleaning because of the labor and equipment requirements, off-site cleaning may be the only alternative to prematurely replacing the membrane elements, and can be a cost-saving practice.

Off-site cleaning can also be useful if cleaning on-site has not been effective at restoring the original performance of membrane elements. Higher crossflow rates can frequently be achieved when cleaning individual elements, than when attempting to clean an entire membrane housing or full system.

Some of the membrane service companies have the capability of removing and replacing the fiberglass outer wrap of spiral-wound elements. This enables them to physically remove heavy foulants or scale lodged between the membrane leaves. A new fiberglass wrap is then placed around the element.

For RO systems that use full-fit spiral-wound elements (elements that do not use an outer wrap), this process of unrolling and physically cleaning the membrane can be performed on-site. Care should be taken not to damage the membrane surface while cleaning. (Consult with the membrane manufacturer.) A vacuum pump can be used to remove any air inside the membrane envelope. This will make it possible to tightly roll the element so that it will fit back into the housing.

An off-site cleaning service can be useful when investigating a fouling or scale formation problem with a large RO system. If it is not clear what cleaning regimen will be most effective in cleaning the particular foulant or scale, various cleaning solutions can be tested on a single element by the service company. What is learned from the single element can be used with the full-scale system. This can save time and expense over using the full-scale RO system to experiment with various cleaning solutions.

RO and Pretreatment System Installation and Start-up

The RO skid should be mounted on a 4-inch-high (minimum) flat concrete pad that corrects for any drainage slope of the main floor. If possible, it is suggested that any chemical injection be located away from the RO skid and pad. Chemical containment may be required around the chemical injection day tanks and pumps.

If located outdoors, the RO system should be covered. Direct sunlight should not be allowed to contact any part of the RO system, even by the changing angle of the sun during different seasons of the year. This protection will reduce the amount of deterioration of the system components due to UV radiation, and will also reduce the amount of heat that the system might pick up while it is shut down.

Coordinating the start-up. Timing is typically critical when a new RO system is started up. Frequently, manufacturers' representatives, who have limited time available, are present for the commissioning. Proper coordination of the start-up events will insure that everyone's time is used to best advantage.

The RO start-up should be scheduled so that the rest of the process is ready for the RO operation. Unfortunately, it is common for a new RO to be started up only to have the system sit stagnant for extended periods of time because the rest of the system is not ready for the water. It may be necessary to fill the RO system with a biostatic storage agent just after its initiation.

An important aspect in coordinating the start-up is to check to make sure that all equipment is on-site and accessible. The system manufacturer should provide a checklist of everything that should be present for the start-up. This should include cartridge prefilters, a breakdown of media to be loaded into the media filters, any chemicals intended to be injected, and O-ring or gasket lubricants as required.

Leak testing. The system should be plumbed in and checked for leaks by pressurization of the connections. This should be done well before scheduling the start-up. It is common to have a few leaks in the plumbing that will need repair prior to start-up, and these repairs will need time for the glue to set up.

Electrical connections. Permanent electrical connections should be made to the RO system prior to start-up. This should include all necessary disconnects, breakers, and panels. A lot of time can be wasted attempting to rig temporary electrical connections, which can be dangerous as well.

System sanitization. With the membrane system isolated from the pretreatment and distribution piping, all new piping should be disinfected. This can be performed using a 200 ppm chlorine solution, a 10% hydrogen peroxide solution, or a 400-ppm peracetic acid solution (also containing 2,000 ppm of hydrogen peroxide). If the RO system has not yet been loaded with membrane elements, the RO housings and pipe manifolding should be included in this disinfection.

(Note: The use of chlorine to disinfect the RO or the pretreatment plumbing is not recommended for PA membrane systems even if the system is not yet loaded with the new membrane. Chlorine has a tendency to be absorbed into plastic system components, which results in the need for an exceptionally long rinse-up before the system will be safe for the installation of PA membrane.)

Media filters should also be included in the disinfection, with the exception of activated carbon and ion-exchange beds that have already been loaded with media. If the beds are already loaded, care should be taken to avoid contact of the disinfectant with the beds and with the RO membrane during the disinfection and rinse-up.

Media filters. Before the RO system is initiated, the upstream pretreatment equipment must be ready to provide water for the RO. Media filters must be loaded with their media and backwashed, and prefilter housings must be loaded with filters.

Before loading the media filters, their flowmeters and flow control orifices on the media filters should be checked, and calibrated if necessary. This is most accurately performed with a calibrated bucket and a stopwatch.

It is easiest to load the media filters if the units are partially filled with water. This will help to reduce the amount of dust generated. Even so, appropriate safety equipment such as breathing gear should be worn during the loading. If it is necessary to enter a tank, local (and plant) safety procedures should be reviewed. It may be necessary to use self-contained breathing apparatus.

Care should be taken to load the correct amount of media, and in the correct order according to the manufacturer's recommendations. Frequently, bags of media are not well identified. Any discrepancy should be clarified prior to loading the filters.

After the media filters are loaded, they should be put through numerous backwash cycles. Unless otherwise specified by the manufacturer, a minimum of five backwash cycles is suggested. A silt density index measurement should be used to verify that the quality of water coming out of the filters is better than what is going in. (See Chapter 2.) The filter from the SDI measurement should be visually inspected after the test to check if the media filters are still sloughing media fines. If so, those filters should be further rinsed or put through another full backwash cycle.

Cartridge filter housings should be rinsed out prior to installing new cartridge filters. Any pressure should be bled from the filter housing prior to opening it.

After the filters have been on-line for a few hours, the system should be shut down. The filters should be removed and inspected. If the upstream media filters are still sloughing media, they should be further backwashed. If it appears that the cartridges are not sealing well in the housing (i.e., if the imprint of the housing's knife-edge seal is not apparent on the ends of the filters), appropriate action should be taken to remedy this situation. (See Chapter 2.)

An SDI measurement should be taken downstream of the prefilters prior to starting the RO system. Since this is the water that will be coming into contact with the high-pressure pump, it should give a reading that is within the membrane manufacturer's guidelines. If it is visually apparent on the SDI filter that large particles are getting through the prefilters, this situation should be resolved prior to starting the RO.

Chemical injection pumps. Day tanks should be filled with the correct amount of purified dilution water for their particular chemical dilution. Since the RO is not yet operating, it may be difficult to obtain purified water. Consult with the system manufacturer if purified water is not available. Dilutions should be performed according to directions provided in Chapter 2.

Prior to the addition of the concentrated chemical, the injection pumps should be primed with the dilution water. To aid in priming the pumps, the inlet tubing should be manually filled with dilution water. The back pressure on the outlet tubing should be minimized either by reducing the pressure in the main pipe line (by shutting off the water upstream), or by disconnecting the tubing from the main line. The injection port should include a check valve to prevent water from spraying out of the main pipe line. (Some injection pumps have a priming device available on the pump outlet that bypasses the effluent line.) Once the inlet and outlet lines are filled and the pump is operating normally, the chemical can be added to the day tank.

Removal of old spiral-wound membrane elements. Before new spiral-wound membrane elements are loaded into an existing RO system, it may be necessary to remove the old elements. If the system has no biological problems, this can be accomplished simply by pushing the new elements in on the feed end of the housing and thereby pushing the old elements out of the other end of the housing.

If biological activity has been a concern with the system, it is a good idea to sanitize the housings and piping after removing the old elements, and prior to loading the new elements. In this manner, the system can be sanitized with a more aggressive biocide without concern about damaging the new membrane. This should slow the bacterial recolonization of the new membrane.

If the old membrane elements are not too badly fouled, they can be pushed out with a "two-by-four" piece of lumber. Sometimes it is easier to unload elements by removing the downstream end caps of the downstream housings and slowly applying water pressure to the system. At some point, the pressure will have enough force to push out the old membrane elements. Then the elements of the housings directly upstream can be removed using the same method.

Any visually apparent scale or foulants should be cleaned out of the RO housings and piping. If the foulants or scale are particularly heavy, it may be necessary to use a high-temperature cleaning of the empty system. Without the membrane, temperatures up to 140 °F can be used to clean out the empty manifolding and housings. After this procedure, the system should be rinsed of the residual cleaning solution.

Calibration/verification of the RO instrument readings. If possible, all monitoring instruments should have their readings verified in the field. If spiral-wound membrane housings have not yet been loaded with the new elements, it should be easy to vary the flowrate through the various meters and verify their readings using a calibrated bucket and a stopwatch.

Pressure, temperature, conductivity, and pH readings should be verified using a calibrated field instrument. If possible, this meter should offer ten times the accuracy of the RO instruments. If any reading is appreciably off, the instrument should be further investigated to insure its integrity. (Cohen 1993).

Loading of new membrane elements. New spiral-wound membrane elements may not yet have their brine seal and O-rings (on the element interconnecting devices) installed. Although there may be a location for a brine seal on both ends of the element, only one brine seal should be used per element. It is usually placed in the slot at the lead end (feed end) of the element. It should be placed such that the water flow going through the element forces the edge of the brine seal against the inside of the membrane housing (Figure 5-11).

The brine seal should be lubricated. Glycerin is suggested as a lubricant for the brine seals as well as for the interconnector O-rings. The glycerin is completely water-soluble, which means it will completely rinse out of the system. Unfortunately, this may make it more difficult to later remove the elements when they need replacement. There will also be a greater risk of breaking O-rings after the glycerin has rinsed out.

If silicone grease is used, it should be used sparingly, as it can foul the membrane elements. Dishwashing soap should not be used because it can cause the O-rings and gaskets to swell.

The elements should be loaded into the housings in the same direction as that of the feed flowrate for each housing. Care should be taken to record the serial number of the element along with its exact position in the array. The loading should be performed slowly, paying attention that brine seals face the correct direction and that all O-rings and interconnectors are correctly installed. Common loading mistakes can affect RO system performance, and will take time to correct.

Downstream end caps should be attached to the housings prior to the upstream ones. For end-entry end caps, the high-pressure manifolding will need to be aligned with the end cap connections prior to setting the end cap firmly into place.

The elements should be pushed firmly into the downstream end caps. Prior to connecting the upstream end caps, the lead end elements should be shimmed to prevent movement within the housing. Shimming involves placing slices of plastic pipe around the outside of the end cap interconnector as a means of taking up any slack between the element and the end cap. There should be some resistance to installing the end cap if it is properly shimmed.

Filling the RO system. Once the system is filled with new elements, it can be filled with low-pressure water. The high-pressure pump should not yet be engaged. First the system should be checked for leaks using the low pressure supplied by the city water or by a system transfer pump. Then the system should be checked for valves that have been inadvertently closed. There should be flow through the concentrate and permeate lines. The permeate pressure should be negligible.

Checking high-pressure pump rotation. If there are no leaks, the high-pressure pump rotation should be checked. With the RO system full of water, the pump should be "bumped" by quickly energizing and then disengaging the pump. The correct pump rotation is usually marked on the pump. With totally-enclosed fan-cooled (TEFC) motors, it may be difficult to see the rotation, but there is usually a way. If the rotation is incorrect, two of the three phase leads powering the pump will need to be switched. If a high-pressure pump is operated with incorrect rotation, it will still create a significant pressure. Typically, this pressure is about half of what it should be at the desired flowrates.

Starting the RO system. Everything should now be ready to start the RO system. If a throttling valve is used on the output of the high-pressure pump, it should be opened about halfway before starting the pump. Upon initiation of the high-pressure pump, the concentrate throttling valve should be immediately adjusted for the desired concentrate flowrate. This flow can be slightly higher than the design rate until the system has stabilized. The pump discharge valve should now be adjusted for the desired membrane pressure. This adjustment will probably affect the concentrate flow, such that adjustment of its throttling valve will be required. Fine-tuning between the two valves may be necessary. Starting up RO systems that do not use pump throttling valves will require only adjustment of the concentrate throttling valve.

As soon as flowrates have stabilized in a desired range, chemical injection concentrations should be checked. With acid injection, the pH should be measured in the RO feedwater. Adjustment of either the injection pump setting or the control settings of the pH meter will probably be necessary.

With PA membrane systems, the concentration of free and total chlorine should be checked. If either of these is present to any appreciable extent, the RO should be shut down until the reason for the biocide's presence is discovered. It may be that the injection pump setting is too low, or the day tank dilution is wrong. If activated carbon filtration is being used for the oxidizing agent removal, bypass may be occurring around the filter valves.

Data collection. After the RO system has operated for at least 30 minutes with stable performance (i.e., flow readings, pressure readings, temperature, pH, and chemical concentrations are all constant), a full set of performance readings should be recorded. This should include individual housing permeate concentrations and flowrates, as well as interstage pressure readings if feasible (as discussed earlier in this chapter).

If any readings are questionable, the instrument should be recalibrated. If an individual housing permeate flowrate or concentration is inconsistent, it should be investigated. If the permeate quality is low for an individual housing, it may be necessary to probe the housing to pinpoint the location of the source of the poor rejection. When starting up a new RO system, it is common to find broken or missing O-rings or brine seals. It may be necessary to further "shim" up the elements in a housing to prevent movement. (See Chapter 7.) Occasionally, an element may be found that does not meet the manufacturer's minimum performance specifications.

If any modifications are made to the system, the operating data should be recorded again. This final data should be copied (twice) and stored in a safe location. A copy should be sent to the membrane manufacturer, and one to the RO system manufacturer. The data can then be used for comparison purposes if a problem should ever arise.■

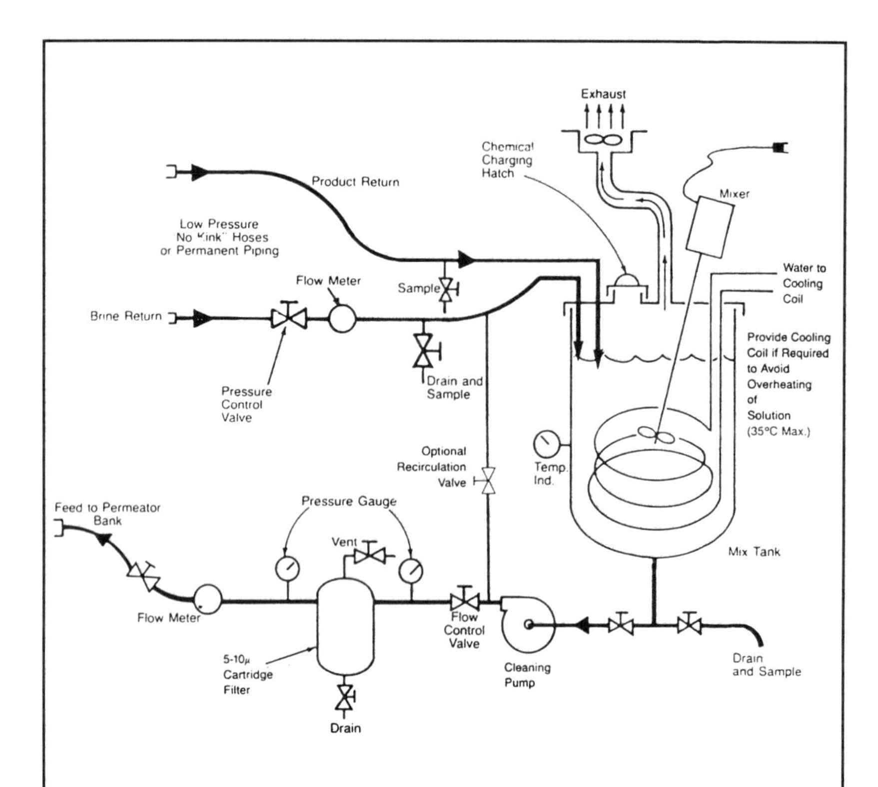

Figure 5-10. Equipment for cleaning, sterilization, and posttreatment.

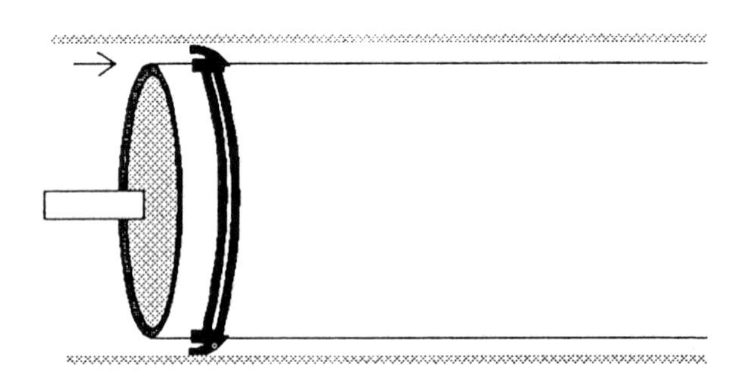

Figure 5-11. Water flow forces the brine seal against the membrane housing.

ADDENDUM TO FIGURE 5-10

The flow requirements for effective permeator cleaning, sterilization and posttreatment vary. In many cases, plants may be treated stage by stage with the flow balancing tubes in place. Procedures which require high brine flows may necessitate a bypass around the flow balancing tubes.

PT-A POSTTREATMENT

The patented PT-A treatment can usually improve the salt rejection of B-9 and B-10 permeators following storage or a cleaning operation where rejection did not improve. Since PT-A and PT-B will react with each other, the permeators should be thoroughly flushed between treatments if both chemicals are used.

Preparation of PT-A Stock Solution

Because the required PT-A dosage for treating permeators is high and the 50 wt. percent concentration is difficult to handle and is slow in dissolving, the PT-A is best added as a diluted aqueous stock solution containing 3 wt. percent PT-A which has been prepared carefully beforehand.

To prepare a supply of 3 wt. percent PT-A stock solution, add 65 grams of the 50 wt. percent PT-A concentrate per liter of good quality chlorine-free cold water. Stir at low speed for 3 hours to ensure complete dissolution. Monitor the temperature to avoid exceeding 35 °C.

Off-line PT-A Treatment Procedure

a. Flush permeators with chlorine-free product or good quality water prior to posttreatment using a once-through flush (brine and product to drain) of 38 liters (10 gallons) of water per 4-inch permeator.*

b. In the mix tank, prepare a well mixed solution containing 80 mg/liter PT-A (active ingredient) plus 530 mg/L NaCl taking into account the volume of water in piping, hoses, and permeators (assume 9.5 liters [2.5 gallons] of water per 4-inch permeator). Use chlorine-free product water.

c. Recirculate the PT-A solution through the permeator. A
brine flow of about 7 L/m (2 gpm) per 4-inch permeator is recommended using a pressure of 345 to 517 kPa (50 to 75 psig). The temperature must not exceed 35 °C. This temperature maximum protects the permeator and prevents precipitation of PT-A (cloud point equals 37 °C).**

d. Monitor the conductivity of the product every 15 minutes. Terminate the PT-A posttreatment when product conductivity is constant for 30 minutes. Usually, constant conductivity is obtained after one hour of treatment.

---

* At higher concentrations, the cloud point is lower; at 3 wt. percent or above, it is 30 °C. Thus, these solutions should be stored at temperatures lower than 30 °C.

**Flows and volumes stated are based on 4-inch-diameter standard length permeators. For 0420 flow and volumes are one-half and for 0410 flows and volumes are one-fourth of the standard 0440 rates. For 8-inch-diameter single bundle permeators, flows and volumes should be increased by a factor of three. When cleaning 10-inch-diameter and two bundle 8-inch-diameter permeators, flows and volumes are increased by a factor of six.

conductivity is obtained after one hour of treatment.

e. When the posttreatment is complete, stop recirculating. Drain the Mix Tank solution to waste. Flush the residual PT-A solution from the permeator with RO product water by operating at 345 to 517 kPa (50 to 75 psig) and at a brine rate of 17 L/m (4.5 gpm) per 4-inch permeator for five minutes in the normal feed direction. During this flush, the brine and product both go to drain. The units may then be flushed using regular feedwater at a reduced pressure (less than 1,380 kPa [200 psig]) for 15 minutes. Both the brine and product go to drain.

f. After the system is stabilized, check the performance to determine the effectiveness of PT-A treatment.

On-line PT-A Treatment Procedure

B-9 and B-10 permeators can be PT-A treated while in operation. The on-line treatment is performed by injecting a PT-A solution into the feed stream. The procedure is as follows:

a. In a mix tank prepare a solution containing up to 2,000 mg/L PT-A using RO product water. The water temperature must be less than 30 °C to ensure that the cloud point of the 2,000 mg/L PT-A solution is not exceeded.

b. Inject the PT-A solution into the RO feed using the following expressions to determine injection rate and volume of solution required:

$$\text{Injection rate, m}^3\text{/hr} = (\text{Feed Flowrate, m}^3\text{/hr})(20\text{ mg/L}) \div 2{,}000\text{ mg/L}$$
$$\text{Volume, m}^3 = (\text{Injection Rate, m}^3\text{/hr})(\text{time, 1.5 hr max.})$$

c. Injection of the PT-A solution at the rate shown above will result in a feed containing 20 mg/L PT-A. Depending on the feed flowrate and injection pump rate, a PT-A solution of less than 2,000 mg/L can be used.

d. During on-line PT-A treatment, measure the product conductivity every 10 minutes. When the product conductivity is constant after three successive readings, stop the PT-A injection. However, the maximum injection time should not exceed 1-1/2 hours.

e. During the PT-A treatment, the feedwater temperature must not exceed 35 °C to ensure that the cloud point of the mg/L PT-A solution is not exceeded.

PT-B Posttreatment

After the initial flush, new B-10 permeators and replacement bundles usually must be treated with the patented PT-B before placing them on-stream. B-10 permeators MUST also be retreated with PT-B after any cleaning operation. B-9 permeators can also be treated with PT-B. Permeators are normally treated with PT-B by shutting down the portion of the system requiring treatment, treating the units and returning them to service. However, under certain conditions, permeators can be treated while in operation. These procedures are discussed below.

Off-line PT-B Treatment Procedures

a. Flush the permeators with chlorine-free product or good quality water (TDS less than 5,000 mg/liter) prior to posttreatment. Use a once-through flush (brine and product to

drain) of 38 liters (10 gallons) of water per 4-inch permeator.**
b. In the Mix Tank, prepare a solution containing 1 wt. percent citric acid plus 80 mg/liter of PT-B, taking into account the volume of water in piping, hoses and permeators (assume 9.5 liters [2.5 gallons] of water per 4-inch permeator). Use RO product or other good quality water (TDS less than 5,000 mg/liter). Since the PT-B is a solid which does not dissolve immediately, the PT-B is best added as a dilute aqueous stock solution containing 3 wt. percent of PT-B.
c. Recirculate the PT-B solution through the permeators. A brine flow of about 7 L/m (2 gpm) per 4-inch permeator is recommended using a pressure of 690 to 1,034 kPa (100 to 150 psig). The temperature must not exceed 35 °C.
d. After one hour, stop recirculating. Drain the mix tank solution to waste. Flush the residual PT-B solution from the permeators with product water at 345 to 517 kPa (50 to 75 psig) and a brine rate of 17 L/m (4.5 gpm) per 4-inch permeator for five minutes in the normal feed direction. During this flush, the brine and product water go to the drain. The units may then be flushed with regular feedwater at a reduced pressure (less than 1,379 kPa [200 psig]) for 15 minutes. The brine and product water go to drain.
e. After the system operation has stabilized, check the performance to determine the effectiveness of PT-B treatment.

On-line PT-B Treatment Procedures

a. Introduction.

B-10 permeators can be PT-B treated while in operation provided the feed pH during treatment is less than 5.0. At a pH greater than 5.0, PT-B is relatively insoluble in seawater and forms a brown precipitate, even at a PT-B concentration of only 20 mg/L. However, if the seawater pH is less than 5.0, no precipitate is formed even at PT-B concentrations of 80 mg/L. The on-line PT-B treatment is performed by injecting a PT-B and citric acid solution into the feed stream. A mix tank and injection system are required. If operating conditions permit on-line treatment, "Permasep" Products Licensees can provide design assistance and operating procedures.

b. Procedure

1. In a mix tank prepare a solution containing 2.7 wt. % citric acid using RO product water. Then add sufficient PT-B to the citric acid solution to obtain a PT-B concentration of 525 mg/L. Since the PT-B is a solid which does not dissolve immediately, the PT-B is best added as a dilute aqueous stock solution containing 3 wt. % PT-B. Other concentrations of the injection solution can be used. Consult "Permasep" products for design assistance.
2. Inject the citric acid/PT-B solution into the seawater feed for 90 minutes using the following expressions to determine injection rate and volume of solution required:

** Flows and volumes stated are based on 4-inch-diameter standard length permeators. For 0420 flows and volumes are one-half and for 0410 flows and volumes are one-fourth of the standard 0440 rates.
For 8-inch-diameter single bundle permeators, flows and volumes should be increased by a factor of three. When cleaning 10-inch-diameter and two bundle 8-inch-diameter permeators, flows and volumes are increased by a factor of six.

Injection Rate, m^3/hr = (Feed Flowrate, m^3/hr)(20 mg/L)/525 mg/L
Volume, m^3 =(Injection rate, m^3/hr)(1.5 hr)

3. Injection of the PT-B/citric acid solution at the rate shown above will result in a seawater feed containing 20 mg/L PT-B and 500 mg/L citric acid. During on-line treatment, feed and brine pH will be 3.0 to 3.5, and product water pH will be 5.0 to 5.5 (200 mg/L citric acid will lower the feedwater pH to approximately 4.5).
4. During treatment, the salt rejection will not improve due to passage of some citric acid into the product water. Therefore, in order to determine the effectiveness of the treatment in increasing salt rejection, continue RO operation for at least 15 minutes after the treatment before obtaining any product water measurements.
5. If HCl, H_2SO_4, and/or $NaHSO_3$ are used in the pretreatment, their use should be continued because the on-line PT-B treatment is effective with these chemicals present. If a scale inhibitor is used instead of acid, the inhibitor should be continued during treatment unless it has been shown to be compatible with on-line PT-B treatment.

c. Other Comments
For small B-10 permeator systems, an injection solution containing 2.7 wt. % citric acid and 525 mg/L PT-B is usually satisfactory based on feed flowrates and injection pump capacity. However, for larger B-10 systems, an injection solution of higher concentration is highly desirable to obtain a reasonable size injection tank and a reasonable rate of injection. The following injection solutions can be used for on-line PT-B.

INJECTION SOLUTIONS

WT. % CITRIC ACID	*+ mg/L PT-B*
2.7	525
5.4	1,050
1 0.8	2,100
21.6	4,200

The solutions given above are prepared by first dissolving the citric acid in water and then adding the PT-B which has been dissolved in about 100 mL of water. Good quality water (chlorine-free, e.g., RO product water) should be used to prepare any of the above solutions.
When the above solutions are used, the injection rate should be adjusted to give 500 mg/L citric acid and 20 mg/L PT-B in the RO feed during on-line PT-B treatment. During treatment the RO feed pH must be measured to be certain the pH is <5.0.

Tannic Acid Quality
a. Specification for Tannic Acid
The suitability of Tannic Acid for membrane posttreatment can be determined by titration with base (NaOH). If the % of the first acid (end point at a pH of approximately 5) is <= 6.0 of the total acid titrated, the tannic acid is of acceptable quality and can be used to PT-B treat "Permasep" permeators.

Example: 1st acid = 0.11 m Eq/g of tannic acid
Total acid = 5.25 m Eq/g of tannic acid

$$1.1 \div 5.25 \times 100 = 2.1\ \%\ \text{(Quality is acceptable)}$$

b. Procedure for Titration of Tannic Acid

1. Weigh 1 g (to nearest .0001 g) sample of tannic acid into a 250-mL beaker.
2. Add 100 mL water and a magnetic stirring bar.
3. Stir into solution.
4. Titrate with 1.0 N sodium hydroxide using a combination glass electrode (Metrohm EA-120) and an automatic titrator (such as Metrohm E-436).
5. Calculation: Neutral Equivalent (m Eq/g) = volume × Normality/sample wt.

NOTE: When two end points are observed, they occur at pH 5 and 10. When only one end point is observed, it occurs at 10.

CHAPTER 6

TROUBLESHOOTING RO SYSTEMS

Isolating the Location of a Decline in Salt Rejection

Salt rejection is probably the most monitored parameter for reverse osmosis systems. Not only is it an indicator of the RO membrane performance, it also directly affects the RO effluent water quality. When the salt rejection of an RO declines, it becomes an immediate concern.

Frequently, engineers and technicians would rather believe that a loss in salt rejection in their particular RO system is caused by something uncommon, something new and mysterious. In most cases of salt rejection decline, however, the particular problem has been experienced by other RO systems, and documented. The potential problems discussed in this chapter will account for better than 95% of the problems encountered in the field.

In some cases, it may be necessary to make improvements without being certain that the problem has been pinpointed. It is suggested that all possible improvements for the potential problems be made anyway. Then the system should be closely monitored to determine if the true problems have been resolved. This chapter will not take on the task of listing all the potential problems that can result from using an RO system in a water reclaim or waste separation system. However, these troubleshooting techniques will offer insight into how to approach such problems.

A loss in salt rejection across the RO membrane barrier is one of the most common and obvious distress signals experienced by RO technicians. The cause of a loss in salt rejection should be determined and corrected as soon as possible for the following reasons:

- The quality of the process water may drop below specifications.
- Even a small drop in salt rejection may indicate the beginning of a serious membrane deterioration, fouling, or scaling problem.
- Organic and colloidal material could start to pass through the RO system and irreversibly foul downstream ion-exchange resins.
- The length of the service cycle of deionizers located downstream of the RO

Editor's note: Most of this chapter is adapted from a series of articles written by Wes Byrne and Michael Bukay and published in Ultrapure Water *journal between March and December 1986.*

system will be reduced because of the additional ionic loading of the deionizer feedwater.

A loss in percent salt rejection may or may not be accompanied by changes in other RO performance indicators such as the normalized (corrected) permeate flowrate[1] or differential pressure (system feed pressure minus concentrate pressure). It depends on what is causing the change in membrane performance.

Reverse osmosis membrane elements are rated according to their ability to reject dissolved salts. *Percent salt rejection* refers to the percentage of salts that are rejected by the RO membrane as the permeate water passes through it.

As RO membrane elements foul or deteriorate, percent salt rejection is often affected. Most of the time it decreases. However, certain foulants can act as a dynamic membrane on top of the actual membrane surface, and increase the percent-salt-rejection reading.

Salt rejection may increase with new membrane. New membrane may have small imperfections that foulants may partially plug during the initial days of operation. As the imperfections plug, the rejection improves. (McPherson 1981).

Percent salt rejection can be monitored continuously with on-line instrumentation, or it can be calculated by using the following formula:

$$\text{percent salt rejection} = (\text{feed TDS} - \text{permeate TDS}) \div (\text{feed TDS}) \times 100$$

where total dissolved solids (TDS) is a term used to refer to the concentration of total salts.

A true TDS measurement by definition also includes the concentration of non-filterable organics. However, TDS is usually obtained on-site by measuring the water conductivity. The monitoring instrument converts the reading to TDS using a ratio based upon the conductivity of a particular salt. Since most dissolved organics do not significantly affect conductivity readings, their impact on TDS and rejection is usually neglected.

A salt rejection calculated by this equation does not take into account the increasing salt concentration exposed to the membrane elements as the feedwater is concentrated through the membrane system; thus, such calculations will result in a value that will usually be lower than the design rejection rating for the membrane element.

An arithmetic average of the feed and concentrate TDS can be used in place of the feed TDS in the preceding equation. This new equation will then take into

[1] The normalized permeate flowrate refers to the permeate flowrate adjusted to what it would be at standard conditions of feedwater temperature and pressure. Usually this is 25 °C and the start-up pressure. It has also been called the "corrected" permeate flowrate. The calculation of this value is explained in detail in an article by Bukay (1984).

account the effect of RO system recovery and its possible variation on the salt rejection:

percent salt rejection = (average TDS - permeate TDS) ÷ (average TDS) × 100

where:

$$\text{average TDS} = (\text{feed TDS} + \text{concentrate TDS}) \div 2$$

Typical rejection values for RO systems with new membrane elements range from 90% to 99%. This varies with the particular membrane, operating conditions, and water composition. Increasing the RO system feed pressure will increase the percent salt rejection of the RO membrane because of the production of additional permeate flow. This additional permeate further dilutes the passage of salts through the membrane.[2]

Divalent ions are rejected better than are monovalent ions (Bukay, 1984). Reverse osmosis systems operating on water sources that have a greater concentration of divalent ions than of monovalent ions will have better salt rejection characteristics. Systems that use sulfuric acid injection for pH control will demonstrate better overall rejection than will systems using hydrochloric acid injection. The sulfate ion in sulfuric acid is divalent, whereas the chloride ion in hydrochloric acid is monovalent.

When a loss in percent salt rejection is noticed, the RO technician should take the following steps:

- Check the instrument calibrations.
- Isolate the location of the decline.
- Investigate potential causes of the problem.
- Correct the potential causes of the problem.

Check instrument calibrations. Instrument calibrations are the first thing to check in any troubleshooting operation. In an RO system, instrumentation can mask a loss in salt rejection. It can cause a false alarm. It can even be the cause of a loss of salt rejection if the error results in operation of the equipment out of specification. The more sophisticated the instrumentation becomes, the more important it is to keep it in calibration.

Conductivity and percent-salt-rejection meters. To obtain the data necessary to determine the percent salt rejection of an RO unit, conductivity readings are obtained in both the feed and permeate streams. Frequently, this is performed on-

[2] Water passage through the membrane is a function of the feed pressure, whereas the salt passage is relatively independent of the pressure over normal operating conditions. Therefore the dilution effect of producing more water at higher pressure results in a lower permeate salt concentration and thus a higher system salt rejection.

line, either by two conductivity instruments, or by one with two probes. The output is expressed in terms of micromhos/cm (or microsiemens/cm). This data can then be used to calculate a percent salt rejection based on conductivity.[3] In-line TDS meters will measure water conductivity and then convert it to a TDS reading. An in-line salt rejection meter electronically performs the percent-salt-rejection calculation based on TDS values converted from conductivity readings.

Accuracy of the instrumentation can be verified by measuring the feed and permeate water TDS with a separate handheld meter. The percent salt rejection can then be calculated and compared to the value obtained from the on-line instrumentation. The two values should be relatively close.

The percent-salt-rejection calculations should agree with each other even though the actual feed and permeate readings from the handheld meter may not agree with the readings from the on-line meter. The variation in readings can occur because each instrument may use a ratio between conductivity and TDS that is based upon different ions or salt solutions. (One may be calibrated to a sodium chloride solution while the other one may be calibrated to a calcium carbonate solution, which have different conductivity to concentration ratios.) This is not a significant problem, however, because much of the variation will cancel out in the percent-salt-rejection calculation.

If a discrepancy arises, recalibrate the on-line meter following the instructions of the manufacturer. Before recalibrating, inspect the probe for the possible accumulation of foreign material that may interfere with the reading. Also, verify that the probes are mounted properly according to the manufacturer's specifications. Improperly mounted probes may trap air bubbles or have inadequate water circulation through the probe.

Flowmeters. Perhaps the most neglected calibrations in water systems are those of the flowmeters. During installation, if the flowrate reading looks reasonable, it is often assumed to be accurate, never to be confirmed unless there is an obvious failure.

Flowmeter calibration is of extra importance in an RO system. The correct measurement of the permeate and concentrate flowrates is critical to the successful operation of the system. For instance, the concentrate flowrate may directly affect the rate of fouling and scale formation of the membrane elements.

The normalized permeate flowrate is a critical parameter to monitor in order to determine trends in fouling or membrane deterioration (see "Normalized Permeate Flowrate" in Chapter 5). A flowmeter inaccuracy can either exaggerate or camouflage a change in RO membrane performance.

Frequently, flow monitors that use a paddlewheel-type transducer are used in

[3] This number will normally be slightly different than a TDS-based rejection value because of the nonlinear relationship between TDS and conductivity.

the water treatment industry. One method of calibrating these flowmeters is to electrically disconnect the paddlewheel sensor portion from the meter and feed an electrical signal to the meter (see Figure 6-1). Using tables published by the manufacturer, the meter can be calibrated to the flowrate that corresponds to the signal. This is an easy calibration because it is quick and there is no risk of getting wet. However, it is an incomplete calibration because it does not include the paddlewheel (transducer), and tells you nothing about the accuracy of the sensor.

The paddlewheel may not be reflecting the true flowrate within the pipe because of turbulence or some other problem. It could be jamming from debris, or have worn parts. This would not be evident when using the calibration method described above.

There are two other methods of calibration that are more accurate. One is the bucket-and-stopwatch method. With this method, the water stream is directed to a vessel of known volume (the bigger the vessel, the greater the accuracy). The water supply is turned on and directed to the ground or a drain until the flowmeter reading stabilizes. Then the water is directed into the vessel, and the time it takes to fill the known volume is measured. The volume in gallons is divided by the time in minutes to obtain the actual flowrate in gallons per minute. This should be repeated several times for accuracy. If the variation is slight, the values can be averaged.

The other method is to calibrate a flowmeter to a known flowrate. Then it can be used to verify the calibration of the other flowmeters. As an example, this method could be used to calibrate the permeate flowmeters of a bank of four RO arrays as well as the flowmeters of four primary deionizers, as shown in Figure 6-2.

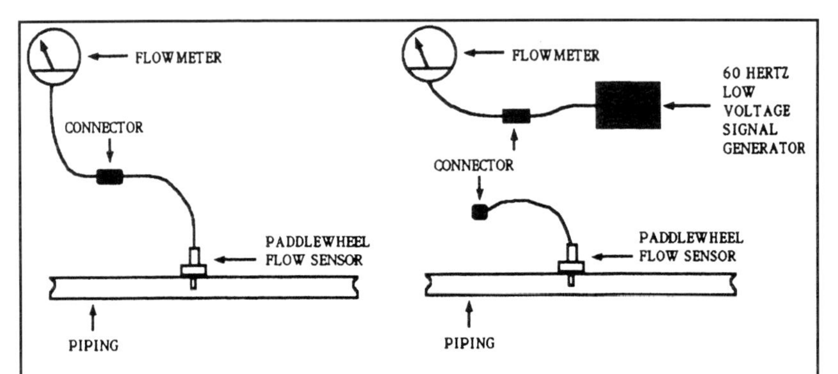

Figure 6-1. One way to calibrate flowmeters that use a paddle wheel as a sensor is to disconnect the sensor from the meter and feed an electrical signal to the meter. Using tables published by the manufacturer, the meter can be calibrated to the flowrate that corresponds to the signal. Unfortunately, this calibration procedure is incomplete because it does not take into account the accuracy of the flow sensor.

The RO permeate flowrate of one array is first calibrated with the bucket-and-stopwatch method. The other three RO arrays are shut down and the valving is set so that all the permeate flow from the first RO goes through one of the primary deionizers. The flowmeter of the deionizer is calibrated to the known flow. This is repeated with each of the other three primary deionizers. The permeate streams from each of the RO units are then directed, one at a time, through one of the primary deionizers, and their flowmeters are calibrated to the known flowrate.

The concentrate flowmeters of the system could be each calibrated individually by the bucket-and-stopwatch method. The RO feed flowrate could then be determined by adding the readings of the permeate and concentrate flowrates.

Flowmeters are most accurate when operating near the middle of their range; thus, an effort should be made during both calibration and operation to keep the meter operating near the middle of its effective range.

A quick method for verifying the accuracy of the feed, concentrate, and permeate flowmeters is to perform a mass balance calculation (Bukay, 1984). This is discussed in Chapter 4.

When more than one technician records flowmeter data, a consistent method should be dictated for reading the flowmeters. If the RO uses digital meters with readings that fluctuate excessively, how the reading is interpreted can vary

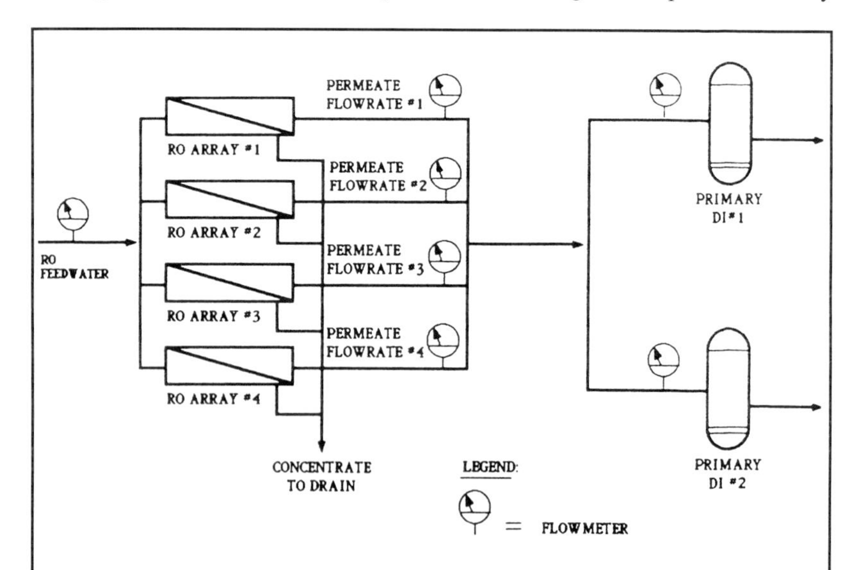

Figure 6-2. An easy way to calibrate many of the flowmeters in this system is to first calibrate one of the RO permeate flowmeters by using the bucket-and-stopwatch method. The calibrated meter is then used to calibrate one of the deionizer flowmeters, which in turn is used to calibrate the remaining permeate flowmeters. See text for details.

dramatically. One method might be to watch the meter for a minute and write down the highest flow reading observed during that time. Some digital flowmeter circuits can be modified so that the readings are averaged over a longer period of time before displaying the reading.

When using rotometers (in-line flowmeters that employ a float), check with the manufacturer to determine whether the flowrate is read in the middle of the float, at the shoulder, or at the top of the float. Record on the rotometer how it should be read.

Pressure sensors. Correct pressure measurements are critical to the successful operation of an RO system. Their measurements allow for the monitoring of the buildup of deposits on the membrane surface, or in the membrane element flow channels. Inaccuracy in the pressure readings will be directly reflected in the normalized permeate flowrate calculation.

If mechanical pressure gauges are used, their accuracy should be periodically verified using a calibrated pressure gauge. Electronic pressure sensors have the potential for greater accuracy; however, they are subject to sensor drift and damage resulting from vibration of the high-pressure pumps or other sources.

It is suggested that an effort be made to avoid mounting electronic pressure sensors directly on the high-pressure stainless steel manifolds in locations where vibration is extreme. To reduce the effects of vibration and pulsation, the sensor can be mounted remotely and connected to the high-pressure piping with a length of 1/4-inch stainless steel or high-pressure nylon tubing.

The location of pressure transducers is critical to the measurement of useful performance data. The taps should not be located immediately downstream of throttled valves or any other type of flow restriction, as these can have a dramatic effect on the reading.

A calibration pressure gauge mounted with a quick-connect fitting is useful for calibrating pressure sensors. A deadweight tester can be used to to calibrate the calibration gauge. The deadweight tester creates an accurate test pressure for a gauge using mechanical weights.

pH. Reverse osmosis systems are designed to operate within a specific pH range. If they are operated outside this range, the membrane elements may be subject to deterioration or scaling. Cellulose acetate membrane is especially sensitive to hydrolysis when operated outside the pH range of 3 to 7. The result is often a loss in percent salt rejection.

The RO pH meters should be regularly calibrated using buffer solutions of a known pH. If the RO system normally operates with an acidic feedwater, a pH buffer solution with a value of 7.0 should be used to zero the instrument, and a buffer with a value of 4.0 should be used to set the gain.

Temperature. Small variations in feedwater temperature do not significantly affect the percent salt rejection of an RO. The temperature readings are important, however, because they are used to determine the normalized permeate flowrate. The accuracy of the feedwater temperature readings should therefore be regularly verified with a laboratory thermometer.

Once the instruments on the RO unit are calibrated, there is greater confidence as to the accuracy of the performance data. If there still appears to be a loss in membrane rejection, then where that loss of rejection is occurring should be pinpointed within the RO membrane system.

Isolate the location of the rejection decline. A loss in salt rejection may be uniform throughout the system, or it could be limited to the front or to the tail end of the system. It is frequently limited to an individual housing or element. The location of the decline can be isolated by following these steps:

- Check the individual housing permeate TDS values.
- Probe the housing (in spiral-wound systems).
- Individually test each of the elements.

Check the individual housing TDS values. The RO membrane elements are housed in a series of tubular vessels (housings) manifolded together with high- and low-pressure piping. A well-designed system contains a sample port located in the permeate stream from each housing. To measure the individual housing TDS values, the water from each sample port is tested for its content of dissolved solids with a handheld TDS or conductivity meter.

For example, suppose that in a 4:2 array, such as the one illustrated in Figure 6-3, the overall percent salt rejection has suddenly dropped from 93.4% to 92.3%. Then the TDS profile might reveal the following:

First-Stage Housing Profile:

Housing #	*Permeate TDS (ppm)*
1	25
2	22
3	49
4	20

Second-Stage Housing Profile:

Housing #	*Permeate TDS (ppm)*
5	36
6	34

There is an apparent problem in housing #3 of the first stage of the array. Its

TDS reading is significantly higher than housings #1, 2, and 4 of the same stage. This data suggests that the loss in percent salt rejection is due to a problem somewhere in housing #3. When the problem is identified and repaired, the overall percent salt rejection will be restored.

Notice that most of the permeate TDS readings in the second-stage housings are higher than the readings from the first stage. This trend is normal. It occurs because the housings from the second stage receive the waste (concentrate) stream from the first stage. Even though the individual second-stage elements may have the same localized percent salt rejection as the elements in the first stage, the second stage has a higher permeate TDS because its specific feedwater contains a higher TDS.

As a routine monitoring procedure, it is useful to obtain a TDS profile of an RO system each month. This allows the operator to establish baseline data for each housing. When a loss of percent salt rejection is experienced, the new

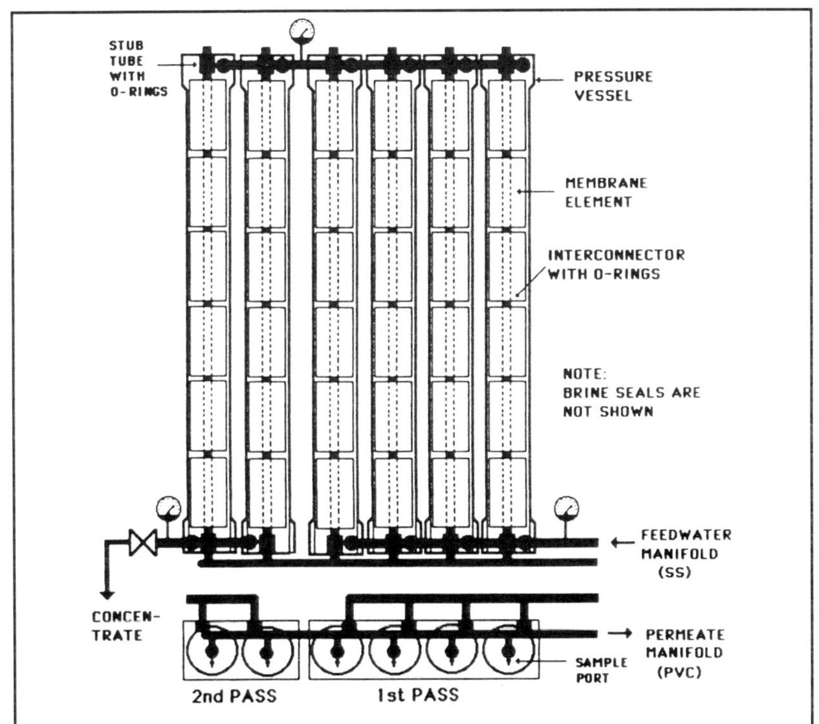

Figure 6-3. Typical arrangement of a spiral-wound RO system in a 4:2 array configuration. The RO membrane elements are housed in a series of tubular vessels manifolded together with high- and low-pressure piping. Sample ports located in the permeate stream from each vessel allow the troubleshooter to isolate a salt rejection problem to individual vessels. See text for details.

profile can be compared to past data for easier interpretation.

Probe the housing (in spiral-wound systems). In spiral-wound membrane systems, the loss in percent salt rejection can now be further isolated within the problem housing, using a technique called probing. Probing involves the insertion of a length of 1/4-inch plastic tubing into the full length of the membrane elements' permeate tubes (Figure 6-4). This can be accomplished by removing the permeate piping manifolds from the housing, or by removing the permeate plug from the opposite end cap (if the membrane element on this end is attached by an interconnector with the end cap).

While the RO system is at normal operating pressures, water is diverted from the permeate stream of the housing(s) in question. After inserting the tubing, a few minutes should be allowed to rinse out the tubing and allow the RO system to equilibrate. The TDS (or conductivity) of the permeate sample from the tubing is then measured with a handheld meter and the data is recorded. This measurement should reflect the TDS of the permeate water being produced by the membrane element at that location. The tubing is then pulled out to the next membrane element and another sample is taken. Results of a housing profile might look like this:

Housing #3 TDS Profile:

Membrane Element #	*Permeate TDS (ppm)*
1 Lead end	25
2	22
3	26
4	20
5	38
6 Tail end	54

These data clearly indicate that the source of the loss in percent salt rejection is the result of the two elements (#5 and #6) located in the tail end of the housing.

Probing the housing allows the troubleshooter to identify the membrane elements that are failing. The technique is limited, however, in that the individual element readings are affected by the flow of water from other membrane elements whose permeate tubes are connected. An element located at the end of a housing may also transport all the permeate from the other elements in the housing. If required, greater accuracy can be obtained by individually testing the elements.

Individually test the membrane elements. This analysis requires the use of an individual element test stand (see Figure 6-5). The test stand consists of a vessel that houses only one membrane element. It allows the troubleshooter to adjust the feed pressure, percent recovery, and feedwater pH (on some units). The recovery setting is particularly important in obtaining a differential pressure that can be compared to other elements. The exact settings should be obtained from the manufacturer's test specifications.

The individual element test stand can use its own separate high-pressure pump, or the unit can be plumbed into the RO feedwater manifolding so that the system RO pump can be used. With the latter arrangement, the acid injection pump stroke setting should be reduced when using the test stand.

Now that the location of the loss in percent rejection has been isolated, the task of finding its cause is greatly simplified. This will be discussed in the next section.

Causes and Prevention of Front-End Salt Rejection Decline

The front end (lead end) of an RO system refers to the particular membrane elements located in the system where the feedwater first enters the element housings. The tail end of the system refers to the elements located where the concentrate water is leaving the system.

This section discusses the causes of front-end rejection decline. Emphasis is placed on corrective actions that can be taken to eliminate these causes of premature RO element failure.

Now that the location of the loss in RO rejection has been isolated (discussed earlier in this chapter), the task of finding the cause is greatly simplified. Similar to signs of a disease in the human body, any reverse osmosis concern will have

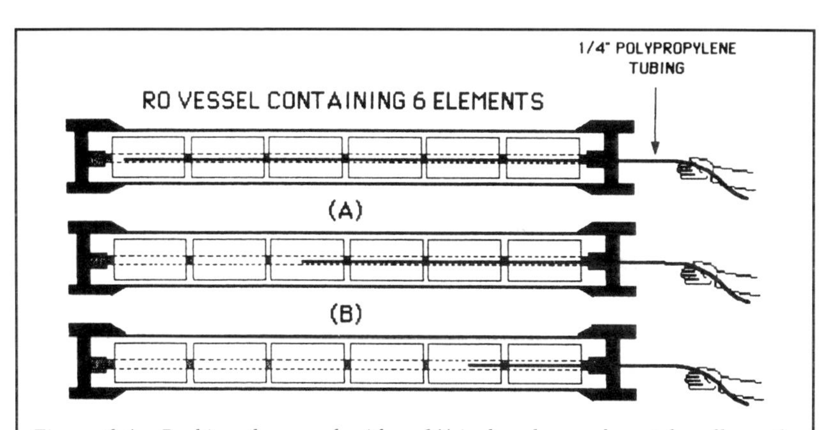

Figure 6-4. Probing the vessel with a 1/4-inch polypropylene tube allows the troubleshooter to identify the membrane elements that are failing. The tubing is inserted into the full length of the permeate tube (A). The TDS of the permeate sample from the tubing is measured with a handheld TDS meter. The tube is withdrawn (B and C), and the procedure is repeated until a TDS profile of each element is obtained. See text for details.

its own set of symptoms. For instance, if an RO system were to be exposed to an excessive amount of some chemical in the feedwater, several things could happen. There might be a loss of salt rejection in the front end where the membrane is first exposed to the chemical, or in the tail end where the chemical is concentrated. In some cases a uniform loss throughout the membrane system would occur. This all depends on the particular chemical, its concentration, and the length of exposure. Knowing the location of the rejection loss can lead to a determination of the specific cause.

Causes of lead end rejection decline. Following are typical causes of lead end rejection decline:

- Excessive acid introduction
- Oxidative attack in polyamide membrane systems
- Hydraulic imbalances
- Heat exchanger inadequacies
- "Compound X" degradation in cellulose acetate membrane

Excessive acid introduction. Acid is typically injected into an RO feedwater for two possible reasons. One is to prevent carbonate scale formation in both CA and PA thin-film elements. The other is to reduce the hydrolysis rate of CA membrane.

In CA membrane systems, injecting the correct amount of acid is critical. Excessive acid concentrations will dramatically increase the hydrolysis rate of CA membrane (see Figure 6-6). This is one of the most common causes of performance deterioration in systems that are otherwise well maintained. Excessive acid can also cause deterioration in a PA-type membrane system, but the

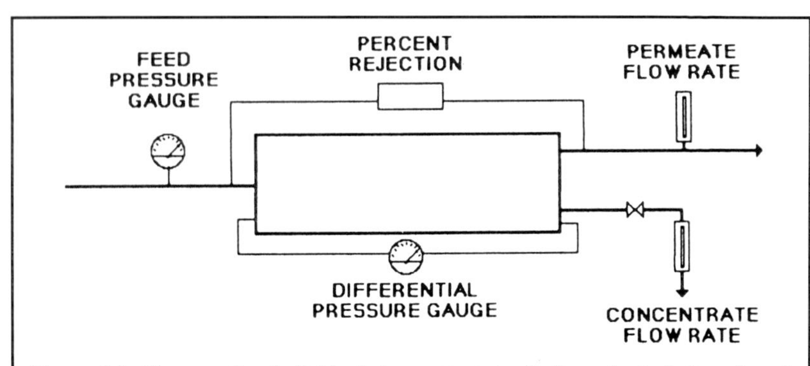

Figure 6-5. The use of an individual element test stand allows the isolation of a salt rejection problem to individual spiral-wound elements.

concentrations usually need to be much greater or the exposure times much longer.

Hydrolysis of CA membrane is the natural breakdown of the cellulose acetate. It is accompanied by a reduction in the salt rejection characteristics of the membrane. When operated within the membrane manufacturer's recommended guidelines, the hydrolysis rate of the CA membrane will not cause a significant loss in the RO system salt rejection during the warranted membrane life. Excessive acid concentrations will speed up this rate of hydrolysis, which can result in a significant loss in salt rejection.

As the excess acid travels through an RO system from the feed to the concentrate, hydrogen ions will permeate through the membrane. As this occurs, the pH of the feed/concentrate stream will increase. Most of the deterioration caused by excessive acid in the feedwater will occur in the lead end of the membrane system where the low pH feedwater first comes in contact with the membrane elements.

Excessive concentrations of acid can result from a poorly adjusted injection control system. The pH control instrumentation may be out of calibration. (Procedures for calibration are described by Wiens [1985].) Sometimes an unauthorized person will incorrectly adjust the controls, or someone may accidentally bump the controls. For these reasons it is suggested that control adjustments be accessible only to authorized personnel.

In some systems it is possible to shut down the high-pressure RO pump without automatically disengaging power to the acid injection pumps. When this occurs, excessive acid can be injected into the feedwater line while the RO is shut down. Upon

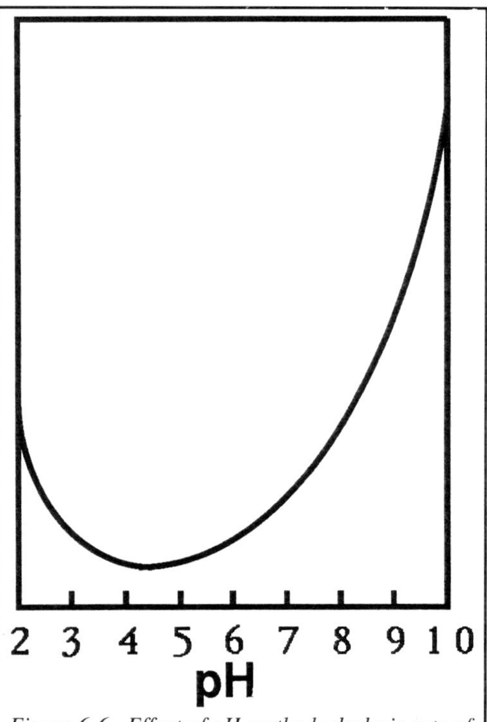

Figure 6-6. Effect of pH on the hydrolysis rate of CA membrane. Note that the hydrolysis rate dramatically increases at both high and low pH. Proper pH control is therefore critical to maintaining the salt rejection characteristics of CA membrane (Vos et al., 1966). See text for details.

start-up of the system, a slug of acid may pass through the elements.

One facility experienced acid-related deterioration of its RO membrane elements after a technician had repaired a high-pressure pump. The technician had turned off the RO power at the quick-disconnect located on the RO unit. In this case, the acid injection pumps were powered by a separate electrical circuit and continued to pump acid.

The RO controls should be wired so that shutting down of the RO system is not possible without shutting down the chemical injection systems. Injection pumps that can be switched to an internal pulse signal should be completely disengaged from power when the RO is shut down.

Concentrate or permeate pipelines are frequently plumbed so that their exit points are below the RO unit. It matters little if these pipelines travel a great distance or not. As long as the point at which the line is exposed to atmosphere is below the unit, a vacuum will be pulled on the RO system while it is shut down (see Figure 6-7).

This vacuum may cause the RO system to drain. It may also siphon the injection chemicals through their connection into the system. These can be chemicals located either upstream or downstream of the RO unit. This siphoning may occur even if a check valve separates the RO elements from the chemical injection day tanks. All it takes is for the valve to leak under vacuum.

It is important that this vacuum be eliminated, either with a vacuum breaker or with an inverted check valve as shown in Figure 6-7. It should be installed at the highest point of the pipeline in question.

If the RO system is draining down while shut down, an aspiration/vacuum

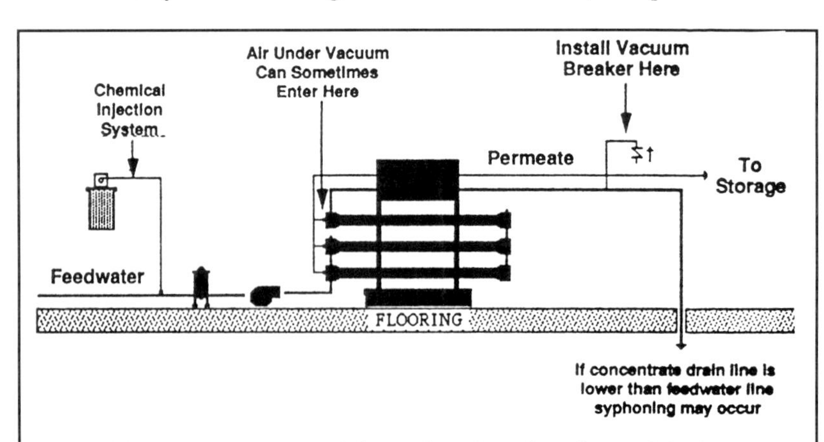

Figure 6-7. Excessive amounts of chemical can be siphoned into RO feedwater if the concentrate or permeate lines are plumbed so that their exit points are below the RO unit. The vacuum can be eliminated by installing a vacuum breaker or an inverted check valve at the highest point of the pipeline in question.

effect can occur when the high-pressure pump is restarted. In refilling a partially empty RO system, the pump may behave as if it had little or no back pressure. It will suck water at great velocities, thus creating a vacuum on the inlet and aspirating the chemicals as the water quickly passes the injection points for the chemical injection. Not only will this introduce excessive chemicals, but it can also mechanically damage the membrane elements.

An automatic shutdown valve can be installed between the chemical day tanks and the RO high-pressure pump. This will reduce the possibility of problems caused by the shutdown. The valve should be inspected on a routine basis to ensure that it is not leaking.

Shutdown valves should automatically open some time prior to start-up of the high-pressure pump. A conservative delay might be 30 seconds between opening the valve and starting the pump. This will allow the RO system membrane housings to fill with low-pressure water. If the RO uses more than one high-pressure pump, there should be a time delay before the sequenced start-up of each pump.

Oxidative attack in PA membrane. The presence of free chlorine or some other oxidizing agent in contact with a PA membrane RO system will cause uniform membrane deterioration. It is a relatively common occurrence, however, to see a similar type of deterioration of PA membrane occurring in just the lead end membrane elements. In most of these cases, the pretreatment system includes a means of removal for the oxidizing agents; therefore, it is commonly assumed that membrane deterioration caused by oxidizing agents should not be occurring anywhere in the system.

There are two factors behind the lead end attack on these systems. First of all, there is a very slight residual concentration of oxidizing agent still present in the feedwater. It could be that either an insufficient concentration or an inadequate contact time has been allowed for complete reaction of an injected reducing agent with the residual oxidizing agent. It might be that an activated carbon filter is undersized, or that its carbon media is mostly oxygenated and requires replacement.The PA membrane in a thin-film spiral-wound membrane element can normally tolerate such slight residual concentrations of chlorine or other oxidizing agents. If iron is present in the RO feedwater, however, the effects of the residual oxidizing agent can be magnified greatly.

Since the iron has been in the presence of oxidizing agents, it will be in the oxidized ferric state (Fe^{+3}), which is insoluble. As it enters the membrane system, it will tend to fall out of suspension in the lead end membrane elements. When the residual oxidizing agent is exposed to the iron on the membrane, its oxidative aggressiveness will be enhanced by the ability of the iron to change valence states. It is as if the iron wants to revert to its reduced state, and in the process lends greater oxidative characteristics to the oxidizing agent.

The solution to the problem is to correct either of the two discussed parameters. One way would be to use better filtration upstream to remove all of the residual iron, manganese, zinc, or copper. The other way to correct the problem would be to insure that all oxidizing agents are removed such that the oxidative/reductive state of the water is greatly reductive.

Hydraulic imbalances. The possibility of the RO system draining during shutdowns has already been discussed. If the high-pressure pump is started before a drained-down system has time to fill, the lead end membrane elements will be exposed to higher-than-normal water velocities. This can hammer the elements, causing mechanical damage to them. In spiral-wound systems, it can cause an effect known as *telescoping,* where the outer membrane layers of the element unravel and extend downstream past the remaining layers (see Figure 6-8A).

Most spiral-wound membrane element manufacturers now seal an outer fiberglass wrap to a plastic anti-telescoping device (ATD) on the element ends. It assists in giving the membrane envelopes more structural support.

If a valve is used to throttle the high-pressure pump, its setting may be critical in preventing telescoping. If the valve is open too far, the excessive flowrates can also cause telescoping. The higher flowrates through these lead end elements create an excessively high pressure drop (from feed to concentrate) across the elements. This pressure drop acts as a great force, compressing the membrane elements. Frequently in spiral-wound systems, this compression will cause movement of the lead end elements that may break the permeate interconnector O-rings of the lead end elements.

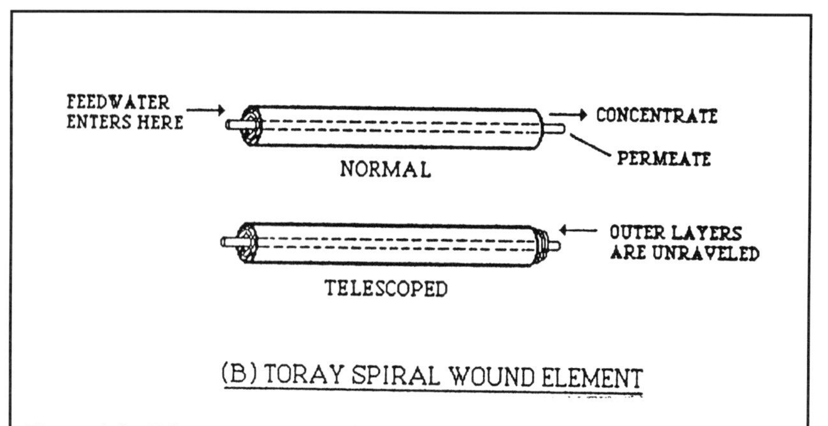

Figure 6-8. Telescoping in typical spiral-wound elements occurs when the outer membrane layers of the element unravel and extend downstream past the remaining layers.

Heat exchanger inadequacies. Any RO system with a heat exchanger as part of its pretreatment system is a candidate for high temperature spikes. The severity of these spikes will depend on the design of the heat exchange system, and on the temperature of the water or steam that is being used in the heat exchanger. In some cases, PVC pipe plumbed downstream from the heat exchanger has actually melted from temperature spikes. Until these slugs of hot water cool down as they pass through the RO elements, they will be causing a dramatically increased rate of hydrolysis in CA membrane elements, or mechanical damage in either CA or PA elements.

The high water temperature can cause failure of plastic parts of the membrane elements; telescoping of spiral-wound elements; breakage of the element interconnector O-rings; and can cause what is known as glue-line failure, a mechanical leaking of feedwater into the permeate water through the end of the membrane envelope where the two layers of membrane and support material have been glued.

Usually these high temperature spikes are created when the RO system starts up after being shut down for a short amount of time. While shut down, the RO feedwater comes to an equilibrium temperature with the hot water or steam stagnant in the heat exchanger (or possibly still flowing through the exchanger). When the RO feedwater valves open, this hot slug of water is introduced to the lead end of the RO system.

One way to prevent high temperature excursions is to use a control system that regulates the amount of steam or hot water that passes through the heat exchanger. A portion of the feedwater can be recirculated around the heat exchanger to assist in smoothing out temperature variations during RO operation. Part of this recirculation system should contain a dump valve that will dump the RO feedwater if the temperature becomes excessive.[4] All these precautions are particularly critical if the heat source is steam or hot water (above 120 °F).

An economical suggestion is to employ a source of waste heat, which is usually available at most industrial facilities that use cooling towers. With a highly efficient plate-and-frame heat exchanger, the feedwater temperature to the RO system can be increased, which will dramatically increase the permeate output of the RO system. This will enable the system to make more water during the colder months of the year at reduced operating costs.

One concern with this or any other type of heat exchanger is the possibility of leaks developing between the hot water/steam side and the RO feedwater. Normally, the hot water/steam side will contain a corrosion inhibitor. These agents can cause damage to the RO membrane if they enter the feedwater stream. Any heat exchanger should be routinely checked to verify that the hot water/

[4] Sometimes an automatic dump valve will be used immediately prior to the RO unit. When the system starts up, this valve always dumps the initial feedwater to drain, to assist in preventing high temperature spikes or chemical slugs.

steam side will maintain its pressure if completely valved off.

"Compound X" degradation. Predominantly found in well waters located in certain areas of the southwestern United States, a type of CA membrane deterioration has been documented and named "Compound X." It is characterized by a relatively rapid lead end loss in salt rejection only in the presence of free chlorine. It has been speculated that "Compound X" deterioration may be caused by some chlorinated species created by the reaction of some unknown substance in the feedwater with free chlorine. Another speculated cause is attack by a type of bacteria that thrives only in the presence of low concentrations of free chlorine. (See "Membrane Compatibility" in Chapter 3.)

Information from several sources indicates that the "Compound X" problem is caused by chlorine attack catalyzed by the presence of iron (or other transition metals). It is known that the oxidation potential of chlorine increases in the presence of oxides of iron. This increase in oxidation potential can occur with either PA thin-film or CA membrane systems when operated in the presence of low concentrations of chlorine and iron.

If "Compound X" is suspected, ammonia or ammonium sulfate can be added to the RO feedwater to convert the free chlorine into chloramines (Mark Wilf,). Another common solution involves simply removing all the free chlorine either by carbon filtration or by the injection of sodium bisulfite into the feedwater. With these solutions, and with the increased popularity of PA thin-film membrane (which necessitates the removal of the free chlorine), verified "Compound X" problems have become a rarity.

Front-end rejection decline can be caused by excessive acid introduction, oxidizing agents, hydraulic imbalances, excessive feedwater temperatures, and possibly "Compound X." A properly monitored and maintained RO system will allow the early identification of the specific causes of front-end deterioration so that corrective measures can be taken. The preventive measures described above allow the RO technician to greatly reduce the risk of premature membrane element failure due to front-end deterioration.

Causes and Prevention of Tail End Salt Rejection Decline

The cost effectiveness of RO is greatly dependent on the permeate flow recovery of the system. The permeate recovery is defined as the percent of feedwater that is converted into permeate water. Reverse osmosis system operating costs can be reduced by modifying the RO to operate at higher recovery. This can save in city water cost, pretreatment cost, energy consumption, and reduced neutralization requirements. With a higher permeate recovery, however, the dissolved salts and foulants become further concentrated within the RO membrane elements, thus increasing the risk of scale formation and/or fouling.

Extreme scaling and fouling can result in a decline in salt rejection. The location of such a decline is most likely to occur in the tail end of the system where the concentrations are the highest.

Severe fouling of the tail end of the system can occur from either suspended solids that are present in the RO feedwater, or from heavy biological activity. Another common cause of tail end salt rejection decline is high feedwater pH with CA membrane, resulting in a greater rate of membrane hydrolysis.

System recovery too high. As the feedwater passes through the RO elements from the feed side to the concentrate end of the RO system, and the permeate water is removed, the feedwater salts become more concentrated. In a 75%-recovery system, the concentrate water contains almost four times the concentration of salts that were present in the feedwater.

A small error in operating the RO at the desired recovery can have dramatic consequences. Suppose a valve is not working correctly, or a flowmeter is not calibrated, such that an RO system is operating with a 5-gpm concentrate flowrate instead of the desired 10-gpm flowrate. If the 10-gpm flow was desired to produce a permeate recovery of 75%, this difference in flowrate will result in the RO operating at 85.7% recovery. This means that now the concentrate water contains almost seven times the concentration of salts in the feedwater.

In high-rejecting systems, this concentration factor is calculated by the following equation:

$$\text{concentration factor} = 1 \div (1 - \text{recovery})$$

With this higher concentration of salts, the potential for scale formation is greatly increased. Also, the lower concentrate flowrate reduces the amount of crossflow turbulence, which will increase the concentration polarization occurring at the membrane surface. This in turn also increases the potential for scale formation.

Concentration polarization is a term used to describe the increased salt concentration that occurs at the surface of the RO membrane (see Figure 6-9). As the permeate water passes through the membrane, the concentration of the rejected salts builds up at the high-pressure side of the membrane surface. The amount of this increase of concentration over that in the bulk stream will depend on how quickly the salts diffuse back into the bulk stream. This rate of diffusion depends on the turbulence of the bulk stream.

A higher salt concentration at the membrane surface results in an increase in salt passage through the membrane. In extreme cases, the increase in local salt concentration can lead to saturation of one or more of the solution components and can result in precipitation (scale formation) on the membrane surface.

Common causes of operating an RO system with too high a system recovery rate (for a given feedwater composition) are operator error, flowmeters out of calibration, and unnoticed changes in the feedwater composition. Preventive measures include daily verification (and readjustment if necessary) of the recovery rate, periodic calibration of flowmeters (see the beginning of this chapter), performing mass balance calculations (Byrne, 1985), and obtaining a periodic analysis of the feedwater composition.

Extreme fouling or scaling. For the purposes of this discussion, scaling is defined as the precipitation of slightly soluble salts, whereas fouling is defined as the buildup of suspended solids and organics. Fouling is considered a normal occurrence in the operation of an RO system in that it can be controlled with proper monitoring and cleaning maintenance (Bukay, 1984).

Scaling may or may not be controllable, depending on the nature of the scale formation. Some types of scale, such as sulfates or silica, can be extremely difficult to clean if allowed to precipitate. Such scale formations can rapidly cause the demise of the RO system. The potential for this sort of scale formation should therefore be controlled with proper pretreatment, design flowrates, and chemical injection systems.

In spiral-wound systems, scale and/or foulants can block off portions of the material separating the membrane envelopes. This material is typically a layer of crisscrossed plastic threads designed to increase the crossflow water turbulence. This turbulence increases the rate of salt diffusion back into the bulk stream.

Scale and foulants can reduce the effectiveness of the spacing material. They can cause a reduction in the flow turbulence, which will result in an increase in the concentration polarization at the membrane surface. The membrane will see a higher concentration of salts and pass a proportionally higher concentration of salts into the permeate water.[5]

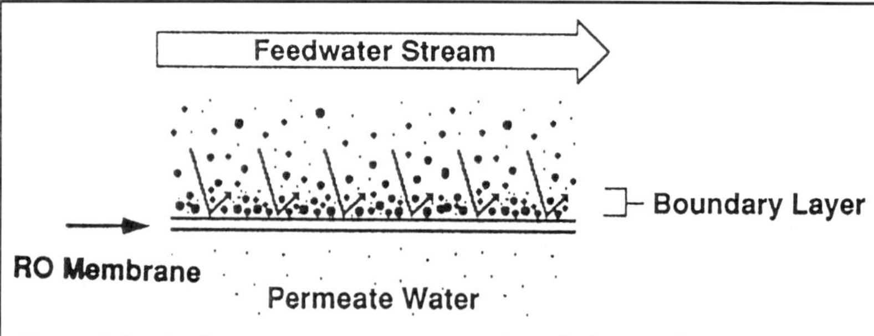

Figure 6-9. As the permeate water passes through the membrane surface, the concentration of the rejected salts builds up at the high-pressure side of the membrane surface and exceeds the bulk concentration. In extreme cases, the increase in local salt concentration can lead to saturation of one or more of the solution components and result in precipitation (scaling) on the membrane surface. See text for details.

Once this condition of reduced turbulence exists, it becomes more difficult to clean the membrane system. There is a reduced ability to force the cleaning solution into these blocked areas of the spacing material.

This blocking off of flow channels can be of even greater concern between the fibers of hollow-fiber membrane elements. Hollow-fiber elements lack mechanisms (such as the spacing material found in spiral-wound elements) to assist in keeping the salts and foulants churned up and in solution.

When fouling or scaling is allowed to proceed without cleaning, RO elements will physically plug to the point that the crossflow pressure drop from feed to concentrate increases dramatically. These high pressure differentials create a force across the membrane elements that can mechanically damage the elements. Element telescoping may be evident in spiral-wound systems. With severe mechanical damage, leakage of water into the permeate can occur that will result in a lower system salt rejection.

Some scale formations can be effectively cleaned out of membrane elements if the cleaning is performed before the pressure drop becomes excessive. A 15% increase in pressure drop is usually a safe limit to use before cleaning carbonate or iron scale.

It should be possible to prevent the formation of scale using the proper system recoveries and chemical injection. Scale inhibitors have been known to control carbonate scale formation up to a Langelier Index value of +2.0. Sulfate and silica scale formation is a much slower precipitation process and is even easier to control using the proper scale inhibitor concentration.

If the system suddenly experiences scale formation when this was not a problem previously, the operating conditions should be investigated. It is possible that the feedwater composition has changed. The solution to preventing further scaling may be as simple as injecting a higher concentration of inhibitor.

Another possible solution is to modify the system so that the high concentrations of salts are rinsed out of the RO system when it shuts down. Scale inhibitors only slow down the precipitation process. An RO system containing supersaturated concentrations of salts should not be allowed to sit dormant for extended periods of time without flushing out the salts. This flushing operation can be accomplished by running pretreated feedwater through the RO system at low pressure (less then 60 psig) until the concentrated water has been displaced.

When sulfuric acid is used for pH control in conjunction with a scale inhibitor, it is wise to ensure the presence of scale inhibitor prior to the point of acid injection. Otherwise, the localized areas of high sulfate concentrations (which are present before the sulfuric acid thoroughly mixes) can accelerate the

[5] Salt passage across an RO membrane is largely a function of the concentration of salt at the rejecting membrane surface. Anything that causes an increase in the concentration polarization effect (i.e., scaling, fouling, or reduced concentrate flowrate) will also result in an increase in salt passage that is proportional to the increase in concentration polarization.

precipitation process upstream of the point at which the inhibitor is injected. The inhibitor should be injected in the same vicinity as or upstream of the sulfuric acid injection.

Sodium hexametaphosphate (SHMP) is one of the more inexpensive scale inhibitors. Unfortunately, it is difficult to dissolve in the day tank. It also breaks down into a less effective orthophosphate with time. In fact, the breakdown phosphate products can themselves precipitate as insoluble salts. Particularly if located in the sun, the sodium hex solution should be dumped and remixed every 3 days.

On feedwaters with low scaling potential, SHMP is usually injected in the RO feed stream at about 5 ppm. Its effectiveness drops off rapidly at lower concentrations.

There are also available commercial substitutes that are more effective than SHMP at lower concentrations. (See "Prevention of Scale Formation" and "Scale inhibition and dispersion" in Chapter 3.) The most common of these will use polyacrylates, organophosphonates, or a combination of the two. The organophosphonates utilize more stable phosphate groups than those of SHMP. (Obtain the supplier's recommendations for desired concentration before using.)

In most cases, simple in-house tests are not available to check for the correct concentration of the scale inhibitor. Samples can be sent away to a qualified laboratory, possibly to the inhibitor supplier.

The injection pump output should be verified. Some chemical injection pumps will not pump their publicized volume with viscous scale inhibitors or in the presence of excessive RO feedwater pressures. This flow verification can be accomplished by checking the rate at which the pump draws diluted chemical from a graduated cylinder. Also, the day tank chemical usage should be recorded daily and used to verify correct chemical consumption as a function of the daily RO feed flow volume.

Feed pH too high. For CA membrane systems, it is necessary to reduce the feedwater pH for most water sources by injecting acid. This decreases the rate of membrane hydrolysis. As shown in Figure 6-6, the rate of hydrolysis dramatically increases as the pH decreases from 4 to 2 and as the pH increases from 6 to 10. The target pH range for CA systems is usually from 5.0 to 7.0. Too much acid (low pH) may result in front-end membrane deterioration, as described earlier in this chapter. Insufficient acid (high pH) can cause tail end deterioration.

With PA thin-film membrane systems, hydrolysis is not a concern. This makes it possible to operate on some waters without acid injection (Kraft, 1985). In these cases, the system relies on low permeate recoveries, softening, or a scale inhibitor to prevent calcium carbonate scale formation.

For many systems that have been relying solely on acid injection to control scale precipitation, savings can sometimes be achieved by also injecting a scale inhibitor. Not only will there be a reduction in the amount of acid required, there will be a dramatic reduction in the amount of carbon dioxide produced. Since carbon dioxide will readily permeate an RO membrane, it can deplete anion exchange resin that may be located downstream to polish the water.

If CA membrane systems are operated for an extended time period at a feedwater pH much higher than 6.0, they will first show a loss in salt rejection in the tail end of the membrane elements. Because of the poor rejection of carbon dioxide (in equilibrium with carbonic acid, H_2CO_3) relative to the rejection of the bicarbonate ion (HCO_3^-), it is common in a 75%-recovery RO system to see a concentrate (tail end) pH of about 7 when the feed pH is 6. Therefore, if the feedwater pH of a CA membrane system is too high, the increased hydrolysis rate of the membrane will be most apparent in the tail end elements of the system.

As previously discussed, another reason for injecting acid in both CA and PA membrane systems has to do with the potential for calcium carbonate scale formation. This potential will depend largely on the concentration of the particular ions, calcium and bicarbonate; as well as on the TDS, the pH, and the presence of scale inhibitor. Acid injection reduces the pH, which increases the solubility of the carbonate scale. It also reduces the concentration of bicarbonate ions by converting a significant portion of the ions into carbon dioxide according to the following reaction:

$$HCO_3^- + H^+ \longrightarrow CO_2 + H_2O$$

If the tail end of an RO system becomes scaled with calcium carbonate,[6] the scaling (if not extreme) may be removed by recirculating a low pH cleaning solution through the affected housings. If the degree of scale formation is too severe, it may not be possible to get the cleaning solution into all the flow channels of the membrane elements, and the elements will require replacement.

The following discrepancies, easily corrected by a vigilant RO technician, are common factors in operation of an RO system at an excessively high pH:

- The pH instrumentation is failing or out of calibration.
- An air lock is present in the acid injection pump.
- The diaphragm in the acid pump may be ruptured or leaking.
- The acid injection pump may be undersized.

Excessive bacteria. In CA or PA membrane systems where little or no biocide is present in the feedwater, the possibility for excessive biological growth exists. This is compounded if the system does not operate continuously and lies stagnant

[6] The tail end will scale first because of the increased concentrations of calcium and bicarbonate and because of the higher pH caused by the poor rejection of carbonic acid by RO membrane.

for extended periods of time. If bacterial growth is allowed to proceed without cleaning, the membrane elements will physically plug to the point that the crossflow differential pressure increases. If this continues, the element flow channels in the spacing material may become blocked, and it will not be possible to clean the elements. Eventually, the differential pressure will increase to the point that physical damage occurs.

There are some types of bacteria that can destroy CA membrane. If allowed to sit stagnant in an RO system without sufficient biocide, they can quickly destroy the membrane elements.

It was previously thought that these bacteria types were cellulose-degrading, and used the cellulose acetate membrane as a food source. It is more common, however, to find bacteria that create low pH conditions around their biogrowths. In these cases, the membrane damage near the bacteria colonies is actually the result of hydrolysis due to the low pH around the biogrowths. This hydrolysis can cause a dramatic decline in salt rejection and an increase in permeate flowrate (Luss, May 1993).

Once introduced into an RO system, bacteria may attach themselves anywhere in the system. From this point they will propagate downstream until all downstream housings and/or elements are infested. Because of this downstream growth, membrane deterioration due to biogrowths frequently occurs first in the tail end stages of the RO system, though the bacteria may have originated upstream.

If it appears that heavy biofouling is occurring with a CA membrane system, the system should be cleaned and disinfected as soon as possible. Unfortunately, during the process of cleaning out the biogrowth, the salt rejection may decline further. The bacterial growth may sometimes serve as a dynamic membrane that improves the salt rejection of the damaged membrane. The full extent of the damage may not be apparent until the biogrowths are cleaned from the system.

To prevent biological problems from reoccurring, several steps may be necessary:

- For CA membrane RO systems, increase the concentration of the biocide in the feedwater. It may be necessary to continuously inject enough sodium hypochlorite to raise the level of free chlorine to higher than 1 ppm to stop certain strains of bacteria. Consult the membrane manufacturer before using high chlorine concentrations.

- If the preceding step is not possible, then shock treatments with high concentrations of chlorine (for CA membranes) or peracetic acid (for CA or PA memmbranes) for short periods of time may be used. Consult the membrane manufacturer for the desired concentration and frequency.

- Increase the running time of the system. It is best if an RO system operates more than 70% of the time. Never allow the system to sit stagnant for 2 days or longer without the presence of a biocide.

- Sanitize stagnant areas of the RO pretreatment system, such as storage tanks and media filters. Remove tanks and filters that are in use only part-time, or use recirculation loops around the equipment (Collentro, 1985).
- With municipal water sources, investigate whether the city is adequately purifying the water before dispensing it. Possibly insufficient biocide is being added to the water at the treatment plant.

Waste backing up the concentrate stream. In some installations, the RO concentrate pipe is plumbed directly into a chemical waste line. This should be avoided if possible by having the concentrate line open to atmosphere before draining by gravity into the waste line. This prevents the possibility of the waste line backing up into the RO system, and thus removes the potential for some waste chemical attacking the tail end membrane elements.

Tail end rejection decline is usually the result of the deposition of scale and/or foulants within the RO elements. This is commonly caused by operating the system at too high a recovery rate, providing improperly adjusted feedwater pH, adding insufficient amounts of scale inhibitor, or by the presence of excessive amounts of bacteria.

A properly monitored and maintained RO system will facilitate the early identification of the specific causes of tail end deterioration so that corrective measures can be taken. The preventive measures described above allow the RO technician to greatly reduce the risk of premature membrane element failure due to tail end deterioration.

Causes and Prevention of Isolated and Uniform Salt Rejection Decline

Although the location of poorly rejecting RO membrane elements can often be traced to the lead or tail end of the system, as discussed earlier in this chapter, it is possible for a loss in salt rejection to occur in one or more isolated elements that may be located anywhere in the system. This may be due to a manufacturing problem, a defective or missing mechanical seal, the initial stages of bacterial infestation, or other causes.

There are some problems, such as exposure to excessive concentrations of oxidizing agents or improper cleaning techniques, that can result in a uniform loss in salt rejection in all the elements throughout an RO system. Sometimes numerous problems occur throughout the system that may create the appearance of a single cause of overall rejection loss. Understanding the various causes of salt rejection decline will insure that efficient and cost-effective actions are taken.

Causes of isolated rejection decline: *O-ring or brine seal damage.* In spiral-wound RO systems, O-rings are used to seal the permeate water tube interconnectors of adjacent elements. They prevent the intrusion of the high-pressure feedwater into the low-pressure permeate water. Damaged O-rings will result in an increase in the permeate water salt concentration in that section of the system.

If an O-ring breaks during operation, it usually happens in the front end of a housing. This is where most of the movement of elements within the housing will occur as the system starts up and shuts down. It is therefore important that this movement be minimized.

Thin slices of plastic pipe, called shims or spacers, can be placed around the male section of the end element's end cap adapter fitting (see Figure 6-10). The

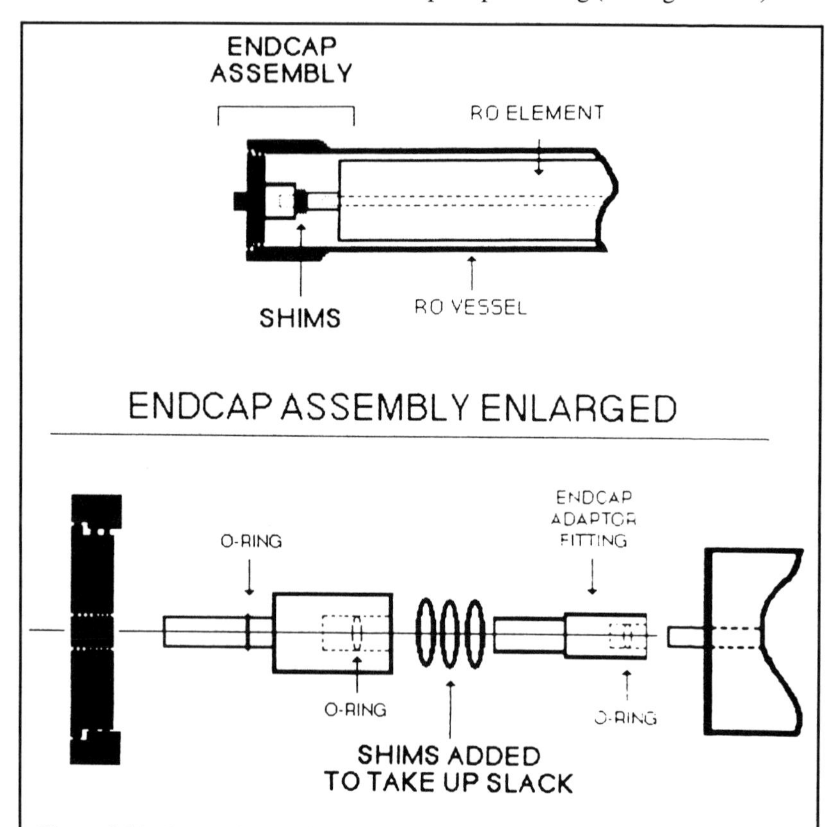

Figure 6-10. In spiral-wound systems, O-ring breaks are usually found in the front end of the vessel, where most of the movement occurs. Thin slices of plastic pipe, called shims or spacers, can be placed around the male section of the end-cap adapter fitting to take up slack and minimize movement of the elements within the vessel.

shims are added until there is some difficulty replacing the end cap of the housing.

Brine seals are the plastic or rubber devices that seal the outside of one end of the spiral-wound elements against the wall of the RO housing. They prevent the bypassing of feedwater around the element.

If a brine seal is forgotten during installation of an element, or if it rolls over during installation or operation, part of the feedwater will bypass the element (see Figure 6-11). Yet the membrane will still be exposed to the same pressure and will be permeating the same amount of water. This will leave less crossflow through the element to create the turbulence through the spacing material. The result will be an increase in concentration polarization, and possibly an increase in the fouling and scaling rate of that particular element. As the affected element becomes more plugged, it may reduce the flow to elements downstream and increase their fouling and scaling rate.

Although there may be a location on each end of a spiral-wound element for a brine seal, only one seal should be installed per element. Occasionally, brine seals will be inadvertently installed at both ends of a spiral-wound element. In one case, this created a problem when combined with a multitude of design deficiencies that were built into this particular system. While shut down, the system would drain through a concentrate line that was plumbed to drain below the RO unit. (For a diagram of this common error, and procedures to correct it, see Figure 6-7.)

In addition, there was no time delay between the start-up of the high-pressure pump and the opening of the RO feed isolation valve. When the system started up, the elements were exposed to high velocities of water. The brine seals on each end of the element prevented the water from quickly reaching the outside of the element. The high-pressure feedwater on the inside of the elements subsequently

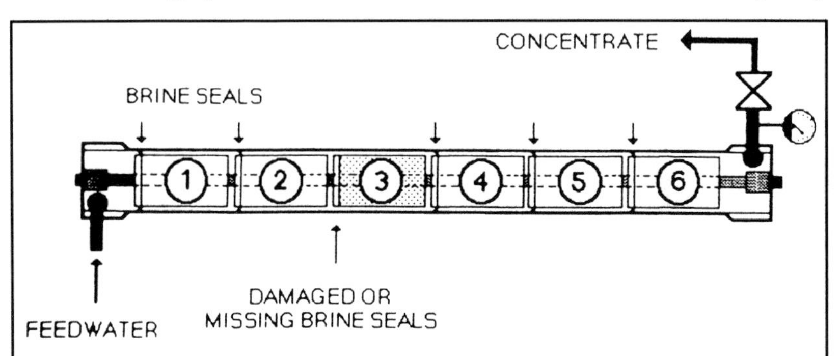

Figure 6-11. Brine seal loss or damage in element #3 allows part of the feedwater stream to bypass the element. Because there is less crossflow through the element to create the necessary turbulence to keep the salts in solution, element #3 is likely to become scaled. As time passes and it becomes more plugged, it will reduce the flow through all elements in the vessel and greatly increase their tendency to scale.

blew out the outer fiberglass wrap on many of the elements.

Excessive permeate pressure. The permeate pressure should never exceed the feed pressure in spiral-wound systems; otherwise, permanent damage to the structure of the membrane elements may occur.

In some water treatment systems, the RO permeate line is plumbed directly into an ion-exchange system, thus creating a back pressure.[7] As long as this pressure can bleed itself off through the filter or DI unit before the feed pressure bleeds off through the concentrate line, there is no cause for alarm. But if there is a blockage in the filter or DI unit, and the RO is shut down, this permeate pressure could exceed the feed pressure. It takes very little permeate back pressure to blow out the envelope leaves of spiral-wound elements.

Hollow-fiber membrane elements are more resistant to back-pressure damage. The fibers in their elements can reportedly withstand as much as 50 psig of permeate back pressure (Permasep, 1982).

If an operator notices that an RO system is operating with a closed permeate valve or some obstruction in the permeate flow, causing the permeate pressure to increase, the permeate pressure should be released before shutting down the system. As long as the system is not shut down, the membrane elements will not be damaged.

If a valve is closed on the permeate line, the permeate pressure will increase until it equals the feed pressure. This poses a safety hazard with plastic permeate lines that are likely to burst. However, the breaking of the permeate line will relieve the permeate pressure and prevent damage to the membrane elements.

In standard spiral-wound systems, it is suggested that the treatment system be designed such that the permeate pressure never exceeds 20 psig. If valves have been plumbed into the permeate line, then an automatic high permeate pressure shutdown switch should be installed with a conservative setting (20 psig or less). A pressure relief valve can also be installed into a tee in the permeate line. If possible, avoid the installation of isolation valves in the permeate line.

Bacteria attack has been discussed earlier in this chapter as a cause of tail end rejection decline. Bacteria can start infesting an element anywhere in the system. If caught in its earliest stages, the downstream movement of bacteria and potential membrane deterioration can be prevented by disinfection of the membrane system. Be sure to follow the recommendations of the membrane manufacturer when disinfecting.

Manifold obstruction. Sometimes an object will work its way into the entrance

[7] This back pressure means that additional feed pressure will be required to make the same permeate flowrate as if there were no back pressure. The additional feed-pressure requirement will be equal to the amount of permeate back pressure.

or exit of a housing, or into the high-pressure manifolding. Possibly a rag was accidentally left in the manifold, or perhaps a hole was incompletely drilled during construction of the manifold. In these cases where the flow through the housing or manifolding is partially blocked, the membrane elements will still be making permeate water. However, because the obstruction has prevented the normal flowrate from passing into or out of the housing, the crossflow rate will decrease. This will increase the concentration polarization, and extreme fouling and/or scaling may occur.

System draining when shut down. The cause and prevention of an RO system draining down when off-line has been discussed earlier in this chapter. This section will describe its possible effects on the top housing elements of an RO system.

In RO systems suffering from drain-down, it is common that the systems will drain down to approximately the same height of water each time the high-pressure pumps are disengaged. For instance, a system may drain all its water down out of its highest housings and down to the middle of its second highest housings (see Figure 6-12). This system may then develop more rapid biological growth in the housings whose elements are exposed to air. The oxygen in the air combined with the dampness left behind by the water make for excellent breeding of bacteria.

With CA membrane systems where free chlorine was initially present in the feedwater, there will be minimal chlorine left behind when the water drains down. Therefore, biological growth in systems that drain down can be a problem for either CA or PA RO elements.

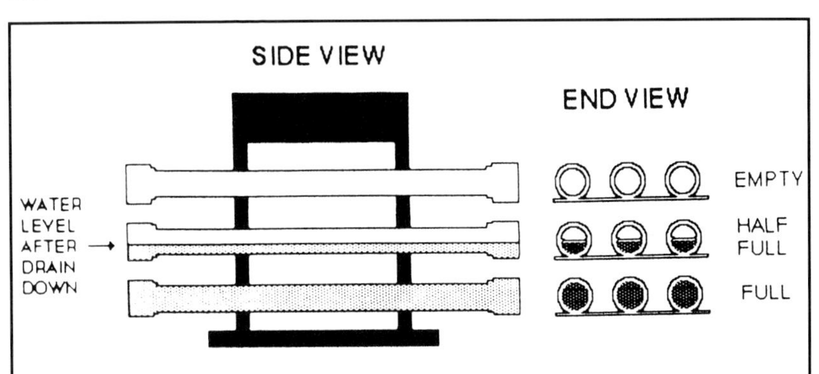

Figure 6-12. An improperly plumbed RO system may drain down to approximately the same height of water each time the system shuts off. In addition to creating hydraulic imbalances, the elements exposed to air are likely to become biologically fouled. See text for details.

The empty or partially empty housings may also be subjected to hydraulic damage when the system starts up. The feedwater will rush in to displace the air. The velocity of this water will depend upon whether or not the system has been allowed to fill before the high-pressure pump is engaged. In any case, it is best if the system is prevented from draining while in the shutdown mode.

RO element manufacturing problem. Occasionally a client will discover that he has received one or more "bad" elements. If an element has a fault, it will usually be evident immediately upon start-up (if the system is properly checked out) or soon afterwards. Element manufacturing problems are rare, however, because of the extensive quality assurance most membrane manufacturers perform.

When a manufacturing problem does happen, it may be mechanical, such as a poor glue line where the two sides of a membrane envelope are not perfectly glued together. It may also be a problem with materials. The plastics may not be able to withstand the hydraulic forces and/or temperatures. The manufacturer should be contacted if any of these problems are suspected.

How new PA thin-film membrane elements are stored prior to start-up is critical. Some membranes have been known to change their performance characteristics during storage, usually by losing permeate flow and increasing in salt rejection. New PA membrane may contain residual amines that can react with biocides such as formaldehyde, glutaraldehyde, hydrogen peroxide, or peracetic acid solutions, also resulting in a loss of permeate flow.

Elements moved from their original position. Moving RO elements from their original position can cause much confusion. Days may be spent troubleshooting an RO system, trying to understand why the tail elements of its first stage have dropped in salt rejection. The RO technician suddenly realizes that these particular elements had been moved out of their original front-end position a year ago last Christmas.

It is suggested that when new elements are installed, their serial numbers are recorded according to their position in the array. If the elements are removed at some time during their operation, they can be installed back in their original positions. If it becomes necessary to move or replace all or some of the elements, the new serial numbers should be noted according to their position.

Sometimes it may be desirable to move RO elements around. For example, suppose permanent scaling were experienced in an RO system to the extent that the second-stage elements of a two-stage system required replacement. If the second-stage housings were reloaded with the new elements, they might be quickly contaminated by residual scale particles possibly present in the first-stage older elements. Also, any residual scale present in the first stage can catalyze the precipitation of new scale that may fall out in the second-stage elements.

It would be best to install the new elements in the first stage. It is also necessary to keep the flowrates balanced. If these new elements were all placed in only one housing (of a spiral-wound system) of two parallel first-stage housings, there is a chance that this housing would experience more crossflow than the other parallel housing, which may contain slightly "scaled" elements. This could then increase the rate of scaling in those already scaled elements.

In these situations, it is recommended that the new elements be split up and installed with the same number of elements in each front end of all the lead housings in the system. The remaining elements should then be installed downstream according to how much relative scaling each has probably experienced.

For example, the front-end elements are probably less scaled than the tail end elements, and should be reinstalled downstream of the new elements. An example of this rearranging is shown in Figure 6-13.

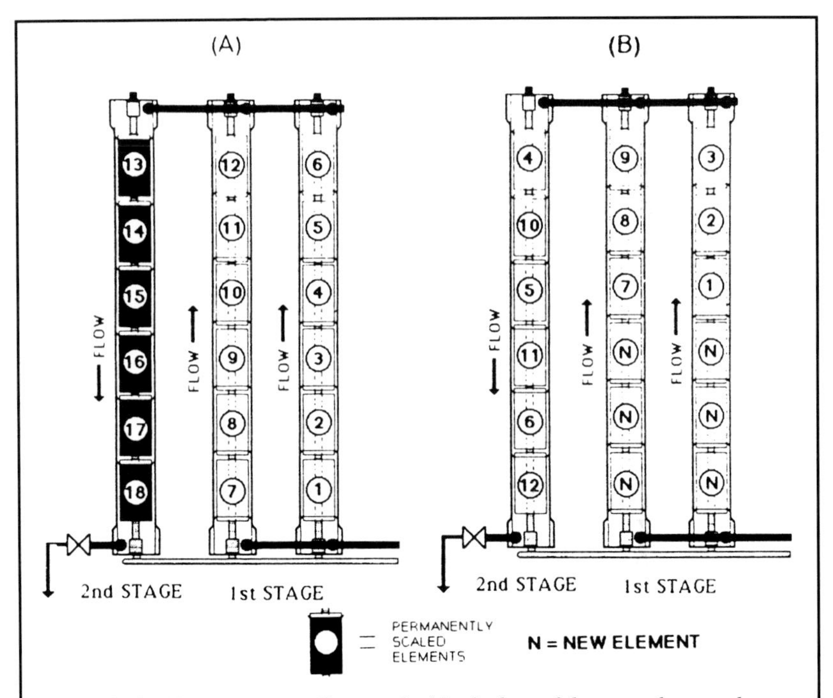

Figure 6-13. If permanent scaling resulted in the loss of the second-stage elements #3 to #18 as illustrated in (A), it would be best to install three new elements in the front end of each of the first-stage vessels (B) to minimize hydraulic imbalances and reduce the risk of contaminating the new elements. The remaining worn elements would be distributed according to the numerical sequence illustrated in (B).

Causes of uniform rejection decline: *Oxidizing agents.* In the absence of metals such as iron or manganese, an excessive concentration of oxidizing agent, such as chlorine, will tend to react uniformly throughout CA and PA membrane systems, the extent depending upon its concentration. Most CA membrane manufacturers publish a concentration limit on continuous free chlorine of 1.0 ppm. If approved by the manufacturer, the system can be exposed to higher concentrations of these agents, particularly short-term, without damage. Yet it is possible with an injection pump problem, siphoning, or aspiration problem, similar to those discussed in previous sections of this chapter, to subject the membrane to high enough biocide concentration to do damage.

Compared to CA membrane systems, PA membrane systems are much more sensitive to trace concentrations of oxidizing agents. Sodium bisulfite or some other reducing agent recommended by the membrane manufacturer is frequently used to break down the free chlorine.

Sometimes a facility will use activated carbon filtration upstream of the RO system. Unfortunately, activated carbon will efficiently remove the free chlorine in the first few inches of filter media. This leaves the bottom of the filter free of biocide, making it an excellent breeding area for bacteria. This bacteria will eventually be shed into the RO system feedwater (Collentro, 1985). Carbon fouling of the RO elements can also occur if the carbon is too soft or if the filter is not operated correctly.

It is recommended that oxidizing agents such as free chlorine (sodium hypochlorite) or iodine not be introduced directly into the RO permeate of PA membrane systems. Even though the injection is downstream of the RO membrane, the oxidizing agent can diffuse upstream when the RO is shut down. This has been known to cause deterioration of PA membrane systems.

If an oxidizing agent is required downstream of the RO, the introduction point should be moved to a location where upstream diffusion of biocide cannot occur, even when the system is shut down.

Check valves are not effective at preventing this sort of upstream diffusion. It is best if the RO permeate flows into the top of a storage tank. The chemical can then be injected through a separate line into the tank.

Exposure to direct sunlight. If an RO system is allowed to sit in the sun for extended periods of time while shut down, temperatures within the system can increase dramatically. In CA systems, the membrane hydrolysis rate will subsequently increase. In both CA and PA membrane systems, the temperatures can increase, possibly to a point of creating problems similar to those discussed earlier under heat exchangers. The increased temperatures can also result in increased biological growth, with increased possibilities of fouling and bacteria-induced membrane deterioration.

An RO system installed outdoors should be covered. The RO housings should be painted a light color to avoid absorbing excessive heat from the sun. Also, RO elements should not be stored in direct sunlight. These simple precautions will assist in extending the useful life of the RO membrane elements.

Heat exchanger leaks. Some of the anticorrosion chemicals used in hot water systems will cause uniform membrane deterioration throughout the RO system if allowed to contaminate the RO feedwater. Heat exchangers should be routinely inspected and pressure-tested to insure integrity.

Cleaning mistakes. During the RO cleaning operation, many mistakes are possible that can quickly affect the performance of (and possibly destroy) an entire set of membrane elements. One of the most common errors is to forget to measure and adjust the pH of the cleaning solution before recirculation. If the components of alkaline cleaning solutions are mixed without adjusting the pH with acid, the cleaning solution may have a pH exceeding 11. Although this may be compatible with PA membrane, it will quickly destroy CA membrane.

Polyamide membranes are particularly sensitive to the charge characteristics of the surfactants used in cleaning solutions. Only cleaning solutions that are approved by the membrane manufacturer should be used.

Positively charged membranes are incompatible with negatively charged (anionic) surfactants, and may be incompatible with nonionic surfactants. Most PA membranes are negatively charged and are incompatible with most cationic and nonionic surfactants. Care should also be taken to verify that disinfecting and storage agents are compatible with the particular membrane.

Sometimes the RO cleaning skid may be borrowed for use with some other maintenance operation. If an incompatible chemical is not adequately rinsed out of the skid before cleaning of the RO membrane elements, membrane deterioration or fouling can occur.

In one facility, a cleaning skid was not thoroughly rinsed of a rejection restorative known as "size" (also known as Colloid 189 from Colloids of America), which had been used to treat a different RO system. This was accidentally mixed with a cleaning solution and recirculated through the RO system. As a result, this system lost about 20% of its normal permeate flowrate. This loss was never regained in subsequent cleanings.

Severe membrane deterioration or multiple problems. When troubleshooting an RO system, one can easily become confused by the signs of deterioration if they are being created by several problems occurring simultaneously, or if one particular problem has been allowed to proceed to extremes. For instance, an acid siphoning problem is easy to spot in its early stages because of its pronounced effect on the front-end membrane elements. If neglected, however, the problem will proceed throughout the system until the performance of all the membrane elements have deteriorated. Then it is much more difficult to determine what caused the damage, since the symptoms now fit those of many other concerns.

Causes and Prevention of an Increase in Differential Pressure

In order to maintain the high standards of performance of which RO systems are capable, a conscientious maintenance program is required. In this chapter on RO troubleshooting, common concerns related to RO system design and maintenance have been discussed as a means of educating the RO technician as to potential pitfalls. These discussions should be helpful in preventing these problems from occurring, or in diagnosing them quickly should they occur.

This section discusses the causes of an increase in differential pressure. This knowledge will make it possible to notice and understand differential pressure concerns and thus prevent their occurrence.

The feed-to-concentrate differential pressure is a measure of the resistance to the hydraulic flow of water through the system. It is very dependent on the flowrates through the element flow channels. An increase in the permeate or concentrate flowrate will increase the system differential pressure. If the feed pressure is changed, it will not necessarily change the differential pressure if the permeate and concentrate flowrates remain the same (and no fouling or scaling occurs). It is therefore suggested that the permeate and concentrate flowrates be maintained as constant as possible in order to notice and monitor any element plugging that is causing an increase in differential pressure.

An increase in differential pressure at constant flowrates is usually caused by the presence of debris, foulants, or scale within the element flow channels. In hollow-fiber membrane systems, this might occur between the membrane fibers. In spiral-wound systems, this may occur within the turbulence-promoting spacing material between the membrane leaves.

This increase is not necessarily a cause for alarm. Most of the time, it is caused by normal fouling tendencies. These can be reversed by simply cleaning the RO elements before the guidelines from the membrane element manufacturer are exceeded. If these guidelines are unavailable, it is suggested that the system be cleaned before the feed-to-concentrate differential pressure drop increases by 15%. If cleaning solutions are found to be ineffective at restoring the baseline (original) performance, the membrane element manufacturer should be notified.

Finding the cause of an irreversible increase in differential pressure may be as simple as inspecting the lead end elements and noting the presence of debris that has entered and plugged the system. In more difficult cases, certain salts may be exceeding their solubility limits and precipitating within the membrane system. The latter can usually be controlled with the proper pretreatment of the RO feedwater, or by reducing the RO permeate recovery. Sometimes, however, the feedwater composition can change so that the original design, pretreatment, and/or operating conditions may need to be modified.

The solution to a problem involving debris may be as simple as properly installing and maintaining the cartridge prefilters. Solving a scaling problem, however, may be more complex than simply finding one aspect of concern and altering it.

Stopping the precipitation of a salt may involve reducing the concentration of that salt in the RO system by reducing the permeate water recovery. It may also require the injection of greater concentrations of scale inhibitor. It may require greater pH adjustment, or possibly changing from sulfuric acid injection to hydrochloric acid injection (to reduce the sulfate ion concentration). Possibly, the precipitation is occurring while the system is shut down, in which case the system should be rinsed at low permeate recovery prior to shutdown. In some cases, all these steps may be necessary to prevent scale formation.[8]

To troubleshoot an increase in differential pressure, it is important to locate where in the RO system the blockage in the element flow channels is occurring. This may be in the front (lead end), tail end, at a random location, or it may be occurring uniformly throughout the membrane system.

Figure 6-14 shows the recommended locations for pressure gauges in an RO system. Note that it is useful to monitor the differential pressure across each stage of the array, as well as the overall feed-to-concentrate differential pressure.

Individual stage differential pressures allow the technician to have greater precision in pinpointing, within the RO system, the location involved in the increase in differential pressure. This can then be further pinpointed to individual elements, using an individual element test stand (to be discussed later in this chapter).

This section will emphasize potential situations that may arise in RO systems due to operational factors, or to changes in the feedwater composition. Some of the common causes and prevention of an increase in the feed-to-concentrate differential pressure will be discussed below, according to the location of that increase in differential pressure.

Front-end increase in differential pressure: ***Bypass in cartridge filters.*** One of the most important purposes of cartridge prefilters upstream of an RO system is to protect the system from large debris that can physically block the flow channels in the lead end membrane elements. Unfortunately, it is common to find cartridge filters loosely installed in their housing.

Sometimes a technician will stack 10-inch (long) cartridge filters to fill a filter housing designed for 20-, 30-, or 40-inch filters without using interconnectors between the filters. In these cases, relatively large debris can pass between the 10-inch filters, possibly lodging within the RO elements.

[8] There have been cases where scaling was inhibited just long enough to protect an RO system, only to scale up the waste treatment pipelines before the RO concentrate had a chance to be diluted with other wastewaters.

Sometimes filters will be left in service for months because they never show an increase in their differential pressure. A good indication that the filters are bypassing, however, is that their differential pressure increases slowly or not at all. In any case, filters should be changed out at least every 2 months to reduce bacteria growth, or when the differential pressure reaches 15 psid.

Sometimes cartridge filters will deteriorate while in operation because of hydraulic shock or the presence of incompatible materials. Cellulose-based filters should be avoided in RO systems with pH control and/or free chlorine. These filters may deteriorate and plug the RO elements.

Pretreatment media filter breakthrough. Occasionally, some of the finer media from sand, multimedia, carbon, or diatomaceous earth pretreatment filters may be shed into the RO feedwater. Cartridge filters should catch most of the larger particles such as sand and anthracite. However, certain types of fine media, such as carbon fines and diatomaceous earth, can pass right through a nominally-rated 5-micron (µm) cartridge filter. In severe cases, these media can permanently plug the element flow channels, resulting in the required replacement of the affected elements.

Diatomaceous earth filtration can offer excellent filtration for an RO system feedwater, or it can be one of the worst threats to an RO system. Although it is capable of removing very fine foulants (down to 1 µm in size), if the diatomaceous earth precoat breaks through its retaining sock, it can plug an RO system. Diatomaceous earth filters that had been installed upstream of RO systems were later taken off-line permanently for this reason. It is therefore important, when considering the use of diatomaceous earth filters upstream of an RO, to evaluate

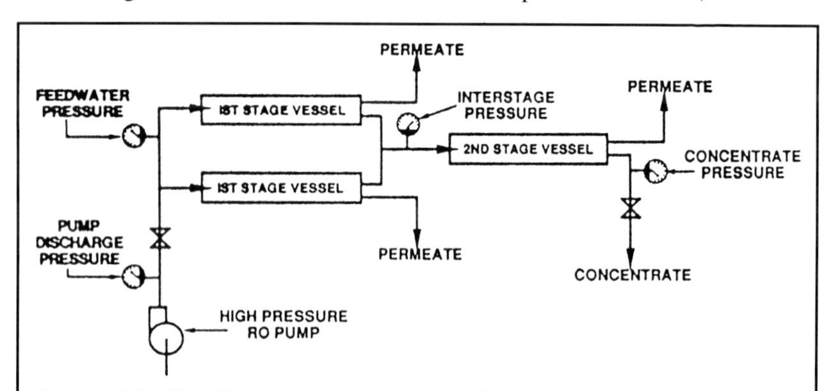

Figure 6-14. This illustrates the recommended locations for pressure gauges in a typical 2:1 array RO system. Note that it is useful to monitor the differential pressure across each stage of the array as well as the overall feed-to-concentrate differential pressure. This allows the technician to have a greater precision in locating the membrane elements involved in the increase in differential pressure.

whether the excellent filtration is worth the risk of fouling the RO elements.

With time, activated carbon can break down and shed a large quantity of carbon fines. This is especially true of soft carbons made from coal. Carbon made from coconut shell with a hardness rating of 95 or better is suggested.

When replacing activated carbon, it is important to remove any residual fines that may be in the gravel underbedding. Backwashing of the gravel with high flowrates may be required. It may be easier just to replace the gravel when replacing the carbon.

New carbon should be sufficiently backwashed to remove fines before the bed is put into service. Multiple backwash cycles may be necessary before silt density index (SDI) values drop to normal. The SDI filter should be visually inspected for the absence of carbon prior to placing the RO system back into service.

Pump impeller deterioration. Many of the multistage centrifugal pumps employ at least one plastic impeller. When a pump problem such as misalignment of the pump shaft develops, the impellers have been known to deteriorate and throw off small plastic shavings. The shavings can enter and physically plug the lead end RO elements.

Many high-pressure pumps are offered with an optional discharge screen. This screen will catch most of the shavings that might be shed by the impellers, and it is therefore recommended. Pump discharge screens should be checked regularly for shavings or other debris. If the output pressure of a high-pressure pump is dropping off at constant flowrates, the discharge screen should be one of the first things investigated. The solution may be as simple as the cleaning or replacing this screen. As part of a routine maintenance schedule, monitoring the discharge pressure of RO pumps prior to any throttling valves (see Figure 6-14) is suggested as a way to see if the pump is maintaining its output pressure.

Telescoping. The longitudinal unraveling of spiral-wound elements, known as *telescoping,* can cause an increase in differential pressure. When the membrane leaves extend beyond the spacing material between the leaves, they can close off areas of the element to flow. This may or may not affect the system salt rejection, depending on the severity.

As discussed in "Hydraulic imbalances" and "Heat exchanger inadequacies" in this chapter, telescoping can be caused by hydraulic surges or by temperature extremes. It is physically damaging to the construction of the membrane element. Most membrane element manufacturers install anti-telescoping devices on their elements to help prevent this from occurring.

Tail end increase in differential pressure: *Insufficient scale inhibitor.* The importance of controlling scale formation in an RO system was discussed in

"Causes and Prevention of Tail End Salt Rejection Decline," earlier in this chapter. If the feedwater composition changes, reducing the permeate recovery or injecting higher concentrations of scale inhibitor may be required.

An interesting example of a potential scale is silica. Homogeneous silica precipitation on the membrane can be extremely difficult to clean. A hazardous cleaning solution, such as one containing ammonium bifluoride, may be required. However, silica usually falls out of solution first either as a silicate salt (at an alkaline pH), or as a colloidal complex containing other organic and ionic species.

Scale inhibitors can slow the precipitation of silicates by binding up the silicates' cationic partner, such as calcium or magnesium (Nusbaum 1991). Some scale inhibitor solutions can assist in keeping colloidal complexes suspended in solution by acting as a dispersing agent.

If silica scale is suspected, it should be cleaned as soon as possible. If allowed to stay on the membrane surface, with time it can cross-link into a more homogeneous silica. It will then be more difficult to remove from the membrane (Luss, May 1993).

The precipitations of the various forms of silica are all time-related. If silica solubilities are being exceeded within an RO system, the system should be flushed of the high silica concentrations when the RO shuts down for any reason. It is suggested that an automatic procedure be used for operating at low pressure or low recovery just prior to shutdown. The system should never be allowed to sit stagnant for extended periods of time if it contains a supersaturated salt.

Most system manufacturers will attempt to design RO systems with a permeate recovery that prevents the precipitation of silica. If the silica concentration increases dramatically in the feedwater such that silica is precipitating even in the presence of scale inhibitor, then it may be necessary to reduce the permeate recovery.[9]

System recovery too high. As has been emphasized throughout this chapter, properly calibrated instrumentation is essential to maintaining an RO system. A slight change in recovery due to instrumentation inaccuracies can make a substantial difference in the concentration of potential foulants and scale in an RO system. Most foulants, and many scale-forming salts such as iron oxides[10], can be removed if the system is cleaned according to the manufacturer's recommended guidelines. However, some scales such as homogeneous silica and barium sulfate may not be effectively removed by cleaning. It is therefore important to maintain an accurate permeate recovery rate.

[9] Development is currently being done on scale inhibitors that are more effective at slowing the silica precipitation process.

[10] Iron in its ferrous state (Fe^{+2}) is soluble and will normally rise out of the system in the concentrate stream. If the iron is ferric (Fe^{+3}), or if ferrous iron has been exposed to air or some oxidizing agent, such as free chlorine (converting it to ferric), it may fall out of solution in the RO membrane elements.

Random increase in differential pressure: ***Brine seal damage.*** Brine seals are rubber or plastic devices that seal the outside of spiral-wound elements against the housing wall. They can be damaged or "turned over" during installation, or by hydraulic surges. This results in a certain amount of feedwater bypass around the element, leaving less flow and velocity to pass through the element. When this occurs, the element is more prone to scale formation. As a scaled element in one of several parallel housings becomes more physically plugged, there is a greater tendency for scale formation to occur within the other housing elements due to insufficient flowrates within that housing.

Contaminated cleaning skid. It is common for facilities to make better use of their RO cleaning skid by also using it in other applications. Unfortunately, this opens up the possibility of contamination. In some cases, carbon or ion-exchange resin from a cleaning skid has been found in just one pass of an RO system (usually the first pass that was cleaned). This pass then demonstrates a higher pressure differential after cleaning.

A cartridge filter housing should be installed on the cleaning skid to assist in preventing large debris from being introduced into the RO elements by way of the cleaning skid. A clean and chemically resistant cartridge filter such as an all-polypropylene filter is recommended in the housing. A cellulosic cartridge is not recommended, since it may not be able to withstand the pH extremes and chemical action of the cleaning solution. Cellulosic filters have been known to break down and plug up the RO elements.

Biological fouling. Under certain conditions, biological growth can quickly foul an RO system. This is especially a concern in chlorine-sensitive PA-type systems where chlorine or iodine cannot be injected into the feedwater. In these cases, it is important to frequently clean out the growth and disinfect the system. Proper performance monitoring (Bukay, 1984) will make it apparent when to clean and disinfect. It is also suggested that bacteria samples be taken and analyzed on a regular basis from the feed, permeate, and concentrate streams.

Uniform increase in differential pressure: ***Leaking valves.*** Scaling could occur uniformly throughout an RO system if the system were shut down with a closed concentrate valve and a leaking feedwater valve. In such a case, the water has nowhere to go but through the membrane into the permeate line, leaving behind its salts on the membrane surface. This is not a common occurrence since the concentrate throttling valve is usually left open when the RO unit shuts down.

Unresolved concerns. Many of the previously discussed concerns can lead to damage or blockage throughout the entire RO system if left unchecked. This might also occur if normal fouling is allowed to proceed beyond the recommended guidelines. In such severe cases, it is unlikely that the system can ever be returned to its original performance because of the permanent blocking off of the element flow channels.

Testing suspect elements. Individual elements can be tested in an single-element test stand to determine their differential pressure drop, permeate flowrate, and percent rejection. The test stand consists of a vessel that houses one element at a time (see Figure 6-15). It allows the technician to adjust the feed pressure, concentrate flowrate, and feedwater pH (on some units). The concentrate flowrate is particularly important in obtaining a differential pressure that can be compared to other elements or to the original element test data.[11] The exact settings should be obtained from the manufacturer's test specifications.

The individual element test stand can use its own separate feed pump, or the unit can be plumbed into the RO feedwater manifolding to use the RO high-pressure pump. If the latter arrangement is used, remember to readjust the acid injection pump settings.

Summary of troubleshooting steps for loss of rejection. The causes, prevention, and correction of a loss in salt rejection and an increase in differential pressure drop have been discussed in this chapter. In summary, the key steps in troubleshooting an RO system are as follows:

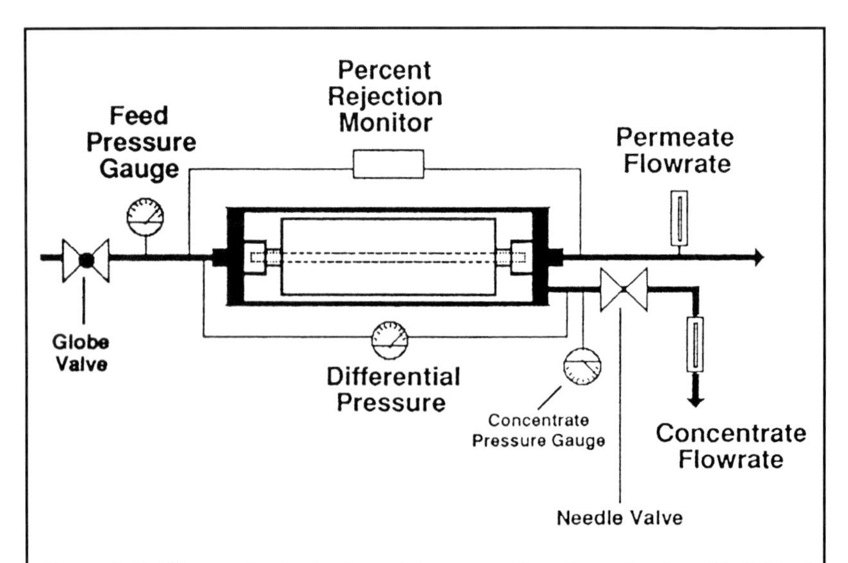

Figure 6-15. The use of a single-element test stand allows the evaluation of individual spiral-wound elements and possible determination of which elements are responsible for an increase in system differential pressure. See text for details.

- Implement an effective monitoring program.
- Check the instrumentation calibrations (including pH, temperature, pressure, conductivity, flowmeters, and percent rejection meter) for accuracy.
- Isolate the location of the RO elements that are losing performance (lead end, tail end, isolated to individual elements, or uniform loss throughout the system).
- Investigate the potential causes of the performance loss (including pH too high or low, hydraulic imbalances, heat exchanger problems, system recovery too high, fouling, scaling, too much or not enough biocide, and system sitting idle too long).
- Correct any concerns or problems as soon as possible to prevent future complications.

Having a broad knowledge of how to operate, maintain, and troubleshoot RO systems is essential to maintaining high-quality performance. The guidelines presented so far in this chapter reflect the personal experiences of the authors.

Analytical Methods for RO Troubleshooting

In the investigation of a membrane fouling, scaling, or deterioration problem, it is helpful to analyze the membrane and foulant as a means of knowing more exactly the nature of the problem. For example, if acidic and alkaline cleaning solutions have been unsuccessful at restoring lost RO permeate flow, knowing more about the nature of the membrane and foulant would provide valuable information from which methods can be devised for resolving the problem.

Visual inspection of a membrane element (by cutting off the outside wrap and the end supports to gain access to the element fibers or the membrane leaves) may be quite informative. There have been cases of lost permeate flow where an element was opened up to find nothing unusual, no foulant or scale. In one case, the appearance of lost permeate flow was the result of a malfunction of a flowmeter. (This points to the importance of double-checking instrument readings before proceeding, particularly before destroying a membrane element.) (See Figure 6-16.)

More commonly in this situation, the foulant will be visually apparent. The appearance of the foulant will provide clues to its nature and to the difficulty in its removal. Large filter media particles (like activated carbon) will be evident to the sight. A biological foulant will have a different appearance than scale, and frequently will smell differently.

[11] When purchasing new or replacement RO elements, it is a good idea to obtain a copy of the membrane manufacturer's test data for each element. The test data can then be used as a baseline for future comparisons.

It sometimes is possible to see mechanical problems with the element that might have contributed to RO system performance problems. This might be the case if the glue line has come apart on a spiral-wound membrane envelope. Another example of a visually apparent problem would be if the permeate carrier material within the membrane envelope has become mushy, instead of rigid. This could be allowing the membrane envelope to close under pressure and to restrict the flow of permeate water into the central permeate collection tube.

Dissolution in acid. If the material on the membrane consists of crystals that are obviously some sort of scale, a simple test can be performed to better define the nature of the scale. If the scale consists of any carbonates, the carbonate will readily dissolve in an acidic solution of pH 3 to 4. There may even be some bubbles formed as some of the carbonate reacts to become carbon dioxide, and degasses.

Sulfate or silica scale will not go back into solution except at extremely low pH, if at all. Carbonate scale can catalyze the precipitation of other types of scale. Therefore, even though the major scale is carbonate, not all scale may go back into solution in the slightly acidic solution. However, the test will still indicate if carbonate is the major component of the scale. Preventing the formation of

Figure 6-16. Dissected spiral-wound membrane element to allow visual inspection.

carbonate scale may then prevent the formation of other scale.

If the scale will not dissolve in a standard acidic solution, the scale can be tested with a 0.1-molar hydrofluoric acid (HF) solution. (Be sure to wear proper safety equipment when working with any acid, particularly hydrofluoric.) Hydrofluoric acid solutions are effective at dissolving silica scale (Luss, April 1993). The success of any particular acid solution in dissolving the scale will give a good indication of what type of solution will be needed to clean the scale out of the membrane system.

Dye test. Intact RO membrane will reject a high percentage of any dye greater in molecular weight than 300 (Comb, 1986). Certain types of membrane deterioration will affect the ability of the membrane to reject dye. As this occurs, the membrane and/or backing material will tend to absorb more of the dye, and dye will pass into the permeate water. Dye will also pass through any mechanical lesions in the membrane or membrane element. Thus, dyes can be extremely useful as troubleshooting guides to learn more about the nature of the degradation, and also to isolate the location of the degradation.

Methylene blue dye is a commonly available dye. Although it may not be the best dye for finding membrane imperfections, its ready availability makes it handy for on-site investigations. Other dyes are available that may offer better sensitivity.

The dye can be introduced into the feedwater of the test element as it is operated on an individual element test stand. It can also be injected into the feedwater of a small operational RO system. (Be certain to isolate the system from the process prior to rinsing the dye from the RO.) The dye will permeate the membrane through areas of degradation, or through mechanical leaks in the system. Its presence in the permeate water can be noted visually, or measured using a spectrophotometer.

The membrane, after it has been exposed to the dye, can be inspected to visually isolate the specific location of the dye passage. If areas of the membrane have deteriorated, those areas may pick up more dye than unaffected areas, depending on the nature of the degradation. This location will provide evidence as to the nature of the problem.

Dye can also be used with a membrane element that has been dissected. The dye can be manually spread lightly over the surface of the membrane in an unrolled element. Areas of membrane damage may tend to absorb the dye more readily than undamaged areas, resulting in a pattern being created across the membrane surface, and offering clues to the nature of the deterioration.

A scattering of spots across the membrane, possibly matching the pattern of the spacing material, would tend to indicate a bacteria attack of a CA membrane. If the bacteria attack has progressed to the point of severe membrane deterioration, the dye pattern will have become more uniform.

Chemical attack or high water-temperature-induced hydrolysis of the membrane will tend to result in uniform absorption of the dye. If the problem is caught before an acid or high water temperature attack is too widespread, the effects may be limited to just the lead end of the element. Attack from a high concentration of an oxidizing agent, such as free chlorine, will tend to result in dye absorption uniformly across the membrane surface.

Optical microscopy. When visual examination can not reveal enough information about the nature of a foulant, optical microscopy can provide additional information. It is usually clear under a microscope if a foulant is biological. Optical microscopy can also provide information about the crystalline structure of a scale formation.

In one situation, a gelatinous material full of fine filaments coated the lead end of a spiral-wound membrane element. Visually, it appeared that the foulant was biological, since some types of bacteria are known to grow filaments. However, under a microscope it was clear that the filaments were simply fine fibers, and were not a fundamental part of the gel structure that surrounded them. Since the RO system had recently had problems with the breakdown of its high-pressure pump impellers, it was easy to surmise that the source of the fibers was the pump impellers.

The clear gel in this example lacked any crystalline structure. It was found that the gel would dissolve in 0.1-molar hydrofluoric acid, but not in sulfamic acid at a pH of 2. For these reasons, it was deduced that the gel was silica-based. The silica used the surface area from the impeller fibers to catalyze its precipitation. (When silica precipitates in its various forms, it tends to be attracted to available surfaces.)

The use of polarized light in optical microscopy can provide additional insight into the nature of a particle. The crystalline structure of different salts will appear differently under polarized light. For example, calcium sulfate has more than one refractive index, which gives it a unique appearance under the microscope (Isner and Williams, 1993). *In the case of foulants that are difficult to clean, it can be extremely useful to analyze a foulant layer. Frequently, the components and the order of the foulant layers will provide direction for the best order in using acidic and alkaline cleaning solutions. It may be that a layer of scale is present directly on the membrane surface, but is covered by a layer of organic solids and biological foulants. In this case, it might be most effective to remove those organic and biological foulants with an alkaline cleaning solution, prior to attempting to remove the scale with an acidic solution. (It should be noted that exposing polyamide RO membrane to an acidic cleaning solution with a pH less than 3.0 can temporarily cause higher rejection and lower permeate flow

*following material contributed by by Zahid Amjad

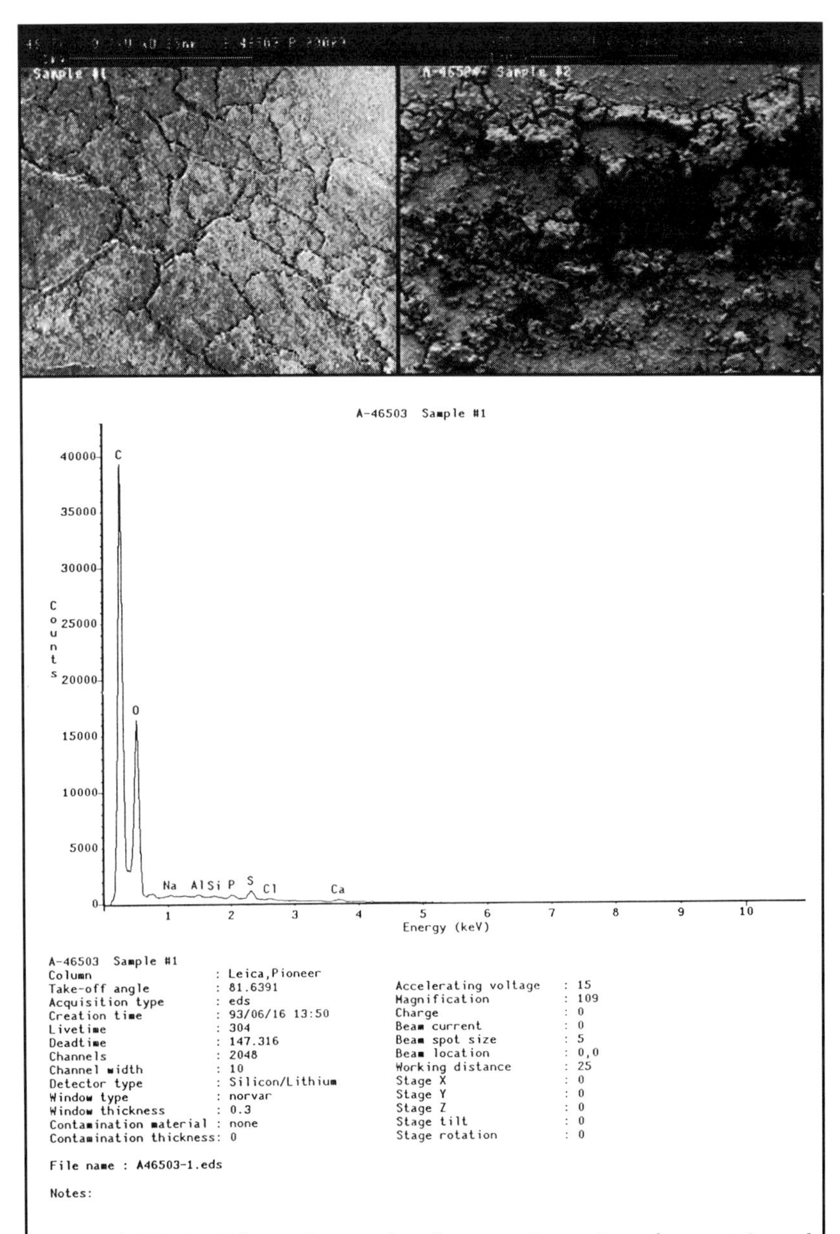

Figure 6-17. An RO membrane taken from a unit running whey protein and concentrating it from 6% to 12% solids. The unit experienced heavy fouling from protein and other material. a.) The SEM show the heavy protein mat on the membrane; b.) X-ray analysis.

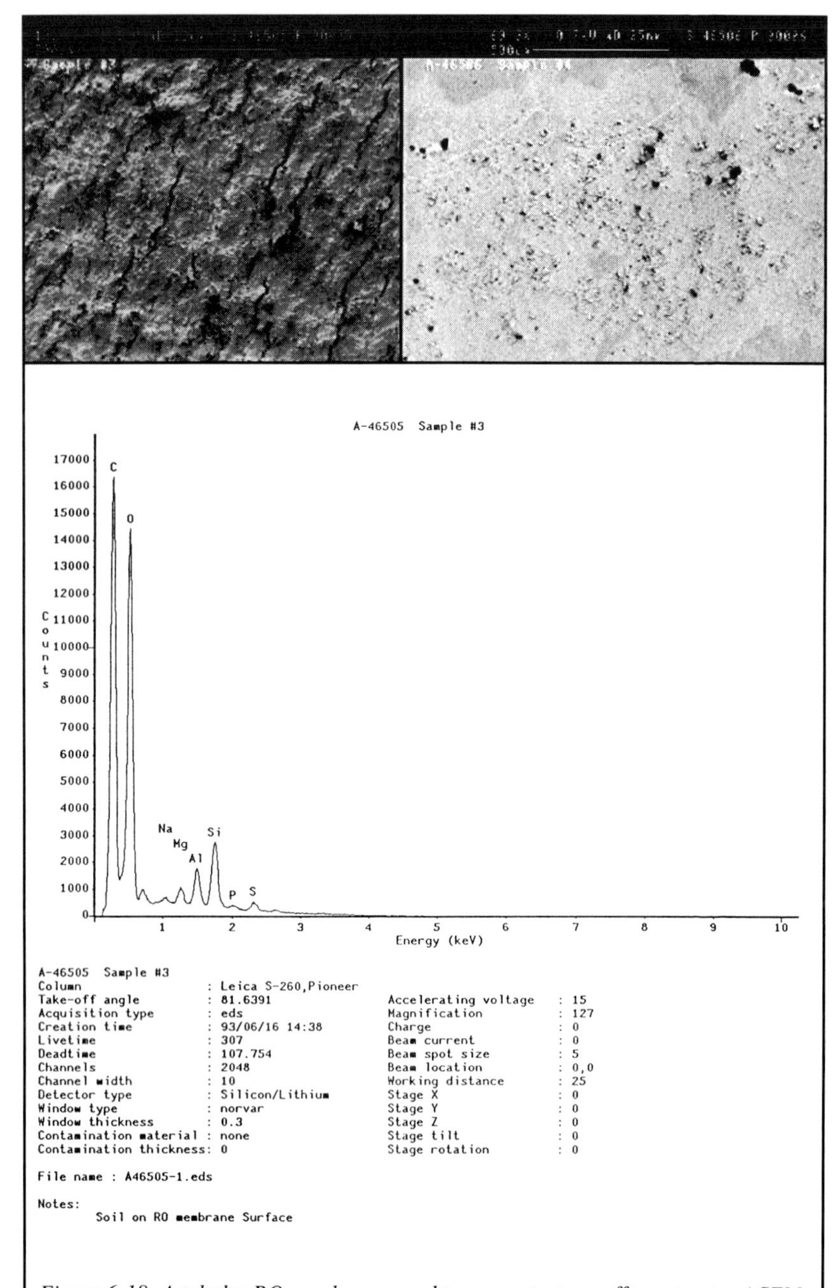

Figure 6-18. A tubular RO membrane, used to concentrate a coffee extract. a.) SEM; b.) X-ray analysis.

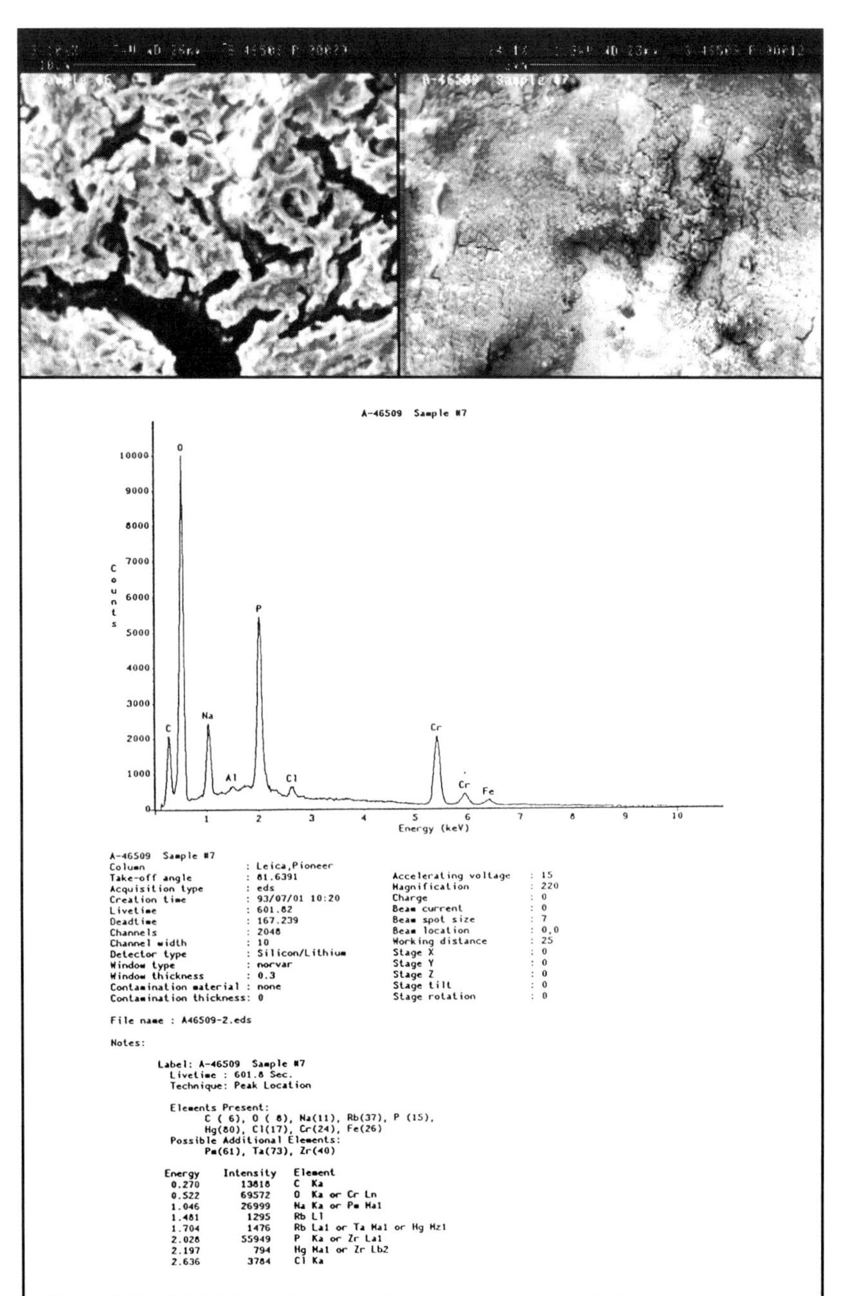

Figure 6-19. A CA RO membrane used to process a chrome plating waste stream. a.) SEM; b.) X-ray analysis.

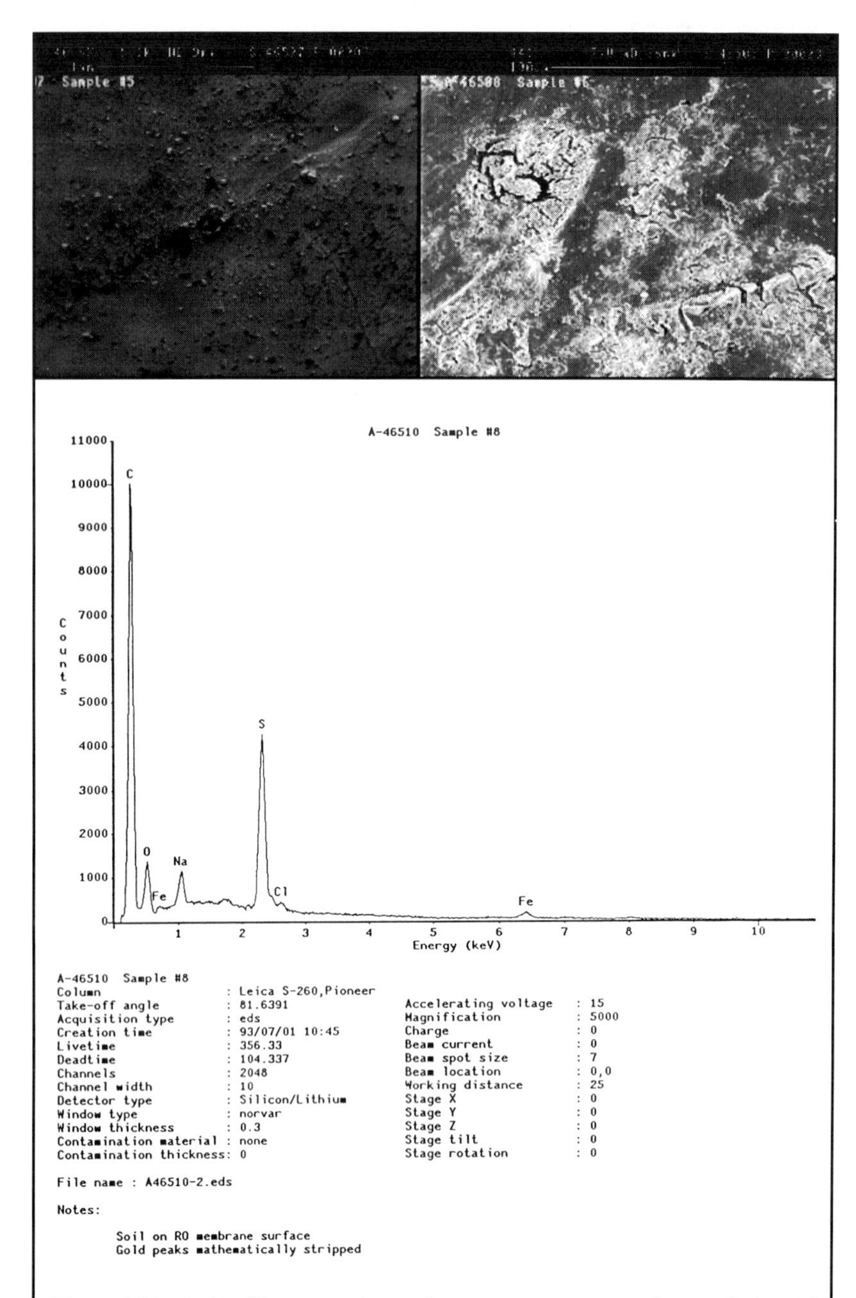

Figure 6-20. A thin-film composite used to concentrate wastes from an industrial laundry. a.) SEM; b.) X-ray analysis.

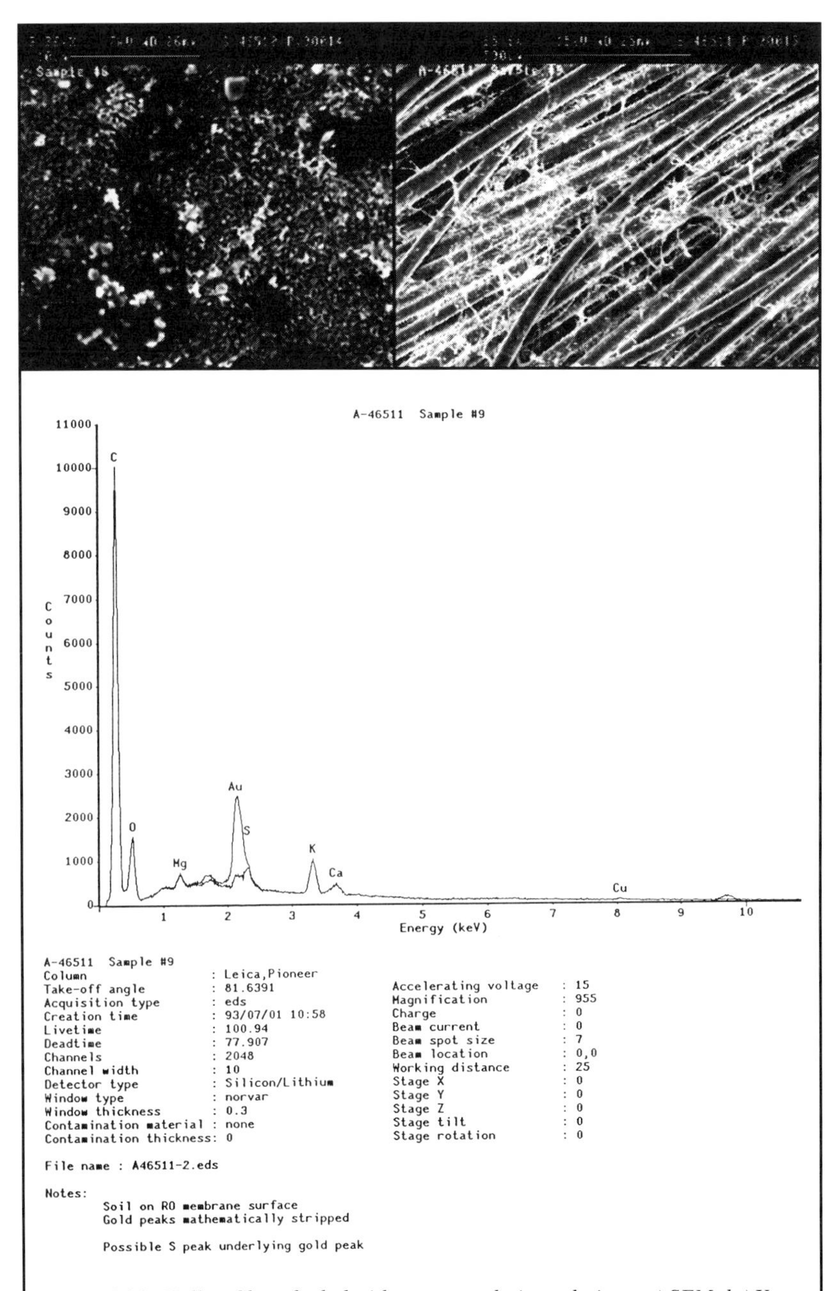

Figure 6-21. Hollow fibers fouled with a copper plating solution. a.) SEM; b.) X-ray analysis.

characteristics. This effect is reversed with a high pH cleaning solution.)

Fourier transform infrared spectroscopy (FT IR) is a method for determining the structure of organic compounds and many inorganic compounds. An infrared spectrum is a type of fingerprint of the foulant molecule, showing the chemical bonds of which the molecule is composed. Nitrogen-carbon bonds will generate peaks that are very different from carbon-hydrogen bonds. The peak intensities will give an idea of the number of those particular bonds present in the molecule. The method is accurate for samples as small as 50 μm (AT&T Analytical Service).

For best accuracy, the FT IR spectrum of peaks should be compared with that generated by a new RO membrane. In this manner, the peaks caused by the membrane material itself will not be confused with the foulant.

An FT IR analysis will also allow the evaluation of the membrane structure. If peaks are missing or reduced as compared to new membrane, this can be indicative of membrane degradation. The bonds that are specifically being attacked will provide information as to the nature of the attacking chemical (or bacteria).

An FT IR analysis can be used to identify the organic foulants that are present on the membrane surface, although it is a surface technique only. To evaluate what is under the foulant surface would require removal of the surface foulant. This might be desirable in the case of a multilayer foulant.

Scanning electron microscopy with X-ray diffraction (SEM w/ XRD). With SEM, devices bombard the sample with electrons and measure how the electrons bounce off the sample. This creates an extremely fine picture of the sample, offering clear photographs of particles as small as 0.01 μm. X-ray radiation (wavelengths shorter than 100 angstroms) is given off by the sample as its electrons return to a lower energy state from the excited state caused by the bombardment. The wavelengths correspond to the nature of the substances in the sample, particularly with crystalline substances (Fulford, May 1990).

Some examples of SEM photographs and corresponding X-ray spectrums are given in the figures. Figures 6-17a and 6-17b are analyses of an RO membrane used to concentrate cheese whey. Figures 6-18a and 6-18b are analyses of a PA tubular RO membrane being utilized to concentrate a coffee extract. Figures 6-19a and 6-19b are from a CA RO membrane used to concentrate a chrome plating waste stream. Figures 6-20a and 6-20b are from a PA thin-film (spiral-wound) membrane used in an industrial laundry. Figures 6-21a and 6-21b are of a PA hollow-fiber membrane element used in concentrating a copper plating solution. The XRD spectrums are useful when attempting to analyze an inorganic scale. They may not be of much value when the foulant is organic in nature.

Obtaining analytical services. Most membrane manufacturers have the analytical capabilities described above. Some specialized private laboratories and cleaning chemical companies also have these capabilities.■

CHAPTER 7

PRELIMINARY STUDIES FOR SPECIALTY APLICATIONS

When a new RO application is being considered, parameters to evaluate include materials compatibility, achievable solute separation, pretreatment and maintenance requirements, and any remaining design considerations for scaling up the system. The most practical evaluation of a new application includes researching the desired application, running a bench-scale test, and running a field pilot study.

Researching the Desired Application

Before large amounts of time and money are invested in an RO pilot system, it is best to first determine whether one of the currently available RO membranes even has the capability of meeting the needs of the application. For instance, there is not much point in running a pilot study on the separation of isopropyl alcohol from ethyl alcohol if it can be determined on a bench scale that no available compatible membranes are capable of achieving a significant separation. Frequently, information already exists regarding the ability of readily available membranes to process various chemicals, and possibly regarding their ability to remove the desired component (or at least something similar in its characteristics).

For a new RO application, typical issues that require consideration usually include the following:

- Membrane and system component compatibility
- Desired membrane rejection characteristics
- Acceptable or controllable fouling/scaling potential
- Desired solute concentration and osmotic pressure

The first step when considering a new application is to contact a major membrane manufacturer and find out what is already known about that application. Even if there is no specific information available, the membrane manufacturer may have enough experience with similar applications to estimate the probability for this application's success.

Membrane and system component compatibility. A general guideline for membrane compatibility is given in Table 7-1. Also, the choice of elastomer material to be used for O-rings and spiral-wound element brine seals (gaskets) should be considered for the application. It should be noted that an excellent rating by this chart does not guarantee compatibility.

Desired membrane rejection characteristics. For most applications, a relatively high solute rejection is required. Either the high rejection is desired to minimize the concentration of solute in the permeate water, or it is necessary to enable the concentration of the solute without losing a significant concentration into the permeate.

If CA membrane is required for the application, it may be necessary to use a specialty membrane designed for especially high rejection. Such a membrane will typically suffer from poor permeate flux. If high rejection is necessary, the RO system may need substantially more membrane elements, which will add appreciably to the system cost.

If the system requires PA membrane, a seawater membrane may be desirable for its higher rejection characteristics. This also will result in the requirement for

Table 7-1
Membrane Compatibility

	Membrane		*O-rings & Gaskets*		
	PA	*CA*	*Buna N*	*EPDM*	*Viton*
Aliphatic solvents	E	E	G	P	G
Alcohols (aliphatic)	G	F	E	E	F
Esters & ketones	E	P	P	G	P
Chlorinated solvents	G	F	P	P	G
Aromatic solvents	E	G	F	P	G
Halogens	E	G	P	F	E
Weak bases	P	--	E	E	G
Strong bases	P	P	G	E	F
Weak acids	P	F	G	E	E
Strong acids	P	P	F	E	E
Strong oxidants	P	F	P	E	E

E: Excellent; no chemical effect on membrane material
G: Good; slight effect such as swelling or distortion may occur
F: Fair; may attack and degrade membrane material
P: Poor; strong chemical attack on membrane material possible

(Courtesy of Osmonics, Inc., and Seelye Plastics, Inc.)

more membrane elements with a resulting higher system cost.

If the objectives of the application can be achieved using standard elements, the RO system cost will be significantly reduced. Operating costs will also be significantly less because of the reduced membrane replacement costs.

The rejection characteristics of some common salts are given in Table 7-2. These values might be typical of a high-rejecting CA membrane. Rejection for PA membrane will tend to be higher across the list. It is also possible to obtain CA membranes with higher rejection characteristics.

The rejection by PA membrane of some common solvents is given in Table 7-3.

Acceptable or controllable fouling/scaling potential. Certain waste streams have been known to have the constituency of molasses. Their suspended solids concentration is so high that severe membrane fouling is an inescapable aspect of the application. It may be a feasible RO application, but only at high capital

Table 7-2
Typical Salt Passage for a High-Rejecting CA RO Membrane

Salt		%
Calcium sulfate	$CaSO_4$)	99
Calcium bicarbonate	$Ca(HCO_3)_2$)	98
Sodium chloride	($NaCl$)	97
Sodium bicarbonate	($NaHCO_3$)	97
Sodium sulfate	(Na_2SO_4)	99
Calcium chloride	($CaCl_2$)	98
Magnesium bicarbonate	($Mg(HCO_3)_2$)	98
Magnesium chloride	($MgCl_2$)	98
Potassium chloride	(KCl)	97
Magnesium sulfate	($MgSO_4$)	99
Potassium bicarbonate	($KHCO_3$)	97
Potassium sulfate	(K_2SO_4)	99
Calcium nitrate	($Ca(NO_3)_2$)	85
Sodium nitrate	($NaNO_3$)	85
Magnesium nitrate	($Mg(NO_3)_2$)	85
Potassium nitrate	(KNO_3)	85
Calcium fluoride	(CaF_2)	97
Sodium fluoride	(NaF)	94
Magnesium fluoride	(MgF_2)	97
Potassium fluoride	(KF)	94

(Courtesy of Osmonics, Inc.,)

and operating costs. If the economics of the application are marginal, it should not be pursued.

If the solubility of salts is to be exceeded in an application, a means of controlling the precipitation of those salts is required. The membrane manufacturer and the suppliers of scale inhibitors should be consulted. If the solubilities are so dramatically exceeded that the suppliers question the application's feasibility, proceeding with bench-scale testing would be of dubious value.

Desired solute concentration and osmotic pressure. As an economically feasible technology, RO is limited by the osmotic pressure of the solute, which is a function of its concentration. If the osmotic pressure of the desired solute concentration exceeds 800 psig, less expensive technologies are probably available for the application. Operating at pressures exceeding 1,000 psig will require special pumps, housings, and fittings that will dramatically increase the cost of the system. Some juice concentration applications may still be feasible in spite of such osmotic pressures. These are discussed in the next chapter.

Bench Test

If a membrane manufacturer is supportive of the application, it has passed the first test. The next step is to run a bench-scale test. It is usually easiest to have the membrane manufacturer run this test at their facility (generally for a fee). The manufacturer should have the capability of testing several different elements to find the most likely membrane types and configurations. There should be more than one membrane flow channel material available to test for an application that has a high potential for fouling.

A successful bench test will have demonstrated that one or more membranes tested were capable of achieving the following:

- The desired separation (i.e., the right components were sufficiently rejected by the membrane).
- Sufficient flux to make a full-scale system practical.
- Good membrane performance with respect to rejection and nearly stable with respect to flux.
- No changes in the appearance of the element's materials of construction (i.e., no visible discoloration should occur).

As discussed above, it is sometimes useful to test more than one element configuration. If it can be shown that a standard spiral-wound configuration can perform without fouling at a significantly faster rate than a tubular or a special-spacing spiral-wound element, this will offer substantial savings in a full-scale system. This is because additional membrane can be packed into a standard spiral-wound element, resulting in greater individual element permeate flowrates. Standard elements also do not require as much concentrate flow to achieve the same crossflow velocities. If an application's economics are questionable, then it is probably better to use a standard element in the pilot test even if its performance is not quite as good.

If all the membrane elements fail the bench test, possibly the pretreatment of the test solution should be reconsidered. It may be possible to eliminate an undesirable component of the application solution.

More often than not, new potential RO applications fail during the bench test and are dropped at that point. Their failure may be caused by catastrophic problems with material compatibility, by insufficient solute rejection, or by high fouling rates. The applications are sometimes dropped because permeate flowrates are too low to justify the potential cost of a full-scale system.

If the bench test passes and the membrane manufacturer is still supportive of the application, the next step is to determine the long-term compatibility of the materials of construction, as well as the pretreatment and maintenance requirements of the application.

Pilot Study

A correctly designed pilot study will duplicate as closely as possible the operating conditions of the proposed full-scale RO system. In fact, in an ideal study the only differences between the pilot system and the full-scale system are the size and the total flowrates (Figure 7-1).

Figure 7-1. Pilot reverse osmosis system.
Courtesy of Osmonics, Inc.

For spiral-wound systems, a smaller-diameter membrane element can be used in a pilot system to emulate the performance of a full-scale RO system. The flow dynamics are fairly similar between different-diameter membrane elements. In fact, the primary difficulty in using a small pilot unit may be in the unavailability of pretreatment equipment that can be accurately scaled up from the flowrate of the smaller system to that of the larger one. For example, it is difficult to find a media filter designed for low flowrates that has the same media grading and depth as a full-scale media filter.

With the pilot RO unit, it is frequently necessary to use some sort of concentrate recycle (back to the feed) as a means of duplicating the desired recovery of the full-scale system. Otherwise, a full train of elements might be required to achieve the desired permeate recovery. This recycle should not have a great impact on the accuracy of pilot study results.

The average membrane concentration will be higher when using concentrate recycle than when not, due to the solute that is being fed back to the beginning of the membrane system. The permeate quality will also be proportionally higher in concentration. Thus, the rejection calculated using the average concentration of the solute in the feed and the concentrate should be similar regardless of whether recycle is used or not.

The osmotic pressure will be greater, and will have a greater effect on permeate flowrate, when concentrate recycle is used. With the higher salt content that will be present in the lead end elements, the lead elements will have a lower permeate flowrate. As long as this higher osmotic pressure is taken into account when sizing the full-size system, this should not present a problem.

The pilot RO pretreatment system should be considered as part of the pilot study. If several options are being considered in the pretreatment scheme, they should each be tested to determine the effects on the performance of the RO

Table 7-3
Rejection of Solvents and Permeate Flux Loss with Polyamide Thin-Film Membrane

Solvent	*Concentration Molar*	*Rejection %*	*Flux Loss %*
Acetone	0.1	70	>25
Ethanol	0.1	40	>25
Phenol	0.1	40	>25
Methylethyl ketone (MEK)	0.1 Molar	75	20
Isopropyl alcohol (IPA)	Independent	90	0

(Courtesy of Osmonics, Inc.)

system.

The operating data to record during the pilot study should include those parameters discussed in Chapter 4, "Related Water Treatment Equipment." Data that would normally be collected intermittently for a full-scale system should be collected regularly for the pilot study. If there is more than one housing in the pilot system, this data would include individual-housing permeate TDS concentrations and flowrates.

The RO performance parameters to calculate and monitor as a means of judging the pilot RO and its pretreatment system performance are the usual:

- The normalized permeate flowrate
- The normalized system differential pressure
- The membrane solute rejection

As discussed in Chapter 5, "RO Maintenance," these parameters will indicate whether and how significantly the RO system is fouling, or if some sort of membrane deterioration is occurring. The values should be recorded on a frequent basis, daily as a minimum. They should be recorded before and after any major change in operating condition.

Pilot study test parameters. Certain parameters are usually not worth the time and trouble of testing because they have typically been tested by the membrane manufacturer. It is best to obtain a conservative value from the membrane manufacturer for these items, and spend more time evaluating the other parameters.

A good example of this would be the minimum element concentrate-to-permeate ratio. It is easiest to decide on a conservative value for the minimum ratio and maintain it while testing other parameters. An attempt to optimize the individual element concentrate-to-permeate ratio will lead to a system design that runs on the edge of its scaling potential. If any variable happens to change, the system may be in trouble.

Another example of a variable possibly not worth testing would be the membrane permeate flux (i.e., the permeate flow per membrane surface area). For applications where fouling is a concern, time can be saved by deciding on a conservative guideline for the maximum membrane flux and using its value in the full-scale RO design calculations. Membrane flux in the full-scale system is typically going to vary dramatically within the system from its lead end to its tail end because of the system pressure, the differential pressure, and the solution osmotic pressure.

High permeate flux will typically result in higher rates of membrane fouling. However, over a range of conservative flux (as provided by the membrane manufacturer), the differences in fouling rate will probably not be significant. Experimenting with the membrane flux might be considered after other variables have been tested. (Note: The membrane manufacturers do not recommend exceeding their design individual element permeate flowrates on applications that could be subject to fouling or scale formation.)

Removing the minimum individual element concentrate-to-permeate ratio and the membrane flowrate from the list of test parameters dramatically simplifies the pilot study. If a particular type of membrane has been chosen based upon the bench test, the number of variables still left to be defined are a manageable quantity.

Typical parameters that will be studied for most RO pilot systems would include one or more of the following:

Phase I - Initial RO performance evaluation

1. A working membrane rejection of the feed solutes as a function of concentration and pH.
2. A working osmotic pressure relationship with solute concentration as a function of concentration.

Phase II - RO maintenance evaluation

3. Finding a cleaning regimen that is effective at restoring the original membrane performance.

Phase III - Optimizing RO recovery

4. The effect of the RO permeate recovery on the RO rejection and fouling rate.

Phase IV - Long-term RO performance

5. The effect of the RO pretreatment equipment on the RO fouling rate and the membrane/system compatibility.
6. The effects that variation in the quantity and quality of the water source has on the RO performance.
7. The long-term compatibility of the RO membrane and system materials of construction with the process chemicals and cleaning regimen.

Phase V - Reducing RO cleaning frequency

8. The extent of fouling and/or scale formation allowable while still allowing the capability of restoring original membrane performance by cleaning.

The study of some of these parameters will be performed actively, whereas others will be studied as an indirect aspect of the ongoing testing.

Phase I - Initial RO performance evaluation. Immediately after starting up the pilot system, its operating data should be recorded. Since it is likely that some aspect of the future testing may affect the membrane performance, a complete evaluation of its operating characteristics should be performed initially.

If the application source varies in quality, it is useful to operate the pilot RO first on a less variable water supply. The city water supply may serve this purpose. In this manner, if the RO performance changes are questionable because of the variation in the pilot application source water, membrane performance changes can be evaluated by operating the RO on the standardized water source. Be certain that the city water is properly treated so that it is not a contributor to fouling or membrane deterioration.

During this phase of the pilot study, conservative operating conditions should be used. Nothing should be attempted that might affect the membrane integrity. Conservative permeate recoveries, feedwater pH, and operating pressures should be used. The pretreatment system should include sufficient equipment to minimize the potential for fouling or scale formation.

Testing can be performed to characterize the membrane solute rejection and the osmotic pressure of the solute. One variable should be altered at a time. After an operating parameter is changed, sufficient time should be allowed for the RO system to stabilize prior to recording the operating data. A minimum of 20 minutes is suggested.

1) Solute rejection: *a) Vary recovery.* Solute rejection should be tested as a function of concentration. This can be performed by varying the RO system recovery, which will alter the solute concentration. An average solute concentration should be calculated using the feed and concentrate concentrations. This can then be used with the permeate concentration to determine the rejection at that particular average concentration.
b) Vary pH. If the pilot system includes a means of pH adjustment, the effect of pH on the solute rejection should be tested. Be certain that the pH is stable prior to recording the operating data.

2) Solute osmotic pressure*: a) Vary membrane pressure at constant permeate recovery.* The permeate flowrate should be recorded as a function of the average membrane pressure (average of the feed and concentrate membrane pressures). Pressures should be chosen that are within a normal operating range. The concentrate flowrate should be adjusted to maintain a constant permeate recovery. Average membrane pressure should be graphed against permeate flowrate (normalized for any temperature variation), as shown in Figure 7-2. If a line is drawn through the points on the graph, the Y-axis intercept will be the sum of the permeate pressure and the osmotic pressure difference between the average membrane solute concentration and the permeate solute concentration. If there is negligible permeate pressure, this Y-axis intercept is just the osmotic pressure differential.

b) Vary recovery and retest. To determine the effect of the membrane solute concentration on osmotic pressure, the permeate recovery should be altered and the previous testing repeated with different operating pressures. A graph of pressure as a function of permeate flow will need to be drawn for each value of the permeate recovery. The relationship (if any) between solute osmotic pressure and solute concentration can be determined from the various osmotic pressure values.

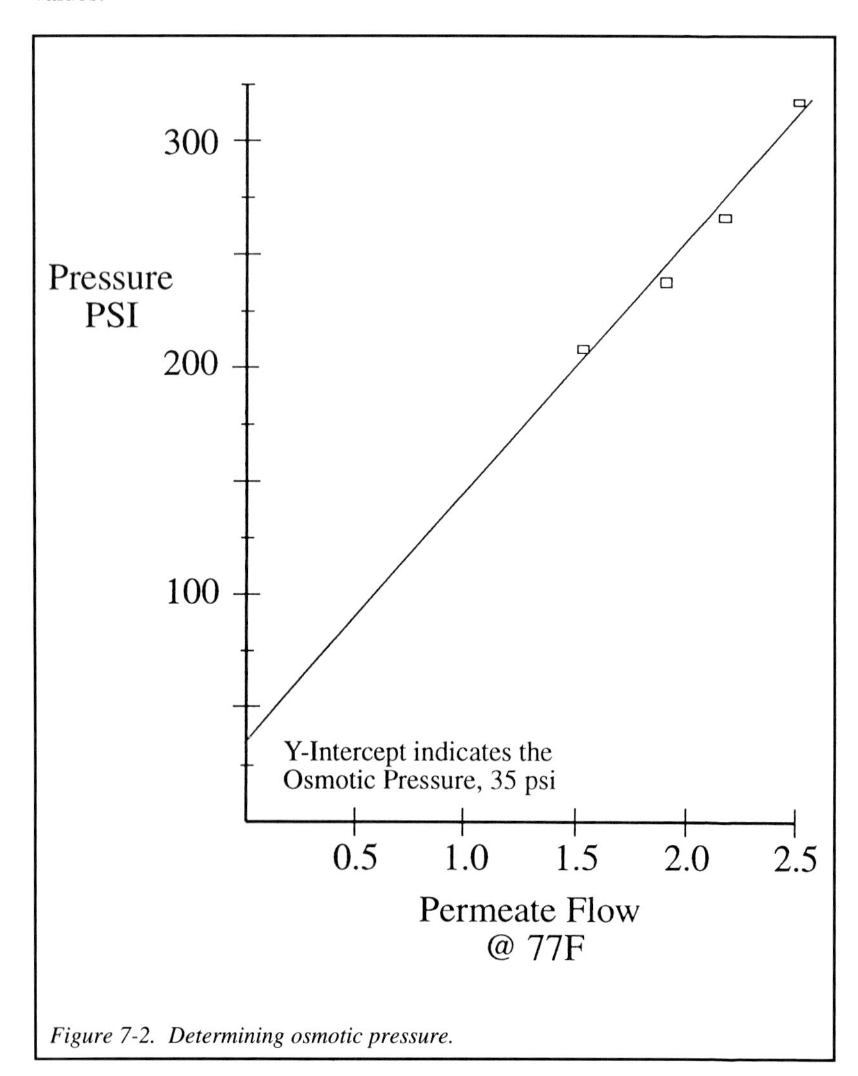

Figure 7-2. Determining osmotic pressure.

Phase II - RO maintenance evaluation. It is possible that the membrane has experienced some fouling and/or scale formation during the testing that was just performed. This phase of testing will better determine the RO system cleaning and maintenance requirements.

If the membrane has already fouled, as indicated by a greater than 10% decline in the normalized permeate flowrate, the membrane should be cleaned. The suggestions given in Chapter 5, "RO Maintenance," can be used in cleaning the pilot system. If the suspected foulant does not fit the characteristics of those discussed, it is suggested that the foulant be analyzed if some can be collected without having to dissect the membrane element. Otherwise, analyzing a piece of fouled membrane will give a better indication of the constituents of the foulant layers. This requires sacrificing a membrane element, but can save a lot of time and expense in not having to guess which cleaning solution might be effective in cleaning the membrane elements.

Once a good idea is available of what will be effective in cleaning, the potential cleaning solution can be tested on the pilot system. Comprehensive operating data should be recorded before the cleaning, and again after the system has rinsed and the performance stabilized. (Keep in mind that the pH extremes present in some cleaning solutions will affect the short-term performance of polyamide thin-film membrane. Acidic solutions with pH below 2.5 can cause a temporary decrease in the permeate flowrate and an increase in solute rejection. Alkaline cleaning solutions with pH greater than 10.5 can cause a temporary decrease in solute rejection and increase in permeate flowrate.)

If it can be determined with confidence that the original membrane performance has not been restored, effort should be made to find a cleaning regimen that will be effective at restoring performance. There is no point in continuing the pilot study if a means is not available to clean the membrane (unless frequent membrane replacement is a viable economic option). It may be necessary to have cleaning studies done on the element by a company specializing in membrane cleaning.

Once a successful cleaning regimen has been found, the pilot study may continue.

Phase III - Optimizing RO recovery. Higher permeate recoveries can now be tested to determine their effect on RO system performance. The effect on membrane fouling and/or scale formation can be tested as a function of the permeate recovery. If the increase in fouling/scaling is minimal as the recovery is increased, it may be possible that the upper limit on recovery will be more a function of the permeate quality.

Although the solute rejection has already been tested at lower recoveries, it is a good idea to check it also at the higher concentrations present at the higher recoveries. The solute osmotic pressure relationship with concentration can also be checked at the higher concentrations.

Phase IV - Long-term RO performance. For most succesful applications, membrane or RO system compatibility has not yet become an issue in the pilot study. Most applications require the long-term compatibility of the RO membrane and the RO system materials of construction with the process fluid and its components, though this is not true in every case. Some applications can live with a relatively short membrane life expectancy because of the potential savings to be reaped by the RO. In this case, the pilot system will allow the determination of how soon the membrane performance is likely to fall below desired levels; thus, the anticipated membrane changeout frequency can be estimated as a means of evaluating the economics of the application.

Sometimes the compatibility problem is with nonmembrane materials within the membrane elements. One example is the possible attack of the polyester material on which the membrane is extruded (in spiral-wound elements), or of the interconnector O-rings, the brine seals, or the melamine resin binder used in the permeate carrier material (of spiral-wound elements).

Occasionally, a compatibility issue will arise from some component in the RO system, independent of the membrane elements. An example is a case where metallic high-pressure pump seals became plated with copper when processing a copper sulfate solution. The pump would subsequently start leaking. In this case the original seals had to be replaced with nonmetallic seals.

Compatibility has to be a concern with severe cleaning solutions if these are required for the application's success. In the dairy industry, the concentration of whey by ultrafiltration requires such strong cleaning and sanitizing chemicals that membrane life expectancy is shortened. Yet these chemicals are required to remove the particular membrane foulants. Shortened membrane life is an accepted aspect of the application.

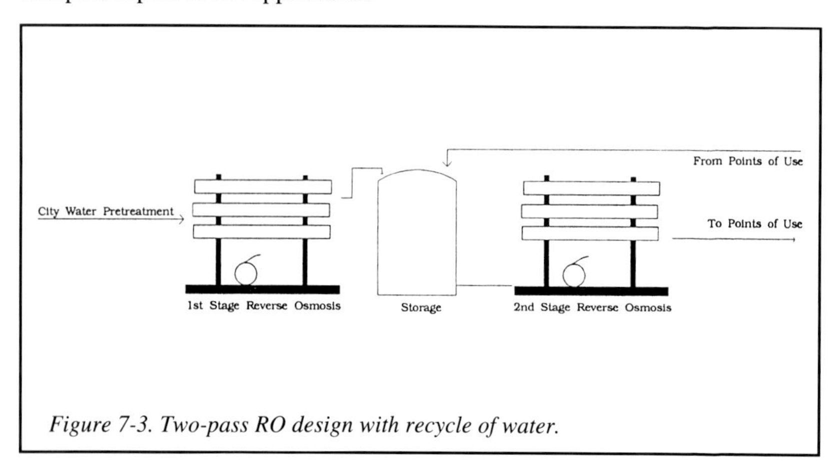

Figure 7-3. Two-pass RO design with recycle of water.

The long-term phase of the pilot study is less intensive than other phases in that the parameters being investigated can be tested only over an extended period of time. During this phase, the successful performance of the RO system while operating under the desired operating parameters will be confirmed. The system should be operated at the recovery that was determined to be best during the optimization phase of the pilot study. The effective cleaning regimen should be used when necessary, as dictated by a loss in normalized permeate flowrate, or by an increase in normalized differential pressure (as suggested by the membrane manufacturer), or possibly by a change in solute rejection.

Operating data should continue to be well monitored during this phase. Changes in the quality and quantity of the water source can thus be noted. If these changes affect the RO performance, this should also be noted. It is rare that the quality of the water source is consistent in terms of the concentration of the desired solute, or the concentrations of potential foulants. For this reason, it is important to develop some history with a new application, even when compatibility is not an issue.

The membrane fouling rate may vary seasonally with the system. Additional pretreatment equipment may be needed only at certain times during the operating season. The RO feedwater quality may decline to the point that the RO may not be capable of meeting permeate quality requirements without substantial changes being made in the system recovery, or in some other parameter.

If the process flowrate feeding the RO and pretreatment system varies dramatically, this can potentially affect the RO performance. If it is intended to operate the RO only when there is sufficient water to feed the system, the RO may end up sitting stagnant for extended periods of time. Biological or scaling problems may ensue.

It is suggested that the pilot study be performed over a period of at least 1 month if minimal fouling and source quality variation occurs, and if no problems are encountered. As this is usually not the case with a new application, most pilot studies should be performed for more than 2 months.

During this time, it may be desirable to try out modifications in the pretreatment system. Changes in pretreatment may result in modest improvements in the membrane fouling rate. Nothing should be attempted during this phase of the pilot study that might endanger the RO membrane.

Phase V - Reducing RO cleaning frequency. For applications that require frequent cleaning, the ability to go longer between cleanings may seriously impact the economics of the application. This may be accomplished by simply letting the membrane foul beyond the recommended guidelines. In allowing the membrane to foul, the risk is being taken that it may not be possible to restore membrane performance. Therefore, this testing should not take place until the long-term success of the application with the approved cleaning frequency has been documented.

In this pilot study phase, membrane fouling is allowed to continue between cleanings with increasing percentage losses in the normalized permeate flow, or percentage increases in the normalized system pressure differential. If difficulty is encountered in cleaning back to original membrane performance, the test is over. The maximum fouling/scaling that was allowed prior to the previously successful cleaning will be used as a guideline for future cleaning.

Scaling Up from the Pilot System

The pilot study can be considered successful if the following occurred:

- The pilot study met minimum permeate recovery and quality requirements.
- The pilot study demonstrated the long-term compatibility of the membrane and system.
- Membrane fouling and scale formation were controllable at the desired RO system recovery.
- A successful cleaning regimen was found to restore the original permeate flow and system differential pressure after fouling/scale formation.

The only remaining issue as to the success of the application with a full-scale RO system is its economics. To determine its financial feasibility requires an estimation of the system cost, which will depend on the system design.

With the data generated during the pilot study, it is now possible to design a full-scale RO system optimized for the application. If the osmotic pressure of the solute or the rejection characteristics are primary factors in the system design, the performance of each element in the full-scale RO system should be evaluated independently, starting at the concentrate end and working backward. This is most easily performed using a computer (see Chapter 9. "Computer Programs"). Permeate recovery and operating pressure can be altered to determine the optimum system recovery as based on RO permeate quality or RO system cost.

Solvent Recovery Using RO

Many of the issues discussed in the previous sections become critical when dealing with a solvent-recovery application. Many solvents are not compatible with certain RO membranes, or with the plastics that are used in the membrane element construction or in the RO system construction. The rejection by the RO membrane is typically substantially less than that of dissolved salts. Also, the desired concentrations for solvent recovery are nearly always in a range where the solute osmotic pressure is a critical parameter in RO system performance.

The first step is to check the compatibility of the solvent with the available RO membranes. Guides are given in Tables 7-1 and 7-3. Most solvents will quickly be eliminated as a candidate for RO because they are incompatible with RO membrane elements. Even if the tables indicate compatibility, the membrane manufacturer should be contacted to obtain their approval prior to further investigation of the application. Existing information about similar applications should be requested from the membrane manufacturer.

Next, if the membrane manufacturer offers an application test program, a sample of the process fluid should be sent to them. This will provide a better idea of rejection characteristics of the solute, and will give an indication of the sample's fouling potential.

Example:

The potential application involves the purification of waste ethylene glycol (antifreeze) in water. This application is unusual in that it is desirable to remove salts from the solution while permeating the ethylene glycol. In fact, extremely poor rejection of the ethylene glycol is essential to the feasibility of the application. The concentration of the glycol could be as high as 50%. Any rejection of it could result in excessive osmotic pressures on the high-pressure side of the membrane, and subsequently poor permeate flowrates.

The salt concentration in the dilute ethylene glycol is 2,500 mg/L. Although it is not necessary to remove all the salts, it would be desirable to remove as much of the salts as possible. The trade-off will be the percentage recovery of the ethylene glycol. The higher the recovery of glycol, the greater the passage of salts there will be into the RO permeate.

What would be the approach to take with this application?

Answer:

When the application is researched, it is found that the membrane manufacturers already have some experience with similar applications. Several have tested a nanofiltration membrane (i.e., a membrane that primarily rejects the multivalent ions, and not monovalent ions like sodium or chloride). Another manufacturer has tested a poorly rejecting CA membrane in the application. The nanofiltration membranes achieved about 50% sodium chloride rejection, whereas the alternative CA membrane had about 85% sodium chloride rejection.

The higher rejection characteristics of the latter membrane make it possible to achieve greater glycol recovery rates for a given product quality. However, this membrane rejected a slight percentage of the ethylene glycol. Its use results in a greater osmotic pressure differential across the membrane. The nanofiltration membranes passed virtually all of the glycol.

The additional osmotic pressure with the CA membrane means that higher operating pressures are necessary to overcome the osmotic pressure of the RO concentrate solution. Another drawback for this membrane is the limitation on the pH of the solution. The solution pH may exceed 8.0, which can lead to an increased rate of hydrolysis of the membrane.

In this situation, there is no obvious answer as to which membrane to use. Since the membrane choice dramatically affects the design of the full-scale RO system, a pilot study should be performed using each membrane.

The pilot study should more accurately reveal the passage of salts and ethylene glycol as a function of concentration (which is directly affected by the RO permeate recovery.) A relatively accurate relationship between the osmotic pressures of the salts and of the ethylene glycol can also be determined. These values can be used in the array design program in Chapter 9 to determine, for a full-scale system, the system design and the overall salt passage as a function of permeate recovery.

It will be evident in the RO performance data whether the RO is experiencing any fouling or scale formation during the pilot study. It will also be apparent if compatibility is a concern for the ethylene glycol solution. If an increased rate of CA membrane hydrolysis occurs because of the high pH, the membrane will experience increase permeate flow and reduced rejection characteristics with time.

Two-Pass RO Design

In most RO systems, the permeate water has passed through the RO membrane only once in achieving the desired solute removal (i.e., in achieving the salt rejection required for the application). Some applications require better salt (solute) removal than can be accomplished using a standard membrane in a typical RO configuration. Ion exchange or distillation is frequently used for this polishing purpose downstream of the standard membrane system.

Ion-exchange systems require regeneration using strong acids and caustic. The waste stream from this process requires neutralization prior to disposal (in most areas in this country). Although less expensive than RO for the capital equipment, ion-exchange systems tend to be maintenance-intensive, and place a greater burden on plant neutralization systems. Some facilities get around the problems of ion-exchange regeneration by using portable exchange tanks supplied by an off-site water treatment vendor. When the exchange tanks are depleted in capacity, they are replaced and taken to an off-site regeneration facility. This service can be economical only if the loading on the portable exchange tanks is minimized.

Distillation systems are expensive to purchase and to operate as a means of polishing the RO permeate. In some Water-for-Injection (WFI) applications, the reliability of distillation in the removal of bacteria and endotoxin still makes it desirable as the final water treatment process. In this case, RO is commonly used upstream to reduce the possibility of scale formation in the distillation equipment.

In many applications, an RO system that utilizes two membrane passes can offer certain advantages, either in addition to other polishing equipment or as an alternative to other equipment. In these systems, the permeate from the first set of membrane elements is passed through a second set of membrane elements. In water treatment systems that produce deionized water, two RO membrane passes can be used to reduce the loading on downstream ion-exchange resins such that the utilization of portable ion-exchange systems is practical. This can sometimes eliminate the need for a waste neutralization system at the facility.

Two-pass RO has been validated as a final process in achieving Water for Injection (WFI) water in the pharmaceutical industry. To date, this process has not replaced distillation as the popular method of producing WFI for drugs used on humans. Reverse osmosis is not considered as reliable as distillation for bacteria and endotoxin removal, and so the process tends to be scrutinized by the Food and Drug Administration (FDA).

Two-pass RO systems are becoming popular in other pharmaceutical and biotech applications. In many of these applications, two-pass RO systems are capable of achieving the desired water purity without ion-exchange polishing. One example is the growing use of two-pass RO systems as pretreatment to WFI distillation systems as a means at reducing scale formation in the WFI stills. Reverse osmosis has the advantage of being biologically "cleaner" than ion exchange, particularly if the second-pass RO runs continuously in a recirculation loop, and is regularly sanitized.

In seawater desalination for potable water, it is fairly common to use two RO membrane passes in series to achieve a desired purity of water. Such a system is useful in other applications as well. Most commonly for these systems, a second high-pressure pump is used to raise the pressure between the two RO passes. This can occur with or without a collection tank between the passes. If a collection tank is not used, care should be taken to insure adequate flow to the second RO system so that its high-pressure pump receives sufficient flow to prevent cavitation.

Membrane choice for each of the passes depends on the priorities of the particular application. If biological control is the priority, then cellulose acetate membrane can be used for both passes because of its compatibility with chlorine. Sometimes CA is used for the first pass, with degasification downstream; and a PA thin-film membrane is then used in the second pass. The degasification removes most of the carbon dioxide released when acid injection is used to control scale formation in the first-pass RO system.

There has been some concern expressed that CA membrane can shed trace organics that can foul PA membrane located downstream. (Nickerson and McClain 1994). As such it is neccessary to over design the second pass PA membrane system when located downstream of CA membrane. This may require nearly doubling the PA membrane elements. It is documented, also, that the nonionic surfactants present in many of the CA-cleaning solutions can temporarily foul PA membrane, if not adequately rinsed out of the CA membrane system prior to putting it back on line.

With the ability of the newer PA membrane systems to achieve high flux at lower pressures, it is practical to stage a second set of RO membranes directly off a first pass (with no repressurization pump between). Reverse osmosis systems that use two membrane passes directly coupled in series are called permeate-staged RO systems. They need only a single high-pressure pump, typically operating around 450 psig. The concentrate of the second pass is valved such that the first pass permeate is back-pressured from 250 to 300 psig; thus, the effective driving pressure of the first pass is actually 200 to 250 psig, and the permeate flux for the first-pass membrane is similar to that in a standard single-pass PA RO system.

In a permeate-staged RO system, the second-pass membrane can be operated at slightly higher pressure than the first pass since the potential for fouling/ scaling is substantially reduced (i.e., the majority of potential foulants and hardness has been removed by the first pass). This is advantageous since it means that fewer membrane elements will be required, and that the second pass will get slightly better salt rejection at the higher driving pressure. (See Figure 7-4.)

Removal of most potential foulants and hardness by the first membrane pass also allows the second pass to operate safely at higher recoveries. It is common for second-pass systems to be designed with a maximum element recovery of 25% (compared to around 14% for standard systems). (Filteau 1993). The membrane manufacturer should be consulted for specific recommendations for the application.

Some of these permeate-staged RO systems use a patented process to achieve better salt removal by raising the pH between the two passes. This is accomplished by injecting caustic between the two stages. Because most of the bicarbonate buffering agents have been removed by the first-pass RO, very little

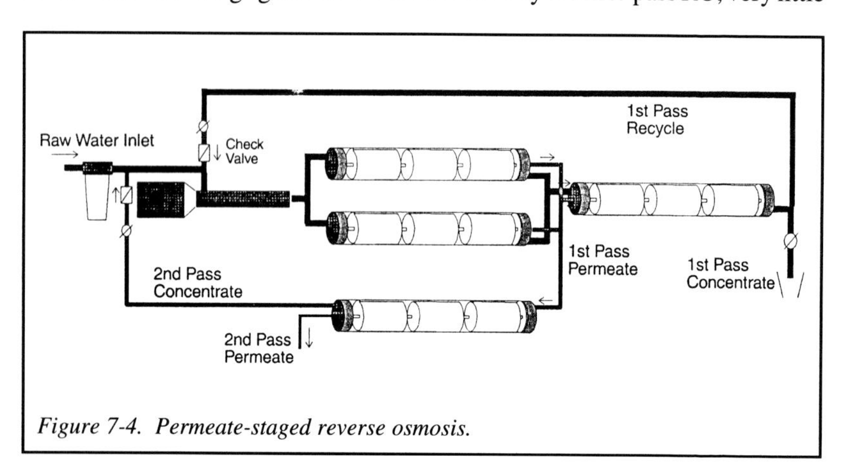

Figure 7-4. Permeate-staged reverse osmosis.

caustic is required to create the change of pH. This higher pH converts most of the dissolved carbon dioxide present into bicarbonate ions, a high percentage of which can then be rejected by the second-pass RO. The higher pH also increases the ionic nature of any dissolved silica, allowing it to also be better removed.

In two-pass RO systems, the concentrate from the second pass is usually recycled back to the feed of the first pass since its purity should be greater than that of the feedwater. This helps to reduce the total dissolved solids (TDS) of the feedwater, and thus helps the system to obtain better overall salt rejection. In fact, the overall rejection of such a permeate-staged system can be as high as 99.9% or more. In one example, feedwater with 300 ppm of TDS was purified using a permeate-staged RO with an effluent water quality registering a resistivity of greater that 2 megohm-cm (which is less than 0.3 ppm TDS).

The first step in designing a permeate-staged RO system begins with the second pass. Most of the same calculations are used as the ones used to design a standard RO system (see Chapter 2). The permeate recovery of the second pass will be determined by the minimum concentrate flowrate required by the housing(s) in its last stage. It is desirable to minimize the second pass concentrate flowrate in order to minimize the number of membrane elements required by the first stage of the system. The second stage can typically operate with an extremely high recovery due to the low potential for scale formation.

Choose a desired driving pressure, then make the following calculations:

Estimate the individual element permeate flowrate:
= driving pressure/design pressure × fouling factor × design permeate flowrate

Calculate the minimum number of membrane elements required:
= desired permeate flow ÷ individual element permeate flow

Calculate the number of housings required (rounding up if necessary)
= number of elements ÷ elements/housing

Recalculate the number of membrane elements
= number of housings × elements/housing

Calculate the individual element permeate flowrate
= desired total permeate ÷ number of elements

Calculate the minimum element concentrate flowrate
= individual element permeate flow ÷ maximum element recovery
- individual element permeate flow

Calculate the total system flow per housing
= individual element permeate flow × elements/housing

Calculate the total stage concentrate flowrate
= number of last stage housings × minimum element concentrate flow

Calculate the last-stage total feed flowrate
= individual housing permeate × housings/stage + concentrate flowrate

Determine a number for the housings in the second-to-last stage.

Calculate the feed flowrate for the second-to-last stage.

Repeat until housings are allocated.

Attempt to optimize the array by removing or adding housings to downstream stages.

Find the temperature correction factor.

Calculate the average osmotic pressure (negligible for systems other than seawater systems) = (feed TDS + feed TDS ÷ (1 - recovery)) ÷ 2

Calculate the array hydraulic pressure differential.

Determine the permeate pressure.

Calculate the required feed pressure
= element permeate flow × psi_{dsgn}* + psi_{prm} + psid ÷ 2 + psi_{osm}
÷ (element design flow × fouling factor + temperature factor)

Now, utilizing the second-pass feed flowrate as the first-pass permeate flow, and the second-pass feed pressure as the first-pass permeate pressure, follow the same procedures to determine an array for the first pass. The desired recovery of the permeate-staged RO is calculated by the second pass permeate flowrate divided by the sum of the second pass permeate and the first pass concentrate flowrate. The feed TDS for the first pass will be diluted by the concentrate of the second-pass array. This can be approximated as follows:

First-pass feed TDS = raw water TDS × raw feed flow
÷ (raw feed + 2nd pass concentrate)

Choose a desired driving pressure.

Calculate the minimum number of membrane elements required.

Calculate the number of housings required.

Recalculate the number of membrane elements.

Calculate the individual element permeate flowrate.

Calculate the minimum element concentrate flowrate.

Calculate the total permeate flow per housing.

Calculate the total system concentrate flowrate based upon a recovery calculated with the second pass permeate flow.

Determine a number for the quantity of housings in the last stage.

Calculate the feed flowrate for the last stage.

Determine a number for the housings in the second-to-last stage.

Calculate the feed flowrate for the second-to-last stage.

Repeat until housings are allocated.

Attempt to optimize the array by removing or adding housings to downstream stages.

Find the temperature correction factor.

Calculate the average osmotic pressure (negligible for systems other than seawater systems).

Calculate the array hydraulic pressure differential.

Determine the permeate pressure.

Calculate the required feed pressure.

Choose the high-pressure pump.■

CHAPTER 8

SPECIAL APPLICATIONS

Reverse Osmosis as Deionized Water Final Filtration

While RO may best be known for removing dissolved salts, its ability to remove fine particulates and total organic carbon is well documented. This ability makes RO attractive as final filtration for high-purity deionized water systems. These water systems are sometimes used in manufacturing semiconductors (Nakagome and Brady, 1986).

Semiconductor manufacturing is mostly a chemical process whereby layers of fine conducting and partially conducting areas are chemically etched out on the semiconductor chips. High-purity water is used to rinse residual chemicals and trace contaminants away from the chips.

It is critical that the water contains a minimal concentration of its own contaminants. Particulates can create defects in the semiconductor chips, as can TOC, which will tend also to support biological activity in the water system.

Particulates and TOC are introduced into a high-purity treatment system by the original water source, or by components of the water treatment equipment. Ion-exchange resins have been reported to be a significant contributor of TOC contamination in high-purity water polishing systems (Ammerer, 1989). Final filtration of the high-purity DI water is required to remove the fine particulates and TOC that either made it through the upstream treatment equipment, or were shed by it.

Cartridge filters. Replaceable cartridge filters have been the primary means used by the semiconductor industry to filter trace particulates from DI water. These filters typically will be constructed using a flat sheet membrane with a submicron porosity. They rely upon the water system line pressure for pushing water through the membrane pores. To minimize the pressure loss across the filters, it is desirable to fit as much flat sheet membrane into the cartridge filter as possible. This is typically accomplished by pleating the flat sheet membrane (Figure 3-27), or by constructing a membrane envelope (as with a spiral-wound RO membrane element) and rolling the envelope into a filter cartridge.

(Editor's note: The first section of this chapter was written by Wes Byrne of American Fluid Technologies Inc. and by Mark L. Maki of Hydranautics, Inc.)

Most water treatment systems are not designed to handle more than a 6- to 10-psid loss of pressure across the cartridge filters. Even when a maximum amount of flat sheet membrane is fit into a filter cartridge, there is still a limitation on how small the membrane pores may be unless a substantially greater pressure drop is acceptable. Usually the tightest cartridge filters that can be practically used in these systems have ratings from their manufacturer of 0.1 μm.

Some semiconductor facilities may use filters that have tighter ratings, but in order to do so, these facilities will have had additional filter housings designed into the system. This reduces the filter pressure drop by decreasing the flow that has to be accommodated by each 10-inch-equivalent filter.

The ratings given the membranes for many of these filters are not based upon the actual size of the pores in the membrane, but rather on the ability of the filter to remove bacteria with the given size. In fact, it is common for the actual pore sizes to be as much as 5 times greater than the manufacturer's rating. The filter is able to remove bacteria and particulates as small as the rating because of the depth of the membrane and the tortuous path the water has to take through the membrane layer.

Reverse osmosis has the potential to be superior to submicron cartridge filtration as a means of particulate removal. It can remove particulates that are much smaller than 0.1 μm because of the near absence of physical pores. This allows RO to remove key particulate contaminants that are common in most high-purity DI water systems. These typically include silica-based particles that are generated in the ion-exchange resins, particles shed by the piping and valves, and stainless steel particles that are shed by stainless steel pumps.

Reverse osmosis is able to correct for contaminant surges in high-purity water systems. Its ability to remove TOC and particulates is relatively independent of the RO feedwater concentration. Reverse osmosis will also remove a high percentage of any ionic contaminants, particularly trace metals, that might bleed from any upstream equipment. A final-filtration RO system thus serves as an excellent safety net against potential problems with upstream equipment.

Cartridge filters are not capable of removing ionic contaminants, and have not been able to achieve TOC reduction in a high-purity DI water system. In fact, many cartridge filters tend to leach organic contaminants into the DI water, particularly when first put into service (Fulford, June 1993).

A limitation with fine membrane filtration of any sort has to do with its ability to pass water. The finer the submicron pores, the more difficult it is to wet out those pores, unless the membrane has an affinity for water. Membranes of this type are called *hydrophilic,* meaning water-loving. Their chemical structures usually have polar groups that have a natural attraction for water, which is also chemically polar. Examples of these types of membranes include nylons (polyamide-based membrane) and cellulose acetates.

The concern with hydrophilic membranes is with their stability. Because of their attraction to water, some may have a tendency to break down with time in the DI water, subsequently shedding organic groups into the DI water; thus, they become a source of trace organic contaminants. Sometimes this breakdown can be sufficiently significant to change the ability of the filter to remove fine particulates according to its micron pore size rating.

Many cartridge filters use more stable membranes that tend to be *hydrophobic,* meaning water-repelling. Examples of these include polysulfones, polycarbonates, and Teflon®s. In order to wet out the pores of these hydrophobic membranes, a wetting agent is required. Such wetting agents have the ability to be attracted to the membrane, as well as to water; thus, they act as a pore lubricant and allow water to pass through the membrane pore.

Wetting agents are intended either to be permanently attached to the membrane, or to completely rinse out of the membrane. Wetting agents that are designed to be permanently attached to the filter membrane surface suffer from the same concerns as hydrophilic membranes. They tend to either break down in the DI water with time, or detach themselves from the filter membrane, thus ending up as a trace organic contaminant.

Teflon® filters are typically wetted out with isopropyl alcohol on-site when they are first put on line. The alcohol rinses out completely from the filter, thus removing the possibility of being a source of organic contamination. Unfortunately, with the absence of any wetting agent, the hydrophobic membrane will tend to attract supersaturated gases to its membrane pores. This phenomenon is more common with some of the tighter hydrophobic membrane filters. When it occurs, the ability to pass water is reduced, resulting in an increase in pressure differential across the filter. This problem is frequently confused with fouling of the filter (Brinda, 1993).

Improvements in RO manufacturing techniques. It is probably not possible to completely eliminate organic shedding from RO membranes, which are hydrophilic. If the amount of organic shedding from the RO membrane is minimized, however, the shedding that remains can be dealt with by the ability of the RO to remove incoming organic contaminants. The overall result of using a properly designed and built RO for DI final filtration would thus be a reduction in total organic contaminants.

As with organic shedding, both cartridge filters and final filtration RO systems will tend to shed some particulates. If the filters or the RO systems are properly designed and manufactured, this shedding can be minimized such that the particulate removal ability of the filter or the RO more than compensates for the particulates it may be adding to the DI water. The overall result is a reduction in particulates.

Some of the first RO systems to be used for final filtration were not well designed and manufactured. Mistakes were made with the manufacturing of the membrane elements, and/or with the RO system design. During an extended period of time after some of these systems were first started up, the quality of RO effluent water was worse than the influent. Eventually, the ionic and TOC concentrations in the RO permeate would return to acceptable levels. However, these long rinse-up times were unacceptable in several cases. (Wong 1990).

Recently, there have been RO systems designed and built that have been able to reduce ionic, organic, and particulate contamination concentrations even when fed with high-purity DI water after a relatively short rinse-up time. The actual performance of a final filtration RO system is shown in Table 8-1.

The choice of materials used in the membrane elements is critical to minimization of contaminant shedding. The glues, spacer materials, and permeate carrier tubes must be constructed of special materials with low particulate and TOC shedding characteristics. The membrane extrusion and element construction must be performed under clean-room conditions to avoid environmental contamination of the materials.

Membrane elements should not be tested with sodium chloride solution. This will add ionic contaminants to the elements. In fact, the elements should be rinsed with low-particulate DI water with a resistivity of 18 megohm-cm (i.e., extremely low salt content) prior to shipment.

The RO system should be designed for a relatively high permeate flux. The permeate flow should be designed for 25 to 30 gallons per square foot per day (gfd) of membrane area, versus closer to 15 gfd for more standard RO applications. With DI water as the feedwater source, there is little concern about a dramatic increase in fouling rate caused by the higher flux, as there would be in other applications.

The amount of contaminants being leached into the permeate water, as well as the percentage of ionic contaminants passing through the membrane from the feedwater, are both relatively independent of permeate flux. Therefore, a high permeate flowrate will result in better dilution of these trace contaminants. The end result is that the RO is able to remove more contaminants than what is being reintroduced by the RO.

Table 8-1
Particle Counts and TOC of RO Feed and Permeate Water

Size	*Feed*	*Permeate*
>0.05 μm	30/mL	5-7/mL
>0.1 μm	8/mL	1-2/mL
TOC	25-30 ppb	2-5 ppb

Note: The extent of TOC removal will depend on the molecular weight, charge, and physical characteristics of the particular organic molecule.

The condition of RO system components is also critical to minimizing the reintroduction of contaminants. All piping, as well as any other system components, should be thoroughly cleaned prior to use. Machining lubricants must be removed from stainless steel. This can be performed using a detergent cleaning, followed by an acidic cleaning step (McPherson and Bedford, 1986).

Low-pressure plumbing (less than 100 psig) is typically constructed using a clean and inert PVDF plastic pipe. Higher-pressure piping is constructed of 316L-grade stainless steel, ground and electropolished to a 2B-rated finish (American Iron and Steel Institute).

The membrane housings (pressure vessels) are fiberglass-reinforced plastic (FRP). Its end caps and internal components are machined PVDF plastic. O-rings and connectors are minimized where possible to reduce potential areas of stagnant water and bacteria growth.

Dead legs, which are defined as lengths of piping with no flow, should be nonexistent. This means that pressure gauges should not be used without a diaphragm gauge guard. Even flowmeters in the permeate lines should be avoided because of their tendency for particle shedding.

Deionized-water RO systems are designed for between 85% and 95% permeate water recovery. Because of the purity of the feedwater, scale formation is not a concern with the higher recoveries. Since DI water is expensive to produce, the RO concentrate water is typically returned to somewhere in the makeup water system. It is not suggested to return this water to a polish DI loop storage tank, as this may result in the buildup of particles and TOC feeding the RO system.

Maintenance of a DI water RO consists mostly of periodic sanitization. These are required at least once per year to minimize the buildup of bacteria in the system. Because the RO is a fundamental part of the DI water polishing system, it may be difficult to schedule such sanitizations, and it will probably require shutting down the DI water polish system.

Sanitization can be performed with up to a 1% hydrogen peroxide solution (semiconductor-grade peroxide), or with a 400-ppm semiconductor-grade peracetic acid solution (which also contains 2,000 ppm of hydrogen peroxide). The solution should be recirculated through the system for a minimum of 4 hours (if possible). Consult with the membrane manufacturer prior to use.

The cost of final filtration RO is dramatically more expensive than cartridge filtration in both capital and operating costs. Capital cost would be roughly 10 times more for the RO. Operating cost would be approximately 4 times more for the RO, based upon annual replacement of the cartridge filters, and replacement of the RO membrane elements every 5 years. The power consumption for the RO high-pressure pump accounts for most of the additional operating cost.

The semiconductor industry continues to move toward products with finer geometries. These semiconductors are sensitive to finer contaminants; thus, the newer devices are driving the need for finer filtration and may justify the higher cost of RO.

Summary. It has been demonstrated that RO systems and their membrane elements can be constructed so as to minimize the shedding of contaminants. These contaminants can be minimized to the point that they are insignificant compared to the ability of RO to remove the contaminants in the RO feedwater. Reverse osmosis can thus be successfully used as final filtration for high-purity DI water.

The use of RO for high-purity DI water final filtration is currently not popular, although it is anticipated that its popularity will increase. This increase will occur as the semiconductor industry requires finer filtration, and membrane cartridge filter technology is not able to meet this need.

Metal-Recovery Applications

Nickel

The recovery of metal salts from plating rinse water for return back to the plating tank was one of the first common applications of reverse osmosis. In plating operations, metal parts are dipped into a plating tank before being dipped into a rinse tank to remove the residual plating solution. In this process, the rinse water picks up the metal contaminants from the plating solution. Removal of the majority of these metal contaminants is required by most sanitary treatment facilities prior to disposing of the water in the sewer. Reverse osmosis can serve this purpose while recycling both the water and the plating salts (Figure 8-1).

Many plating applications are ideal for RO in that there is no waste stream generated. The RO concentrate containing the plating salts is returned to the

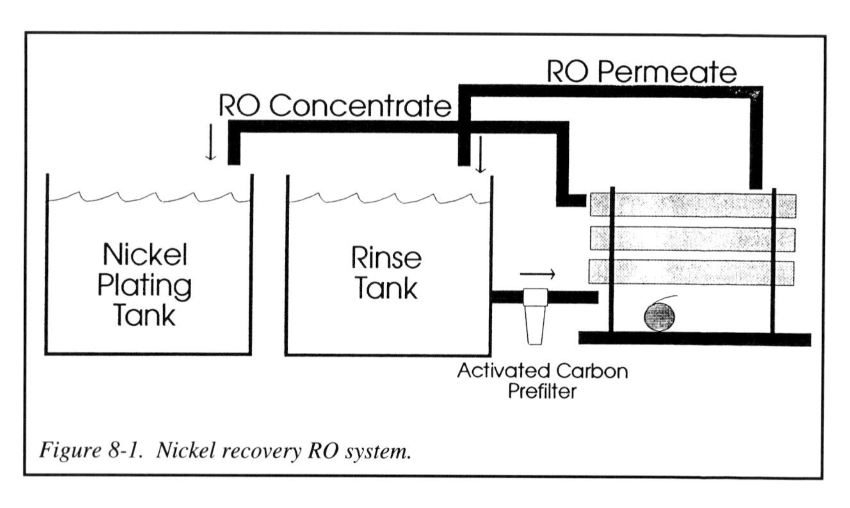

Figure 8-1. Nickel recovery RO system.

plating tank, offering significant savings by recycling the salts. The RO permeate is returned to the rinse tank, offering savings in water usage. The RO permeate water is also of a quality that is typically better than most city waters, thus offering a better rinse.

When RO is used to recover nickel from the rinse water, some savings occurs from not having to treat and dispose of the waste nickel. Most localities require that nickel concentrations in the plant discharge water be at minimal levels. Depending on this limit, standard hydroxide precipitation systems may have difficulty achieving sufficient purity. Achieving this low concentration by diluting the waste water with city water is heavily discouraged (and may be illegal).

In some areas, there are companies that offer off-site treatment of nickel wastes. They pick up the liquid wastes, treat it, and dispose of it. This service tends to be fairly expensive.

The most common means of on-site treatment of nickel has been hydroxide precipitation. The pH of the water is raised by injecting caustic (sodium hydroxide) in the water. The solubility of nickel decreases at higher pH, and the nickel falls out of solution as nickel hydroxide. Most of the residual liquid is removed with a filter press. The solid waste is then hauled off to a landfill. This process is becoming more expensive as approved landfills become more scarce.

Nickel-recovery application history. Many of the original plating shops that tried RO to recycle nickel would not consider the application a success. Due to the high permeate water recoveries required by this application, the control of scale formation was critical to the success of the application. How this concern was addressed by the original equipment vendors had a great deal to do with the success or failure of the particular system.

Treatment of the makeup water to prevent scale formation was essential. Reverse osmosis, softening, or ion exchange was needed to remove from the makeup water the ions that might precipitate in the nickel-recovery RO.

Nickel-recovery RO systems operate with permeate water recoveries of 96% or greater. This means that the concentration of salts in the feedwater increases by a factor exceeding 30 within the RO systems. Any hardness (calcium or magnesium) present in the system will likely fall out of solution as a carbonate or other salt.

Most of the original spiral-wound element RO systems sold in the application used cellulose acetate membrane. Any residual hardness that might make its way into the rinse water tanks would tend to fall out of solution as carbonate scale on the CA membrane surface. The carbonate scale would cause a dramatic increase in the hydrolysis rate of the membrane, often resulting in its demise.

Most of these RO systems were in the field only a few months before the membrane stopped performing adequately. Many of the plating customers did not know enough about how the system should work to know that the system was losing performance. Eventually, the performance would become so inadequate that the rinse water would contain nearly as much nickel as the plating tank. The membrane elements would then have to be replaced. After replacing the elements a few times within a short period of time, the plating shops would find it less expensive to get rid of the RO and treat the nickel rinse water using some other means.

Polyamide membrane elements. The recovery of nickel from a nickel rinse water is one of the easier metal-recovery applications, yet it is also one of the most critical since the nickel rinse usually takes place immediately prior to a very sensitive chrome-plating process. A poor rinse due to poor-quality rinse water will nearly always result in poor-quality chrome plating.

A nickel-plating bath will not normally contain any oxidizing agents (unless introduced by the makeup water); therefore, PA thin-film spiral-wound membrane elements are compatible. This offers the end user more maintenance flexibility, since the operator does not have to be concerned about calcium carbonate hydrolyzing the membrane. Because the thin-film membrane also gets better nickel salt rejection at lower pressures, a smaller pump and less membrane elements are required.

Even as compared to well-maintained CA membrane systems, the thin-film membrane will tend to last longer between membrane changeouts. The overall advantages of the thin-film membrane system for this application are that the system costs less to buy, less to operate, and requires less attention to maintain it.

Reverse osmosis pretreatment. One requirement of a PA thin-film RO system (which is not the case for a much-lower-rejecting CA membrane system) is the need for activated carbon filtration of the RO feedwater. The carbon serves two purposes. First, it removes any chlorine that might have been introduced in the city water makeup (and could otherwise oxidize the thin-film membrane). Secondly, it removes the organic brighteners used in the nickel-plating bath that have been dragged out into the rinse water.

The thin-film membrane is capable of rejecting and concentrating some of the brighteners from the plating solution. These should not be recycled back to the plating tank because this will alter the ratio of the organic components of the brighteners. Without the capability of analyzing for the exact concentrations of organic constituents in the plating tank, it is impossible to know which organic components need to be added (to make up for drag-out losses). Therefore, it is easiest to remove all the brighteners that have been dragged out into the rinse tank, and simply replace them in the plating tank with the original formulation.

It is critical that the brightener concentration not be allowed to get too high in the plating tank, and that the organic constituents not be allowed to get out of balance. Typically, if either of these conditions exist, organic films are visually apparent on the plated parts. It appears as if the nickel rinse water is not rinsing off the brighteners, although the problem may not necessarily be with the rinse water.

The high permeate water recovery is required of the nickel-recovery RO so that the RO concentrate does not overflow the nickel-plating tank. This would defeat the purpose of the RO. The RO concentrate flow must be less than the combined evaporative losses and carryover of the plating tank. Fortunately, the nickel-plating tank normally operates at around 140 °F, which aids in the evaporation.

Most nickel-recovery systems operate in the 96% to 98% recovery range. This is achievable with RO by recycling most of the last housing concentrate flow back to the feed of the high-pressure pump. For example, the desired flow leaving a single 4-inch membrane element might be 5 gpm (chosen to meet the maximum individual element recovery guidelines as set by the membrane element manufacturer). At 97% recovery, a small RO system might have a system concentrate flow as low as 0.09 gpm. This would mean that 4.91 gpm would need to be recycled back to the feed of the high-pressure pump.

The concentration of salts going by the membrane will be roughly 30 times the concentration of salts in the feedwater. Because nickel is a divalent ion, it is very well rejected by the thin-film membrane. In fact, its rejection is about 99.5%, depending on other components of the feed rinse water. This means that the overall system rejection will typically be greater than 70% in spite of the extremely high recovery rate.

Even with PA membrane elements, the makeup water quality is critical to the performance of the RO. Any calcium or magnesium hardness present in the rinse water feed to the RO will result in precipitation of salt on the membrane surface. With PA thin-film membrane systems, the leakage of hardness into the rinse water will result in scale formation that will require cleaning. Removing this hardness in the makeup water is important to avoid having to clean the RO every time water is added to the rinse tank.

The amount of makeup water to the system that needs to be treated will normally be minimal. The majority of the water losses from the system will be from the evaporation in the plating tank. With PA membrane it is possible to operate a system without treating the makeup water, but it will mean frequent cleanings of the RO system (at least once a week).

Treating the makeup water can be as simple as putting an automatically regenerated softener on the makeup line. Reverse osmosis can also be used, although it may not provide as good a method of removing hardness as softening, or using two-bed (a cation bed followed by an anion bed) ion exchange. Treating the makeup water will have the added benefit of creating a better-quality rinse water.

If a water softener is used to treat the makeup water but is operated past the point of exhaustion, hardness ions will be shed into the RO feedwater. The result will be a fairly rapid decrease in RO permeate flow and a decrease in permeate quality (mostly in the tail end membrane elements). If these symptoms are noted in an RO system with softener pretreatment, the softening system should be investigated. The RO permeate flow and quality can usually be recovered if an acidic cleaning solution is recirculated through the membrane system before the problem becomes too severe. (Consult the membrane manufacturer.)

Cleaning. Even with proper treatment of the makeup water, scale formation within the membrane elements should be anticipated. Monthly cleanings with an acidic cleaning solution will probably be necessary. The system should be designed to facilitate the frequent cleanings. The cleaning should not be avoided, as its omission may result in the permanent blockage of flow channels within the membrane elements.

Some means of disposing of the nickel-laden cleaning solution will be required. If no means of treating the cleaning solution is available on-site, it may be necessary to haul it to an off-site treatment service. This can significantly add to the operating cost of a nickel-recovery system. Another option is to replace the fouled elements and send them to an off-site membrane cleaning service (one set up for dealing with hazardous wastes).

Water temperature control. With the high recovery rate of the RO, there tends to be a significant amount of heat added to the rinse water as it recirculates. Given

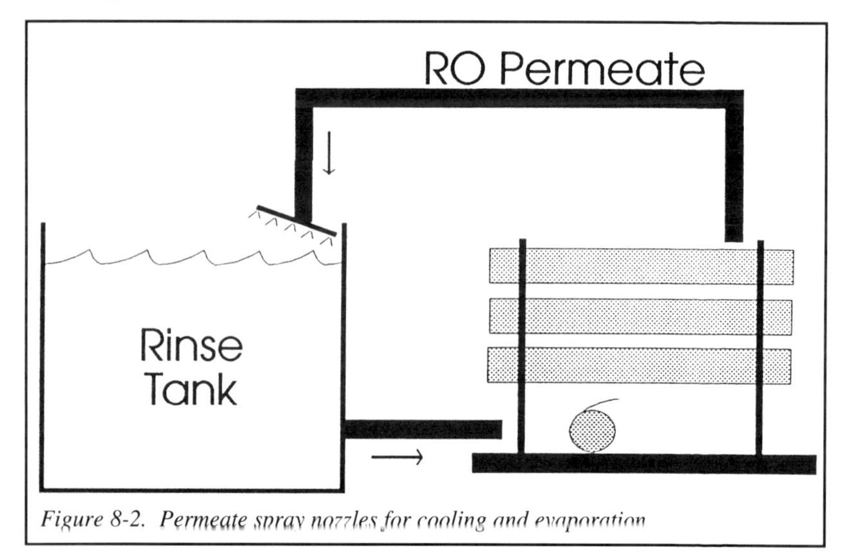

Figure 8-2. Permeate spray nozzles for cooling and evaporation

that the rinse water is already warmed by the drag-out from the plating tank (140 °F), the rinse water temperature can sometimes exceed 100 °F. With time, this can permanently affect the performance of the RO membrane.

Particularly in warmer climates, it may be necessary to install some means of cooling the rinse water. This can be as simple as installing spray nozzles (Figure 8-2) on the RO permeate return to the rinse tank. The evaporation of the permeate water as it is sprayed over the rinse tank will usually drop its temperature enough to keep the rinse tank water temperature below 100 °F. When designing the spray nozzles, be sure not to place too much back pressure on the RO permeate. This pressure would take away from the driving force across the membrane, and could reduce the permeate flowrate.

Nickel-recovery RO cost comparison. With proper treatment of the makeup water and cleaning of the RO membrane, the membrane life can be expected to exceed 2 years. As shown in the following economic calculations, RO can be attractive as a means of saving money, depending on the requirements of the facility. In some situations, RO can save as an alternative to the installation of an expensive hydroxide precipitation system. In areas where it is desirable to achieve zero liquid waste discharge, the nickel-recovery RO application may offer significant advantages compared to other technologies.

The costs of operating an RO system in a metal-recovery application are similar to those discussed for a standard RO system in Chapter 2. Maintenance requirements are somewhat higher due to the increased fouling rate of the application.

Power. The amount of energy to drive a pump can be calculated using the following:

$$\text{kwh/1,000 gal permeate} =$$
$$\text{pump pressure} \times 0.00728 \div (\text{pump efficiency} \times \text{motor efficiency} \times \text{recovery}^{*})$$
$$= 400 \text{ psig} \times 0.00728 \div (0.52 \times 0.95 \times 0.33^{*)}$$
$$= 17.9 \text{ kwh/1,000 gal permeate}$$

$$17.9 \text{ kwh} \times \$0.08/\text{kwh} = \$1.43/1{,}000 \text{ gal permeate}$$
$$\$1.43 \div 1{,}000 \text{ gal} \times 4 \text{ gpm} \times 1{,}440 \text{ minutes/day} = \$8.24/\text{day}$$

Cartridge prefilters. The RO uses a 20-inch activated carbon cartridge that is replaced weekly at a cost of about $20.00.

$$\$20.00 \div 7 \text{ days} = \$2.86/\text{day}$$

*The recovery fraction to be used in determining electricity demand should be based on the permeate flowrate divided by the total feed flowrate (including any concentrate recycle flow).

Membrane replacement. Using a projected life of 2 years for the RO membrane elements, a cost per day can be determined:

$$\text{membrane cost/day} = (\$450 \times 3) \div (365 \text{ days} \times 2) = \$1.85/\text{day}$$

Cleaning chemicals and hauling. In the case study used, cleaning was required approximately every 3 weeks. Phosphoric acid was used at a cost of about $15.00.

$$\$15.00 \div 21 \text{ days} = \$0.71/\text{day}$$

Cleaning generates about 30 gallons of waste solution, which has to be hauled to an off-site treatment facility. Cost of this hauling and treatment is estimated at $2.00/lb.

$$30 \text{ gal} \times 8.3 \text{ lb/gal} \times \$2.00/\text{lb} \div 21 \text{ days} = \$23.71/\text{day}$$

Total cleaning cost is therefore $0.71 + $23.71 = $24.42/day

A less expensive alternative to hauling the cleaning solution might be to batch treat it by hydroxide precipitation on-site.

Maintenance labor. It is estimated that 1 hour of maintenance was required per day at a cost of $20.00/hour.

Summary:

Power	$8.24
Prefilters	$2.86
Membrane	$1.85
Cleaning chemicals	$24.42
Labor	$20.00
Total/day	$57.37

The cost of the makeup water, as well as the cost of softening the makeup water, will be considered negligible because of its low volume (90 gallons per day).

Hydroxide precipitation operating cost. The cost of using a hydroxide precipitation (conventional treatment) system and not reclaiming the metal salts would include the following:

Makeup nickel salts. It is estimated that about a pound per day is saved in both nickel sulfate and in nickel chloride salts. At about $1.50/lb:

$$2 \text{ lb/day} \times \$1.50/\text{lb} = \$3.00/\text{day}$$

Sludge hauling. When treated with caustic, approximately 25 pounds of sludge would be generated (50% solids) per day. At a hauling cost of $1.00/lb:

$$25 \text{ lb/day} \times \$1.00/\text{lb} = \$25.00/\text{day}$$

Caustic (sodium hydroxide) expense. The volume of caustic required to treat the nickel would include the volume required to react with the nickel salts as well as that required to raise the pH of the water to a range of 10.5 to 11.0. This would be estimated to be about 7 lb/day of 100% caustic. Its cost is about $0.08/lb of 50% caustic. Total cost:

$$7 \text{ lb/day} \times 2 \times \$0.08/\text{lb} = \$1.12/\text{day}$$

Water and sewer. A total cost of $0.01/gal will be used for both city water and sewer costs.

$$4 \text{ gpm} \times 60 \text{ minutes} \times 24 \text{ hours/day} \times \$0.01/\text{gal} = \$57.60/\text{day}$$

Labor. A rough estimate of the average time to control and maintain a hydroxide precipitation system would be 1 hour per day at $20.00/hour.

Total:	
Nickel salts	$3.00
Sludge handling	$25.00
Caustic	$1.12
Water and sewer	$57.60
Labor	$20.00
Total/day	$106.72

Many of the costs that are given in this example for both the RO and the hydroxide precipitation will vary dramatically, depending on the facility's location, the quantity of parts processed (i.e., the amount of nickel drag-out), and the number of plating/processing shifts per day. It is interesting to note that the highest costs of operating the RO system are the cost of the maintenance labor and the cost of disposing of the RO cleaning solution. This cost will be extremely dependent on how well hardness is removed from the rinse tank makeup water. Its importance might justify the use of several leased or purchased softening units in series to insure the complete removal of the hardness ions.

The highest costs of operating the precipitation system are the water and sewer costs, the maintenance, and the cost of hauling the waste sludge to a landfill. It is likely that the cost of the sludge hauling and the sewer costs will increase in the future as the result of tighter government restrictions.

The capital cost of a complete hydroxide precipitation system will typically be much greater than the cost of an RO system. However, the precipitation system may also be required to treat other waste streams resulting from the plating processes. If this is the case, the RO may be just an added capital expense that would need to be justified by the operating savings.

The savings in the previous example would be about $49.35/day. If the RO and makeup water treatment system had an installed cost of $30,000, it would take 608 days for the RO to pay for itself. This would probably be considered a good return on investment, particularly considering the political benefits of recycling the salts versus hauling them to a dump.

Chromic Acid

Chromic acid (CrO_3) is also known as hexavalent chrome, so named because the chrome atom in the acid is in the hexavalent valence state (charge of +6). It is considered an acid because in low pH water, the CrO_3 molecule combines with water to form H_2CrO_4. The extent of its dissociation will depend on the pH of the solution.

Although the chrome atom is highly charged, which might suggest that it would be well rejected by an RO membrane, at a pH of less than 6 the chromate molecule (CrO_4^{-2}) begins to associate with hydrogen ions to form $HCrO_4^-$. As a monovalent ion, $HCrO_4^-$ is not well rejected. At extremely low pH, chromic acid associates completely with hydrogen ions to form H_2CrO_4, which is very poorly rejected by an RO membrane. Thus, the rejection of chromic acid by RO is a very strong function of pH. (Comb 1986).

At a feedwater pH greater than 4, the RO membrane will tend to pass a greater percentage of acidic ions while rejecting the divalent chromate anion, resulting in a lower permeate pH. Since this water is returned back to the rinse tank from which it came, the rinse water will drop in pH with time until it stabilizes between a pH of 3 and 4. At this pH, the rejection of hexavalent chrome is going to be in the low 90% range, at best. It is therefore critical that a membrane with high rejection characteristics be used for the application. It is suggested that the membrane have minimum sodium chloride rejection characteristics of 98.5%.

Unfortunately, chromic acid is an oxidizing agent. With time it will attack a PA thin-film membrane. Since the charged polysulfone membranes do not have sufficient rejection characteristics for this application, the only choice available for the membrane is a high-rejection cellulose acetate.

Chrome-plating baths operate at lower temperatures than nickel-plating baths. This means that less water will be lost to evaporation from the bath, as compared to nickel-plating baths. This in effect reduces the flow of RO concentrate that can be put back into the plating tank without overflowing the tank; thus, a chrome-recovery RO will have to operate at an even higher recovery than a nickel-recovery RO. Yet the recovery is limited by the pH of the concentrate. Excessive recoveries will result in a concentrate pH approaching 2, which will result in reduced life for the CA membrane because of an increased rate of hydrolysis.

The effect of all these parameters is that the chrome-recovery application is somewhat of a tightrope. It requires close monitoring to insure that the chrome tank is not going to overflow while making sure that the RO permeate is of sufficient quality that it can be used for rinsing.

Increasing the number of membrane elements in the RO system will slightly improve the quality of the permeate water. However, the RO concentrate flowrate and the subsequent concentrate salt concentration are dictated by the makeup flowrate that can be tolerated by the plating tank. Increasing the RO permeate flowrate by increasing the number of elements will not change the limit on the concentrate flowrate, or the resulting RO concentrate salt concentration.

Adding more membrane elements will result in better life expectancy from the lead end elements. Yet the tail end elements that see the greater concentration of salts and lower pH will still experience a reduced life expectancy because of the faster rate of CA membrane hydrolysis.

The only way to reduce the concentration of salts in the RO concentrate is either to reduce the drag-out of salts into the rinse tank, or to use some means of further concentrating the RO concentrate stream prior to sending it back to the chrome-plating tank. In some areas (depending on air quality restrictions), evaporation can be used to further increase the concentration of the chrome salts from the RO concentrate. For example, the RO concentrate can be run through an evaporation tank containing heating coils. This allows the RO system to operate with a higher concentrate flowrate, resulting in better permeate quality and longer membrane life expectancy. However, the energy demand of the evaporation will affect the economics of the application.

If short-term performance improvement is desired from the chrome-recovery RO, this can be achieved by raising the pH of the rinse water. As the pH approaches 6, the rejection of chrome will increase dramatically as the chrome dissociates into the divalent chromate anion.

This method of getting better chrome rejection can be only short-term for two reasons. If the RO concentrate is being returned to the plating tank, the additional sodium introduced by the caustic injection is undesirable in the plating tank. The second reason is that the chromic acid is a more aggressive oxidizing agent at the higher pH and may cause an increased rate of breakdown of the CA membrane.

If the concentrate is not being returned to the plating tank for recycling, but is instead being concentrated prior to some other form of treatment and disposal, the hexavalent chrome can be chemically reduced to trivalent chrome (Cr^{+3}). The trivalent chrome now behaves as a trivalent metal that is well rejected by the RO membrane (independently of the RO feedwater pH). This chemical reduction can be accomplished by injecting sodium bisulfite (or some other reducing agent) into the RO feed stream.

Case study. The chrome-recovery RO application has been used successfully. In a shop specializing in plating automotive bumpers, one RO was used to recycle the nickel salts, and another was used to recycle the chrome salts. This shop already had a conventional hydroxide precipitation system. Yet the RO systems enabled the shop to reduce their water consumption by two-thirds, as well as reduce their chemical and sludge hauling requirements by approximately two-thirds.

This shop needed to perform metal-stripping operations on old bumpers prior to plating them; therefore, the hydroxide precipitation system was still required to remove the metals from these operations. Because of the lower waste water flowrates, however, and because metal concentrations were reduced, there was a dramatic reduction in the chemical requirements of the system, as well as increased ease in meeting the local sanitary sewer discharge guidelines.

Chrome-recovery system design. The chrome-recovery RO system was set up

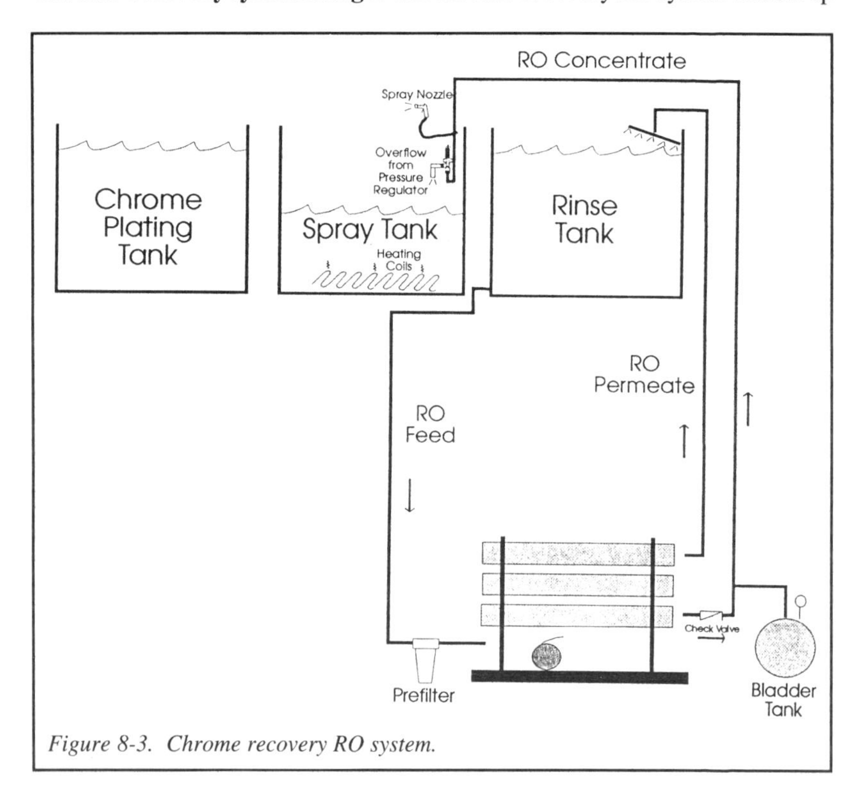

Figure 8-3. Chrome recovery RO system.

as shown in Figure 8-3. The RO concentrate was plumbed to a bladder-type pressure tank. This tank was plumbed to a pressure relief valve whose effluent went into a tank located after the chrome-plating tank. This tank contained heating coils that assisted in further concentrating the chrome solution via evaporation. A hose with a spray nozzle was teed into the line with the pressure relief valve.

The spray nozzle was used to wash out chrome-plating drag-out that had been left in pockets within the bumpers. This was performed over the middle tank. The spraying made a tremendous difference in reducing the amount of chrome that was dragged into the rinse water tank. Since the spray water contained a high chrome concentration at a pH of about 2.0, great care had to be taken to ensure that the chrome solution did not get on anyone's skin or eyes.

When the spray was not being used, the pressure tank would fill. At the desired set pressure of about 25 psig, the RO concentrate water escaped through the pressure relief valve into the middle tank.

If the chrome-plating tank level dropped, the solution from the middle tank was used to make up for the losses. The temperature in the middle tank was adjusted so that the solution was not overly concentrated to the point that there was not enough solution available to make up for the losses in the plating tank. It was also important to keep the temperature of the solution below 140 °F to prevent the breaking down of constituents in the solution.

If the chrome rinse water was allowed to get too concentrated in chrome by excessive drag-out, it would turn a yellow color. A hazy film would be apparent on the automobile bumpers. This film could be rubbed off, although this extra labor-intensive step was undesirable. Usually, the rinse water had this problem only if the parts were not being adequately sprayed prior to dropping them in the rinse water.

One nice feature of the chrome-recovery application was that the RO membrane never fouled or scaled. It therefore required little or no maintenance. However, the concentrate bladder tank was incompatible with the oxidants in the solution and required periodic replacement. Since the solution was under pressure, care had to be taken to place the bladder tank in a location where a catastrophic failure would not create a spill or allow the solution to come into contact with any personnel.

The chrome-recovery RO system was on-line for only 1 year prior to the shop shutting down their plating operations (for reasons independent of the waste water treatment systems). Over that year, there was no significant loss in membrane performance. It is suspected that it would have been possible to obtain at least 1 more year of life from the RO membrane before poor permeate quality would necessitate the membrane replacement.

Summary. When RO is used to recycle nickel and chrome solutions back to their plating tanks, it becomes very possible that no other waste treatment system is required for the plater (depending on their particular situation and how well the solutions are isolated). Platers who must perform metal-stripping operations prior to plating will likely need some sort of treatment for the rinse water from the stripping operations. (See the next section, "Oily Waste Water.") Also, some means of handling spent plating, stripping, or cleaning baths is required. This may mean hauling the solution to some independent treatment facility, or treating it on-site by some conventional means (such as hydroxide precipitation).

Oily Waste Water

Sometimes a waste stream that requires the concentration of a salt may also happen to contain a high concentration of grease and oil. This tends to be the case in various wash water recovery applications.

In localities of water scarcity, RO is commonly used to recover a low-TDS water from oily wash waters. In car-wash applications, this water can sometimes be purified to the point of being used for a "spot-free" rinse. A rule of thumb in the car wash industry is that the water has to contain less than 60 ppm of total dissolved solids for the water to not visibly spot when it dries.

In metal-cleaning applications, rinse water can become contaminated with heavy metals, which may prevent its discharge into sanitary sewers. Frequently these rinse waters will also contain oils and grease rinsed off the metal parts.

In both of these RO applications, membrane fouling due to the oils and grease is almost a certainty. Cellulose acetate RO membrane tends to be more resistant to oil fouling than is polyamide thin-film membrane. However, cleaning solutions with a pH greater than or equal to 11 tend to be most effective at stripping oily foulants. These cleaning solutions would be compatible only with PA or charged polysulfone RO membrane. While all membrane types might be used in an oily-waste application, none is ideal.

In designing RO systems for oily waste applications, it is critical to well overdesign the system. Because it is certain that the system is going to lose permeate flux due to fouling, there should be more than enough membrane in the system to compensate for losses in permeate flow. The objective is to have enough membrane in the system so as to allow a reasonable time between membrane cleanings. Cleaning a system several times a day to meet the permeate flow requirements of the application is not practical.

An excess of membrane also makes it possible to operate at lower permeate flux and lower operating pressures. This will reduce the rate of membrane fouling. (It will also reduce the overall salt rejection characteristics of the membrane and thus may not be feasible for some applications.)

Even an over-designed RO system may have to be cleaned as frequently as once per day. Processing oily waste therefore tends to be a relatively expensive RO application as compared to other RO applications, and should be used only when the alternatives are even more expensive. This might be the case if the alternative is hauling off all the waste water to an external treatment facility.

Case study. In one plating application, old bumpers are stripped of oil and grease in an alkaline cleaning solution, and then stripped of their metal finish in a hydrochloric acid solution. They are then plated with nickel and chrome.

The nickel- and chrome-plating operations were already set up at this facility to recover the metals from their respective rinse tanks using evaporation. The rinse water was concentrated by evaporating water off the solution using heating coils. It was then added back into the plating tank. (It should be noted that using RO instead of evaporation in these systems would have resulted in significant operational savings because of the high energy demand of evaporation.)

The facility wanted to achieve a zero waste discharge status. It was therefore necessary to do something with the rinse water from the cleaning and acid stripping operations. It was decided to preconcentrate this water with RO and then evaporate the RO concentrate. The salts left by the evaporation would be hauled to a hazardous-waste landfill.

It was visually evident that the water would contain a high loading of oils and grease. Spiral-wound PA thin-film membrane elements were chosen for the application to enable the use of aggressive alkaline cleaning solutions to remove the oils and grease from the membrane. The PA membrane also had the advantage of being compatible with the anionic surfactant that was used in the facility's alkaline cleaning tank.

The RO was overdesigned by a factor of 8. In other words, when the membrane was clean, the RO contained enough membrane elements (four each 4-inch by 40-inch elements) to permeate 8 times the flow required to keep up with the flow from the rinse tanks (which was only 0.4 gpm). The RO was designed to operate at 98% recovery, but also had the flexibility to operate at lower recoveries if the osmotic pressure of the feed solution was extremely high (i.e., if the salt concentration was so high that there would be very little permeate flow at the 98% recovery rate).

The waste water was first collected in a 3,000-gallon conical tank. Water was decanted off the top of this tank into another 3,000-gallon tank. The water from this tank was then pumped to the pretreatment system.

Pretreatment of the RO feedwater consisted of two parallel multimedia filters that fed a diatomaceous-earth (DE) precoat filter. This system constantly ran in a loop, returning back to the second waste water storage tank. The RO system pulled water from this loop downstream of the precoat filter. Just prior to the RO, the water passed through an oversized cartridge filter housing containing 1-μ depth-type prefilters. Permeate water from the RO was collected in a third tank, from which it was distributed back to the rinse water tanks in the process (Figure 8-4).

Makeup water was automatically added to the system when the level was low in the third tank. This water made up for water lost via the RO concentrate stream, evaporation losses, and for water that was used as makeup to the chrome and nickel rinse systems (whose water was not returned to the waste treatment system). The makeup water consisted of city water that was deionized via two-bed portable ion-exchange columns. By eliminating the salts that were added to

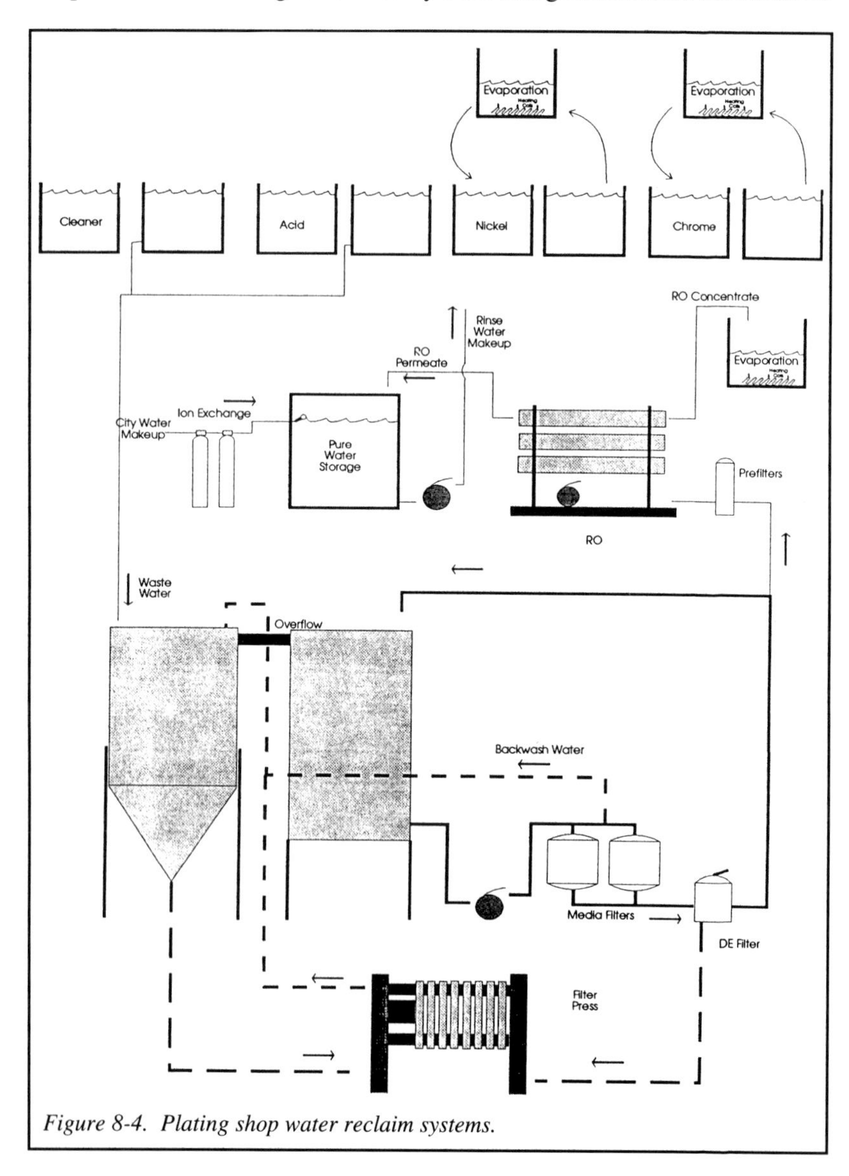

Figure 8-4. Plating shop water reclaim systems.

the system, the RO would be capable of achieving a higher permeate recovery before scale formation would occur. The ion-exchange columns were regenerated off-site by a local water service company.

The multimedia filters were shallow-bed filters designed mostly for high-flow surface filtration. They were backwashed several times a day for periods of 2 minutes each. The backwash water was sent to the conical storage tank.

The DE precoat filter was designed to have the DE mechanically bumped off the filter socks. This bumped DE would then recoat the socks when the filter was put back into service. This allowed the filter to achieve a longer run before having to precoat the socks with new DE. When bumping was no longer effective at extending the life of the precoat, the filter was bumped and the DE was sucked out of the filter to a filter press. New DE was then sucked into the inlet of the precoat filter. The DE that was sucked into the filter press would assist the press in achieving better filtration and forming a better filter cake.

Once a day the bottom sediment of the conical tank was sucked out and through the filter press. The effluent of the filter press was plumbed back to the conical tank.

System operation. As expected, the RO would rapidly lose its permeate flowrate because of the oil and grease still present in the feedwater after the pretreatment system. It would lose about 75% of its flow in 3 to 4 days and would then be cleaned. An alkaline cleaning solution with pH between 11 and 12 was quite effective at restoring the original permeate flowrate. The spent cleaning solution (approximate volume of 25 gallons) was sent directly to the evaporation tank.

The rejection characteristics of the membrane on the waste water were poor except immediately after cleaning. The system would typically run with an overall salt rejection between 70% and 80%. The quality of this water was still sufficient for use in the rinse water tanks.

One of the first problems encountered with the system was that the concentration of metal salts in the return water from the rinse tanks was much greater than anticipated. This resulted in a much higher osmotic pressure in the RO system. With the high osmotic pressure, the permeate flowrate was much lower than its design rate. It was necessary to operate the RO at lower recovery. This in turn, made it impossible for the evaporation tank to keep up with the concentrate flow from the RO system.

Upon investigation, a leak was found in a pump that recirculated the nickel-plating solution through a filter and back to the plating tank. The reclaim water system had been set up to recover all water that dumped into a local collection area on the floor of the facility. This collection area also picked up the leaking nickel solution.

To prevent the possibility of a leak causing a similar problem in the future, all the systems were isolated so that there was no way for leaks or spills from the concentrated plating tanks to automatically make it into the reclaim system.

Another problem was encountered with the pH control of the water in the reclaim storage tanks. If for some reason there was a high concentration of metals in a storage tank, and water from the alkaline rinse was introduced into the tanks, metal hydroxide precipitation would occur. The alkaline rinse water was high enough in pH to exceed the solubility of the metal hydroxides. This problem was resolved by manually adjusting down the pH of collected water with hydrochloric acid if it happened to be greater than 7.0 during the daily maintenance of the system. This was only infrequently necessary after the plating tanks were isolated.

Overall maintenance requirements of the RO and pretreatment equipment stabilized to less than 2 hours per day even though it was necessary to clean the RO every 3 to 4 days.

The RO reclaim system had met the goal of zero discharge. No industrial water was ever released to the sanitary sewer. This resulted in a dramatic improvement in relations between the facility and the municipal authorities.

Sugar and Juice Concentration

The use of RO to recover and concentrate sugar solutions from various food processes is an application of growing popularity. Both PA and CA RO membranes can achieve rejection percentages of sucrose and fructose in excess of 99%. Reverse osmosis can be used as a replacement for or in conjunction with evaporation. In applications where either RO or evaporation could be applicable, RO is lower in both capital and operating expenses.

Standard RO systems are limited in how far they can concentrate a sugar solution by the osmotic pressure of that solution. For economic reasons, this limit tends to be about 24 °Brix.* To operate beyond this concentration requires pressures in excess of 600 psig, thus requiring special membrane elements and materials of construction, at which point it is usually less expensive to use evaporation.

Fruit juice concentration. The concentration of fruit juice is commonly performed in the food industry. Reverse osmosis is attractive for preconcentration of the fruit juice prior to evaporation if the feed stream has a concentration of sugars less than 14 °Brix. In this range, concentration can be achieved prior to evaporation with a substantial savings in operating costs. If the evaporator has not yet been purchased, the use of RO can also result in capital equipment savings if a smaller evaporator can now be purchased.

*Refractometers, frequently used to determine the sugar concentration in water, operate by measuring the bending of light caused by the organics (sugars) in the solution. The bending of light corresponds to the organic concentration, and each degree Brix is comparable to a percentage concentration by weight as sugar. Thus, 24 °Brix is about 24% sugar.

It should be noted that evaporation will also tend to change the flavor characteristics of juices. Volatile components can be lost, and certain lipids can be oxidized (Koseoglu and Guzman, 1993). In most cases, there is less flavor damage when using RO to concentrate juice than when using evaporation (Paulson et al., 1985).

Concentration of maple syrup. In the production of maple syrup, the sap from the maple tree is collected and concentrated. It is taken from its original sugar concentration of 2% to 2.5%, to between 8% and 12% as sugar. Previously, this was performed using only evaporation. The sap was boiled at atmospheric pressure in an open-pan thermal evaporator (Willets et al., 1967).

By preconcentrating the sap, reverse osmosis has become popular as a means of reducing the energy demand of straight evaporation. The most common membrane used for the application is CA membrane in a spiral-wound configuration. Its advantage over PA membrane is its tolerance to chlorine (hypochlorous acid created by the dissociation of sodium hypochlorite), which is used to clean and sanitize the membrane elements (Mohr et al., 1988).

The maple flavor of the sap is brought out by heat. It is therefore still necessary to use an evaporation step in the concentration of the sap. Typically, RO is used to remove 75% of the water, with evaporation providing further concentration as well as the flavor enhancement. The use of RO reduces the cost of processing the sap by 33% (Gekas et al., 1985).

Recovery of second-press apple juice. One of the most appropriate applications of RO in the food industry is in the recovery of second-press apple juice. Apple juice will typically have a sugar (fructose) concentration between 12° and 13 °Brix. Many presses are limited in their ability to remove all the residual sugars from the apple pomace. Water has to be added to the pomace for a second pressing. This second-press juice may have a sugar concentration somewhere around 6°Brix.

Reverse osmosis is well suited to take this diluted juice and concentrate it to match the original juice concentration. Actually, it is common for juice manufacturers to vary the RO recovery to produce the particular concentration desired to stabilize the final juice product after remixing. Because the sugar concentration from the apples will vary throughout the processing season, the concentration of the second-press juice can be varied as required to standardize the final juice sugar concentration.

It is important to remix the RO-produced juice with first-pass apple juice in order to maintain the desired flavor of the juice. It has been reported that only 25% of the low-molecular-weight flavor components are retained with CA membrane, but that PA thin-film membrane elements retain 86% of these low-molecular-weight components (Sheu and Wiley, 1983). This provides an advantage for PA thin-film membrane in this application, although sanitizing the membrane will be more difficult due to the membrane's intolerance to chlorine.

Countercurrent RO for sugar solutions. A method is currently being developed to achieve higher sugar concentrations with RO while using normal pressures. This method, called countercurrent reverse osmosis, uses a sugar water flush on the permeate side of the membrane. The osmotic pressure limitation is thus overcome by creating an osmotic pressure on the permeate side of the membrane with the sugar; thus, the osmotic pressure differential across the membrane is reduced.

The countercurrent RO system operates using two RO passes. The first pass concentrates the sugar solution from low concentrations to a concentration of about 20%. This preconcentrated sugar solution then feeds a second-pass RO system. Part of the concentrate of this second pass, which may have a sugar concentration approaching 30%, is fed into the permeate side of the second-pass tail end membrane elements. It is diluted by the permeate water and exits out of the lead end membrane elements. From here it is returned to the feed stream of the first-pass RO (Figure 8-5).

This type of system design is currently limited in its ability to achieve higher sugar concentrations by the existing membrane element configurations. Hollow-fiber elements have an excessive pressure drop through their fibers, while spiral-wound elements have poor flow distribution inside the membrane envelopes. Tubular elements suffer from high concentration polarization effects at the membrane surface (Ray et al., 1986).

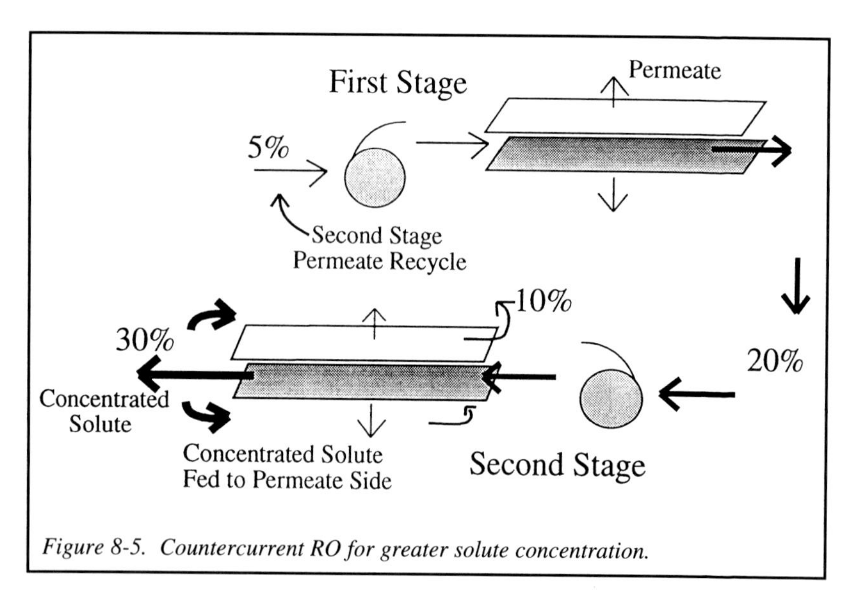

Figure 8-5. Countercurrent RO for greater solute concentration.

Biological control. Preventing biological activity is a primary concern with sugar-recovery RO systems. Such activity can occur very quickly in a contaminated system, possibly leading to the premature demise of the membrane elements. For lower temperature solutions (less than 100 °F), the compatibility of CA membrane with free chlorine offers a significant advantage over PA thin-film membrane.

It is recommended that an element configuration be used that has minimal stagnant areas. For example, outer wraps that seal the outside of spiral-wound elements off from continuous flow should be avoided. Many of the membrane manufacturers that sell to the food industry offer membrane elements that are constructed to completely fill the RO housing diameter. This element configuration is commonly called a *full-fit* element. If an outer wrap is used for the element, it is designed to allow flow around the outside of the membrane element.

Daily sanitizations with up to 100 ppm of chlorine are sometimes necessary to prevent biological activity in CA membrane systems, even though long-term membrane life may be affected. Some sort of sanitizing procedure should definitely be performed before any system shutdown. The membrane manufacturer should be consulted as to specific operating instructions.

Specialty membranes for the food industry. Some of the membrane manufacturers offer a high-temperature spiral-wound PA membrane element. This is an advantage when high temperatures have been used to sanitize the sugar solution prior to the RO system. Since the RO concentration step is commonly followed by evaporation, a high temperature membrane reduces the need to cool the solution only to have to reheat it in the evaporator. Unfortunately, these membranes are typically expensive and offer a relatively low flux, around 12 gallons per day per square foot of membrane, as compared to about 20 gfd for standard membranes. Still, the energy savings can sometimes compensate for the additional capital and operating costs.

Reverse osmosis has been used successfully in concentrating a number of different juices, including apple, strawberry, peach, pineapple, and pear juice. The measure of success of the application is usually dictated by how well the juice is filtered prior to the RO. Juice that has been filtered by diatomaceous earth or ultrafiltration has the greatest likelihood of success with RO. Unfortunately, it is becoming more difficult to dispose of used diatomaceous earth, and ultrafiltration can sometimes be very maintenance-intensive.

Development work is being performed with spacing material between the membrane leaves of spiral-wound elements that are more open, so as to allow the passage of larger particles and fibers. Within these membrane elements, there are fewer restrictions on which the juice particles and fibers can get caught. These elements with larger spacing between leaves will contain less membrane than standard elements, so are not as economical. They also require greater flowrates to achieve the same crossflow velocities as standard elements. This requires a larger pump with greater energy consumption.

Concentrating olive waste water. Highly concentrated caustic and salt water solutions are used in the processing of olives. The processing takes place during the 2-1/2 months during and immediately after the olives are picked. In California, this season takes place from October into December.

The waste water from the processing contains high concentrations of salts and organics (from the olives). Its volume, quality, and pH will vary dramatically, depending on the particular process step taking place with the olives. There will also be variation depending on how far it is into the processing season.

The high organic concentration in the waste water places a large demand on municipal waste treatment systems. Biological oxygen demand (BOD), which is a measurement of this loading, can run as high as 15,000 ppm.

Graber Olive House is a small, privately owned olive manufacturing facility that has been located in Ontario, California, (in the far suburbs of Los Angeles) since 1894. It is well known in southern California for producing specialty tree-ripened olives.

The municipal waste district in which Graber Olive is located had recently made dramatic changes in their sanitary sewer disposal guidelines. Graber Olive was given the following restrictions for additional salts (over what was already in the city water supply) that they could discharge into their sanitary sewer:

Sodium (Na^+)	75 mg/L
Chloride (Cl^-)	75 mg/L
Sulfate (SO_4^{-2})	40 mg/L
Total dissolved solids (TDS)	230 mg/L

Although it varies dramatically, the following is a typical analysis of Graber Olive's processing discharge concentrations:

pH	6.0
Calcium	19 mg/L
Magnesium	36 mg/L
Sodium	520 mg/L
Potassium	3,550 mg/L
Alkalinity	2,020 mg/L as HCO_3^-
Sulfate	1,600 mg/L
Chloride	900 mg/L

The water is low in pH and high in sulfate concentration because of the addition of sulfuric acid, which had been required to meet the previous waste discharge pH requirements. The analysis is high in potassium because potassium hydroxide (KOH) was being used to treat the olives, instead of the more commonly used sodium hydroxide (NaOH). This was because of the need to meet previous discharge limits for sodium.

There were not many options available for the removal of salt from the waste stream. The waste water could be evaporated and the salt hauled to a sanitary dump. Another possibility would be to haul the entire waste stream to a municipal waste facility that is capable of dumping its effluent directly into the ocean (where the high salt content is not a concern).

During the olive-processing season, Graber's waste water flowrates would vary between 1,000 and 12,000 gpd. Evaporating all the waste water would be extremely expensive because of the energy demands. Having to haul all of the waste water would also be expensive, costing about $15,000 per month. This would have had a significant economic impact on Graber Olive. If the water could be reduced in volume prior to hauling, there would be substantial savings.

Because this would be a relatively small-volume application, it did not make sense to spend a great deal of money piloting an RO system as a means of perfecting the system design and pretreatment requirements. A comprehensive study would have cost more than the potential savings that could be incurred from the RO application. At the risk of frequent membrane replacement, the approach taken was to make a best guess at the system design. There would likely be system modifications required as more was learned about the demands of the application. The approach taken was to treat the final RO system as though it were the pilot system.

Designing the RO system required characterizing thoroughly the waste stream. The extent of the variation in flowrates and water quality would need to be investigated in order to put together an RO design that would keep up with the flow demands, and would also be capable of producing a permeate water that would meet the sanitary sewer discharge requirements.

There was the possibility of reusing the RO permeate water to process the olives. It was decided by the facility owners not to risk affecting the quality of the olives. Some amount of low-molecular-weight organics would likely permeate the membrane, which could affect the flavor of the olives. The option would always be open to reuse the permeate water in the future, if desired.

It was easy to accurately determine the average flow and salt concentrations in the Graber Olive waste water. There was an exact number of vats that were used when processing olives. Each vat would hold 30 gallons of liquid. The process was well defined, such that it was possible to calculate how much water, salt, and potassium hydroxide would be used during the season. (Figure 8-6)

It was then necessary to determine peak values for flowrates. This was found by tracking the flow during the prior processing season. These numbers are charted in Figure 8-7.

The peak daily flow was about 11,300 gallons. Unfortunately, this flow could occur over several subsequent days. It would not be possible to undersize the RO system and rely on storage as a means to allow the RO to catch up. Also, space was limited for installing additional storage tanks. Graber Olive is located in a residential area. This created space and noise limitations that would affect the system design.

The TDS of the waste water varied from around 2,000 to 40,000 mg/L. The average TDS for the season was around 11,000 mg/L. Such a large variation in the TDS of the RO feedwater meant that the osmotic pressure of the solution would vary dramatically. This would seriously impact the recovery of the RO system, as well as the membrane flux. When the TDS was low, the flux could increase to the point of causing excessive fouling of the membrane, which would alter the flow balance through the RO array. When the TDS was high, it would be necessary to decrease the system recovery in order to get any permeate flow from the tail end membrane elements, and to obtain sufficient permeate quality to be able to discharge the permeate water to the sewer.

The overall system design is shown in Figure 8-8. Two unusual safety measures were designed into the system. First, the RO was designed to operate continuously as a means of reducing biological activity. When the water level

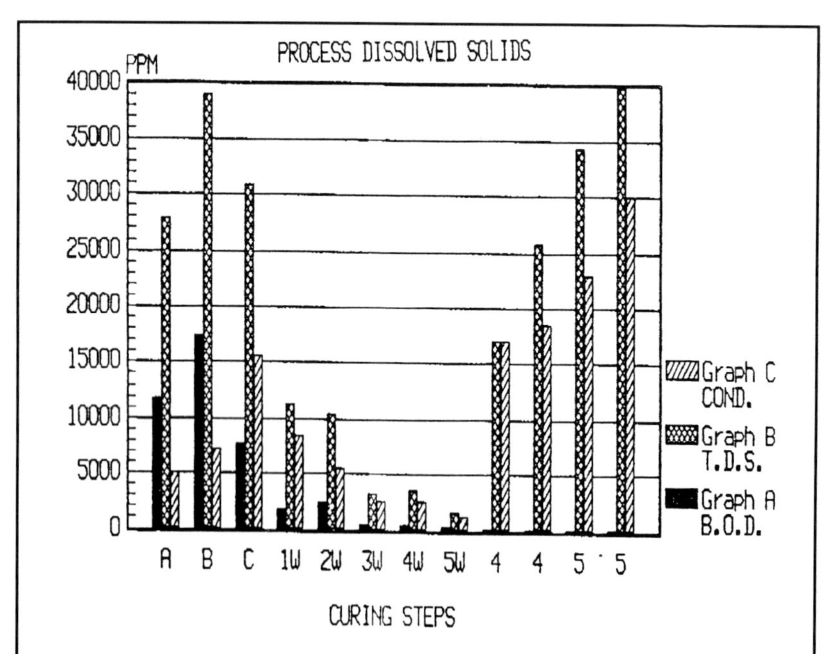

Figure 8-6. Effluent water conductivity, TDS, and BOD as a function of the olive oil curing step.

in the RO feedwater tanks reached a low level, automatic valves would open to divert the RO concentrate and permeate water back to the feedwater tanks. This same diversion system would be used if the RO concentrate tanks were full (in case the truck was late to haul off the waste water).

The second safety measure caused the RO permeate to divert back to the feedwater tanks if the permeate conductivity was not of sufficient quality. The permeate diversion could be used to prevent out-of-specification water from being dumped to the sewer during times of excessively high RO feed concentrations. This dilution of the RO feedwater should assist in improving the quality of the RO permeate so that it could be dumped.

The pretreatment system consisted of multimedia and cartridge filtration. The multimedia filter was backwashed with city water. Because of a high concentration of salts in this backwash, it was necessary to haul most of this water.

Even though the feedwater would normally contain less than 15,000 mg/L of TDS, it was decided to use seawater PA thin-film spiral-wound membrane elements. The seawater membrane uses a thicker polyamide membrane layer than brackish-water thin-film membranes. It offers better salt rejection at the expense of permeate flowrate. In this application, the better rejection characteristics would be necessary to consistently meet the permeate discharge require-

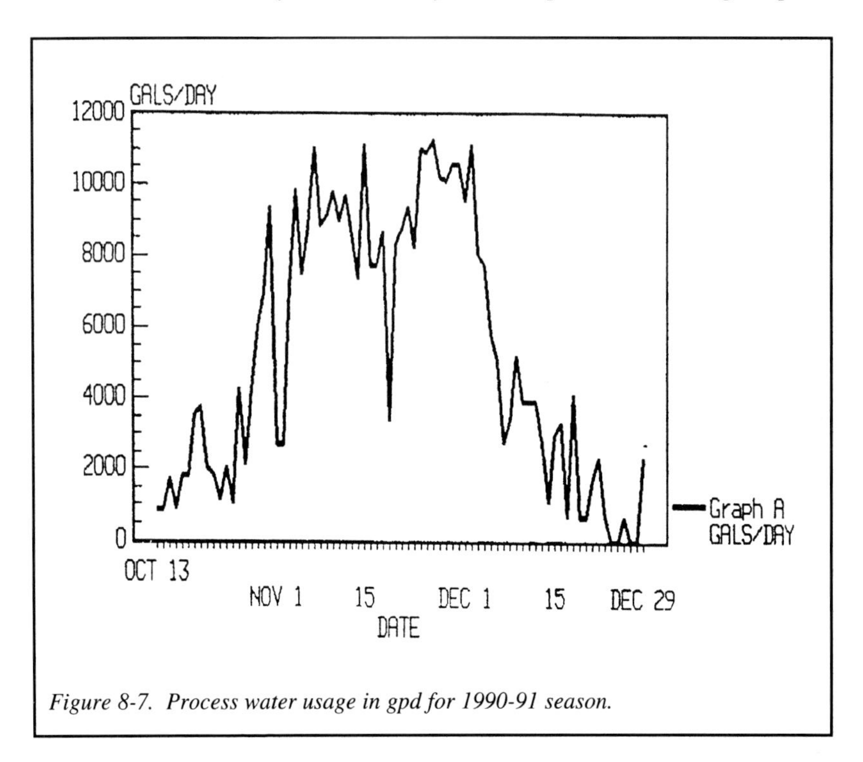

Figure 8-7. Process water usage in gpd for 1990-91 season.

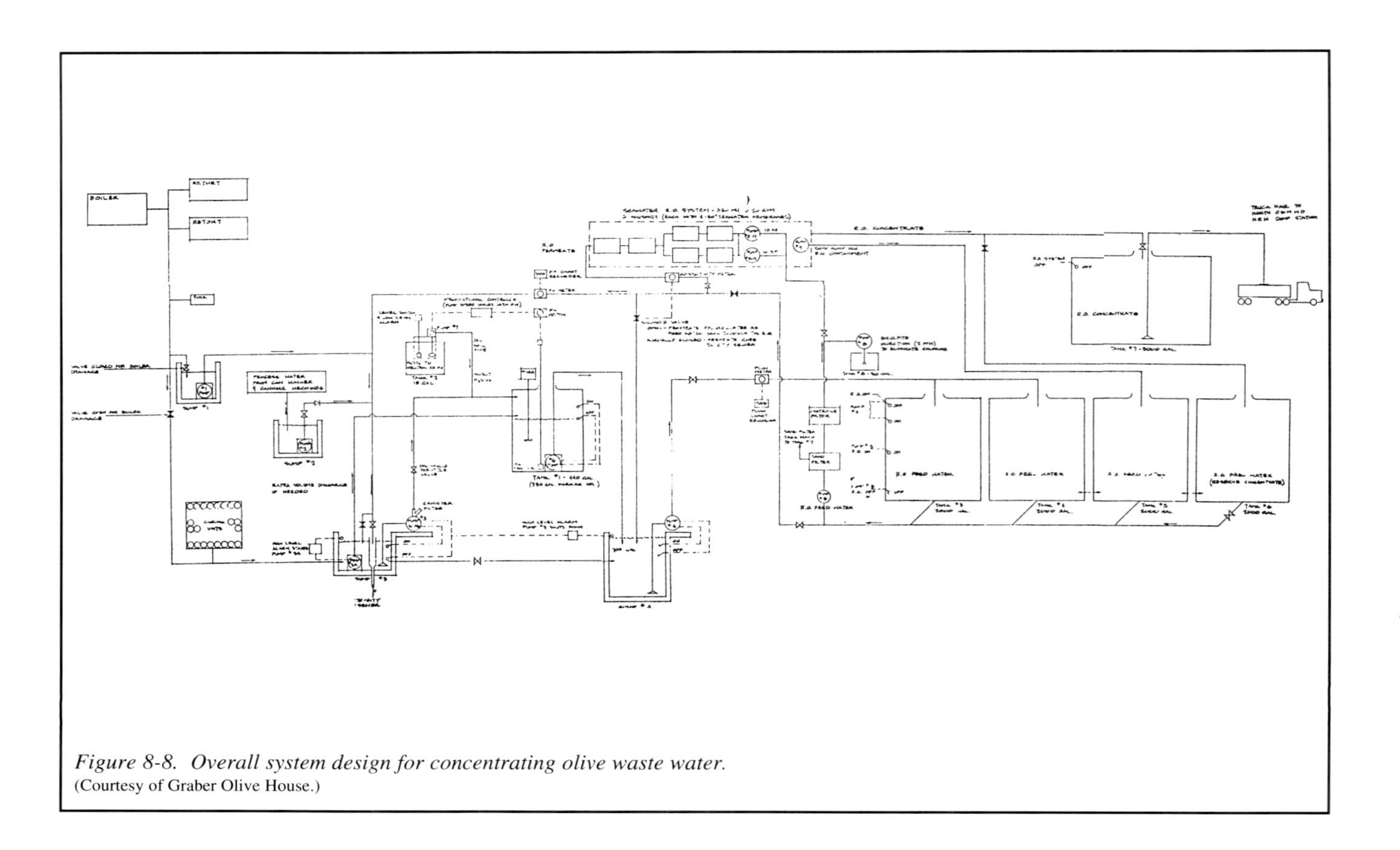

Figure 8-8. Overall system design for concentrating olive waste water.
(Courtesy of Graber Olive House.)

ments. The lower flux would be useful in preventing the overproduction of water in the lead end elements when the TDS was low (due to the low osmotic pressure).

Because of space limitations, the membrane housings for the system were limited in length to those that would hold four elements, 40 inches in length. The housings were plumbed for a 2:2:1:1 array. The concentrate from the first two housings was plumbed to the next two, which were plumbed to a single housing, and so forth. This array design was capable of achieving the desired permeate recovery with flowrates that were relatively balanced (Figure 8-9).

Table 8-2 shows the flowrates and pressures predicted for operation with varying salt concentrations. Concentrate recycle is used when the TDS is less than 10,000 ppm in order to achieve higher permeate recoveries. The final RO system after manufacturing is shown in Figure 8-10.

The RO required frequent cleaning. It would lose 50% of its flux during the day. At night it would need to be cleaned using an alkaline cleaning solution containing sodium perborate and sodium ethylenediaminetetraacetic acid (EDTA). This was effective at restoring the permeate flux to the original design conditions.

One unexpected problem occurred with the inlet seal on the second high-pressure pump. The two pumps had been chosen in the system design to give maximum flexibility for how the system could be operated. The inlet seal of the second pump was expected to handle the 400 psig delivered by the first pump, but failed to do so.

In order for the seal to hold, it was necessary to shim the impellers of the first pump as a means of reducing its effluent pressure. Unfortunately, this reduced the maximum pressure capabilities of the system, which reduced the potential recovery of the RO system.

Table 8-2
Flows and Pressures versus Concentrations

TDS (ppm)	*5,000*	*11,200*	*30,000*
Recovery (%)	92	78	45
Recycle flowrate (gpm)	5	0	0
Membrane pressure (psig)	840	840	840
Permeate flowrate (gpm)	21.7	23.3	12.4
Concentrate flow (gpm)	1.9	6.5	15.2
RO system rejection (%)	92.6	97.6	97.2
System pressure drop (psid)	74	77	140
Smallest individual element permeate flow (gpm)	0.35	0.42	0.15
Largest individual element permeate flow (gpm)	1.2	1.3	.82

The RO was still able to reduce the volume of waste water hauling required by Graber Olive by more than 60%. This substantially reduced the high cost of hauling, and made the hauling operation more workable. Because of the efforts that were made by Graber Olive to meet the waste discharge demands of the city, they are now on excellent terms with municipal authorities.

Seawater Desalination*

When seawater is to be put through the reverse osmosis process, the designer of the system must use special care. The goals may be similar to other RO applications, but seawater desalination offers its own special challenges. The focus of this section is on the unique aspects of RO systems that are designed for seawater desalination. Predominant applications using seawater desalination are offshore oil platforms, ships (from small personal sailboats to large transcontinental vessels), and communities located next to a seawater supply. Drinking water has been the main use for the RO permeate water.

Offshore oil platforms and ocean-going vessels use desalination for some industrial uses as well as for production of drinking water. Compared to other methods of desalination, RO units have a better volume and weight-to-production ratio. Because RO plants are smaller in size, they are finding increased usage aboard ships and offshore platforms. If there are space limitations, RO desalination equipment can be built on one skid or made into modular components.

Reverse osmosis has provided potable water from seawater for communities located in drought regions such as the Middle East and California. Drinking water has also been provided by RO for locations where human development has

This last section of this chapter was written by Kelly Sullivan., Village Marine Tec.

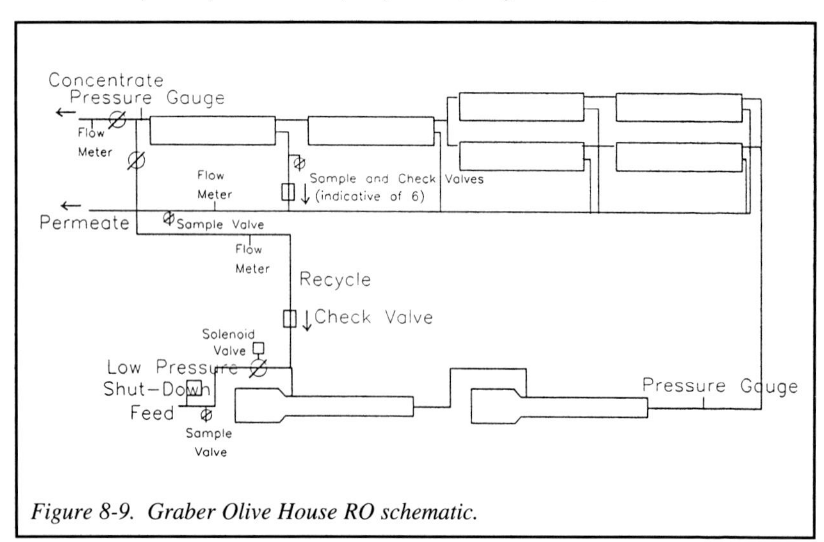

Figure 8-9. Graber Olive House RO schematic.

exceeded the demand on the natural water supply, such as regions of the Caribbean and the Mediterranean.

Pretreatment. The world's oceans and seas vary in temperature, salinity, and turbidity. To determine the proper pretreatment for an RO system, various questions should be asked related to the system design. Of primary concern is whether the installation is mobile or stationary.

If the installation is stationary, special requirements of the location will have to be considered. Will it be necessary to reduce the feedwater concentration of organic material, hydrogen sulfide gas, or other potential fouling agents? A thorough feedwater analysis must be obtained before the pretreatment system can be designed.

If the installation is mobile (e.g., a ship) the temperature and turbidity range of the water in which the ship will primarily be operating must be obtained before the design can be evaluated.

Typical pretreatment devices include:

Media filtration	Bisulfite injection
Settling tank	Chlorination
Cyclone separation	Antiscalant injection
Cartridge filters	Acid injection

Table 8-3 compares these pretreatment devices.

Posttreatment. The water produced from the RO membrane in most cases require some form of posttreatment, be it chlorination or pH adjustment. The need for posttreatment generally depends on several factors. What the water is going to be used for (e.g., potable water, industrial uses) will affect the type and amount of chemicals required.

There are many regulations that apply to potable water. A thorough understanding of the regulations governing the particular application is required before a design evaluation can be performed. Generally, municipalities require some form of chlorination, ozonation, or ultraviolet sanitization. Chlorination or bromination is used to disinfect the potable water aboard ships.

The posttreatment generally must address the aggressive nature (low pH) of the RO product water (permeate). One must also look at the pretreatment (e.g., acid injection) to see if it is going to affect the permeate. The RO permeate water can destroy certain types of piping materials, such as galvanized steel or asbestos-cement. Chemicals must be injected to raise the pH and balance the chemical nature of the water. Calcium carbonate salts are commonly used for this purpose. The permeate may not need to be conditioned if the piping is constructed of PVC, stainless steel, or copper/nickel alloys (CuNi).

Corrosion. Seawater by its very nature is extremely corrosive to many materials. A careful evaluation of RO unit component materials that come in direct contact with seawater must not be overlooked. (See Table 8-4.) There are several factors that go into material choice:

- ***Cost:*** Generally the more corrosion-resistant the material, the greater the expense (but this is not always the case).
- ***Design life:*** Corrosion resistance and durability in a seawater environment.
- ***Fabrication ability:*** The relative difficulty in manufacturing a component, in terms of welding, machining, and assembly.

Quality of produced water. A single-pass RO system will produce water meeting potable water standards, containing less than 500 ppm of sodium chloride (NaCl). Currently, there are two grades of membrane elements used for seawater desalination: standard seawater membrane elements and high-rejection seawater membrane elements. High-rejection elements sacrifice some water quantity for better water quality.

A two-stage RO system takes the permeate from the first-stage RO and

Table 8-3
RO Pretreatment Considerations

Filtration Method	*Target Item of Filtration*	*Land-Based/Ocean-Based*
Media filtration	Large organic material and silt	Mostly used for land-based equipment, limited use
Settling tank	Silt	Land-based only
Cyclone separator	Silt	Mostly ocean-based
Cartridge filtration	Silt and organic material	Land- and ocean-based
Sodium bisulfite injection	Organic material	Land- and ocean-based
Chlorination	Organic material	Land- and ocean-based
Antiscalant	Aids in keeping minerals and organics in solution	Land- and ocean-based
Acid injection	Keeps mineral scale in solution	Land- and ocean-based

processes it through another RO membrane system. The RO permeate will be in the range of 4 to 40 ppm of NaCl. The second-stage permeate water has been used for purposes such as battery water, boiler feedwater, and turbine wash-downs. When even higher purity water is needed, the second-stage water can be passed through a mixed-bed demineralizer and polished to a quality of approximately 0.5 ppm of NaCl. The highly polished water has been used for purposes such as turbine fuel injection, boiler feed, and electronic makeup water.

Experiments on three-stage RO systems are being run. The preliminary data suggests that a product quality of approximately 0.5 ppm of NaCl can be achieved.

Instrumentation. Below is a list of instrumentation that is commonly found on seawater RO desalination plants. The list has been confined to specialized instrumentation, and does not cover the various pressure gauges and flowmeters found on RO plants.

- Feedwater temperature: for monitoring changes in production due to changing feedwater temperature. Shipboard RO plants tend to travel through waters with a large range of temperatures. The temperature indicator allows the

Advantages	*Disadvantages*
Removes large percentage of particles from feedwater	Large and heavy, backflushing required
Easily maintained	Large and susceptible to biological growth
Small, and easy to maintain	Uses pumping energy, which limits its use to small RO plants (less than 100,000 gpd)
Removes particles before the RO membrane	Potentially costly
Removes oxygen from the RO feedwater, preventing biological growth	Constant chemical use, thus more maintenance and cost
Prevents biological activity	Constant chemical use, thus more maintenance and cost. The chlorine must be neutralized before the RO membrane.
Reduces RO membrane scale formation and fouling	Constant chemical use, thus more maintenance and cost
Reduces RO membrane scale formation	Constant chemical use, thus more maintenance and cost

operator to make adjustments to the membrane pressure to compensate for the temperature.

- Feedwater oxidation reduction potential (ORP): If the feedwater is chlorinated, neutralization of the chlorine will be necessary. The ORP instrument will monitor the relative amount of free chlorine left in the feedwater.
- Feedwater pH: If the RO pretreatment includes acid injection, the pH will have to be monitored.
- Permeate conductivity: This is monitored for evaluation of the membrane performance.
- Permeate pH: For RO systems with acid injection, the permeate may need pH adjustment.

Attention should be given to evaluating each gauge and instrument for its merit to the system. Each instrument adds information as well as complexity to the RO plant. Increased complexity will raise the chance of failure. In addition, instrumentation that has no specific purpose can confuse the operator and should be avoided.

Maintenance of RO plants cannot be overemphasized. Preventive maintenance is more likely to be performed with land-based systems. It is common for this maintenance to be overlooked with offshore and ship systems.

It is suggested that RO systems used for offshore and ship applications be designed to require as little preventive maintenance as possible. Crew members have to service a large variety of mechanically related systems; therefore, specialization in one area is rare. Often, concerns are resolved only when problems are encountered, or are simply left to the next shift.

Table 8-4
Materials for Seawater Applications

Material	*Cost*	*Design life*	*Fabrication Ability*
Titanium	High	Long	Difficult
Hastalloy	High	Long	Difficult
Inconel 625	High	Long	Difficult
70/30 CuNi	High	Long	Moderate
316L Stainless steel	Moderate	Moderate	Moderate
Ni-Al Bronze	Moderate	Moderate	Moderate
PVC	Low	Moderate	Easy
GRP/FRP	Low	Long	Easy

The design evaluation must include a serious look at the man-hours required for preventive maintenance. If preventive maintenance is required, how simply can it be performed? Training programs are highly recommended.

Specifically, there are key areas to look at when evaluating a system in terms of maintenance. Access is the key word. The less room around the unit, or the more difficult it is to perform the maintenance, the less preventive maintenance will be performed. Evaluate and consider modularizing if there are space limitations.

Shipboard filters will require more servicing than most types of RO plants due to the high amount of particulate matter that can be encountered in the ocean. They must be readily accessible.

When the pump needs an overhaul, will it be possible to service it on-site? Otherwise it will be necessary to remove it from the ship. Seawater decreases the life of many components. Overhauls should be expected more frequently than for other applications.

The membrane elements should be relatively easy to replace. Particular attention should be paid to the end cap design of the membrane housing. Membrane replacement varies for a number of reasons, such as cleaning frequency, temperature of the seawater, organic loading in the seawater, and the effectiveness of the prefiltration. Average life expectancy is in the range of 3 to 5 years.

Energy recovery. Several factors make energy recovery attractive for seawater applications. The permeate water recovery of the RO system is low, therefore leaving large amounts of brine energy to recover. As the system operates at high pressures (800 to 1,000 psig), energy consumption of the plant is high, and any reduction in the amount of energy reduces the cost of operation.

Energy recovery is simply a recovery of the energy present as pressure in the RO brine. Energy-recovery systems are able to convert the brine pressure into a useful energy form prior to discharge of the brine. There are two primary types of energy-recovery methods: positive displacement (piston type) or centrifugal (turbine type). Positive displacement methods offer greater recovery but increased complexity when compared to centrifugal methods. Various ways of using the recovered energy are possible. Different applications will require customized solutions.

Reverse Osmosis Technology for Pharmaceutical Water*

Historically, water used by a pharmaceutical manufacturer was purified to the level of United States Pharmacopoeia Water for Injection (USP WFI), using a distillation process. The WFI water that was used directly in the pharmaceutical product had low concentrations of salt, organics, and very low levels of bacteria

** This section was written by Leland Comb, Osmonics, Inc.; and Wes Byrne, American Fluid Technologies, Inc.*

and their endotoxin by-products. Rather than provide two different water systems, all water used in the plant (whether intended to be part of the product or not) would be WFI. This included water required only to meet the USP Purified Water quality guidelines.

As the costs associated with distillation have escalated, a demand for less costly methods of producing USP Purified Water has developed, with much of this attention focused on membrane systems. Initially, the use of membranes — more specifically, reverse osmosis — was looked upon as an alternative to distillation to produce WFI. However, since RO is operated more or less at ambient temperatures, there is no assurance that the water — no matter how extensively purified — would retain the microbiological state required by the USP WFI specifications.

Instead, plants began to study their true USP needs, separating the WFI and the Purified Water requirements. At the same time, it was recognized that a good approach to reducing distillation costs was to pretreat the water feeding the still. This reduced the cleaning requirements of the still and extended its life, resulting in a reduction in the overall costs associated with the production of WFI water.

The majority of a typical pharmaceutical plant's water usage is USP Purified Water. The end result of the need for this quality water as well as the need for pretreatment to WFI stills has been an increase in the use of reverse osmosis as a means of producing USP Purified Water.

In most cases, RO has proven itself to be suitable to meet the current USP Purified Water specifications for total dissolved solids of 10 mg/L as measured gravimetrically. However, the TDS figure is generally not the most critical measurement. Instead, the chloride content will often be the most difficult. As currently defined by the USP, the chloride test is qualitative and not quantitative. Nevertheless, it is often the chloride test that determines whether a single RO alone will be appropriate to produce USP water on a given water supply.

In cases where the chloride figure, or any other criteria spelled out in the USP, cannot be met using RO, a deionization process may be added or a two-pass RO may be considered. For most water supplies, two-pass RO will meet the USP requirements, even as these criteria become re-evaluated to more specifically define the measurements.

New pharmaceutical guidelines for water. New standards have been proposed for the pharmaceutical industry that will be published in the *United States Pharmacopoeia XXIII (USP XXIII),* to be effective January 1995. This will include water conductivity measurement as a replacement for qualitative measurements currently used to test for calcium, sulfate, chloride, ammonia, and carbon dioxide concentrations. Testing for heavy metals and dissolved solids will be eliminated. A measurement of water pH will be required only when the water conductivity exceeds 2.4 microsiemens/cm (μS/cm) of a batch water sample after aeration. Also, if the water conductivity measures less than 1.25 μS/cm in-line, pH measurement will not be required (Gethard, 1994). This should greatly simplify the qualification of the process water.

If the water conductivity is greater than 2.4 μS/cm, pH measurement will be required with an aerated batch water sample. The conductivity limit will then be a function of the water pH. This is to take into account the fraction of the conductivity that is caused by the concentration of dissolved carbon dioxide (which becomes carbonic acid) in the water sample (Figure 8-10).

Total organic carbon analysis will be replacing the measurement of oxidizable substances. The TOC limit will be set at 0.5 mg/L (500 ppb). This can be performed either in-line or via batch samples (Pharmaceutical Manufacturers Association, Water Quality Committee, 1990).

Many USP Purified Water treatment systems that were able to consistently meet the old guidelines may have trouble meeting the new ones. They may have to be modified to achieve greater salt removal. This may mean switching to a polyamide thin-film RO membrane from a cellulose acetate membrane, and possibly using two-pass RO. An alternative might be to polish the RO permeate water using ion exchange. Since ion exchange can more easily become a source of bacteria and endotoxin, there are some advantages to meeting the water quality guidelines using just RO.

System design. Reverse osmosis systems have been applied in a variety of designs, depending on the water quality of the feedwater and upon the system design criteria employed by the particular pharmaceutical plants. Some facilities have relied entirely on polished stainless steel piping and skids, whereas others have been willing to consider less expensive alternatives for the materials of construction.

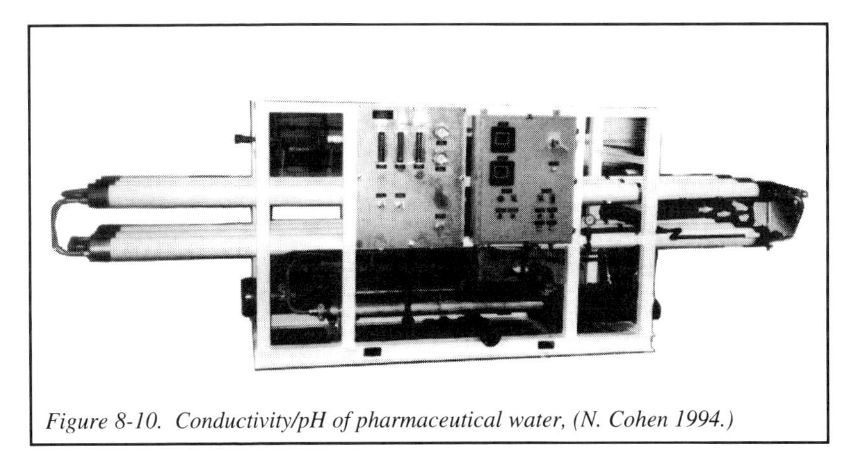

Figure 8-10. Conductivity/pH of pharmaceutical water, (N. Cohen 1994.)

As the quantity of pharmaceutical RO systems in the USA now numbers in the hundreds, however, some common engineering designs are beginning to become established. These are listed below:

Pretreatment. A dual-media sand filter is typically the first piece of equipment in the system sequence. A backwashable filter provides inexpensive protection for an RO system from excessive suspended solids that could foul the system. Such a filter should be designed conservatively with respect to service flowrate per square foot of media, and backwash flowrate per square foot, and with 100% of freeboard. Optimum service flowrate occurs at 3 to 5 gpm/ft^2 of media (McPherson, April 1981). The backwash flowrate should be at least 12 gpm/ft^2 of media.

Typically, the tanks are lined carbon steel, but in more and more cases, fiberglass-reinforced plastic is being considered from a cost-savings and corrosion-resistance point of view.

Sufficient chlorine or chloramines should be present in the feedwater to the sand filter to prevent biological activity within the bed. Otherwise, periodic shock sanitization of the bed will be required.

Softening. Following the sand filters, a water softener is often the choice for pretreating the water to remove hardness, thereby reducing the potential for scale formation on the RO membrane. The same can be achieved with acid addition, but the softening approach keeps the carbon dioxide content low, therefore simplifying the ability to meet the USP carbon dioxide specifications.

One aspect of softeners considered by the new USP guidelines is the proliferation of microorganisms. Recirculation of water through the softener is suggested when usage is low (i.e., when the RO is not operating). Periodic sanitization of the brine tank regeneration system and possibly of the softener itself may be required. The presence of chlorine or chloramines in the feedwater will reduce the biological activity within the softener, but may affect the long-term life expectancy of the resin. Ultraviolet light sanitization can also be upstream of the softener to reduce the biological loading on the system.

The USP guidelines are also concerned with proper flowrate design and the possible fracture of the resin beads. A filter should be used downstream to capture any resin fines shed by the softener.

Chlorine removal. Chlorine needs to be removed ahead of the RO, a task that has typically been assigned to activated carbon (AC) filtration. The AC can be a breeding ground, however, for bacteria that ultimately can end up on the membrane surface. There have been cases where the carbon has been periodically steam-sterilized with reasonable success. Such carbon filters are usually constructed of stainless steel components in order to be compatible with the steam.

As an alternative, the addition of a reducing agent such as sodium bisulfite has been used very successfully for chlorine removal. Bisulfite is biocidal at high concentrations and biostatic at lower concentrations. It can be very effective at removing chlorine and minimizing the growth of bacteria within the RO system.

The new USP guidelines clarify concerns about added substances in USP water systems. If it is necessary to add something to the water within a USP water treatment system, it will also be necessary to monitor its removal. In the case of injecting sodium bisulfite for the removal of chlorine or chloramines, the concentration of bisulfite downstream of any reverse osmosis or ion-exchange equipment will need to be measured regularly and must be nondetectable. If the water system uses only single-pass reverse osmosis for salt removal, some bisulfite may be detectable in the RO permeate. This would not be acceptable.

Reverse osmosis. Two-pass RO systems should be constructed to include a high-rejection polyamide-type membrane, operating in a clean crossflow mode, thereby keeping the fouling to a minimum. Either one high-pressure (400 to 500 psig) or two lower-pressure (200 to 300 psig) pumps (one for each pass) can be used, although the latter has proven to have certain design advantages.

One advantage of working with two pumps is that the system can be designed to compensate for temperature swings in the feedwater by application of more pressure. (During colder temperatures, to achieve constant flowrates, the pumps are run at 350 to 400 psig instead of at 200 to 300 psig.) One could do the same with the one-pump design, but it would require a 600- to 800-psig pump, which would be more expensive.

There are two other reasons to support the two-pump design. The first-pass RO permeate can be plumbed using low-pressure pipe for additional cost savings. Also, the second pass can be more easily designed to operate at higher pressures, thus keeping the number of membrane elements in the second pass to a minimum. The reasoning is that one wants to keep the flux per unit area of membrane low for the first pass, but there is strong logic in running the membrane at a much higher flux rate on the second pass where concerns over fouling are almost nonexistent.

Operating pressures, flowrates, and water temperature should be monitored on the RO system. In addition, the USP guidelines call for the monitoring of microbial levels and TOC. Bacteria counts should be monitored in the RO concentrate stream as an indication of the biological activity on the feed/concentrate side of the membrane. They should also be monitored in the permeate stream as an indication of what may be growing in the permeate carrier within the membrane envelopes (of spiral-wound membrane elements).

Integrity challenges are required for the membrane system. This would be extremely difficult to accomplish using standard methods. There is too much membrane area in most spiral-wound RO systems to be able to bubble-point or to accurately check the diffusive flow of air through the membrane. Higher rates of air passage through the membrane resulting from membrane mechanical imperfections or damaged O-rings would not be detected unless the problem was extremely severe. Dye testing could be used to check the integrity of the system, but would be time-consuming and might result in the contamination of the permeate water.

It is suggested that analyzing the concentration of a divalent ion, such as calcium, be used as a method of checking membrane system integrity. Divalent ions are rejected by most RO systems well in excess of 99%. Even as the RO membrane performance deteriorates, divalent ion rejection will remain high unless there is a mechanical leak, such as a membrane glue line failure or damage to an O-ring, within the system. The calcium concentration could be checked in the permeate of the individual membrane housings as a function of the concentration of calcium in the feedwater. A calcium solution could be injected into the feedwater of second-pass RO systems to check the integrity of its membrane housings. The calcium rejection of each housing should be recorded.

Storage. Once the permeate is produced as USP Purified Water, the product water should be either used within 24 hours, held at 80°C, or ozonated (with downstream ozone removal using UV light). The absence of ozone in the water should be continuously monitored, and on-line instruments are available for this purpose.

Storage tanks should have smooth interiors, and spray heads on the incoming water line(s) to insure water movement in the tank head space. Hydrophobic and microretentive membrane cartridge filters should be used on air vents. All such microretentive filters here or elsewhere in the water system must be sanitized and integrity-tested prior to use, and periodically thereafter. A rupture disk should also be provided on the storage tank, along with a rupture alarm device.

Distribution. Plastic piping is typically used for low-pressure piping, with a preference for fittings that are welded or mechanically joined with a gasket. For the high-pressure piping, 304 stainless steel is entirely suitable. At the point of the second-pass permeate, the piping should become sanitary design 316L stainless steel with either orbital welds, checked by boroscope, or Tri-Clamp type connections. Ideally, while a Tri-Clamp device is considered sanitary, it should be used only where necessary. Also, there is logic in keeping the amount of valving and instrument connections to a minimum in the second-pass permeate lines, to keep cost down and to keep the system as sanitary as possible.

This sanitary concept should be carried back through the internal piping within the RO unit itself, making certain that the permeate tubing within the membranes is designed in a reasonably sanitary manner and that there are no dead legs in the upstream end of the permeate tubing.

Distribution pumps should be of a sanitary design. Valve selection should be diaphragm valves where possible because of their smooth internal surfaces that are continuously exposed to the flowing water. Distribution lines should be sloped and fitted to allow for complete draining. Stagnant dead-leg areas should be avoided. The USP defines a dead leg as a tie-in point with a length-to-diameter ratio of greater than 6. Reverse flow from connections to processes or equipment should be prevented. Sample valves should be located at the storage tank, at the points of use, and in the distribution water return line to the storage tank.

Cleaning. Clean-in-place (CIP) systems should be provided for cleaning and sanitizing the RO system. With two-pass RO systems, it may only be necessary to design in cleaning connections for the first pass. The second pass need only be fitted with a CIP feature to allow the introduction of a sanitizing solution that would be periodically run through the RO system to drain, or in some cases through the entire distribution piping system.

Validation of the system should include documentation that the system is able to meet the USP Purified Water requirements for a given period of time (generally 60 days). It should include standard operating procedures (SOPs) that address routine maintenance and monitoring issues, as well as reaction plans to system failures. Alert and action levels should be set for all parameters. A reaction plan should be documented for each alert and action level.

Validation may also include certain construction documentation criteria, including boroscope track of the orbital welds in the second-pass permeate. Material certification records for these piping materials should be kept as well.

Conclusions. Reverse osmosis has become an important water treatment device in the pharmaceutical industry as an inexpensive alternative to distillation, or as pretreatment to WFI stills. With the new *USP XXIII* guidelines, two-pass polyamide thin-film membrane systems should gain in popularity as additional salt removal may be required from USP Purified Water systems.

As the industry becomes more familiar with RO technology, it is starting to standardize on particular ways of designing the system and components. These system designs take advantage of economical methods available for utilizing RO and RO pretreatment technologies.■

CHAPTER 9

BIOLOGICAL FOULING

This chapter discusses the biological fouling problems that are prevalent at Water Factory 21 in Orange County, California. Water Factory 21 takes secondary municipal waste and purifies it with reverse osmosis treatment prior to injecting it back into the area's aquifer. The RO systems at this installation have been considered to be a worst-case scenario for biological fouling of RO membranes. The fouling rates have been dramatic, requiring frequent cleaning. Restoring original membrane performance has been difficult. Also, the biological fouling has resulted in reduced membrane life expectancy.

The experimental investigations performed at this installation have been important to the RO membrane industry because they address the critical limitation of RO in applications where biological fouling is encountered. The tone of this chapter is different from the rest of the book in that the results and discussions include microbiological theory. However, this work affects practical aspects of reverse osmosis in most nonwaste treatment applications. Specifically, some of the important items to note include the following:

- The availability of nutrients in the Water Factory 21 municipal wastewater resulted in the propagation of bacteria that grew quickly and could readily adhere to membrane surfaces.
- The high concentration of ammonia in the water caused any free chlorine that was injected into the water to quickly react to form chloramines. The chloramines were found to be only partially effective at controlling biological fouling.
- It was easier to clean biofilm from the membrane when the bacteria were in a stressed environment created by the presence of a biocide.
- It was more difficult to completely clean a membrane biofilm as the biofilm aged.
- A good surfactant assisted cleaning solutions in removing membrane biofilm.

Editor's note: This chapter is contributed by Harry F. Ridgway, Orange County Water District, Fountain Valley, California; and Dr. David G. Argo, Black & Veatch Inc., Irvine, California.

Background

During the last decade, much basic and applied research has been conducted at the Orange County Water District (OCWD) in southern California concerning biological fouling (biofouling) of RO membranes (Ridgway et al., 1981). Membrane biofouling is a serious and widespread problem that can disrupt normal plant operations and significantly increase operating and maintenance (O&M) costs (Durham, 1989; Flemming, 1991; Lepore and Ahlert, 1988; Ridgway, 1987, 1988; Schaule et al., 1993; Zeiher et al., 1991). The OCWD has maintained a strong interest in this area because of a growing dependence on RO and related membrane separation processes (e.g., ultrafiltration and continuous microfiltration) to better manage a large groundwater basin that supplies more than 60% of domestic water needs for approximately 2.5 million residents. Membranes are now utilized by OCWD and other water management agencies for enhancing wastewater reclamation and recovering groundwater contaminated with nitrate or other constituents of aesthetic or public health concern (e.g., humic and fulvic acids, trichloroethylene, pesticides, and total organic carbon).

The current principal uses of reclaimed wastewater by OCWD are for injection into a subterranean seawater intrusion barrier system (Argo, 1985; Mills, 1993; Nusbaum and Argo, 1984) and local greenbelt irrigation. Treatment of municipal wastewater (i.e., activated sludge effluent) is done at Water Factory 21, a 0.66-cubic-meter-per-second (m^3/sec) (15 [mgd]) advanced wastewater reclamation facility that incorporates a 5-mgd RO plant (Nusbaum and Argo, 1984). The cost of membrane biofouling at Water Factory 21 is substantial, accounting for nearly 25% of the total O&M costs of the facility.

To learn more about membrane biofouling, a comprehensive long-range research program was embarked upon at Water Factory 21 in the early 1980s. The intent of the program was to analyze the chemical constitution, microbiology, and ultrastructure of the membrane fouling layers on the surfaces of the blend cellulose acetate (CA) membranes installed at Water Factory 21. This work included use of scanning electron microscopy (SEM); energy dispersive X-ray microanalysis; and a variety of other chemical, biochemical, and microbiological analytical techniques. By learning more about the nature and mechanism of RO membrane biofouling, it was felt that specific recommendations could be proposed for the prevention or control of membrane biofouling.

Much of the information presented here was adapted from Argo and Ridgway (1984), which addressed research conducted during the period from about 1979 to 1983. Many of the earlier findings discussed have been presented elsewhere and the reader is referred to the original literature references for additional details (Argo and Ridgway, 1984; Ridgway, 1987, 1988; Ridgway et al., 1981, 1983, 1984a, 1984b, 1984c, 1985, 1986; Ridgway and Safarik, 1991).

Cost of Membrane Biofouling at Water Factory 21

Given the substantial economic impact of RO biofouling, it is remarkable how little is known concerning the biology of the microorganisms that constitute membrane biofilms or the mechanism of biofilm formation. It is estimated that in the United States alone, RO biofouling costs tens of millions of dollars annually because of lost productivity, the need for specialized feedwater pretreatment processes and chemicals, increased O&M and system energy consumption, and decreased membrane life. At Water Factory 21, located in Fountain Valley, California, membrane biofouling is estimated to cost more than $700,000 annually, which represents nearly 25% of the total annual plant operating costs (Table 9-1).

Biofouling is especially prevalent and even more costly in large seawater RO plants, which now provide a substantial share of the available drinking water in many arid regions of the world such as the Canary Islands and the Middle East. If effective preventative measures cannot be developed, it is anticipated that the

Table 9-1
Estimated Costs of Membrane Biofouling at Water Factory 21*

Cost Area/Item	*Unit Cost/Conversion*	*Subtotal ($)*
Membrane Cleaning		
(a) Labor costs	$27/hr; 12 days/yr	7,776
(b) Chemicals	$696/mo; 1 clean/mo	8,352
Pretreatment Processes		
(a) Lime process	60% recycle; 25% for biofilm control	31,536
(b) Cartridge filters	320 filters; 3X/yr; $7/filter	6,720
(c) Polymer addition	3 mg/L; 150 lb/da; $1.30/lb	7,414
(d) Biocide addition	chlorine 8 mg/L; $325/ton; 77 ton/yr	25,084
(e) Pretreatment energy	1,660 kwh/AF; 25% for biofilm control	223,000
Membrane Performance Loss		
(a) Flux decline	6 mgd; 1,694 kwh/AF; $911,000/yr; assume 150% OP for 80% of 4-yr life	242,934
(b) Rejection loss	Economic impact not calculated	
Shortened Membrane Life	$1 million membrane inventory; 4-yr actual life; 8-yr theoretical life	125,000
Testing/Evaluations	Testing of new membranes, cleaners, Pretreatments, etc.	25,000
	Total Annual Added Cost Due to Biofouling	702,816

*All computed costs shown are estimates based on approximated energy consumptions and materials costs (by H. F. Ridgway).

economic impact of membrane biofouling will grow significantly as new large-scale RO facilities are constructed and placed into operation.

Description of Water Factory 21

Water Factory 21 was constructed during the mid 1970s to reclaim municipal wastewater for the purpose of injection into a subterranean seawater intrusion barrier system. Because the injection water was commingled with the domestic groundwater supply, it was necessary for the reclaimed wastewater to meet or exceed all federal and state (California) drinking water standards. The wastewater reclamation facilities at Water Factory 21 were designed to treat 0.66 m^3/sec (15 mgd) of unchlorinated secondary (i.e., activated sludge) effluent from a municipal wastewater treatment plant (County Sanitation Districts of Orange County) by the process diagrammed in Figure 9-1. The "Q" numbers in Figure 9-1 designate various sampling locations (e.g., Q-1 is plant influent).

The original treatment processes include high-pH lime (calcium oxide) clarification with sludge recalcining, ammonia air stripping at elevated pH, recarbonation, prechlorination, mixed-media filtration, granular activated carbon (GAC) adsorption with carbon regeneration, final chlorination, and RO demineralization. Air-stripping of ammonia was discontinued in the early 1980s. Demineralization of a portion (about one third) of the flow at Water Factory 21 is required to satisfy federal and state drinking water regulations regarding the maximum allowable concentration of total dissolved solids.

Description of 5.0-mgd RO Facility

A flow diagram of the 5-mgd RO plant at Water Factory 21 is shown in Figure 9-2. Included in this process are feeding of sodium hexametaphosphate or organic polymer (polyacrylate) antiscalants to inhibit mineral precipitation on

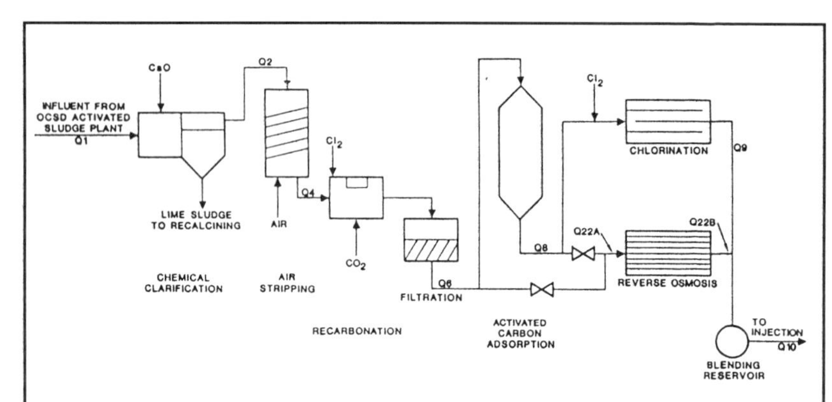

Figure 9-1. Flow schematic and sampling locations for Water Factory 21. Note that Q-6 is currently used as feed for the RO plant.

the membrane surfaces (i.e., mineral scaling); continuous addition of 1 to 3 mg/L chlorine (at Q-5, recarbonation) to control biological growth within the membrane modules; and cartridge prefiltration (25-μm nominal pore size) to remove suspended particulates from the feedwater.

The water is then pressurized by three vertical turbine feed pumps equipped with variable frequency drives to a total dynamic head of 460 psig. Over the years, it has been possible to incrementally reduce the initial RO operating pressure to about 225 psig as higher-flux membrane polymers have become commercially available. Chlorine added at Q-5 (recarbonation) rapidly combines with an excess of ammonia in the water to produce monochloramine; thus, the RO membranes are typically not exposed to a free chlorine residual. Sulfuric acid is injected into the high-pressure feed header to adjust pH to approximately 5.5 before the water is applied to the RO membranes.

The RO system consists of six identical subunits, each consisting of 42 pressure vessels mounted on a structural steel frame in a 4:2:1 array (i.e., 24 pressure vessels in series with 12 more vessels, and these in series with 6 more pressure vessels.) The operating pressure and product water (permeate) recovery are adjustable for each subunit. The plant is designed to provide 90% salt removal while achieving 85% recovery. The demineralized water receives posttreatment in two packed-tower decarbonators to air-strip the dissolved carbon dioxide that results from pH adjustment. A detailed description of the RO plant design criteria has been provided in previous reports (Nusbaum and Argo, 1984).

RO System Performance

Although the feed pressure has been varied over the years depending on

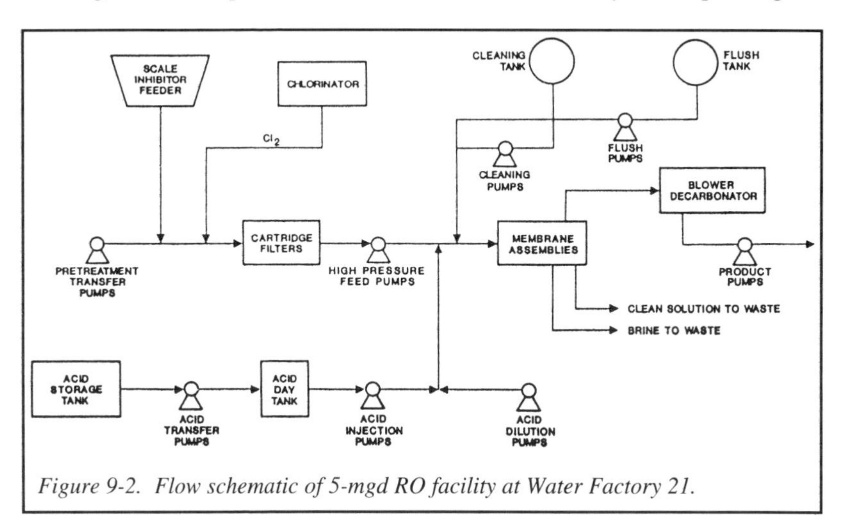

Figure 9-2. Flow schematic of 5-mgd RO facility at Water Factory 21.

membrane type, system recovery has been maintained at 85%. Four of the six subunits were reloaded in the last half of 1979 with Fluid Systems (FSD) Model 8150-HR spiral membranes (CA polymer). Fluid Systems product specifications for these elements indicate a minimum rejection of 96% NaCl and a nominal flux of 12.5 gfd at 420 psig and 77 °F. The other two subunits were reloaded in December 1979 with a different CA type of membrane, but their performance was very similar to that of units loaded with HR type of membrane. Discussion of plant performance in this chapter will be limited to subunit 1A (during the period of January 1979 to September 1981), since it was typical of all subunits and has had the longest period of uninterrupted operation.

Normalized flux and rejection versus time for subunit 1A is plotted in Figure 9-3. As indicated, the subunit was cleaned on numerous occasions during the period from July 1979 to September 1981. The flux was normalized to a referenced temperature of 77 °F and a net operating pressure of 400 psig by applying a temperature and pressure correction factor based on the monthly average for both. The rejection values are weekly averages calculated for the subunit during any given month.

Membrane flux displayed biphasic kinetics, declining rapidly from July 1979

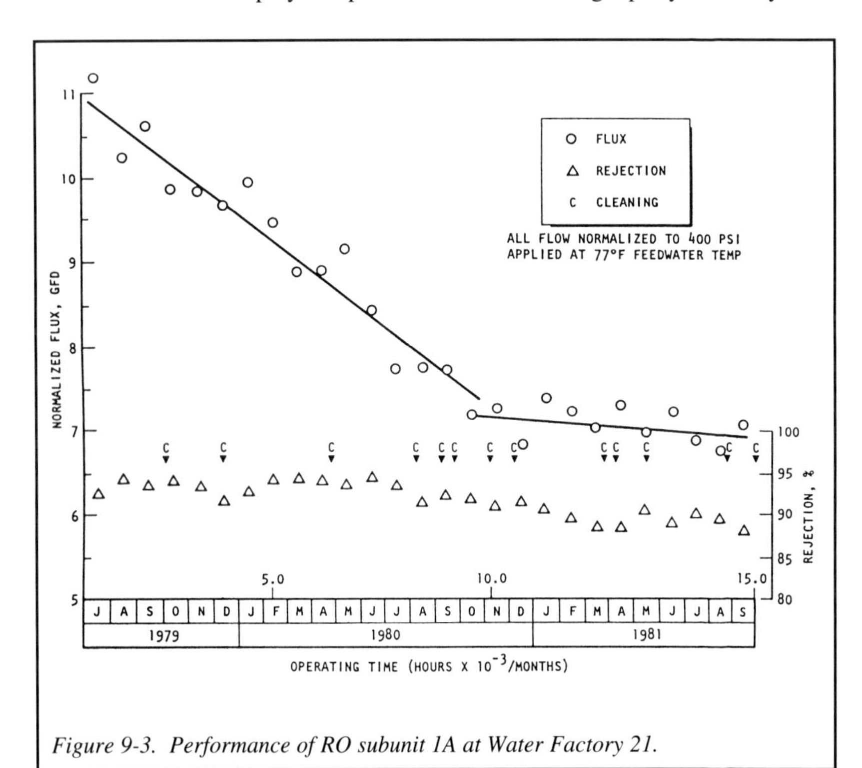

Figure 9-3. Performance of RO subunit 1A at Water Factory 21.

to October 1980 (-0.226 gfd*/month), and more slowly from October 1980 to September 1981 (-0.025 gfd/month). This transition in flux decline appeared after about 9,000 hours of operation. The highest flux was 11.18 gfd, observed in July 1979, immediately after reloading and start-up of the new membranes. The flux declined steadily to a minimum of 6.84 gfd in December 1980, despite cleaning of the membrane eight times. Following a shutdown, the flux increased to 7.39 gfd in January 1981, and the level was maintained between this value and 6.77 gfd through September 1981. During this 9-month period, the unit was cleaned five times. The mean flux maintained for 1981 was 7.09 plus or minus 0.21 gfd.

Inspection of several modules removed after 12,000 hours of operation indicated that the principal cause of the observed flux decline was accumulation

* gfd = gallon per square foot per day

Table 9-2
Analysis of Biofilm from RO Membranes at Water Factory 21

		Membrane		
Parameter	*Unit*	*1st Pass*	*2nd Pass*	*3rd Pass*
Quantity of foulant	mg (wet wt) cm^{-2}	8.2	9.9	12.6
Moisture content	% wet wt	93.7	93.0	93.9
Total organic fraction[1]	% dry wt	92.6	91.6	87.3
Total inorganic fraction[2]	% dry wt	7.4	8.4	12.7
Protein	% dry wt	15.0	24.1	30.1
Carbohydrate	% dry wt	13.2	17.6	13.2
CaO	% dry wt	1.9	2.0	2.65
K	% dry wt	0.03	0.04	0.06
Al	% dry wt	0.08	0.08	0.08
Fe	% dry wt	0.21	0.32	0.35
Cr	% dry wt	0.12	0.09	0.05
C	% dry wt	0.003	0.003	0.003
Cl	% dry wt	0.60	1.10	2.40
SO_4	% dry wt	0.87	0.93	0.90
PO_4-P	% dry wt	0.90	1.00	1.40
NO_3-N	% dry wt	0.01	0.01	0.01
SiO_2	% dry wt	<0.01	<0.01	<0.01

[1]Volatile at 550 °C
[2]Nonvolatile at 550 °C

of fouling material on the surface of the membrane. Several RO elements were removed after 15,000 hours, and the fouling material was scraped from the membrane surface to determine its composition. The results of this testing are shown in Table 9-2. Data are given for typical elements removed from all three arrays (i.e., first, second, and third pass), indicating that between 87% and 92% of the fouling material was organic (Ridgway et al., 1981; Nusbaum and Argo,

Table 9-3
RO Membrane Mineral Rejection Performance
(October 1980 - September 1981)

Parameter	*Conc.*	*Influent*	*Effluent*	*Percent Rejection*
Sodium	mg/L	210	28	86.6
Calcium	“	82	0	100.0
Chloride	“	271	39	85.6
Sulfate	“	218	6	97.2
Ammonia	“	22	3	86.3
Nitrate	“	0.5	0.3	40.0
TOC	“	9.0	1.2	86.6
COD	“	19.3	1.9	90.0
Fluoride	“	0.85	0.18	78.8
Boron	“	0.56	0.53	5.3
Potassium	“	12.6	1.3	89.6
Silver	µg/L	0.30	0.05	83.3
Arsenic	“	<5.0	<5.0	—
Aluminum	“	24.7	7.4	70.0
Barium	“	41.8	1.8	95.7
Beryllium	“	<1.0	<1.0	—
Cadmium	“	0.70	0.04	94.3
Cobalt	“	0.50	0.02	96.0
Chromium	“	2.8	0.5	82.1
Copper	“	8.0	3.7	53.8
Iron	“	41.0	3.5	91.5
Mercury	“	0.2	0.2	—
Manganese	“	1.70	0.07	95.9
Nickel	“	21.0	0.5	97.6
Lead	“	<1.0	<2.0	—
Selenium	“	<5.0	<5.0	—
Zinc	“	<100	<100	—

COD: Chemical oxygen demand <: Detection limit

1984). Further research designed to identify the chemical and microbial composition of the organic foulant was conducted, and the results of these studies are summarized below and in previous publications (e.g., Ridgway, 1987, 1988).

Besides a flux decline, a long-term decrease in rejection was also observed. Rejection, after reloading, was 92.6% based on feedwater and improved to 94.7% the next month, apparently because of the development of a "dynamic membrane" as the fouling layer plugged microscopic holes and other imperfections in the membrane surface. Following this initial improvement, rejection again declined, and after 6 months had fallen to 92.0% with a mean of 92.9 plus or minus 1.3%. After a shutdown in December 1980 when the subunit was out of service for a period of 41 days, there was a decline in rejection from 91.5% to 90.3%. The rejection fell to 89.2% in February 1981 and remained about the

Table 9-4
RO Subunit 1A Cleaning Effectiveness

		Product Flow, gpm		
Date	*Cleaning Solution*	*Before*	*After*	*Percent Change*
11/5/80	Solution A 200 lb Citric acid 13 lb sodium bisulfite pH unknown Solution B 200 lb trisodium phosphate (TSP) 100 lb EDTA pH unknown	482	560	16.2
12/11/80	Same as 11/5/80	440	489	11.1
3/25/81	Solution B 252 lb TSP 100 lb EDTA 12.6 lb Triton X-100 pH 7.5	430	485	12.8
4/13/81	Same as 3/25/81	452	505	12.6
5/19/81	Same as 3/25/81 except pH 7.4	462	485	5.0
8/24/81	Same as 3/25/81 except pH 7.8	460	560	19.6
9/21/81	Same as 3/25/81	468	530	13.2

The above chemicals are mixed in 1,500 gallons of RO product water and the solution temperature is raised to 120 °F after pH adjustment and prior to recirculation in pressure vessels.

same through August 1981. Following a cleaning in August which was unusually effective in flux restoration, the rejection declined to a low of 87.9%, suggesting that the removal of the protective foulant layer had exposed a slightly degraded membrane.

The principal cause of the gradual decline observed in salt rejection is speculated to be a combination of fouling and gradual hydrolysis of the membrane. Typical annual average rejections of other feedwater constituents for the RO system during operation from October 1980 through September 1981 are given in Table 9-3. The membrane had been in service about 9,000 hours prior to October 1980, and by September 1981 had been in service nearly 16,000 hours.

During the 27 months of operation, subunit 1A was cleaned 13 times (see Figure 9-3). During the period from June 28, 1979, to September 30, 1980 (15 months), the subunit was cleaned six times but, as previously reported, these cleanings were relatively ineffective. From October 1, 1980, to September 1981, the subunit was cleaned seven times, and the results of these cleanings are given in Table 9-4. The data shown are from readings taken before and the day after cleaning. Cleanings before 1981 had little effect on monthly average normalized flow. The dramatic change in flux decline between December 1980 and January 1981 corresponds to changes in cleaning procedures and solutions and increased frequency of cleanings; but even during this period, a continued decline in flux occurred because of fouling.

Other explanations for the observed biphasic flux decline for subunit 1A include (1) establishment of an equilibrium condition between foulant accumulation on the one hand and foulant removal by hydrodynamic shear on the other, and (2) unidentified physicochemical or microbiological changes in the RO feedwater composition.

Chemical Analyses of Foulants

Several membranes were removed from the 5-mgd RO plant for analysis on November 3, 1980, and January 26, 1981. Visual inspection of the dismantled membrane elements indicated that after 11,000 hours of operation the membrane surfaces had become uniformly coated with a darkly colored gelatinous fouling layer. The fouling material could be readily wiped or scraped from the membranes and, upon chemical analysis, was found to be approximately 93% water by weight (Table 9-2).

Nearly 90% of the dry (dehydrated) weight of the fouling material was organic in composition, since it could be volatilized at 550 °C. Calcium, phosphorus, sulfur, and chlorine were the major elements constituting the remaining inorganic fraction as determined by conventional chemical analysis and scanning electron microscopy/energy-dispersive X-ray (Ridgway et al., 1983). As much as 30% of the dry weight of the fouling material consisted of protein, while up to 17% consisted of carbohydrate, suggesting that biologically synthesized macromolecules were an important component of the fouling layer. Indeed, bacteriological plate counts indicated up to 5×10^8 colony-forming units (cfu) per gram of native biofilm (Table 9-5). These data suggested that the organic fouling layer was composed primarily of layers of living and dead (i.e., nonculturable) bacterial cells.

Chemical and Microbiological Properties of the RO Feedwater

To better define the role of the feedwater in microbial fouling of the RO membrane surfaces, water samples were collected for analysis on the same days membranes were removed. Chemical analyses of these water samples showed that the data fell within the limits of variability of the average chemical data of the preceding 1-year period (see Table 9-6). Hence, the 11/3/80 and 1/26/81 feedwater data were representative of the long-term water chemistry for various treatment processes at Water Factory 21.

From a microbiological standpoint, there were ample concentrations of dissolved minerals (nitrogen, phosphorus, and others) and organic substances to allow bacteria to proliferate at each treatment process preceding RO. The presence of substantial concentrations of dissolved organic matter in Q-1, Q-2, Q-5, Q-6, Q-8, Q-9, and Q-22A water samples was indicated by measurement of TOC, COD, and UV absorbance. These organic substances serve as potential sources of nutrients (carbon and energy) for the growth of microorganisms at the various treatment processes, including the RO process. Indeed, feedwater nutrients may be viewed as the "potential" biofilm.

Water samples collected at each treatment location were filtered through 0.2-μm pore size polycarbonate Nuclepore membrane filters and examined by scanning electron microscopy (SEM). Figure 9-4 is a series of SEM photomicrographs of the suspended particulate matter found at various sampling locations throughout the treatment train at Water Factory 21. The plant influent water

Table 9-5
Bacteriological Properties of RO Membrane Biofilm Scrapings

Parameter	*Unit*	*Membrane* *1st Pass*	*2nd Pass*	*3rd Pass*
ATP[1]	$\mu g\ g^{-1}$ dry wt	NA[2]	337	69.5
cfu[3] detected using:				
m-SPC medium	cfu cm^{-2}	4.8×10^5	4.2×10^6	3.5×10^6
R-2A medium	cfu cm^{-2}	4.2×10^5	5.3×10^6	5.6×10^6

[1]ATP: adenosine-5'-triphosphate
[1]NA: Parameter not analyzed
[3]cfu: Total number of colony-forming units

(Q-1) contained the highest concentration of particulate debris, which consisted mainly of large aggregations (visible flocs) of rod-shaped and filamentous microorganisms. Direct enumeration of bacteria in the SEM photomicrographs indicated that the total number of bacteria per field (i.e., viable plus nonviable cells: also referred to as *culturable* plus *nonculturable*) remained relatively constant (approximately 5×10^6 cells/mL) from one treatment process to the next, with the exception of Q-1 and Q-22B water, where bacterial numbers were significantly higher and lower, respectively.

In order to determine the viability of bacteria observed in SEM micrographs, several independent tests were performed and the results compared. These tests included (1) determination of the concentration of adenosine-5'-triphosphate

Table 9-6
Chemical Analysis of Water Samples for 1-Year Period[1]

Parameter	*Unit*	*Location* *Q1*	*Q2*	*Q8 (RO feed)*	*Q22B (RO permeate)*
EC	µmho	1,672.0-1,764.3	1,591.2-1,685.9	ND[2]	177.8-193.0
TDS	mg/L	958.23-1,045.29	891.21-978.52	904.80-1,036.25	86.57-113.82
pH	pH	7.43-7.48	10.90-11.00	7.16-7.29	6.80-6.88
Ca	mg/L	85.60-89.68	76.04-81.40	ND	BDL[3]
Mg	"	20.73-23.50	1.33-2.73	1.99-5.23	0.03-0.13
Na	"	196.42-237.60	192.84-235.10	228.29-247.30	26.49-29.01
K	"	12.05-13.62	12.16-13.73	ND	1.19-1.55
Al	"	ND	ND	ND	ND
Fe	"	ND	ND	ND	ND
OH as $CaCO_3$	"	ND	120.24-137.63	ND	ND
CO_3 as $CaCO_3$	"	ND	70.39-77.81	ND	ND
HCO_3 as $CaCO_3$	"	262.50-274.22	ND	ND	18.44-19.88
Hardness as $CaCO_3$	"	286.91-1,045.29	209.46-252.53	187.66-240.02	5.17-8.48
Cl	"	234.92-246.95	239.47-252.69	ND	35.36-39.11
F	"	1.17-1.53	0.77-0.94	0.77-1.02	0.16-0.20
SO_4	"	220.80-239.64	194.85-211.48	ND	3.01-4.06
PO_4-P	"	5.02-5.60	0.01-0.10	ND	BDL
NO_3-N	"	0.02-0.09	0.01-0.18	0.44-0.87	0.16-0.33
NH_3-3	"	21.09-27.82	22.39-29.32	24.15-30.27	2.55-3.38
Org N	"	2.01-3.02	0.92-1.50	0.28-1.15	0.18-0.38
TKN	"	23.81-30.72	21.74-29.82	24.73-32.40	2.93-3.83
TOC	"	13.77-20.24	8.90-11.13	1.30-10.38	0.82-2.46
COD	"	44.69-53.41	28.58-33.90	6.02-22.41	0.98-2.26
SiO_2	"	21.26-23.28	11.88-14.05	12.68-16.24	3.37-3.93
Free Cl_2	"	ND	ND	ND	BDL
Total Cl_2	"	ND	ND	ND	0.35-0.60
Turbidity	FTU	ND	ND	ND	ND
Color	Units	32.60-35.32	18.91-21.66	2.69-9.80	BLD
UV	% abs.	0.178-0.196	0.121-0.157	0.017-0.082	0.021-0.34

[1]Values shown are 95% confidence intervals for the 1-year period from October 1, 1980, to September 30, 1981.
[2]ND: No data available
[3]BDL: Below detection limit

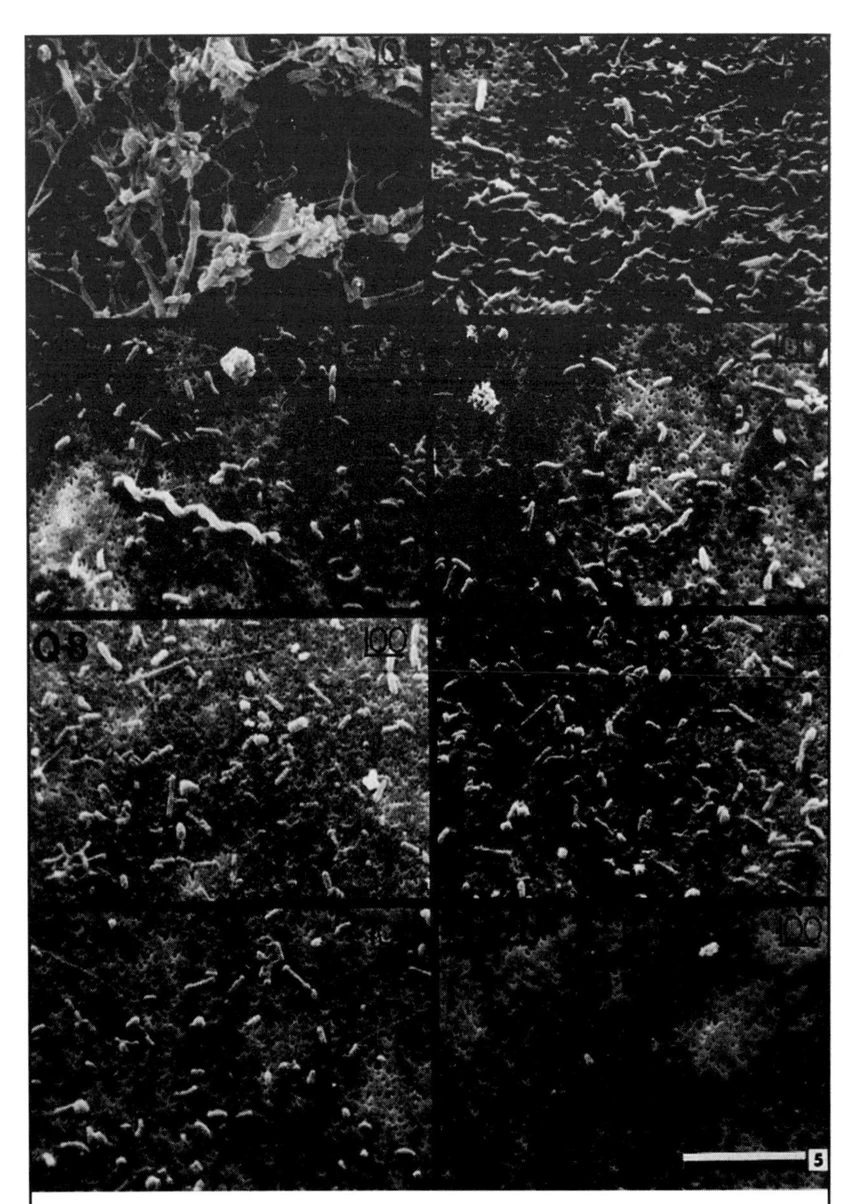

Figure 9-4. Photomicrographs (by SEM) of suspended particulate matter in water samples from Water Factory 21 treatment process locations. The 'Q' numbers in upper left corner of panels indicate sample locations (see Figure 1). Number in upper right corner indicates volume (mL) of sample filtered. Reference size scale in panel Q22b: 5 µm.

(ATP) to give a relative indication of the number of viable bacteria in the water sample, (2) direct microscopic enumeration of total (i.e., culturable plus nonculturable) microorganisms using epifluorescent light microscopy and SEM microscopy of filtered water samples, and (3) enumeration of bacterial colonies on low-nutrient R2A medium (Reasoner and Geldreich, 1985) or on rich m-SPC medium (Means et al., 1981). Figure 9-5 is a graphic summary of the results.

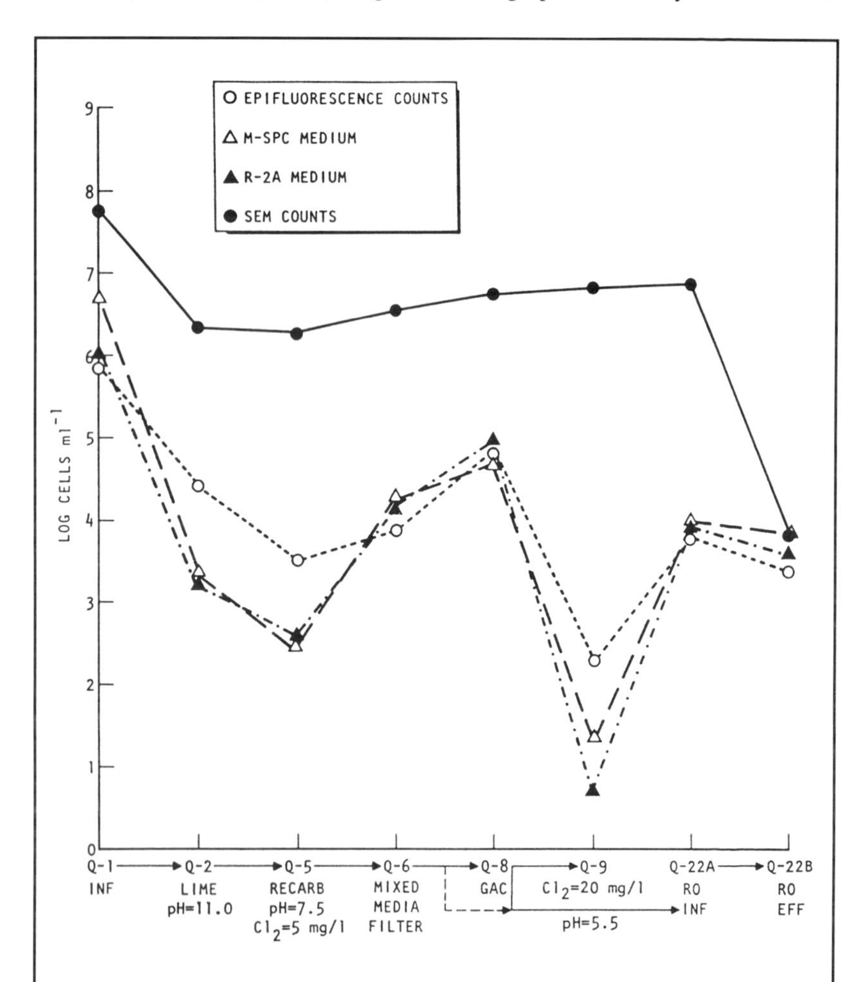

Figure 9-5. Total *(i.e., culturable plus nonculturable cells) and* viable *(i.e., culturable) bacterial populations following treatment processes at Water Factory 21. Total bacteria were directly enumerated by SEM of cells captured on 0.2 -µm filters. Viable cells were enumerated by the number of cfu appearing on R-2A or m-SPC plates (see text).*

The data show an approximate 2-log reduction in total bacteria after lime treatment, probably resulting from flocculation/precipitation processes. Total bacterial numbers (by SEM direct count) remained relatively constant at about 5×10^6 cells/mL throughout the remaining treatment processes. However, viable bacterial counts (expressed as cfu/mL on R2A or m-SPC medium) fluctuated widely, depending upon treatment location, with the lowest counts observed at points in the process where chlorine disinfection was employed (Q5 and Q9).

The graph shows that the RO feedwater (Q-22A) had a total bacterial count of about 7.5×10^6 cells/mL and a viable bacterial count of about 1.0×10^5 cells/mL. Apparently, substantial regrowth of microorganisms occurred during the mixed-media filtration and granular activated carbon adsorption processes immediately preceding the RO plant. Based on this finding, granular activated carbon treatment was eventually omitted as a pretreatment to RO. Microbial isolates of different bacterial genera recovered from the RO feedwater indicated that most of the organisms were *Flavobacterium/Moraxella, Corynebacterium/Arthrobacter*, or *Pseudomonas/Alcaligenes* (also see Ridgway et al., 1981, 1983).

Electron Microscopy of RO Biofilms

The surfaces of the RO membranes were examined using SEM. Figure 9-6 is a series of photomicrographs of the feedwater side of the RO membrane surface, showing the existence of a fouling layer with a surface texture that appears rough at low magnification. At higher magnification, the SEM revealed that the outermost surface of the biofilm consisted of a complex network of fissures and cavities that harbored large numbers of rod-shaped and/or filamentous microorganisms. The rod-shaped bacterial cells were generally on the order of 0.3 to 0.5 μm in diameter and about 0.7 to 0.9 μm in length.

All of the bacteria occupying a single microcolony generally exhibited a similar morphological appearance, suggesting that cell growth and multiplication had occurred in situ. Individual bacterial cells were typically firmly attached to the biofilm surface by means of a network of extracellular polymeric fibrils that radiated outward from the cell surface. Though these extracellular fibrils were not purified and chemically analyzed, it is likely that they consisted of acidic mucopolysaccharides or glycoproteins known to mediate surface adhesion of a wide variety of bacteria (Bryers, 1993; Costerton et al., 1987; Costerton et al., 1985; Geesey, 1982; Marshall, 1985; Marshall and Blainey, 1991).

It was sometimes possible to examine the biofilm in an edgewise orientation where the RO membrane had been sliced and subsequently curled away from the membrane surface during dehydration (Figure 9-7). In such an orientation the overall biofilm thickness was estimated to be from 10 to 20 μm. In addition, the biofilm was seen to possess a distinct laminar construction in edgewise orientation, whereby several individual lamellae of apparently compressed bacterial

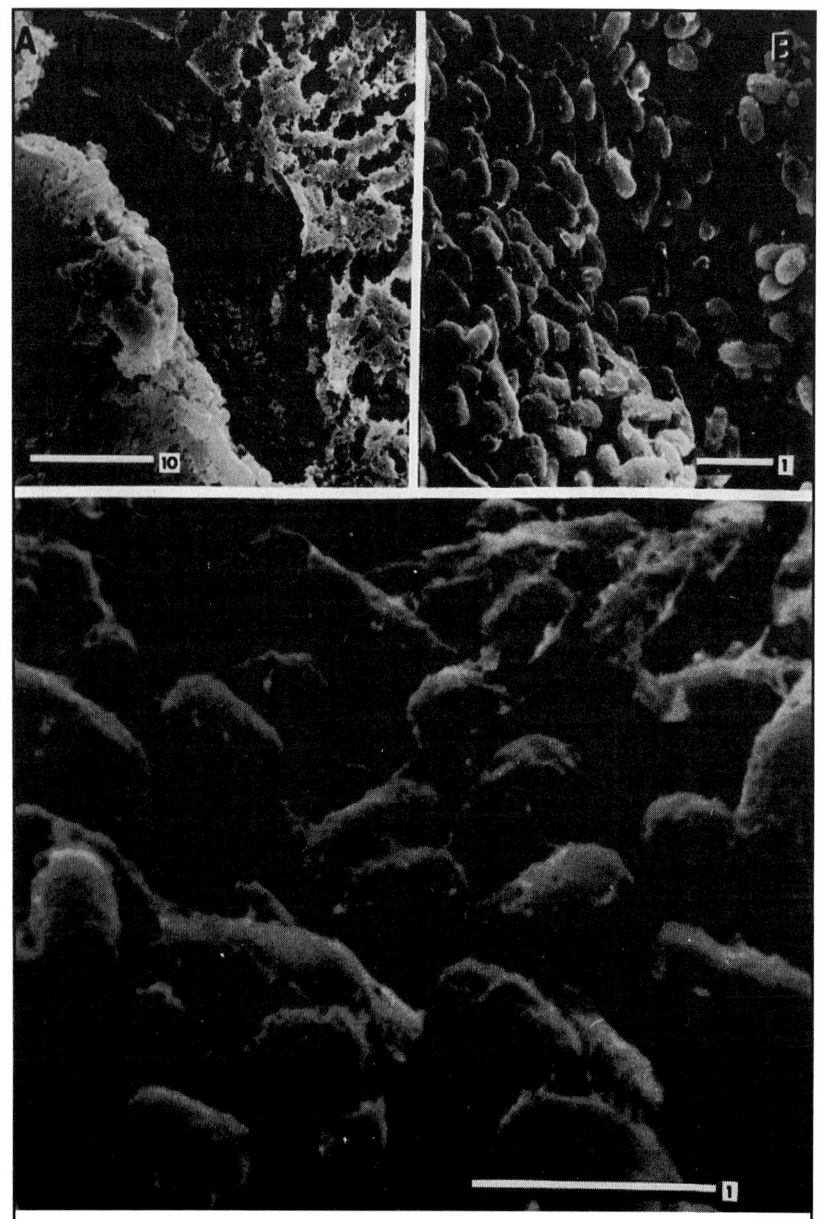

Figure 9-6. Series of SEM photomicrographs of typical RO biofilm associated with the feedwater surface of a CA membrane at Water Factory 21. Membrane was operated for approximately 2 years. Reference scale given in μm.
(From Ridgway et al., 1983.)

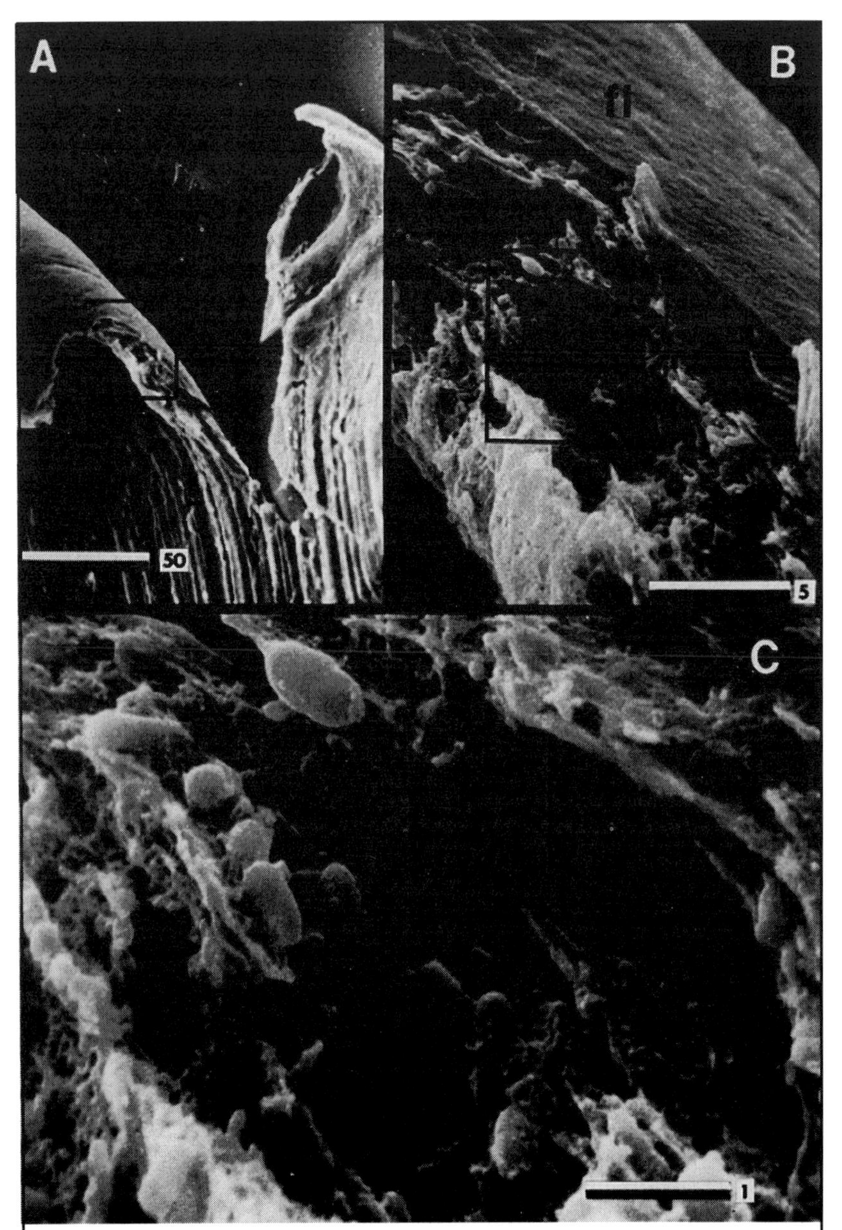

Figure 9-7. Series of SEM photomicrographs of RO membrane biofilm viewed in edge orientation. Note overall biofilm thickness of 10 to 20 μm. Reference scale given in μm.
(From Ridgway et al., 1983.)

cells were layered upon one another, collectively constituting the biofilm. Each of the layers was on the order of 3 to 5 μm in thickness, the outermost layer (on the feedwater surface of the membrane) being the least compacted. Presumably this outermost layer corresponded to the most recently deposited fouling material.

Very sparse (i.e., patchy) microbial colonization was observed on the surfaces of the polyester Texlon support fibers on the permeate surface of the membranes (data not shown; see Ridgway, 1987; Ridgway et al., 1981, 1983). Low nutrient concentrations (essentially oligotrophic conditions) and the presence of a continuous chloramines residual on the permeate surface undoubtedly contributed to poor biofilm development.

Bacterial Enumeration and Identification

The number of viable microorganisms associated with the RO membrane fouling layer was determined by plating samples of the fouling material onto m-SPC and R-2A nutrient media using a membrane filtration technique. The total numbers of colony-forming units detected in biofouling material from the first-, second-, and third-pass membranes are shown in Table 9-5. The number of viable bacteria in the fouling material (expressed as cfu/cm^2 of membrane surface) varied from a low value of approximately 5×10^5 for the first-pass membrane to approximately 5×10^6 for the third-pass membrane. In addition, relatively high concentrations of intracellular ATP were detected in the fouling material from the second- and third-pass RO membranes, indicating the presence of large numbers of metabolically active (i.e., viable) bacterial cells.

Identification of bacterial isolates found in the RO membrane fouling layer are summarized in Figure 9-8. The major bacterial genus (*Acinetobacter*) associated with the heavily fouled feedwater surface of the RO membrane constituted approximately 78% (on R-2A medium) of the total number of isolates examined from the feedwater surfaces of all three elements.

Subsequent studies (not described herein) have revealed that bacteria belonging to the acid-fast genus *Mycobacterium* are most commonly isolated from more recently biofouled membranes at Water Factory 21 (Ridgway, 1987, 1988; Ridgway and Safarik, 1991; Ridgway et al., 1984b; Safarik et al., 1989). Evidently, mycobacteria are among the earliest microorganisms to invade RO membranes at Water Factory 21, colonizing both CA and PA membranes alike (Safarik et al., 1989).

A variety of other microbial genera, including the *Acinetobacter* species mentioned above, appear on the membrane surfaces only after the mycobacteria have established an extensive biofilm (Ridgway et al., 1984a; Ridgway and Safarik, 1991). Thus, membrane biofouling is a successional process. The documented resistance of mycobacteria to antimicrobial agents such as chlorine compared to other microorganisms may partially explain their ability to outcompete other bacteria in the earliest stages of RO membrane biofouling. In addition, because they possess an exceedingly hydrophobic cell exterior, mycobacteria tend to exhibit a higher binding affinity for RO membranes than most other bacteria (Ridgway et al., 1985; also see below).

Rate of Biofilm Formation and the Effect of Chlorination

It was hypothesized that accumulation of the observed biological material on the RO membrane surface is the cause of flux decline. Therefore, experiments were devised to compare the rate of biofilm development with flux decline and membrane performance. Six 2-1/2-inch-diameter RO modules were installed in parallel on a side-stream loop of the 5-mgd RO plant feedwater. The small "loop tester" modules were made with the same blend CA membrane (Fluid Systems Model 8150-HR) as used in the full-scale plant. The membrane elements were pre-compacted by operating at 500 psig on pure water for 48 hours, to eliminate the effect of compaction on flux decline. Because the fouling layer consisted mostly of bacteria, a chlorine feed pump was installed and two of the six elements received further chlorination (15 mg/L). All added chlorine was in the combined form (chloramines), since the feedwater contained a considerable concentration of ammonia (see Table 9-6, above).

During the course of the study, elements receiving either *chlorinated* or *nonchlorinated* (normal plant) feedwater were periodically removed, the membrane examined by SEM, and the foulant material analyzed. The *unchlorinated* feedwater actually contained 1 to 3 mg/L combined (i.e., monochloramine) residual as a result of chlorination during recarbonation pretreatment at Q5. *Chlorinated* feedwater received an additional dose of 15 mg/L of chlorine just

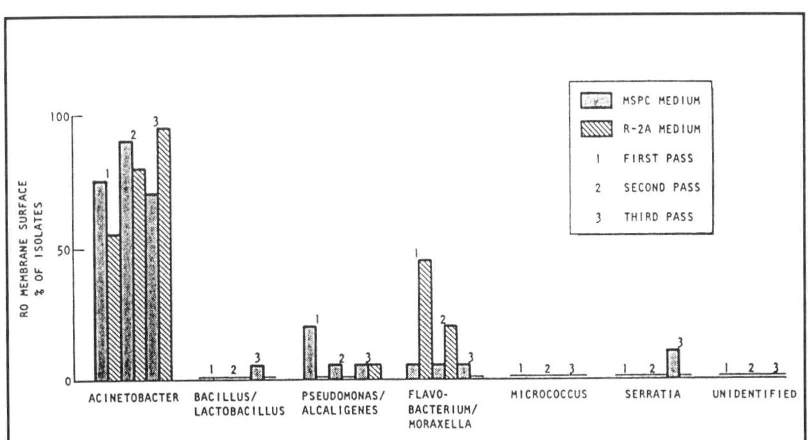

Figure 9-8. Identification of major bacterial groups associated with RO membrane biofilms depicted in SEM photomicrographs, above.
(Data adapted from Ridgway et al., 1981, 1983.)

before entering the RO loop tester modules, yielding a final combined residual of 16 to 18 mg/L).

The performance of all elements was monitored for flow and salt rejection as well as for biofilm development. Biofilm development was determined by total plate counts (on m-SPC or R-2A medium) of biofilm scrapings. Figure 9-9 shows the total number of viable bacteria associated with the different sampling times. The data indicate that microbial growth on the membrane surface occurred in three distinct stages: (1) an initial very rapid exponential growth/attachment phase lasting for about the first 5 days of membrane operation; (2) a much more gradual increase in viable cell numbers lasting for an additional 35 to 40 days; and finally (3) a gradual decline phase lasting for the duration of the

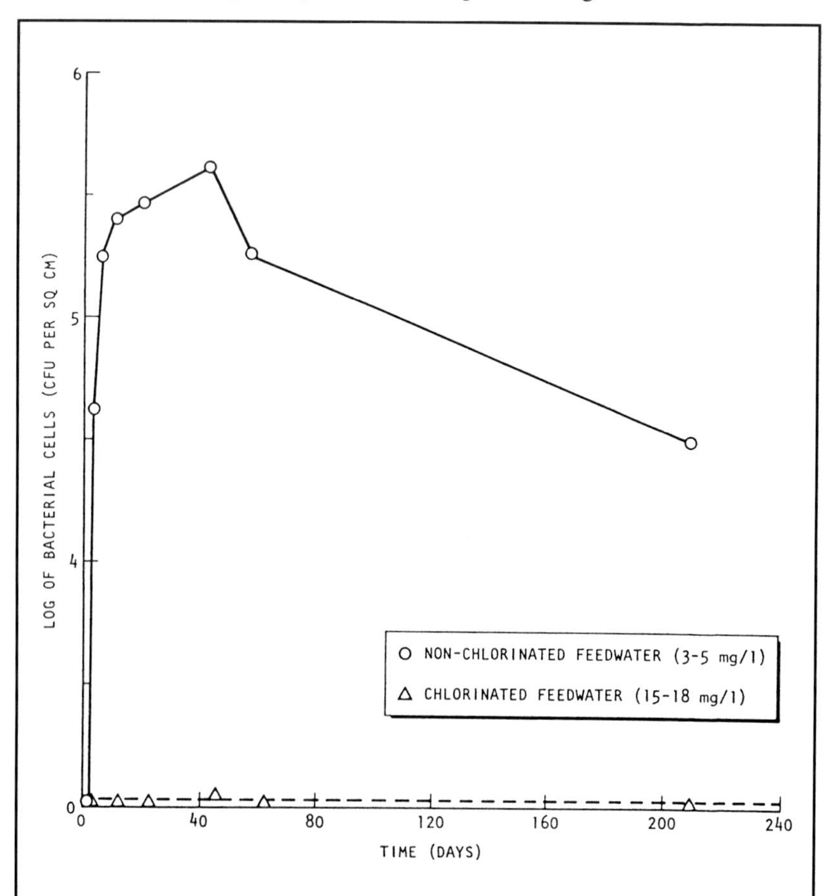

Figure 9-9. Number of viable bacteria per cm² (cfu/cm²) as a function of time of membrane operation.
(From Ridgway et al., 1981, 1984a.)

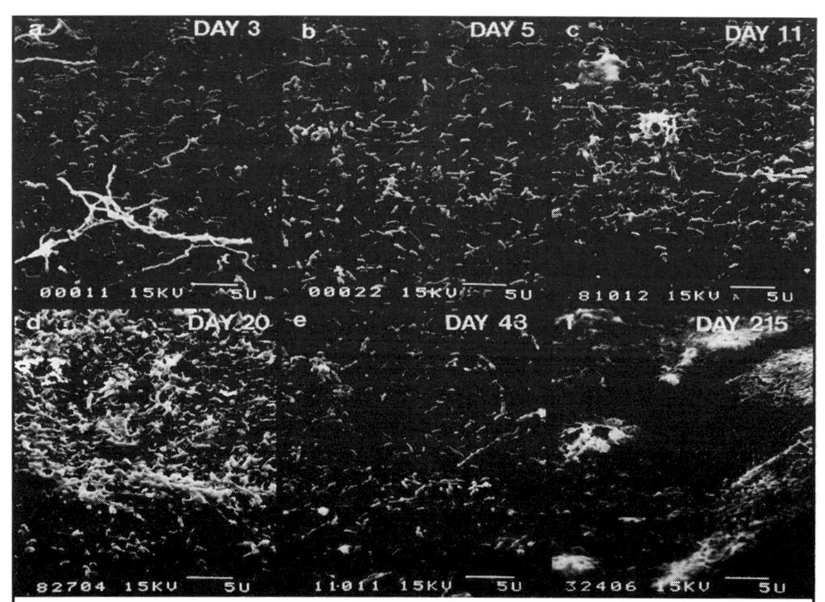

Figure 9-10. Series of SEM photomicrographs depicting time course of biofilm development on CA membrane surface at Water Factory 21. Membranes received nonchlorinated feedwater. (From Ridgway et al., 1981, 1984a.

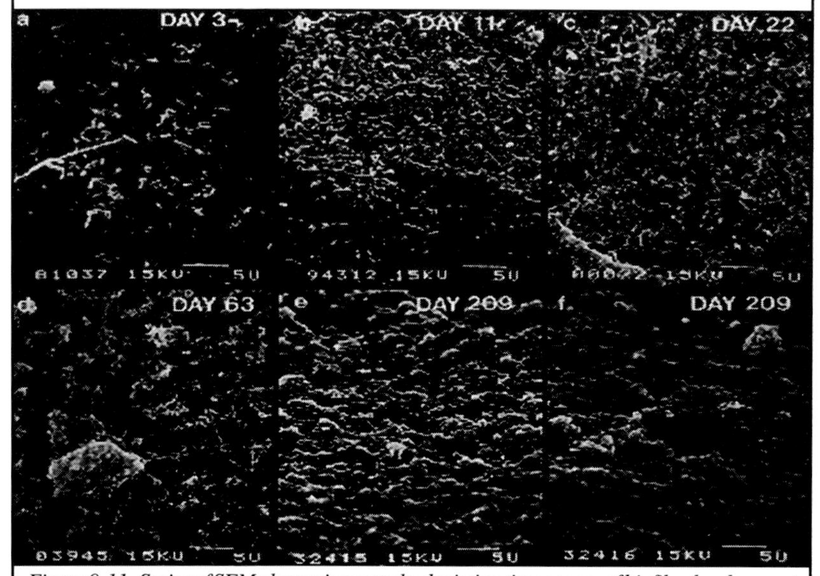

Figure 9-11. Series of SEM photomicrographs depicting time course of biofilm development on CA membrane surfaces at Water Factory 21. Membranes received chlorinated feedwater. Note apparent membrane hole in panel f, *possibly resulting from chlorine oxidation damage to membrane.* *(From Ridgway et al., 1981, 1984a.)*

experiment (209 days). The maximum number of viable bacteria was approximately 4.4×10^5 cells per cm^2 of membrane surface on day 43. With the exception of day 63 of the sampling schedule, attempts to recover viable bacteria from the chlorinated membranes were uniformly unsuccessful. Even on day 63, however, the total number of viable microbes detected was low (less than 100 cfu/cm^2 of membrane surface).

By comparison, direct microscopic enumeration in the SEM indicated approximately 8.7×10^6 and 3.3×10^6 attached total bacteria per cm^2 of the nonchlorinated and chlorinated membrane surfaces, respectively, after only 70 hours of operation (Figures 9-10 and 9-11). Since viable microorganisms could not be recovered from the chlorinated membranes in almost all cases, it was concluded that these attached cells were inactivated or injured by the combined chlorine residual (15 to 20 mg/L) present in the RO feedwater. Hence, continuous addition of a high dose of combined chlorine to the feedwater apparently did little to impede the rate of bacterial adhesion and membrane biofouling.

Bacterial colonization of membrane surfaces attained a nearly confluent state of membrane coverage as early as the tenth or eleventh day of operation (Figure 9-10, panel *c*). These bacteria appeared to be attached to the RO membrane surface (and to other bacteria) by means of extracellular polymeric substances (EPS); (Geesey, 1982), which were frequently so extensive as to occlude individual cells from view.

Further increases in the numbers of organisms contributed principally to growth in the overall thickness of the biofilm. Apart from increases in biofilm thickness and the amount of extracellular polymer, no other changes in the microstructural properties of the fouling layer subsequent to day 22 were observed.

Comparison of the chlorinated and nonchlorinated membrane surfaces indicated a slightly slower accumulation of the biofilm on the chlorinated surface. The chlorinated membrane biofilm appeared to be composed primarily of bacterial cell debris rather than intact living cells (Figure 9-11).

The membrane depicted in Figure 9-11, panel *d* (i.e., day 63) received feedwater containing 10 to 15 mg/L of free chlorine for a period of 2 or 3 days prior to sacrificing that element and acquiring the picture shown. Just before day 63, the ammonia concentration in the feedwater was significantly reduced by enhanced nitrification activity in the biological treatment processes at the sewage treatment plant supplying Water Factory 21. As a result, the 15-mg/L chlorine dose added was mainly in the free residual form. Note the extensive cellular aggregation and apparent destruction of the normal rod-shaped morphology of bacteria after exposure to elevated free chlorine.

Exposure to the high free chlorine residual also resulted in the appearance of numerous holes in the membrane surface (panel *f*, Figure 9-11). The appearance of such holes was correlated in time with sudden deterioration of membrane function (i.e., loss of mineral rejection and increase in membrane flux) (Figure 9-12). However, exposure to the free chlorine residual, while evidently damaging the membrane, did not result in significant biofilm removal.

Extensive microbial fouling of the RO membrane surfaces was reflected by the accumulation of biochemical substances. Accumulation of protein and carbohydrate on the membrane surfaces is plotted in Figures 9-13 and 9-14, respectively. On both the chlorinated and nonchlorinated membranes, these

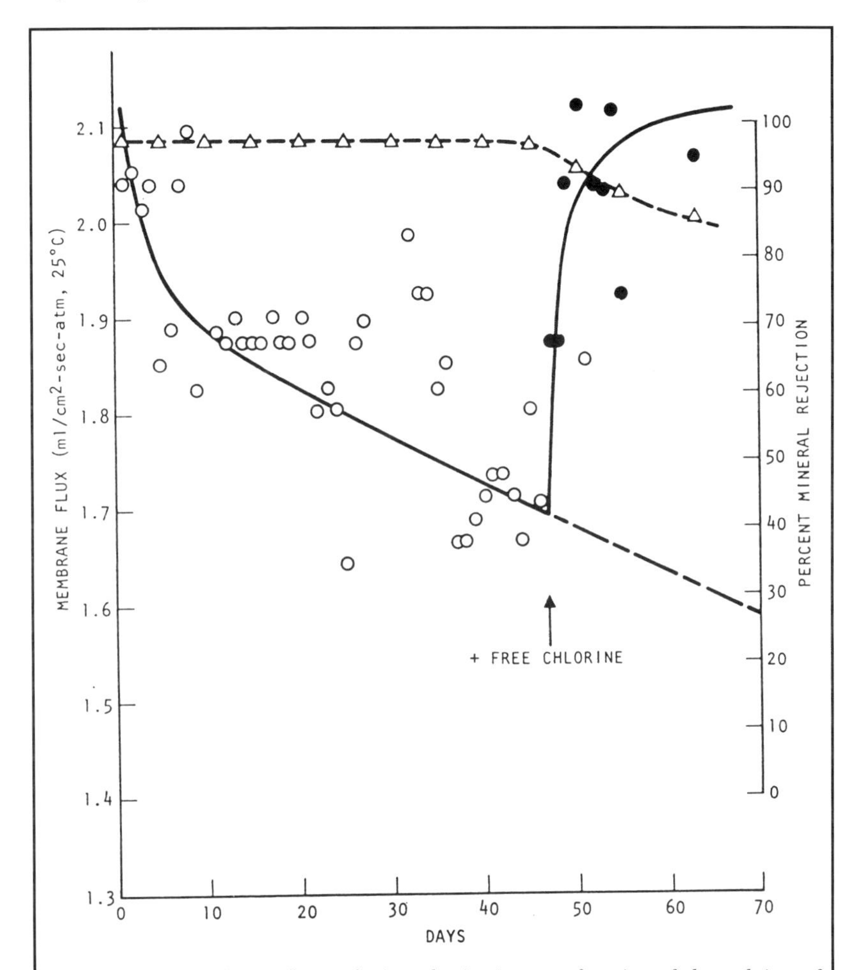

Figure 9-12. Membrane flux and mineral rejection as a function of elapsed time of operation. Membrane received chlorinated feedwater. Note loss of mineral rejection when free chlorine was initiated (see text for details).
(From Ridgway et al., 1981, 1984a.)

parameters underwent rapid increases during the initial 5 to 10 days of operation. Like bacterial growth, this initial rapid increase in the biochemical parameters was followed by a more gradual rate of fouling for the remaining duration of the experimental period.

An objective of the early membrane biofouling studies at Water Factory 21 was to correlate the accumulation of bacteria on the membrane surfaces with the flux decline. The performance for a module receiving feedwater identical to that for the full-scale RO plant (i.e., nonchlorinated) is plotted in Figure 9-15a, while the performance of an identical module receiving the chlorinated feedwater is shown in Figure 9-15b. The operating time for the two membranes was identical (about 115 days).

As previously discussed, the chlorine usually reacted with ammonia in the feedwater (see Tables 9-3 and 9-6) such that the RO membrane received only

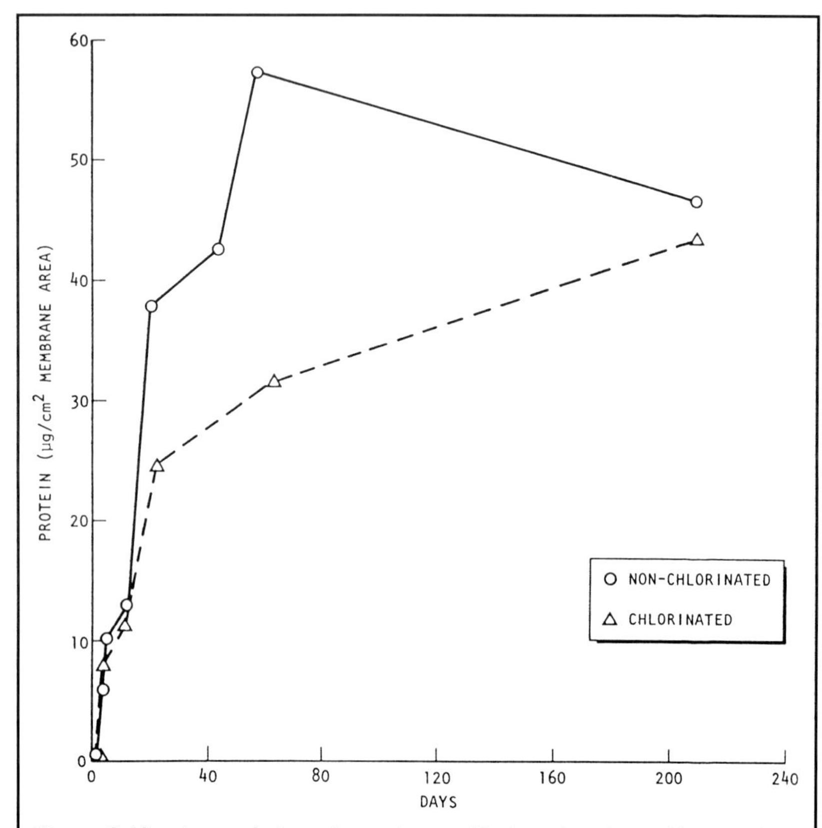

Figure 9-13. Accumulation of protein on chlorinated and nonchlorinated CA membrane surfaces as a function of elapsed time of operation.
(From Ridgway et al., 1981, 1984a.)

monochloramine. A comparison of Figures 9-15a and 9-15b indicates that the rate of flux decline was greatest for the element receiving the chlorinated feedwater. This result was consistent with the SEM and protein/carbohydrate observations, indicating that chlorination did not prevent adhesion of bacteria to the membrane surface. However, based on that observation, one would expect the rate of flux decline for the two membranes to be approximately equal or

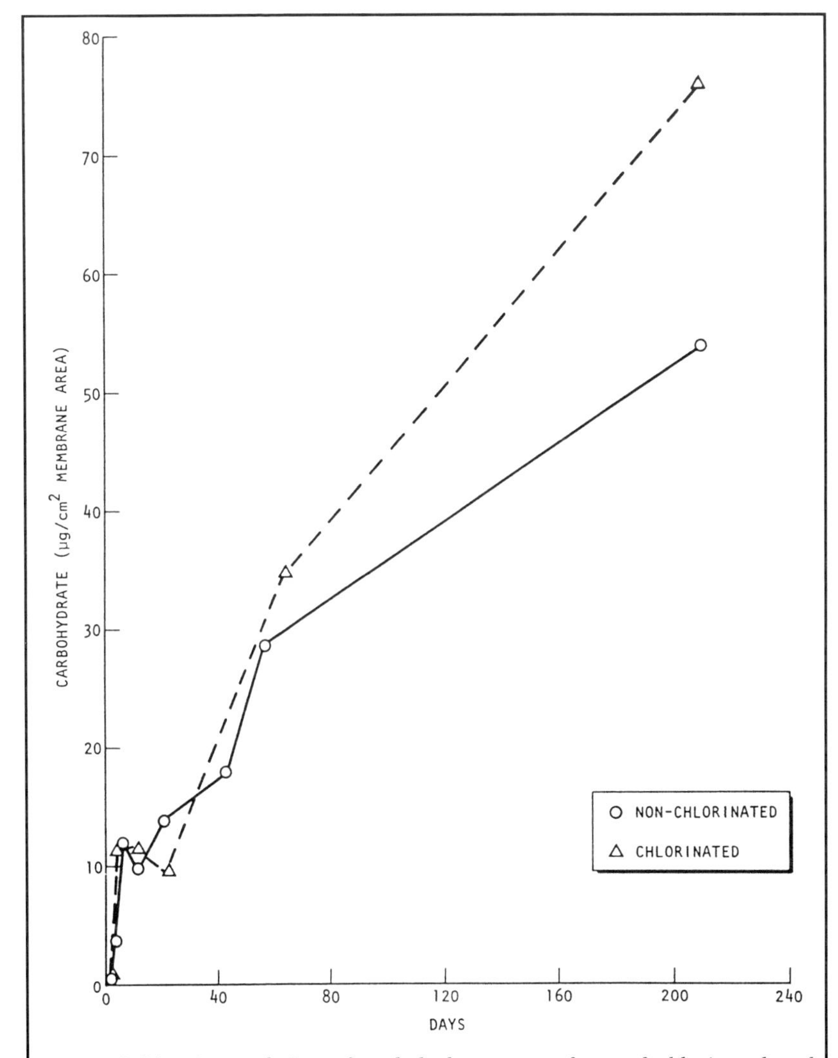

Figure 9-14. Accumulation of carbohydrate on surfaces of chlorinated and nonchlorinated CA membranes as a function of elapsed time of operation.
(From Ridgway et al., 1981, 1984a.)

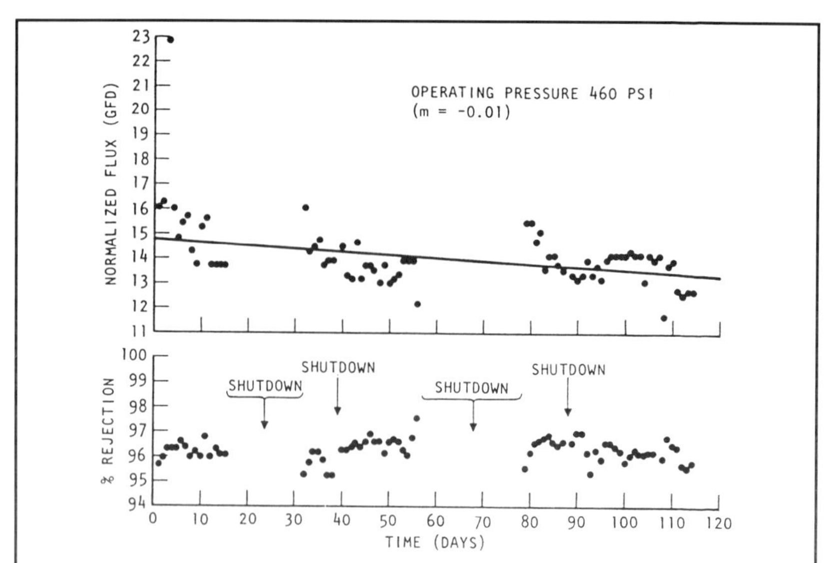

Figure 9-15a. Membrane flux and mineral rejection versus time for nonchlorinated CA membrane receiving normal pretreated feedwater (1 to 3 mg/L chlorine as monochloramine) at Water Factory 21.

(From Ridgway et al., 1981, 1984a.)

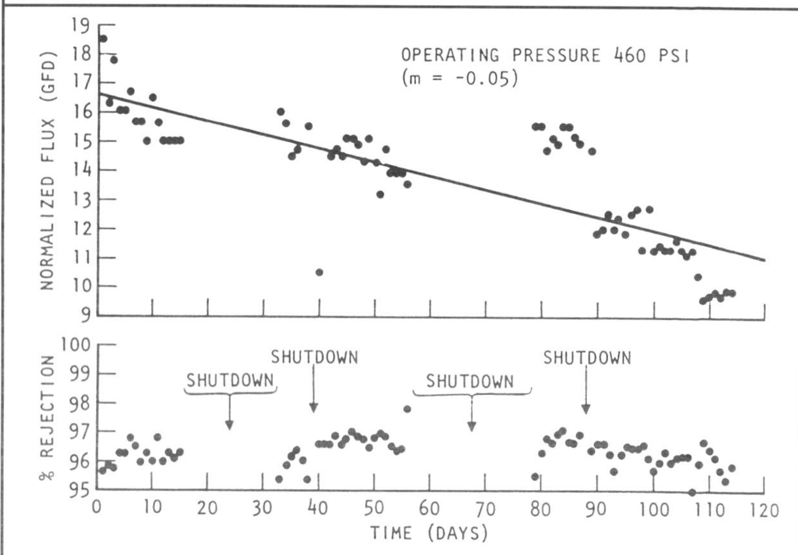

Figure 9-15b. Membrane flux and mineral rejection versus time for chlorinated CA membrane receiving normal pretreated feedwater plus 15 mg/L of chlorine (as monochloramine).

(From Ridgway et al., 1981, 1984a.)

perhaps slightly less for the chlorinated membrane.

Surprisingly, the flux loss (m value) for the chlorinated membrane was about 5-fold greater than that of the nonchlorinated membrane. At least two explanations for the observed difference in performance are possible.

First, the biofilm on the membrane receiving chlorinated feedwater is com-

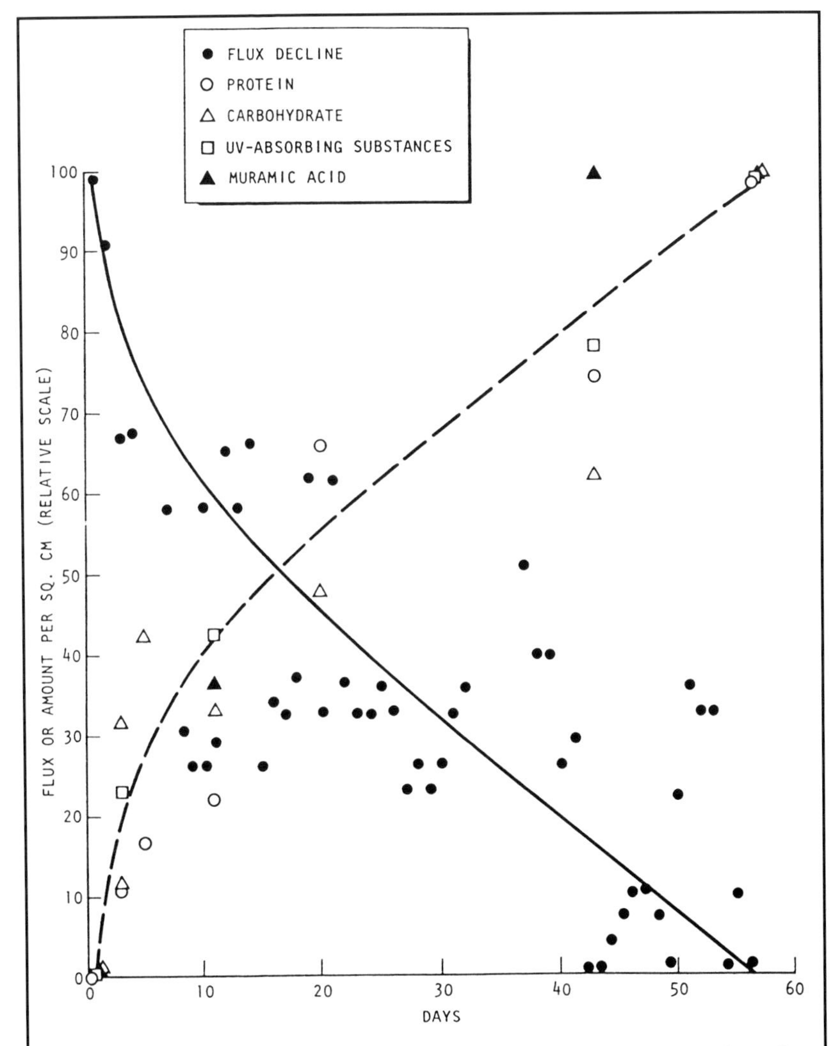

Figure 9-16. Inverse relationship between CA membrane flux and time-dependent accumulation of biologically-derived substances on feedwater surface of membrane. All parameters were normalized to percent.
(From Ridgway et al., 1981, 1984a.)

posed primarily of inactivated or stressed bacteria unable to produce colonies on R-2A medium. Such inactive cells may become more densely compacted (and less permeable to water) than metabolically active cells associated with the nonchlorinated membrane. Second, high dosages of chloramines may react with the CA membrane polymer to reduce its flux characteristics. It was observed that the membrane that had received chlorinated feedwater was more brittle, suggesting that detrimental physicochemical changes might have occurred within the membrane polymer.

The relative (percentage) change in several biochemical parameters such as

Table 9-7
Formulation of Experimental RO Membrane Cleaning Solutions

Cleaning Solution Code	*Treatment*	*Time*	*Dose*
UTC*	Sodium phosphate buffer	60 minutes	0.1M
		pH 7.0	
1	Esterase	30 minutes	100 μg/mL
	EDTA		10 mM
	Trypsin	30 minutes	100 μg/mL
	Triton X-100		0.1% v/v
2	Urea	60 minutes	2M
	TSP	1% w/v	
4	GuHCl	60 minutes	2M
	CTAB		1% w/v
5	Urea	60 minutes	2M
	SDS		1% w/v
6	Urea	60 minutes	2M
	CTAB	1% w/v	
7	Papain	100 μg/mL	
	Triton X-100	30 minutes	0.1% v/v
	EDTA		10 mM
	Protease		100 μg/mL
	Triton X-100	30 minutes	0.1% v/v
	EDTA		10 mM
8	BIZ	60 minutes	1% w/v
9	Pancreatin	60 minutes	100 μg/mL
	Triton X-100		0.1% v/v
	EDTA		10 mM
10	Thermolysin	60 minutes	100 μg/mL
	Triton X-100		0.1% v/v
	EDTA		10 mM
11	GuHCl	30 minutes	2 M
	CTAB		1 % w/v
	Protease		100 μg/mL
	Papain	30 minutes	100 μg/mL
	Esterase		10 μg/mL
	EDTA		10 mM
12	CTAB	60 minutes	1% w/v
13	SDS	60 minutes	1% w/v
14	Triton X-100	60 minutes	1% w/v
15	MBTC	60 minutes	1% w/v
	CTAB		1% w/v
16	ZDDC	60 minutes	1% w/v
	Triton X-100		.01% v/v

protein, carbohydrate, UV-absorbing substances, muramic acid (a bacterial cell wall component), and membrane flux were plotted against time for membranes receiving the nonchlorinated feedwater (Figure 9-16). The rate of accumulation of the biochemical substances on the membrane correlated inversely with the rate of membrane flux decline.

This observation suggests, but does not unequivocably prove, that the loss of water flux through the RO membranes is caused by biofilm formation. Presumably, the viscoelastic gel-like consistency of the biofilm imposes an effective diffusional barrier between the bulk fluid phase and the RO membrane. A thick

Cleaning Solution Code	*Treatment*	*Time*	*Dose*
17	STP	60 minutes	1% w/v
	EDTA		1% w/v
	TSP		1% w/v
18	EDTA	60 minutes	100 mM
19	Trypsin	60 minutes	100 μg/mL
	EDTA		10 mM
	Triton X-100		0.1% v/v
20	Esterase	60 minutes	10 μg/mL
	EDTA		10 mM
	Triton X-100	0.1% v/v	
21	Urea	60 minutes	2M
22	TSP	60 minutes	1% w/v
23	Pancreatin		100 μg/mL
	EDTA	30 minutes	10 mM
	Triton X-100		0.1% v/v
	CTAB	30 minutes	1% w/v
	EDTA		10 mM
	Triton X-100		0.1% v/v
20	Esterase	60 minutes	10 μg/mL
	EDTA		10 mM
	Triton X-100	0.1% v/v	
21	Urea	60 minutes	2M
22	TSP	60 minutes	1% w/v
23	Pancreatin		100 μg/mL
	EDTA	30 minutes	10 mM
	Triton X-100		0.1% v/v
	CTAB	30 minutes	1% w/v

EDTA: the disodium salt of ethylenediaminetetraacetic acid
TSP: trisodium phosphate
GuHCl: guanidine hydrochloride
CTAB: cetyltrimethylammonium bromide
SDS: sodium dodecyl sulfate
BIZ: tradename for a laundry presoaking detergent
MBTC: methylene bisthiocyanate
ZDDC: zinc dimethyldithiocarbamate
*UTC: untreated control
Note: All solutions prepared in 100 mM Na-phosphate buffer, pH 7.0

biofilm would effectively extend the diffusional path length over which a water or solute molecule must travel before reaching the underlying RO membrane. This effect can be offset by increasing the system operating pressure to accelerate water transport kinetics through the biofilm.

Evaluation of Cleaning Solutions in Removal of Biofilm

The membrane elements that were removed from the loop tester to evaluate the development of the biofilm were also used to evaluate experimental cleaning solutions. Strips of biofouled membrane were placed in sterile glass test tubes and different experimental cleaning solutions were added to each (see Tables 9-7 through 9-9; also see Whittaker et al., 1984, for additional details). After a 1-hour incubation at 35 °C, the membrane strips were removed, inspected by SEM, and analyzed by means of a battery of assays including ATP analysis, bacterial enumeration, and biochemical analyses. One strip placed in a phosphate buffer (pH 7.0) served as an untreated experimental control.

The appearance of the treated membranes in the SEM was compared with that of the corresponding untreated control membrane, and the cleaning solution was then assigned a relative grade on a scale of 0 to 4+, based on a subjective visual evaluation of the extent of biofilm removal (4+ = greatest biofilm removal). The

Table 9-8
SEM Chemical Cleaner Grading System

Numerical Score	*Degree of cleaning or biofilm removal*	*Presence of MLB**	*Presence of SLB**
0	No removal of biofilm evident The untreated control for each membrane was arbitrarily assigned a zero value. All treatments were compared to the control preparation.	Yes	Yes
1+	Poor biofilm removal. Generally less than 20% to 30% removal observed	Yes	Yes
2+	Fair biofilm removal. Usually #50% to 70% removal	No	Yes
3+	Good biofilm removal. Usually #70% to 80% removal	No	Yes
4+	Excellent biofilm removal. Usually greater than 90% removal A few scattered cells may remain, and some small patches, but no extensive patches.	No	No

*MLB: Multilayer biofilm (i.e., a biofilm two or more cell layers thick).
*SLB: Single-layer biofilm (i.e., a biofilm no more than one cell layer thick).

grading system is explained in more detail in Table 9-8. The results of these studies have been reported in detail in a report by Whittaker et al. (1984).

Figures 9-17 through 9-19 are series of representative SEM photomicrographs illustrating the effects of selected cleaning treatments on biofilm removal; Table 9-9 gives the relative grades assigned to each cleaning solution. Figure 9-17 shows a series of photomicrographs of membranes cleaned with seven different experimental cleaners after receiving nonchlorinated feedwater for 11 days. The code in the top left-hand corner corresponds to the type of cleaning solution used (see Table 9-7), and the effectiveness of each is noted in the upper right-hand corner according to the grading system outlined in Table 9-8. Figure 9-18 shows a series of photomicrographs of membranes that had received chlorinated feedwater and were then cleaned with the same solutions as in Figure 9-17. Clearly, the relative effectiveness of the different cleaning solutions varied

Table 9-9
SEM Evaluation of RO Biofilm Removal by Experimental Cleaning Solutions

Cleaning	------- Number day ------									
Solution	3	3	5	11	11	20	22	43	57	63
Code	*NC*[*]	*C*[**]	*NC*	*NC*	*C*	*NC*	*C*	*NC*	*NC*	*C*
TC	0	0	0	0	0	0	0	0	0	0
1	2+	3+	2+	3+	3+	2+	3+	2+	2+	4+
2	0	2+	1+	2+	2+	2+	-	1+	3+	1+
4	-	-	-	-	-	-	-	-	-	-
5	2+	3+	3+	3+	3+	3+	4+	2+	3+	4+
6	2+	2+	0	1+	2+	1+	3+	0	2+	4+
7	2+	2+	3+	3+	3+	2+	3+	2+	1+	4+
8	3+	3+		4+	4+	3+	3+	2+	1+	4+
9	2+	3+		2+	3+	3+	3+	2+	1+	4+
10	1+	0		1+	0	1+	2+	-	-	-
11	-	-		1+	1+	1+	2+	-	-	-
12	0	1+		0	1+	1+	3+	0	0	1+
13				3+	3+	1+	3+	1+	1+	4+
14				2+	2+	0+	3+		1+	4+
15				0	0	1+	1+	-	-	
16						0	-	-	-	-
17								1+	1+	4+
18								0	1+	4+
19								1+	1+	4+
20								2+	2+	4+
21								0	2+	2+
22								0	0	1+
23								2+	1+	3+

[*]NC: nonchlorinated membrane
[**]C: chlorinated membrane

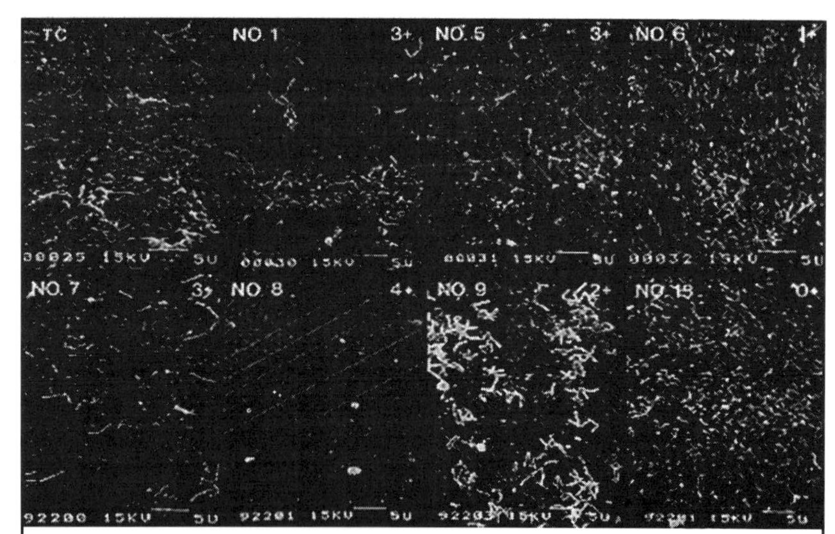

Figure 17. A series of SEM photomicrographs showing relative effectiveness of different experimental membrane cleaning solutions on nonchlorinated membranes. Number in upper left corner corresponds to the cleaning solution code given in Table 9-7. Number in upper right corner refers to relative degree of cleaning achieved by experimental solution (see Table 9-8).
(From Whittaker et al., 1984.)

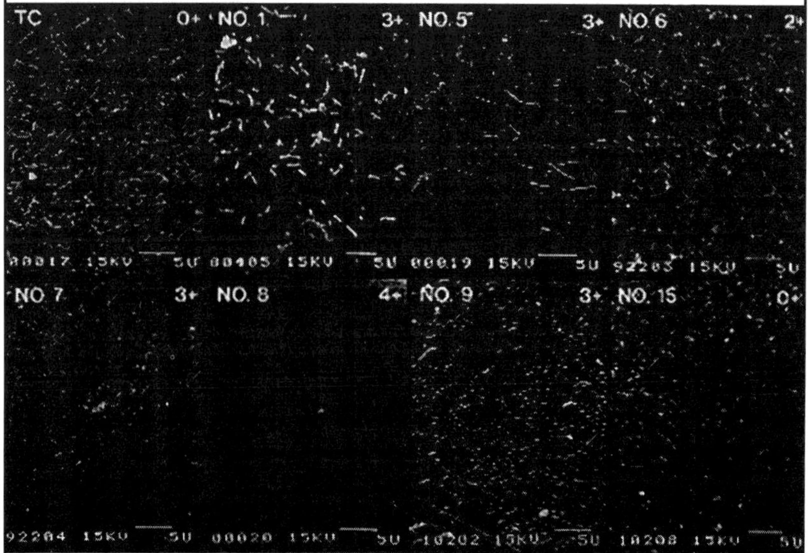

Figure 18. A series of SEM photomicrographs showing relative effectiveness of experimental membrane cleaning solutions on chlorinated membranes. Numbers are as indicated in Figure 9-17.
(From Whittaker et al., 1984.)

considerably.

Moreover, it appeared that most cleaning solutions were less effective as the age of the biofilm increased (see Figure 9-19). A decrease in cleaning effectiveness with biofilm maturity may be accounted for by a number of possible factors, such as increases in biofilm thickness, the degree of biofilm compaction, changes in the type and amount of EPS synthesized by the fouling microorganisms, or the accumulation within the biofilm matrix of heavy metals (e.g., chromium; see Table 9-2) that could inhibit enzyme activity during cleaning.

No experimental cleaning solution tested consistently removed all of the biofilm from all of the membranes. Moreover, some solutions appeared to work well once, but poorly at other times. However, general trends in the data were discernible. For example, inspection of the data in Table 9-9 indicated that biofilm removal from the membranes that received chlorinated feedwater was, in general, more efficient than removal from membranes that received nonchlorinated feedwater. This effect is probably caused by the bactericidal activity of monochloramine (LeChevallier, 1991), and possibly by the partial oxidation of cellular material by the combined residual.

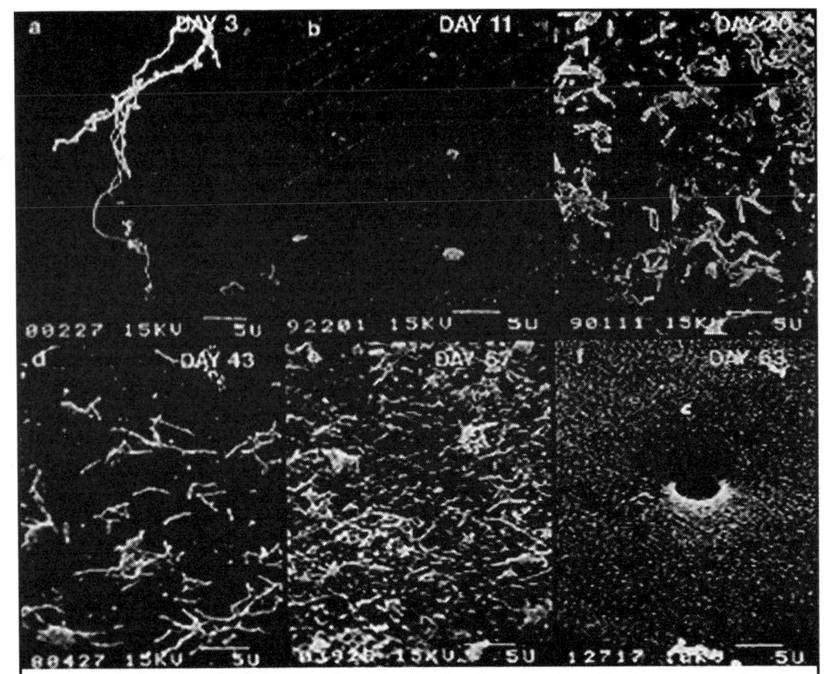

Figure 19. A series of SEM photomicrographs showing cleaning effectiveness as a function of RO biofilm age. Cleaning solution used was the commercial detergent BIZ (see Table 9-7, solution #8). Note comparatively poorer cleaning of older biofilms. All membranes received chlorinated feedwater containing 16 to 18 mg/L monochloramine.

Whereas bacteria inactivated by chlorine are still capable of passively attaching to the RO membrane surface (Ridgway and Safarik, 1991), they might be expected to be dislodged more easily than living microorganisms since they would be unable to manufacture adhesive extracellular biopolymers. Bacterial EPS has been demonstrated to be involved in the irreversible attachment of bacteria to a wide variety of surfaces (Costerton et al., 1985; Marshall, 1985; Marshall and Blainey, 1991).

The highest-rated cleaning solutions in terms of overall biofilm removal were the commercial laundry detergent BIZ, urea/SDS, papain/Triton X-100/EDTA, protease/Triton X-100/EDTA, and esterase/Triton X-100. It is interesting to note that a common feature of all of these solutions is the incorporation of one or more surfactants. Subsequent research has indicated that certain surfactants can be effective in inhibiting the attachment of specific fouling microorganisms to RO membrane surfaces (Ridgway, 1987, 1988; Ridgway et al., 1985, 1986; Flemming and Schaule, 1988a, 1988b; also see below).

Concluding Remarks

Biological fouling of RO membranes results in a number of deleterious effects with regard to membrane performance. Some of the principal effects include the following:

- A decline in the apparent membrane flux due to accumulation of a gel diffusion barrier (i.e., the biofilm) separating the bulk fluid phase from the semipermeable membrane surface
- A gradual increase in the frictional resistance of water flowing tangential to the membrane surface, resulting in an increase in the feed-brine pressure differential (i.e., an increase in the membrane delta-p)
- A reduction of convective fluid flow proximal to the membrane surface that, in turn, results in enhanced concentration polarization and greater solute passage through the membrane (i.e., lowered permeate quality)
- Accelerated deterioration of membrane integrity and/or module components (e.g., glue lines) caused by direct attack (e.g., by bacterial enzymes) or indirect processes (e.g., by local pH changes) associated with normal microbial metabolism and growth processes in biofilms
- An increased health risk to plant workers and the public due to accumulation and concentration of potentially pathogenic microorganisms on the membrane surfaces

The combined result of the above effects is to significantly shorten membrane lifetime and increase plant O&M costs. In addition, expensive pretreatment processes (e.g., continuous biocide dosage, multimedia and cartridge filtration) may be required to combat membrane biofouling.

On a cellular level, biological fouling of RO membranes is an exceedingly complex phenomenon involving many undefined or unidentified variables associated with the RO feedwater, the membrane polymers, and the bacteria. Numerous different microbial genera are involved in membrane biofouling, some of which colonize the membrane surface early (e.g., the mycobacteria; Ridgway et al., 1984a), and others much later (Flemming, 1993; Flemming and Schaule, 1988a, 1988b). Little is known concerning the physiological ecology and genetics of the membrane-fouling bacteria or the mechanism(s) by which they attach to RO membrane surfaces. The kinetics of microbial growth and the relationship of growth to feedwater nutrient availability are also still largely unexplored. Without such critical fundamental information, it is unlikely that elegant and truly cost-effective membrane biofouling control strategies will be developed in the near term; thus, basic research is needed to better delineate microbial attachment mechanisms and define the role of feedwater nutrients (type and availability) in biofilm growth processes.■

CHAPTER 10

COMPUTER PROGRAMS

Many of the calculations discussed in this book are most easily performed by a computer. This chapter will include some simple examples of computer programs that may be of assistance in evaluating or designing an RO treatment system. These programs are written in the BASIC programming language, and should run with most personal computers. Newer DOS operating software includes a Quick Basic program (QBASIC.EXE) that is capable of running this software.

Advanced programs are currently available from the RO membrane manufacturers for predicting RO permeate quality, and for monitoring and graphing normalized RO performance. Commercial programs are also available for monitoring and graphing operating data, and for predicting piping pressure losses.

RO Data Normalization and Graphing

Chapter 5, "RO Maintenance," discussed in detail the importance of monitoring RO operating data in order to follow performance trends. The program that follows takes daily operating data and normalizes it for temperature, pressure, flow, and osmotic pressure variation so as to provide the following three parameters in a form that can be easily graphed and compared:

- Salt rejection
- Normalized permeate flowrate
- Normalized differential pressure

The program will also graph this raw data if so desired:

- Temperature
- Membrane feed pressure
- Concentrate pressure

Note: Advanced editions of the following five programs are available for purchase from American Fluid Technologies (1-800-783-4955). These are written in a "WINDOWS"-style graphics environment.

- Permeate flowrate
- Concentrate flowrate
- Feed concentration
- Permeate concentration

It is suggested that the normalization calculations that are used in this program be incorporated into a more advanced data storage and management system.

```
(note: for commands that are too long to fit on line "»" denotes that
            commands following belong to previous line")
CLS
PRINT "American Fluid Technologies RO/DI Data Monitoring Program"
PRINT
PRINT "                                    Data Entry"
Month:

n = n + 1
INPUT "Month (mm)                    "; Mnth(n)
IF Mnth(n) > 12 GOTO Month
INPUT "Day (dd)                      "; Dy(n)
IF Dy(n) > 31 GOTO Month
INPUT "Year (yy)                     "; Yr(n)
IF Yr(n) > 99 OR Yr(n) < 88 GOTO Month

GOSUB CalcDays
'The number of days is counted from the first data entry

INPUT "Temperature-F                 "; Temp(n)
INPUT "Feed Pressure                 "; FdPsi(n)
INPUT "Concentrate Pressure          "; ConPsi(n)
INPUT "Permeate Flow                 "; PermFlow(n)
INPUT "Concentrate Flow              "; ConFlow(n)
INPUT "Feed Concentration            "; FdConc(n)
INPUT "Permeate Concentration        "; PermConc(n)
PRINT
INPUT "More Data [Y] "; q$

IF q$ = "" OR q$ = "Y" OR q$ = "y" GOTO Month

EndDt = n
graphs:
CLS

Itv = 1

'Data is normalized to the pressure and flow rates of the first entry.

PrmDsFlw = PermFlow(1)
ConDsFlw = ConFlow(1)
FdDsPsi = FdPsi(1)
```

```
'Normalization  Calculations

FOR n = 1 TO EndDt

IF FdConc(n) = 0 THEN Rej(n) = 0
IF FdConc(n) > 0 THEN Rej(n) =(FdConc(n)-PermConc(n))/FdConc(n) *100
DeltaPsi(n)=(PermFlow(n)+2*ConFlow(n))/(PrmDsFlw + 2 * CnDsFlw) ^ 2 *
        » (FdPsi(n) - ConPsi(n))
OsmPsi(n) = (FdConc(n) - PermConc(n)) * .01
TmpCtFct = EXP(2500 * (1 / (((Temp(n) - 32) * 5 / 9) + 273) -
        » 1 / 298))
IF FdPsi(n) = 0 THEN NormPrm(n) = 0: GOTO Movin2
NormPrm(n) = PermFlow(n) * TmpCtFct * FdDsPsi / (FdPsi(n) - OsmPsi(n)
        » - (FdPsi(n) - ConPsi(n)) / 2)

'The maximum and minimum values for each parameter is found in order
to set the  'y-axis for the graphs.

Movin2:
IF Temp(n) > MxTemp THEN MxTemp = Temp(n)
IF Temp(n) < MnTemp THEN MnTemp = Temp(n)
IF FdPsi(n) > MxFdPsi THEN MxFdPsi = FdPsi(n)
IF FdPsi(n) < MnFdPsi THEN MnFdPsi = FdPsi(n)
IF ConPsi(n) > MxConPsi THEN MxConPsi = ConPsi(n)
IF ConPsi(n) < MnConPsi THEN MnConPsi = ConPsi(n)
IF PermFlow(n) > MxPermFlow THEN MxPermFlow = PermFlow(n)
IF PermFlow(n) < MnPermFlow THEN MnPermFlow = PermFlow(n)
IF ConFlow(n) > MxConFlow THEN MxConFlow = ConFlow(n)
IF ConFlow(n) < MnConFlow THEN MnConFlow = ConFlow(n)
IF FdConc(n) > MxFdConc THEN MxFdConc = FdConc(n)
IF FdConc(n) < MnFdConc THEN MnFdConc = FdConc(n)
IF PermConc(n) > MxPrmConc THEN MxPrmConc = PermConc(n)
IF PermConc(n) < MnPrmConc THEN MnPrmConc = PermConc(n)
IF Rej(n) > MxRej THEN MxRej = Rej(n)
IF Rej(n) < MnRej THEN MnRej = Rej(n)
IF DeltaPsi(n) > MxDeltaPsi THEN MxDeltaPsi = DeltaPsi(n)
IF DeltaPsi(n) < MnDeltaPsi THEN MnDeltaPsi = DeltaPsi(n)
IF NormPrm(n) > MxNormPrm THEN MxNormPrm = NormPrm(n)
IF NormPrm(n) < MnNormPrm THEN MnNormPrm = NormPrm(n)

'If there is only one data entry, the minimum and maximum values are
        » the same.

IF n = 1 THEN
          MnTemp = MxTemp
          MnFdPsi = MxFdPsi
          MnConPsi = MxConPsi
          MnPermFlow = MxPermFlow
          MnConFlow = MxConFlow
          MnFdConc = MxFdConc
          MnPrmConc = MxPrmConc
          MnRej = MxRej
          MnDeltaPsi = MxDeltaPsi
          MnNormPrm = MxNormPrm
          END IF
NEXT n
```

```
GrphMenu:
PRINT
PRINT
PRINT "1) Rejection"
PRINT "2) Normalized Permeate Flow Rate"
PRINT "3) Normalized System Pressure Drop"
PRINT "4) Temperature"
PRINT "5) Membrane Feed Pressure"
PRINT "6) Concentrate Pressure"
PRINT "7) Permeate Flow Rate"
PRINT "8) Concentrate Flow Rate"
PRINT "9) Feed Concentration"
PRINT "10) Permeate Concentration"
PRINT
INPUT "Enter a number for the desired graph"; NMBR

'The next section sets up the graph for the particular parameter.

ON NMBR GOTO Rejection, NrmPrmF, NrmDeltaP, Tmp, FdPres, ConcPres,
        » PrmF, ConcF, FdCnc, PrmCnc

Rejection:
CLS
SCREEN 2
LOCATE 25, 20: PRINT "% REJECTION";
MX = MxRej: Mn = MnRej
GOSUB VertAx
FOR j = 1 TO EndDt STEP Itv
          y(j) = 163 - CLNG((Rej(j) - LwrLmt) * 160 / Span)
        NEXT j
GOTO Continue

NrmPrmF:
CLS
SCREEN 2
LOCATE 25, 12: PRINT "Normalized Permeate Flow Rate";
MX = MxNormPrm: Mn = MnNormPrm
GOSUB VertAx
FOR j = 1 TO EndDt STEP Itv
          y(j) = 163 - CLNG((NormPrm(j) - LwrLmt) * 160 / Span)
        NEXT j
GOTO Continue

NrmDeltaP:
CLS
SCREEN 2
LOCATE 25, 6: PRINT "Normalized Differential Pressure Drop";
MX = MxDeltaPsi: Mn = MnDeltaPsi
GOSUB VertAx
FOR j = 1 TO EndDt STEP Itv
          y(j) = 163 - CLNG((DeltaPsi(j) - LwrLmt) * 160 / Span)
        NEXT j
GOTO Continue

Tmp:
CLS
```

```
SCREEN 2
LOCATE 25, 20: PRINT "Temperature - F";
MX = MxTemp: Mn = MnTemp
GOSUB VertAx
FOR j = 1 TO EndDt STEP Itv
          y(j) = 163 - CLNG((Temp(j) - LwrLmt) * 160 / Span)
        NEXT j
GOTO Continue

FdPres:
CLS
SCREEN 2
LOCATE 25, 16: PRINT "Membrane Feed Pressure";
MX = MxFdPsi: Mn = MnFdPsi
GOSUB VertAx
FOR j = 1 TO EndDt STEP Itv
          y(j) = 163 - CLNG((FdPsi(j) - LwrLmt) * 160 / Span)
        NEXT j
GOTO Continue

ConcPres:
CLS
SCREEN 2
LOCATE 25, 18: PRINT "Concentrate Pressure";
MX = MxConPsi: Mn = MnConPsi
GOSUB VertAx
FOR j = 1 TO EndDt STEP Itv
          y(j) = 163 - CLNG((ConPsi(j) - LwrLmt) * 160 / Span)
        NEXT j
GOTO Continue

PrmF:
CLS
SCREEN 2
LOCATE 25, 14: PRINT "Actual Permeate Flow Rate";
MX = MxPrmFlow: Mn = MnPrmFlow
GOSUB VertAx
FOR j = 1 TO EndDt STEP Itv
          y(j) = 163 - CLNG((PermFlow(j) - LwrLmt) * 160 / Span)
        NEXT j
GOTO Continue

ConcF:
CLS
SCREEN 2
LOCATE 25, 18: PRINT "Concentrate Flow Rate";
MX = MxConFlow: Mn = MnConFlow
GOSUB VertAx
FOR j = 1 TO EndDt STEP Itv
          y(j) = 163 - CLNG((ConFlow(j) - LwrLmt) * 160 / Span)
        NEXT j
GOTO Continue

FdCnc:
CLS
SCREEN 2
```

```
LOCATE 25, 16: PRINT "Feed Water Concentration";
MX = MxFdConc: Mn = MnFdConc
GOSUB VertAx
FOR j = 1 TO EndDt STEP Itv
        y(j) = 163 - CLNG((FdConc(j) - LwrLmt) * 160 / Span)
      NEXT j
GOTO Continue

PrmCnc:
CLS
SCREEN 2
LOCATE 25, 20: PRINT "Permeate Concentration";
MX = MxPrmConc: Mn = MnPrmConc
GOSUB VertAx
FOR j = 1 TO EndDt STEP Itv
        y(j) = 163 - CLNG((PermConc(j) - LwrLmt) * 160 / Span)
      NEXT j

Continue:
LOCATE 22, 1:PRINT STR$(Mnth(1)) +"/" + RIGHT$(STR$(Dy(1)),2) + "/" +
      RIGHT$(STR$(Yr(1)), 2)

'The X-axis data point is determined by the ratio of the number of
      » days since the first entry over the number of days over the
      » entire time span.

FOR k = 1 TO EndDt STEP Itv
        x(k) = 42 + CLNG((DifDys(k) - DifDys(1)) * 554 /
      (DifDys(EndDt) - DifDys(1)))
        IF INT((x(k) - 116) / 80) = (x(k) - 116) / 80 THEN
              LOCATE 22, (x(k) - 116) * .125 + 12
           PRINT STR$(Mnth(k)) + "/" + RIGHT$(STR$(Dy(k)), 2)
            END IF
      NEXT k

LOCATE 1, 1: PRINT TpLmt;
LOCATE 6, 1: PRINT ThrdLbl;
LOCATE 11, 1: PRINT ScndLbl;
LOCATE 16, 1: PRINT FstLbl;

Drawing:
LOCATE 21, 5: PRINT CHR$(211);
LOCATE 1, 5: PRINT CHR$(183);
FOR i = 2 TO 20
        LOCATE i, 5: PRINT CHR$(182);
      NEXT i
FOR w = 15 TO 70
IF INT((w - 5) / 10) =(w -5)/10 THEN LOCATE 21, w: PRINT CHR$(194);
NEXT w

LOCATE 21, 6: PRINT STRING$(9, CHR$(196));
LOCATE 21, 16: PRINT STRING$(9, CHR$(196));
LOCATE 21, 26: PRINT STRING$(9, CHR$(196));
LOCATE 21, 36: PRINT STRING$(9, CHR$(196));
LOCATE 21, 46: PRINT STRING$(9, CHR$(196));
LOCATE 21, 56: PRINT STRING$(9, CHR$(196));
```

```
LOCATE 21, 66: PRINT STRING$(9, CHR$(196));
LOCATE 21, 75: PRINT CHR$(191);
FOR p = 1 TO EndDt STEP Itv
        PSET (x(p), y(p))
       NEXT p

LOCATE 25, 45: INPUT "Enter Number for Desired Graph"; NMnu
ON NMnu GOTO Rejection, NrmPrmF, NrmDeltaP, Tmp, FdPres, ConcPres,
      PrmF, ConcF, FdCnc, PrmCnc
END

VertAx:
TpLmt = (INT(MX / 10) + 1) * 10
LwrLmt = TpLmt - 20
IF TpLmt < 20 THEN TpLmt = 20
     LwrLmt = 0
     ENDIF
IF Mn < LwrLmt THEN LwrLmt = TpLmt - 40
IF TpLmt < 40 THEN TpLmt = 40
     LwrLmt = 0
     ENDIF
IF Mn < LwrLmt THEN LwrLmt = TpLmt - 80
IF TpLmt < 80 THEN TpLmt = 80
     LwrLmt = 0
     ENDIF
IF Mn < LwrLmt THEN LwrLmt = TpLmt - 200
IF TpLmt < 200 THEN TpLmt = 200
     LwrLmt = 0
     ENDIF
Span = TpLmt - LwrLmt
FstLbl = Span / 4 + LwrLmt
ScndLbl = 2 * Span / 4 + LwrLmt
ThrdLbl = 3 * Span / 4 + LwrLmt
RETURN

CalcDays:
Dys(n) = Dy(n)
Jan:
IF Mnth(n) < 2 GOTO Total
Feb:
Dys(n) = Dys(n) + 31
IF Mnth(n) < 3 GOTO Total
Mar:
Dys(n) = Dys(n) + 28
IF INT(Yr(n) / 4) = Yr(n) / 4 THEN Dys(n) = Dys(n) + 1
IF Mnth(n) < 4 GOTO Total
April:
Dys(n) = Dys(n) + 31
IF Mnth(n) < 5 GOTO Total
May:
Dys(n) = Dys(n) + 30
IF Mnth(n) < 6 GOTO Total
June:
Dys(n) = Dys(n) + 31
IF Mnth(n) < 7 GOTO Total
July:
```

```
Dys(n) = Dys(n) + 30
IF Mnth(n) < 8 GOTO Total
Aug:
Dys(n) = Dys(n) + 31
IF Mnth(n) < 9 GOTO Total
Sep:
Dys(n) = Dys(n) + 31
IF Mnth(n) < 10 GOTO Total
Oct:
Dys(n) = Dys(n) + 30
IF Mnth(n) < 11 GOTO Total
Nov:
Dys(n) = Dys(n) + 31
IF Mnth(n) < 12 GOTO Total
Dec:
Dys(n) = Dys(n) + 30
Total:
DifDys(n) = Dys(n) + (Yr(n) - 88) * 365 + 1 + INT((Yr(n) - 89) / 4)
RETURN
```

RO Permeate Projection and Scaling Potential

The calculations to predict RO permeate quality were presented in Chapter 2, "Reverse Osmosis Design." Calculations to predict the potential for scale formation, including calculations for the Langelier Index and the Solubility Product constant, were provided in Chapter 3, "RO Pretreatment." The following program predicts the possibility for scale formation and also projects the RO permeate quality for standard water treatment RO applications.

```
1000 CLS
1010 PRINT "RO PERMEATE PROJECTION AND SCALING POTENTIAL"
1015 INPUT "Customer "; cust$
1020 INPUT "Values as ion (I) or as CaCO3 (Ca) "; Q$
1030 Q$ = UCASE$(Q$)
1040 INPUT "Ca++ "; CA
1050 INPUT "Mg++ "; mg
INPUT "Na+   "; na
INPUT "K+    "; k
INPUT "HCO3-"; hco3
INPUT "SO4—"; so4
INPUT "Cl-   "; cl
INPUT "NO3-  "; no3
INPUT "F-    "; f
INPUT "pH    "; ph
INPUT "Water Temperature - F "; t
INPUT "Membrane (PA, CA(SR), CA(HR))"; memb$
memb$ = UCASE$(memb$)

'Solute passage numbers will be based on that of PA membrane.
      » The solute passage for other membrane types will be
      » relative to that of PA membrane.
```

```
IF memb$ = "PA" THEN memratio = 1
IF memb$ = "CA(HR)" THEN memratio = 4
IF memb$ = "CA(SR)" THEN memratio = 8

Recovery:
INPUT "Choose Recovery (as a fraction)"; Recov
IF Recov > 1 GOTO Recovery
INPUT "Enter estimated membrane driving pressure in psi [205]";
      » drvpsi
IF drvpsi = 0 THEN drvpsi = 205

INPUT "Desired pH"; phx
IF phx = ph GOTO CaCO3Conversion
INPUT "Choose acid for pH adjustment (H2SO4, HCl)"; acid$
acid$ = UCASE$(acid$)
IF phx = 0 THEN phx = ph

CaCO3Conversion:
IF Q$ = "CA" GOTO skipconversion
CA = CA * 2.5
mg = mg * 4.1
na = na * 2.18
k = k * 1.28
hco3 = hco3 * .82
so4 = so4 * 1.04
cl = cl * 1.41
no3 = no3 * .81
f = f * 2.5

BalancingAnalysis:
'The total concentration of cations must equal that of anions
      » when expressed as CaCO3.
skipconversion: cattotal = CA + mg + na + k
antotal = hco3 + so4 + cl + no3 + f
IF 100 < ABS(cattotal - antotal) THEN PRINT "Difference between
      » the cation and anion total is greater than 100!"
IF cattotal > antotal THEN INPUT "Cation total exceeds the anion
      » total. What anion should be used to balance"; anbal$
IF antotal > cattotal THEN INPUT "Anion total exceeds the cation
      » total. What cation should be used to balance"; catbal$
catbal$ = UCASE$(catbal$)
anbal$ = UCASE$(anbal$)
iondif = ABS(cattotal - antotal)
IF LEFT$(catbal$, 2) = "CA" THEN CA = CA + iondif
IF LEFT$(catbal$, 2) = "MG" THEN mg = mg + iondif
IF LEFT$(catbal$, 2) = "NA" THEN na = na + iondif
IF LEFT$(catbal$, 1) = "K" THEN k = k + iondif
IF LEFT$(anbal$, 4) = "HCO3" THEN hco3 = hco3 + iondif
IF LEFT$(anbal$, 3) = "SO4" THEN so4 = so4 + iondif
IF LEFT$(anbal$, 2) = "CL" THEN cl = cl + iondif
IF LEFT$(anbal$, 3) = "NO3" THEN no3 = no3 + iondif
IF LEFT$(anbal$, 1) = "F" THEN f = f + iondif
```

```
IF 0 = INT(ABS(CA + mg + na + k - hco3 - so4 - cl - no3 - f))
       » GOTO pHCalculations
GOTO skipconversion

pHCalculations:
co2 = hco3 / (10 ^ (ph - 6.3))
kon = co2 + hco3
co2x = kon / 10 ^ (phx - 6.3) / (1 + 1 / 10 ^ (phx - 6.3))
hco3x = kon - co2x
caf = CA: mgf = mg: nafd = na: kfd = k
hco3f = hco3: so4f = so4: clf = cl: ff = f: no3f = no3

'Concentration adjustments are made for the acid concentration
       » used in pH adjustment.
IF acid$ = "H2SO4" THEN so4 = so4 + hco3 - hco3x
IF acid$ = "HCL" THEN cl = cl + hco3 - hco3x
hco3 = hco3x: clx = cl: so4x = so4
cai = CA / 2.5: mgi = mg / 4.1: nai = na / 2.18: ki = k / 1.28
hco3i = hco3 / .82: so4i = so4 / 1.04: cli = cl / 1.41: no3i =
       » no3 / .81
fi = f / 2.5
tds = cai + mgi + nai + ki + hco3i + so4i + cli + no3i + fi

'The Langlier subroutine determines the pH at which the water is
       » saturated in CaCO3.
GOSUB Langelier
fdlindex = lindex


PRINT temp
INPUT "temp"; lskdj

'Adjustments for membrane pressure, temperature, and osmotic
       » pressure
TmpFctr = EXP(2500 * (1 / temp - 1 / 298))
Ratio = memratio * 220 * TmpFctr / (drvpsi - (1 + 1 / Recov) *
       » tds * .01)
PRINT Ratio
INPUT "Ratio"; zpvi

'The permcalc subroutine associates cations with anions in the
       » order of salt formation preference.
GOSUB permcalc

PermeateQuality:
'The passage of each salt is calculated.

INPUT "salt passage"; oeiryt

caso4 = caso4 * .99 * (1 + 1 / (1 - Recov)) / 2
papca = .008 * Ratio * caso4: papso4 = .008 * Ratio * caso4
cahco32 = cahco32 * .98 * (1 + 1 / (1 - Recov)) / 2
```

```
papca = papca +(cahco32 * .015): paphco3 = .015*Ratio*cahco32
nacl = nacl * .97 * (1 + 1 / (1 - Recov)) / 2
papna = .02 * Ratio * nacl: papcl = .02 * Ratio * nacl
nahco3 = nahco3 * .97 * (1 + 1 / (1 - Recov)) / 2
papna = papna + (.02 * Ratio * nahco3)
paphco3 = paphco3 + (.02 * Ratio * nahco3)
na2so4 = na2so4 * .99 * (1 + 1 / (1 - Recov)) / 2
papna = papna + (.008 * Ratio * na2so4)
papso4 = papso4 + (.008 * Ratio * na2so4)
cacl2 = cacl2 * .98 * (1 + 1 / (1 - Recov)) / 2
papca = papca + (.015 * Ratio * cacl2)
papcl = papcl + (.015 * Ratio * cacl2)
mghco32 = mghco32 * .98 * (1 + 1 / (1 - Recov)) / 2
papmg = .015 * Ratio * mghco32
paphco3 = paphco3 + (.015 * Ratio * mghco32)
mgcl2 = mgcl2 * .98 * (1 + 1 / (1 - Recov)) / 2
papmg = papmg + (.015 * Ratio * mgcl2)
papcl = papcl + (.015 * Ratio * mgcl2)
kcl = kcl * .97 * (1 + 1 / (1 - Recov)) / 2
papk = .02 * Ratio * kcl: papcl = papcl + (.02 * Ratio * kcl)
mgso4 = mgso4 * .99 * (1 + 1 / (1 - Recov)) / 2
papmg = papmg + (.008 * Ratio * mgso4)
papso4 = papso4 + (.008 * Ratio * mgso4)
khco3 = khco3 * .97 * (1 + 1 / (1 - Recov)) / 2
papk = papk + (.02 * Ratio * khco3)
paphco3 = paphco3 + (.02 * Ratio * khco3)
k2so4 = k2so4 * .99 * (1 + 1 / (1 - Recov)) / 2
papk = papk + (.008 * Ratio * k2so4)
papso4 = papso4 + (.008 * Ratio * k2so4)
cano32 = cano32 * .775 * (1 + 1 / (1 - Recov)) / 2
papca = papca + (.15*Ratio*cano32): papno3 = .15*Ratio * cano32
nano3 = nano3 * .73 * (1 + 1 / (1 - Recov)) / 2
papna = papna + (.18 * Ratio * nano3): papno3 = papno3 +
      » (.18 * Ratio * nano3)
mgno32 = mgno32 * .775 * (1 + 1 / (1 - Recov)) / 2
papmg = papmg + (.15 * Ratio * mgno32): papno3 = papno3 +
      » (.15 * Ratio * mgno32)
kno3 = kno3 * .73 * (1 + 1 / (1 - Recov)) / 2
papk = papk + (.18 * Ratio * kno3): papno3 = papno3 +
      » (.18 * Ratio * kno3)
caf2 = caf2 * .94 * (1 + 1 / (1 - Recov)) / 2
papca = papca + (.042 * Ratio * caf2): papf = .042 *Ratio * caf2
naf = naf * .895 * (1 + 1 / (1 - Recov)) / 2
papna = papna + (.07 * Ratio * naf): papf = papf +
      » (.07 * Ratio * naf)
mgf2 = mgf2 * .94 * (1 + 1 / (1 - Recov)) / 2
papmg = papmg + (.042 * Ratio * mgf2): papf = papf +
      » (.042 * Ratio * mgf2)
kf = kf * .895 * (1 + 1 / (1 - Recov)) / 2
papk = papk + (.07 * Ratio * kf): papf = papf + (.07*Ratio * kf)
papcctot = papca + papmg + papna + papk
p(2) = papca: p(3) = papmg: p(4) = papna: p(5) = papk:
```

```
        » p(6) = paphco3
p(7) = papso4: p(8) = papcl: p(9) = papno3: p(10) = papf

c(1) = LOG(paphco3 / co2x) / LOG(10) + 6.3
v(1) = phx: v(2) = caf: v(3) = mgf: v(4) = nafd
v(5) = kfd: v(6) = hco3x
v(7) = so4x: v(8) = clx: v(9) = no3f: v(10) = ff
c(2) = p(2) / 2.5: c(3) = p(3) / 4.1
c(4) = p(4) / 2.18: c(5) = p(5) / 1.28
c(6) = p(6) / .82: c(7) = p(7) / 1.04: c(8) = p(8) / 1.41
c(9) = p(9) / .81: c(10) = p(10) / 2.5
hco3 = (hco3x - p(6) * Recov) / (1 - Recov)
CA = (caf - p(2) * Recov) / (1 - Recov)
mgc = (mgf - p(3) * Recov) / (1 - Recov)
nac = (nafd - p(4) * Recov) / (1 - Recov)
kc = (kfd - p(5) * Recov) / (1 - Recov)
so4c = (so4x - p(7) * Recov) / (1 - Recov)
clc = (clx - p(8) * Recov) / (1 - Recov)
no3c = (no3f - p(9) * Recov) / (1 - Recov)
fc = (ff - p(10) * Recov) / (1 - Recov)
phc = LOG(hco3 / co2x) / LOG(10) + 6.3
tds = CA / 2.5 + mgc / 4.1 + nac / 2.18 + kc / 1.28 + hco3 /
        » .82 + so4c / 1.04 + clc / 1.41 + no3c / .81 + fc / 2.5
GOSUB langelier
CLS
PRINT cust$
PRINT
PRINT "Specified Recovery "; Recov
PRINT "Membrane                "; memb$
PRINT "Estimated Membrane Driving Pressure"; drvpsi
PRINT "                  Raw Feed     Treated Feed     Permeate Projected
        » Permeate Proj."
PRINT "      (as CaCO3)     (as CaCO3)      (as CaCO3)     (as ion)"
PRINT "pH                "; ph
PRINT "Ca++              "; caf
PRINT "Mg++              "; mgf
PRINT "Na+               "; nafd
PRINT "K+                "; kfd
PRINT "HCO3-             "; hco3f
PRINT "SO4--            "; so4f
PRINT "Cl-               "; clf
PRINT "NO3-              "; no3f
PRINT "F-                "; ff
PRINT "CO2              "
FOR n = 1 TO 10
LOCATE 7 + n, 26
PRINT v(n)
LOCATE 7 + n, 44
p(1) = c(1)
PRINT p(n)
LOCATE 7 + n, 62
PRINT c(n)
```

```
NEXT n
LOCATE 18, 62
PRINT co2x
PRINT
PRINT "Langelier saturation index for the pH adjusted feed water
      » is "; phx - fdlindex
PRINT
PRINT "The concentrate pH is "; phc
PRINT "and the Langlier index is "; phc - lindex
INPUT "Do you want a printed version? (Y/N) "; qst$
qst$ = UCASE$(qst$)
IF qst$ = "Y" GOTO Printing ELSE END

Printing:
n = 2
LPRINT
LPRINT "                    American Fluid Technologies"
LPRINT "             Reverse Osmosis Permeate Concentration"
LPRINT "                         Computer Projection"
LPRINT
LPRINT cust$
LPRINT "Membrane Driving Pressure "; drvpsi
LPRINT "Temperature "; t; " F"
LPRINT "Specified Recovery "; Recov
LPRINT "Membrane                  "; memb$
LPRINT
LPRINT "      Raw Feed    Treated Feed    Permeate Projected
      » Permeate Proj."
LPRINT "            (as CaCO3)     (as CaCO3)          (as CaCO3)
      » (as ion)"
LPRINT "pH              "; LEFT$(STR$(ph), 5); "             ";
      » LEFT$(STR$(phx), 5); "
      » "; LEFT$(STR$(c(1)), 5); "              ";
      » LEFT$(STR$(c(1)),    5)
LPRINT "Ca++            "; LEFT$(STR$(caf), 5); "            ";
      » LEFT$(STR$(v(n)), 5); "
      » "; LEFT$(STR$(p(n)), 5); "              ";
      » LEFT$(STR$(c(n)),   5)
n = n + 1
LPRINT "Mg++            "; LEFT$(STR$(mgf), 5); "            ";
      » LEFT$(STR$(v(n)), 5); "
      » "; LEFT$(STR$(p(n)), 5); "              ";
      » LEFT$(STR$(c(n)),5)
n = n + 1
LPRINT "Na+             "; LEFT$(STR$(nafd), 5); "            ";
      » LEFT$(STR$(v(n)), 5);
      » "               "; LEFT$(STR$(p(n)), 5); "
      » "; LEFT$(STR$(c(n)),5)
n = n + 1
LPRINT "K+              "; LEFT$(STR$(kfd), 5); "            ";
      » LEFT$(STR$(v(n)), 5);
      » "               "; LEFT$(STR$(p(n)), 5); "
```

```
        » "; LEFT$(STR$(c(n)),5)
n = n + 1
LPRINT "HCO3-          "; LEFT$(STR$(hco3f), 5); "              ";
        » LEFT$(STR$(v(n)), 5);
        » "                     "; LEFT$(STR$(p(n)), 5); "
        » "; LEFT$(STR$(c(n)),5)
n = n + 1
LPRINT "SO4--           "; LEFT$(STR$(so4f), 5); "            ";
        » LEFT$(STR$(v(n)), 5); "
        » "; LEFT$(STR$(p(n)), 5); "                    ";
        » LEFT$(STR$(c(n)),5)
n = n + 1
LPRINT "Cl-              "; LEFT$(STR$(clf), 5); "            ";
        » LEFT$(STR$(v(n)), 5); "
        » "; LEFT$(STR$(p(n)), 5); "                   ";
        » LEFT$(STR$(c(n)),5)
n = n + 1
LPRINT "NO3-             "; LEFT$(STR$(no3f), 5); "              ";
        » LEFT$(STR$(v(n)), 5);
        » "                 "; LEFT$(STR$(p(n)), 5); "                      ";
        » LEFT$(STR$(c(n)),5)
n = n + 1
LPRINT "F-                 "; LEFT$(STR$(ff), 5); "                ";
        » LEFT$(STR$(v(n)), 5);
        » "                       "; LEFT$(STR$(p(n)), 5); "
        » ";LEFT$(STR$(c(n)),5)
n = n + 1
LPRINT "CO2      "; co2x
LPRINT
LPRINT "Langelier saturation index for the pH adjusted feedwater
        » is "; phx - fdlindex
LPRINT
LPRINT "The concentrate pH is "; phc; "and the Langelier index
        » is "; phc - lindex
LPRINT
'The calcium sulfate concentration is compared to that of
        » saturation.

LPRINT "Calcium Sulfate concentration is"; LEFT$(STR$(((v(2) -
        » Recov * p(2)) * (v(7) - Recov * p(7)) /(1 -Recov) ^ 2/
        » 251000) ^ .5), 5); " times its  solubility limit"
LPRINT "in the concentrate."
END

permcalc:
IF CA > so4 THEN caso4 = so4 ELSE caso4 = CA
CA = CA - caso4: so4 = so4 - caso4
IF hco3 > CA THEN cahco32 = CA ELSE cahco32 = hco3
CA = CA - cahco32: hco3 = hco3 - cahco32
IF na > cl THEN nacl = cl ELSE nacl = na
na = na - nacl: cl = cl - nacl
IF na > hco3 THEN nahco3 = hco3 ELSE nahco3 = na
```

```
na = na - nahco3: hco3 = hco3 - nahco3
IF na > so4 THEN na2so4 = so4 ELSE na2so4 = na
na = na - na2so4: so4 = so4 - na2so4
IF CA > cl THEN cacl2 = cl ELSE cacl2 = CA
CA = CA - cacl2: cl = cl - cacl2
IF mg > hco3 THEN mghco32 = hco3 ELSE mghco32 = mg
mg = mg - mghco32: hco3 = hco3 - mghco32
IF mg > cl THEN mgcl2 = cl ELSE mgcl2 = mg
mg = mg - mgcl2: cl = cl - mgcl2
IF k > cl THEN kcl = cl ELSE kcl = k
k = k - kcl: cl = cl - kcl
IF mg > so4 THEN mgso4 = so4 ELSE mgso4 = mg
mg = mg - mgso4: so4 = so4 - mgso4
IF k > hco3 THEN khco3 = hco3 ELSE khco3 = k
k = k - khco3: hco3 = hco3 - khco3
IF k > so4 THEN k2so4 = so4 ELSE k2so4 = k
k = k - k2so4: so4 = so4 - k2so4
IF CA > no3 THEN cano32 = no3 ELSE cano32 = CA
CA = CA - cano32: no3 = no3 - cano32
IF na > no3 THEN nano3 = no3 ELSE nano3 = na
na = na - nano3: no3 = no3 - nano3
IF mg > no3 THEN mgno32 = no3 ELSE mgno32 = mg
mg = mg - mgno32: no3 = no3 - mgno32
IF k > no3 THEN kno3 = no3 ELSE kno3 = k
k = k - kno3: no3 = no3 - kno3
IF CA > f THEN caf2 = f ELSE caf2 = CA
CA = CA - caf2: f = f - caf2
IF na > f THEN naf = f ELSE naf = na
na = na - naf: f = f - naf
IF mg > f THEN mgf2 = f ELSE mgf2 = mg
mg = mg - mgf2: f = f - mgf2
IF k > f THEN kf = f ELSE kf = k
k = k - kf: f = f - kf
RETURN

langelier:
a = .1 * LOG(tds) / LOG(10) - .1
temp = (t - 32) * 5 / 9 + 273
b = -13.12 * LOG(temp) / LOG(10) + 34.55
cart = CA
IF CA < 3 THEN cart = 3
hco3l = hco3
IF hco3 < 1 THEN hco3l = 1
c = LOG(cart) / LOG(10) - .4
d = LOG(hco3l) / LOG(10)
lindex = 9.3 + a + b - c - d
RETURN
```

RO Array Design

The application of this program is specific to designing RO systems for new and unusual applications. The program determines an optimum array design, and estimates the permeate quality produced by that array. By calculating and adding permeate flowrates and solute passage on an individual-element basis, relatively accurate results are predicted even when the osmotic pressure or solute passage for the application is particularly high.

This program is suggested for use in designing RO systems when dealing with the following issues:

- Physical length constraints for the RO housings
- Permeate recoveries in excess of 75%
- Osmotic pressures exceeding 40 psig in the RO concentrate
- Solute rejections of less than 92%

The program will require some finessing. To obtain reasonable results may demand several iterations while attempting subtle variations in input variables such as membrane pressure or desired concentrate recycle flowrate.

```
CLS
PRINT "American Fluid Technologies Reverse Osmosis Array Design
      » Program"
PRINT
INPUT "RO Membrane (PA, HR(CA), SR(CA), SW-PA)              ";
      »Memb$
Memb$ = UCASE$(Memb$)
INPUT "Four Inch (4) or Eight Inch (8) Housings [4]  "; Diam
INPUT "Number of Elements Per Housing                ";
      » NmElmPrHsng
INPUT "Temperature (F)                                "; Temp
INPUT "Maximum Element Recovery [.143]                ";
      » MaxElmRecov
INPUT "Desired Operating Pressure (psi)               "; FdPsi
INPUT "Permeate Flow Rate (gpm)                       "; PrmFlw
Recv:
INPUT "System Recovery (fraction)                     "; Recov
If Recov=0 GOTO Recv
INPUT "Feed Concentration of Primary Solute           "; FdConc2
INPUT "Osmotic Pressure of Primary Solute (psi/ppm)  "; Osm
INPUT "Rejection of Primary Solute (fraction)         "; Rej
INPUT "Desired Recycle Flow (gpm) [0]                 "; Recycle2
PRINT
RecQ$ = UCASE$(RecQ$)

NewVar:
IF Diam = 0 THEN Diam = 4
IF Rej = 0 AND Memb$ = "PA" THEN Rej = .98
IF Rej = 0 AND Memb$ = "SW-PA" THEN Rej = .98
IF Rej = 0 AND Memb$ = "HR(CA)" THEN Rej = .96
IF Rej = 0 AND Memb$ = "SR(CA)" THEN Rej = .94
IF Memb$ = "PA" AND Diam = 4 THEN DsgnPsi = 205:DsgnElmPrm = 1.1
```

```
IF Memb$ ="SW-PA" AND Diam =4 THEN DsgnPsi =205:DsgnElmPrm = .42
IF Memb$ ="HR(CA)" AND Diam =4 THEN DsgnPsi =400: DsgnElmPrm = 1
IF Memb$ = "SR(CA)" AND Diam = 4 THEN DsgnPsi = 400:
       » DsgnElmPrm = 1.25
IF Memb$ = "PA" AND Diam =8 THEN DsgnPsi = 205: DsgnElmPrm =4.5
IF Memb$ ="SW-PA" AND Diam=8 THEN DsgnPsi = 205: DsgnElmPrm =1.7
IF Memb$ = "HR(CA)" AND Diam=8 THEN DsgnPsi =400: DsgnElmPrm =4
IF Memb$ = "SR(CA)" AND Diam=8 THEN DsgnPsi =400: DsgnElmPrm =5

IF MaxElmRecov = 0 THEN MaxElmRecov = .143
TempC = (Temp - 32) * 5 / 9
TmpFctr = EXP(2500 * (1 / (TempC + 273) - 1 / 298))

IF Osm = 0 THEN Osm = .01
Passg = 1 - Rej
k1 = 1 - Passg / .7
k2 = 1 / (1 - Recov)

'The following are minimum individual element concentrate flow
       » rates based upon the element diameter.

MinElmDesFlw4 = 4
MinElmDesFlw8 = 20

wrt = 0
DeltaPRpt:
j1 = 0: AW = 0: zd = 0: lp = 0: fg$ = "": IntHsPrm1 = 0
LgstIntHsPrm = 0

IF DeltaPsi = 0 THEN EstPsiDrp = 30 ELSE EstPsiDrp = DeltaPsi
IF Diam = 4 THEN MinElmDesFlw = MinElmDesFlw4
IF Diam = 8 THEN MinElmDesFlw = MinElmDesFlw8
Stg = 8
FdConc = FdConc2
Recycle = Recycle2
FOR sf = 1 TO 8: HsngPr(sf) =0: PresDrp(sf) =0: ConcConc(sf) =0
          PrmSlt(sf) = 0: ConcFlw(sf) = 0: NEXT sf

PreliminaryData:
FdFlw = PrmFlw / Recov
ConcFlw = FdFlw - PrmFlw
ConcFlw(8) = ConcFlw + Recycle
ConcConc(8) = FdConc2

'An average membrane solute concentration is determined
       » utilizing values for the permeate recovery and solute
       » passage.

ConcTest:
AvConc = ((FdConc2 * FdFlw + ConcConc(8) * Recycle) / (FdFlw +
       » Recycle) * 2 + ConcConc(8)) / 3
IF Passg * (FdPsi - EstPsiDrp / 2 - AvConc * Osm) / DsgnPsi *
```

```
        » PrmFlw * AvConc + ConcConc(8) * ConcFlw < FdConc2 *
        » FdFlw THEN
            ConcConc(8) = ConcConc(8) * 1.02: GOTO ConcTest
          END IF

PerStg:
Rcycl = ConcFlw(8) - (1 - Recov) / Recov * PrmFlw
IF Rcycl > .01 THEN Recycle = Rcycl
FdConc = (FdConc * FdFlw + Recycle * ConcConc(8)) / (FdFlw +
        » Recycle)
ConcOsmPsi = ConcConc(Stg) * Osm
ElmPrmFlw = (FdPsi - EstPsiDrp + PresDrp(Stg + 1) - ConcOsmPsi)
        » / DsgnPsi * DsgnElmPrm / TmpFctr
MinConcFlwStgPls1 = MinConcFlw
MinConcFlw = ElmPrmFlw / MaxElmRecov - ElmPrmFlw
IF MinConcFlw < MinElmDesFlw THEN MinConcFlw = MinElmDesFlw
IF zd = 0 THEN HsngPr(Stg) = INT(ConcFlw(Stg)/MinConcFlw *1.01)
zd = 0

IF HsngPr(Stg) = 0 THEN HsngPr(Stg) = 1
lp = lp + 1
IF lp > 100 THEN
            INPUT "Program is caught in a loop.  Please change a
        » variable [continue]"; ds:
            OPEN "SCRN:" FOR OUTPUT AS #2: GOTO PrtRslts
          END IF
IntHsConc = ConcFlw(Stg) / HsngPr(Stg)

'If there is insufficient concentrate flow to meet the minimum
        » individual element concentrate flow requirements for a
        » single housing, recycle flow must be utilized to
        » achieve the desired permeate recovery.

IF Stg = 8 AND IntHsConc < MinConcFlw THEN
            IntHsConc = MinConcFlw: Recycle = MinConcFlw *
        » HsngPr(8) - ConcFlw
            GOTO PreliminaryData
          END IF

'If the desired permeate flow is still not allocated by the
        » number of housings already allocated, another upstream
        » stage will be utilized.

IF HsngPr(Stg) * ElmPrmFlw * NmElmPrHsng < 1.01 *
        » (PrmFlw - (ConcFlw(Stg) -  ConcFlw(8))) GOTO KpGoing

'If there is insufficient permeate flow left to be allocated for
        » the number of housings determined by the minimum
        » concentrate flow rate, a number is determined
        » by the flow that is actually available.

IF HsngPr(Stg) * ElmPrmFlw * NmElmPrHsng > PrmFlw -
```

```
        » (ConcFlw(Stg) - ConcFlw(8)) THEN
            HsngPr(Stg) = INT((PrmFlw - ConcFlw(Stg) + ConcFlw(8))
        » * 1.1 / ElmPrmFlw / NmElmPrHsng)

            'If there is not enough permeate flow left for one
        » housing, the array is complete.

            IF HsngPr(Stg) < 1 THEN
                    HsngPr(Stg) = 0
                AW = 1
                    Stg = Stg + 1
                    GOTO ChkCst
                END IF
        END IF
IntHsConc = ConcFlw(Stg) / HsngPr(Stg)

'The following finds which stage can best handle one
        » less housing.

LwFlwR = ConcFlw(8)
LwFlwRStg = 8
FOR k = 1 TO 9 - Stg
z = 9 - k
FlwAtst(z) = ConcFlw(z) / (HsngPr(z) + 1)
IF HsngPr(z) > 1 THEN FlwRtst(z) = ConcFlw(z) / (HsngPr(z) - 1)
        » ELSE FlwRtst(z) = 1000
IF FlwRtst(z) < LwFlwR THEN
            LwFlwRStg = z
            LwFlwR = FlwRtst(z)
        END IF
NEXT k

'If the lead end stage can best handle a reduced number of
        » housings, then the array is complete.

IF ConcFlw(Stg) / HsngPr(Stg) < LwFlwR THEN
        AW = 1
        GOTO KpGoing
        END IF
HsngPr(LwFlwRStg) = HsngPr(LwFlwRStg) - 1
Stg = LwFlwRStg
zd = 1
GOTO PerStg

KpGoing:
ActPrm = 0
ConcConc = ConcConc(Stg)

Calcinterstgflow:
PresDrp = PresDrp(Stg + 1)
prmsltot = 0
IF Stg = 8 THEN
```

```
          IntHsPrm=0: LgstIntHsPrm=0: IntHsPrm1=0: FOR ry =1 TO 7:
            HsngPr(ry) = 0: PresDrp(ry) = 0: ConcConc(ry) = 0:
            PrmSlt(ry) = 0: ConcFlw(ry) = 0: NEXT ry:
          END IF

'The permeate flow, pressure drop, & solute passage is
       » calculated for each element in the stage.

FOR p = 1 TO NmElmPrHsng
f = IntHsConc + IntHsPrm / 2
IF Diam = 4 THEN IntHsPsiDrp = 7.52E-06 * f ^ 4 + .0005
       » * f ^ 3 + .00625 * f ^ 2 + .553 * f - 1.73
IF Diam = 8 THEN IntHsPsiDrp = -3.305E-07 * f ^ 4 + .0000396
       » * f ^ 3 + .00227 * f ^ 2 + .00172 * f + 1.14
IF IntHsPsiDrp < 0 THEN IntHsPsiDrp = 0
PresDrp = PresDrp + IntHsPsiDrp
IntHsPrm = (FdPsi - EstPsiDrp - ConcConc * Osm + PresDrp)
       » / DsgnPsi * DsgnElmPrm / TmpFctr
IF Stg = 8 AND p = 1 THEN IntHsPrm1 = IntHsPrm
IF IntHsPrm > LgstIntHsPrm THEN LgstIntHsPrm = IntHsPrm
PrmSlt = IntHsPrm * Passg * ConcConc * (DsgnPsi /
       » (FdPsi - EstPsiDrp + PresDrp - ConcConc * Osm))
ConcConc = (PrmSlt + IntHsConc * ConcConc) / (IntHsPrm
       » + IntHsConc)
prmsltot = prmsltot + PrmSlt
ActPrm = ActPrm + IntHsPrm
IntHsConc = IntHsConc + IntHsPrm
NEXT p

PresDrp(Stg) = PresDrp
ConcConc(Stg - 1) = ConcConc
PrmSlt(Stg) = prmsltot * HsngPr(Stg)
HsngPrm(Stg) = ActPrm
ConcFlw(Stg - 1) = IntHsConc * HsngPr(Stg)

IF AW = 1 GOTO ChkCst
Stg = Stg - 1
GOTO PerStg

ChkCst:

'If the calculated system pressure drop is not close to the
       » initial assume pressure drop, a new pressure drop is
       » calculated and the calculations are all repeated.

IF PresDrp > EstPsiDrp * 1.2 AND wrt < 2 THEN
            EstPsiDrp = (PresDrp + EstPsiDrp) / 2 * .95: wrt
       » = wrt + 1:
           GOTO DeltaPRpt
          END IF

TotSlt = 0: AW = 0
```

```
FOR rt = 1 TO 8
IF HsngPr(rt) = 0 THEN
          PrmSlt(rt) = 0: ConcFlw(rt - 1) = 0
          ConcConc(rt - 1) = 0
         END IF
TotSlt = TotSlt + PrmSlt(rt)
NEXT rt

AlmostDone:
ActlPrm = ConcFlw(Stg - 1) - ConcFlw(8)
PumpPsi = PresDrp - EstPsiDrp + FdPsi
OPEN "SCRN:" FOR OUTPUT AS #2

PrtRslts:
CLS
PRINT #2,
PRINT #2,
PRINT #2,
PRINT #2,
PRINT #2,
PRINT #2, "American Fluid Technologies Reverse Osmosis Array
      » Design Program"
PRINT #2,
PRINT #2, "1) RO Membrane:                    "; Memb$;
      » "         7) Desired Permeate Flow: "; PrmFlw

PRINT #2, "2) RO Element Diameter:      "; Diam;
      » "         8) System Recovery:"; Recov

PRINT #2, "3) Elements per Housing:     "; NmElmPrHsng;
      » "         9) Solute Feed Concentration: "; FdConc2

PRINT #2, "4) Temperature:                  "; Temp; "      10)
      » Osmotic Pressure: "; Osm; "psi/ppm"

PRINT #2, "5) Max Element Recovery:    "; MaxElmRecov; "    11)
Solute Membrane Rejection: "; Rej

PRINT #2, "6) Desired Operating Psi: "; FdPsi; "      12)
      » Desired Recycle Flow:"; Recycle2; "gpm"

PRINT #2,

FOR gp = 1 TO 7
PRINT #2, HsngPr(gp); "-";
NEXT gp
PRINT #2, HsngPr(8)

PRINT #2, "Lowest Element Permeate Flow Rate:     "; IntHsPrm1
PRINT #2, "Highest Element Permeate Flow Rate:    "; LgstIntHsPrm
PRINT #2,
PRINT #2, "Actual Operating Pressure: "; PumpPsi
```

```
PRINT #2, "Actual Feed Flow Rate:      "; ActlPrm / Recov
PRINT #2, "Actual Permeate Flow Rate:"; ActlPrm
PRINT #2, "Actual Recycle Flow Rate: "; Recycle
PRINT #2,
PRINT #2, "Permeate Solute Concentration: "; TotSlt / ActlPrm
PRINT #2, "Overall Solute Rejection:          "; (FdConc2 - TotSlt /
ActlPrm) / FdConc2

PRINT #2, "Differential Pressure:               "; PresDrp
CLOSE #2
IF UCASE$(fg$) = "Y" THEN OPEN "SCRN:" FOR OUTPUT AS #2:
       » fg$ = "": GOTO PrtRslts
INPUT "Do you want a print-out [N] "; fg$

IF UCASE$(fg$) = "Y" THEN OPEN "LPT1:" FOR OUTPUT AS #2
       » GOTO PrtRslts

ReDo:
INPUT "Enter number of item to be altered, or just enter to quit
       » "; Try
IF Try = 0 AND jl = 0 THEN END
IF Try = 1 THEN
       INPUT "RO Membrane (TFC, HR(CA), SR(CA)) "; Memb$
     Memb$ = UCASE$(Memb$)
     jl = 1
     GOTO ReDo
     END IF
IF Try = 2 THEN
       INPUT "Four Inch (4) or Eight Inch (8) Housing "; Diam
     jl = 1
     GOTO ReDo
     END IF
IF Try = 3 THEN INPUT "Number of Elements per Housing ";
       » NmElmPrHsng
     jl = 1
     GOTO ReDo
     END IF
IF Try = 4 THEN INPUT "Temperature (F) "; Temp: jl = 1
       » GOTO ReDo
IF Try = 5 THEN INPUT "Maximum Element Recovery "; MaxElmRecov:
jl = 1: GOTO ReDo
IF Try = 6 THEN INPUT "Desired Operating Pressure "; FdPsi
       » jl = 1: GOTO ReDo
IF Try = 7 THEN INPUT "Permeate Flow Rate "; PrmFlw: jl = 1
       » GOTO ReDo
IF Try = 8 THEN INPUT "System Recovery (fraction) "; Recov
       » jl = 1: GOTO ReDo
IF Try = 9 THEN
       INPUT "Feed Concentration of Primary Solute "; FdConc2
     jl = 1
     GOTO ReDo
     END IF
```

```
IF Try = 10 THEN
        INPUT "Osmotic Pressure of Primary Solute (psi/ppm) "; Osm
      jl = 1
      GOTO ReDo
      END IF
IF Try = 11 THEN
        INPUT "Membrane Rejection of Primary Solute (fraction)";Rej
      jl = 1
      GOTO ReDo
      END IF
IF Try = 12 THEN INPUT "Desired Recycle Flow "; Recycle2
      jl = 1: GOTO ReDo
IF jl = 1 GOTO NewVar
END
```

RO Array Performance

Similar to the previous program, the permeate and solute passage through individual elements is evaluated to more accurately predict the performance of RO arrays that might be utilized for unusual applications. The particular array (staging) is entered, along with the characteristics of the system (e.g., feed pressure, feed solute concentration, solute osmotic pressure, solute rejection). The program then predicts the system permeate flowrate, quality, and system pressure differential, and estimates the appropriate capital cost of the RO system.

```
CLS
PRINT "American Fluid Technologies Reverse Osmosis Array
       » Performance Program"
PRINT
INPUT "RO Membrane (PA, HR(CA), SR(CA), SW-PA)          "; Memb$
Memb$ = UCASE$(Memb$)
INPUT "Four Inch (4) or Eight Inch (8) Housings [4] "; Diam
INPUT "Number of Elements Per Housing                   ";
       »  NmElmPrHsng
INPUT "Temperature (F)                                  "; Temp
INPUT "Maximum Element Recovery [.143]                  ";
       » MaxElmRecov
INPUT "Desired Operating Pressure (psi)                 "; FdPsi
INPUT "Desired Permeate Flow Rate (gpm)                 "; PrmFlw2
INPUT "Desired System Recovery (fraction)               "; Recov
INPUT "Feed Concentration of Primary Solute             "; FdConc2
INPUT "Osmotic Pressure of Primary Solute (psi/ppm)  "; Osm
INPUT "Rejection of Primary Solute (fraction)           "; Rej
PRINT

EstPsiDrp = 30
PrmFlw = PrmFlw2
FOR ptr = 1 TO 8
PRINT "Number of Housings in #"; ptr;
INPUT " Stage                "; HsngPr(ptr)
IF HsngPr(ptr) = 0 THEN
```

```
            TotStgs = ptr - 1
            GOTO NewVar
            END IF
NEXT ptr


NewVar:
PRINT
PRINT "Working..."
IF Diam = 0 THEN Diam = 4
IF Rej = 0 AND Memb$ = "PA" THEN Rej = .98
IF Rej = 0 AND Memb$ = "HR(CA)" THEN Rej = .95
IF Rej = 0 AND Memb$ = "SR(CA)" THEN Rej = .92
IF Memb$ = "SW-PA" AND Diam = 4 THEN DsgnPsi = 205
        » DsgnElmPrm = .42
IF Memb$ = "PA" AND Diam = 4 THEN DsgnPsi = 205
        » DsgnElmPrm = 1.1
IF Memb$ = "HR(CA)" AND Diam = 2 THEN DsgnPsi = 400
        » DsgnElmPrm = .1
IF Memb$ = "HR(CA)" AND Diam = 4 THEN DsgnPsi = 400
        » DsgnElmPrm = 1
IF Memb$ = "SR(CA)" AND Diam = 4 THEN DsgnPsi = 400
        » DsgnElmPrm = 1.25
IF Memb$ = "SW-PA" AND Diam = 8 THEN DsgnPsi = 205
        » DsgnElmPrm = 1.7
IF Memb$ = "PA" AND Diam = 8 THEN DsgnPsi = 200
        » DsgnElmPrm = 4.4
IF Memb$ = "HR(CA)" AND Diam = 8 THEN DsgnPsi = 400
        » DsgnElmPrm = 4
IF Memb$ = "SR(CA)" AND Diam = 8 THEN DsgnPsi = 400
        » DsgnElmPrm = 5

IF MaxElmRecov = 0 THEN MaxElmRecov = .143
TempC = (Temp - 32) * 5 / 9
TmpFctr = EXP(2500 * (1 / (TempC + 273) - 1 / 298))

IF Osm = 0 THEN Osm = .01
Passg = 1 - Rej
MinElmDesFlw4 = 4
MinElmDesFlw8 = 20
jl = 0: fg$ = "": IntHsPrm1 = 0: LgstIntHsPrm = 0
ActlPrm = 0: Recycle = 0

DeltaPrpt:
IF Diam = 4 THEN MinElmDesFlw = MinElmDesFlw4
IF Diam = 8 THEN MinElmDesFlw = MinElmDesFlw8
stg = 1
FdConc = FdConc2
FOR sf = 1 TO 8: ConcConc(sf) = 0: HsngPrm(sf) = 0
            PrmSlt(sf) = 0: ConcFlw(sf) = 0: NEXT sf

PreliminaryData:
```

```
FdFlw = PrmFlw / Recov + Recycle
FdFlw2 = PrmFlw / Recov
ConcFlw = FdFlw2 - PrmFlw
FinConc = FdConc2

ConcTest:
'The purpose of the following calculation is to estimate the
       » concentrate concentration which is the same as the
       » concentration of the recycle that is being returned to
       » the feed stream.  This will affect the feed
       » concentration.

AvConc = ((FdConc2 * FdFlw2 + FinConc * Recycle) / FdFlw *
       » 2 + FinConc) / 3
FlwFctr = (FdPsi - EstPsiDrp / 2 - AvConc * Osm) / DsgnPsi
IF Passg / FlwFctr * PrmFlw * AvConc + FinConc *
       » ConcFlw < FdConc2 * FdFlw2 THEN
         FinConc = FinConc * 1.02: GOTO ConcTest
        END IF

FdConc = (FdConc2 * FdFlw2 + Recycle * FinConc) / FdFlw

PerStg:
IF stg = 1 THEN
         IntHsPrm = 0: PresDrp = 0: ActPrm = 0: PrmSltot = 0
        ConcConc = FdConc
         IntHsFd = FdFlw / HsngPr(1)
         IntHsConc = IntHsFd
        END IF

FOR p = 1 TO NmElmPrHsng

'Individual element pressure drop calculations.

f = IntHsConc - IntHsPrm / 2
IF Diam = 4 THEN IntHsPsiDrp = 7.52E-06 * f ^ 4 + .0005 *
       » f ^ 3 + .00625 * f ^ 2 + .553 * f - 1.73
IF Diam = 8 THEN IntHsPsiDrp = -3.305E-07 * f ^ 4 + .0000396 *
       » f ^ 3 + .00227 *f ^ 2 + .00172 * f + 1.14

GoOn:
'For each element in a housing, the individual element permeate
       » flow is calculated as well as its contribution to the
       » total permeate salt content.

PresDrp = PresDrp + IntHsPsiDrp
IntHsPrm = (FdPsi - ConcConc * Osm - PresDrp) / DsgnPsi *
       »  DsgnElmPrm / TmpFctr
IF stg = TotStgs AND p = NmElmPrHsng THEN IntHsPrm1 = IntHsPrm
IF stg = 1 AND p = 1 THEN LgstIntHsPrm = IntHsPrm
PrmSlt = IntHsPrm * Passg * ConcConc * (DsgnPsi /
       » (FdPsi - PresDrp - ConcConc * Osm))
```

```
'A mass balance calculation is performed to determine the
       » individual element's concentrate concentration which
       » becomes the next element's feed concentration.

ConcConc = (IntHsConc * ConcConc - PrmSlt) /
       » (IntHsConc - IntHsPrm)
PrmSltot = PrmSltot + PrmSlt
ActPrm = ActPrm + IntHsPrm
IntHsConc = IntHsConc - IntHsPrm
NEXT p

ConcConc(stg) = ConcConc
PrmSlt(stg) = PrmSltot * HsngPr(stg)
HsngPrm(stg) = ActPrm * HsngPr(stg)
ConcFlw(stg) = IntHsConc * HsngPr(stg)

MinConcFlw = IntHsPrm / MaxElmRecov - IntHsPrm
IF MinConcFlw < MinElmDesFlw THEN MinConcFlw = MinElmDesFlw

IF (p = NmElmPrHsng + 1) AND (IntHsConc < .98 * MinConcFlw) THEN
         PresDrp = 0: ActPrm = 0: PrmSltot = 0
       Recycle =(MinConcFlw - IntHsConc) * HsngPr(stg) + Recycle
          GOTO DeltaPrpt
         END IF

stg = stg + 1
ActPrm = 0: PrmSltot = 0
IF HsngPr(stg) = 0 GOTO ChkCst
IntHsConc = ConcFlw(stg - 1) / HsngPr(stg)
GOTO PerStg

ChkCst:
Evaluate:
NmHouses = 0: TotSlt = 0: TotConc = 0

FOR rt = 1 TO 8
IF ConcConc(rt) > TotConc THEN TotConc = ConcConc(rt)
TotSlt = TotSlt + PrmSlt(rt)
ActlPrm = ActlPrm + HsngPrm(rt)
NmHouses = NmHouses + HsngPr(rt)
NEXT rt

'If the calculated permeate flow is more than 1.5% off from the
       » desired flow,an average permeate flow is used and the
       » calculations are repeated.  If the pressure drop is
       » more than 5% off an estimated value, the calculations
       » are repeated.
IF (ActlPrm < PrmFlw * .985) OR (ActlPrm > PrmFlw * 1.015) THEN
         PrmFlw = (ActlPrm + PrmFlw) / 2
        GOTO NewVar
        END IF
```

```
IF (PresDrp < EstPsiDrp * .95) OR (PresDrp > EstPsiDrp
       » * 1.05) THEN
           EstPsiDrp = (PresDrp + EstPsiDrp) / 2
          GOTO NewVar
          END IF

'The cost of the particular system is estimated based upon
       » its components.

IF Diam = 8 THEN
           HsCst = (2000 + NmElmPrHsng * 100) * NmHouses
           ELSE HsCst = (650 + NmElmPrHsng * 80) * NmHouses
          END IF
IF Diam = 8 AND Memb$ = "PA" THEN
           MmCst = NmHouses * NmElmPrHsng * 1665: GOTO AlmostDone
          END IF
IF Diam = 8 AND Memb$ <> "PA" THEN
           MmCst = NmHouses * NmElmPrHsng * 850: GOTO AlmostDone
          END IF
IF Memb$ = "PA" THEN
           MmCst = NmHouses * NmElmPrHsng * 650
           ELSE MmCst = NmHouses * NmElmPrHsng * 280
          END IF
AlmostDone:
Hp = FdPsi * ConcFlw(stg - 1)
PumpCst = -1.61E-18 * Hp ^ 4 + 5.77E-13 * Hp ^ 3 + 2.145E-07
       » * Hp ^ 2 + .0721 * Hp + 6480
IF Diam = 8 THEN
        SkdCst = 2500 + 1225 * (NmElmPrHsng - 2) + 100 * NmHouses
   ELSE SkdCst = 2500 + 1225 * (NmElmPrHsng - 2) + 20 * NmHouses
   END IF
InCst = 9.69E-08 * ConcFlw(stg - 1) ^ 4 - .000152 *
       » ConcFlw(stg - 1) ^ 3 - .0138 * ConcFlw(stg - 1) ^ 2 +
       » 108 * ConcFlw(stg - 1) + 3943
ActlConc = ActlPrm / Recov - ActlPrm
ActlRecyc = ConcFlw(stg - 1) - ActlPrm / Recov
ConcCst = -6.07E-08 * ActlConc ^ 4 - .0001223 * ActlConc ^
       » 3 - .0237 * ActlConc ^ 2 + 41.8 * ActlConc + 884
RecyCst = -.0000439 * ActlRecyc ^ 4 + .00342 * ActlRecyc ^
       » 3 + .641 * ActlRecyc ^ 2 - 5.96 * ActlRecyc + 743
PrmCst = 1.57E-06 * ActlPrm ^ 4 - .000517 * ActlPrm ^
       » 3 - .345 * ActlPrm ^ 2 + 65.85 * ActlPrm + 325
TotCst = MmCst + HsCst + PumpCst + SkdCst + InCst + ConcCst
       » + RecyCst + PrmCst + 5000
OPEN "SCRN:" FOR OUTPUT AS #2

PrtRslts:
CLS
PRINT #2,
PRINT #2,
PRINT #2,
PRINT #2,
```

```
PRINT #2,
PRINT #2, "American Fluid Technologies Reverse Osmosis Array
        » Design Program"
PRINT #2,
PRINT #2, "1) RO Membrane:               "; Memb$; "        7)
        » Desired Permeate Flow:"; PrmFlw2

PRINT #2, "2) RO Element Diameter:    "; Diam; "        8)
        » System Recovery: "; Recov

PRINT #2, "3) Elements per Housing:   "; NmElmPrHsng; " 9)
        » Solute Feed Concentration: "; FdConc2

PRINT #2, "4) Temperature               "; Temp; "     10)
        » Osmotic Pressure: "; Osm; "psi/ppm"

PRINT #2, "5) Max Element Recovery:   "; MaxElmRecov; "   11)
        » Solute Membrane Rejection: "; Rej

PRINT #2, "6) Desired Operating Psi: "; FdPsi;
PRINT #2,
PRINT #2,
FOR gp = 1 TO 7
PRINT #2, HsngPr(gp); "-";
NEXT gp
PRINT #2, HsngPr(8)
PRINT #2, "Concentrate Concentration: "; TotConc
PRINT #2, "Smallest Element Permeate Flow Rate: "; IntHsPrm1
PRINT #2, "Largest Element Permeate Flow Rate:   "; LgstIntHsPrm
PRINT #2,
PRINT #2, "Actual Feed Flow Rate       "; ActlPrm / Recov
PRINT #2, "Actual Permeate Flow Rate "; ActlPrm
PRINT #2, "Actual Recycle Flow Rate:           "; Recycle
PRINT #2,
PRINT #2, "Permeate Solute Concentration: "; TotSlt / ActlPrm
PRINT #2, "Overall Solute Rejection:        "; (FdConc2 - TotSlt /
        »  ActlPrm) / FdConc2
PRINT #2, "Differential Pressure:          "; PresDrp
PRINT #2,
PRINT #2, "Estimated Retail Cost of Reverse Osmosis System:
        » $"; TotCst
CLOSE #2
IF UCASE$(fg$) = "Y" THEN OPEN "SCRN:" FOR OUTPUT AS #2:
        » fg$ = "": GOTO PrtRslts
INPUT "Do you want a print-out [N] "; fg$
IF UCASE$(fg$) = "Y" THEN OPEN "LPT1:" FOR OUTPUT AS #2
        » GOTO PrtRslts

ReDo:
INPUT "Enter number of item to be altered, or just enter
        » to quit "; Try
IF Try = 0 AND i1 = 0 THEN END
```

```
IF Try = 1 THEN
     INPUT "RO Membrane (TFC, HR(CA), SR(CA)) "; Memb$
     Memb$ = UCASE$(Memb$)
     jl = 1
     GOTO ReDo
     END IF
IF Try = 2 THEN
     INPUT "Four Inch (4) or Eight Inch (8) Housing "; Diam
     jl = 1
     GOTO ReDo
     END IF
IF Try = 3 THEN
     INPUT "Number of Elements per Housing "; NmElmPrHsng
     jl = 1
     GOTO ReDo
     END IF
IF Try = 4 THEN INPUT "Temperature (F) "; Temp: jl = 1
       » GOTO ReDo
IF Try = 5 THEN INPUT "Maximum Element Recovery "; MaxElmRecov
       » jl = 1: GOTO ReDo
IF Try = 6 THEN INPUT "Desired Operating Pressure "; FdPsi
       » jl = 1: GOTO ReDo
IF Try = 7 THEN INPUT "Permeate Flow Rate "; PrmFlw
       » jl = 1: GOTO ReDo
IF Try = 8 THEN INPUT "System Recovery (fraction) "; Recov
       » jl = 1: GOTO ReDo
IF Try = 9 THEN
     INPUT "Feed Concentration of Primary Solute "; FdConc2
     jl = 1
     GOTO ReDo
     END IF
IF Try = 10 THEN
     INPUT "Osmotic Pressure of Primary Solute (psi/ppm) "; Osm
     jl = 1
     GOTO ReDo
     END IF
IF Try = 11 THEN
  INPUT "Membrane Rejection of Primary Solute (fraction) "; Rej
     jl = 1
     GOTO ReDo
     END IF
IF jl = 1 GOTO NewVar
END
```

Predicting Pressure Drop in Piping

Based upon the calculations given in Chapter 4, "Related Water Treatment Equipment," this program predicts the pressure drop through a piping loop using different water usage rates. Data is entered into the program from the beginning of the piping loop to the end. Pressure drop is calculated for the flowrate that occurs between nodes in the system.

A *node* is simply a point where some sort of flow or pipe change occurs. Piping diameters, fittings, usage points, and flowrates are entered in the program as required to define the nodes in the distribution piping system. Fittings are converted into relative straight pipe lengths, as a means of calculating a pressure drop over that equivalent piping length. This is performed over each node in the system. The various pressure drop values are then added to find the total pressure loss through the piping system.

The program will predict the pressure changes that occur when flow usage is minimal, average, or at peak usage conditions. It can be used to project the size of distribution pump required for the piping system. It will also give a good idea of the maximum pressure variation that may occur at the usage points.

```
CLS
DIM Flow(1 TO 55), opt(1 TO 55), dp(1 TO 55), d(1 TO 55), nl(1
        » TO 55), n90(1 TO 55), n45(1 TO 55), ntotl(1 TO 55),
        » mnfl(1 TO 55), avfl(1 TO 55), pkfl(1 TO 55),
        » Inflow(1 TO 55), mindptot(1 TO 55), Outflow(1 TO 55)
g = 1

PRINT "This program calculates pressure drop for straight PVC
        » Schedule 80 Piping."
PRINT "It will generate a table of 55 pressure drop values as a
        » function of flow."
PRINT
INPUT "Enter the gpm increment for the distribution flow
        » rate [0.2]"; increment
IF increment = 0 THEN increment = .2
INPUT "Enter the minimum flow rate to enter the loop";Miniflow

IDQ:
INPUT "Enter the pipe's inside diameter (inches)"; d(g)

nodes:
PRINT
INPUT "Enter how much straight pipe before next node - ft";
        »  nl(g)
INPUT "Enter the number of 90 degree elbows before next node";
        »  n90(g)
INPUT "Enter the number of 45 degree elbows before next node";
        »  n45(g)
L45 = .01309 * d(g) ^ 4 - .0322 * d(g) ^ 3 - .09652 *
        » d(g) ^ 2 + 1.557 * d(g) - .174
L90 = .1341 * d(g) - .3602 * d(g) ^ 3 - .9876 * d(g) ^
        » 2 + 5.857 * d(g) - 1.728
ntotl(g) = L45 + L90 + nl(g)

IF j = 1 GOTO complete
options:
PRINT
```

```
PRINT "Following are options for what occurs at the next node"
PRINT "1) A point-of-use occurs at this node."
PRINT "2) A filter or UF module is in-line."
PRINT "3) The inside diameter of the piping changes."
PRINT "4) The loop ends in a pressure regulator, throttled
       » valve, or open line."
PRINT
INPUT "Enter the number of the correct option"; opt(g)
ON opt(g) GOTO usepoint, filter, pipeid, valve

usepoint:
INPUT "Enter the minimum flow rate - gpm [0]"; mnfl(g)
INPUT "Enter the average flow rate - gpm [0]"; avfl(g)
INPUT "Enter the peak flow rate - gpm [0]"; pkfl(g)
d(g + 1) = d(g)
g = g + 1
GOTO nodes

filter:
INPUT "If a filter/UF is on the line, enter psi /
       » gpm ratio [0]"; nfrat(g)
g = g + 1
GOTO IDQ

pipeid:
g = g + 1
GOTO IDQ

valve:
INPUT "If there is a throttled valve, enter a ratio for
       » psid/gpm through it [0]"; nfrat(g)
IF nfrat(g) = 0 GOTO regulator
GOTO complete

regulator:
INPUT "Enter the psi at which the regulator is set, or 0
       » for an open line"; rgpsi

complete:
PRINT
PRINT "Enter the difference in height between where the
       » line opens to the tank"
INPUT "and the average height of water in the tank";
       » ht
max = g
g = 1

FOR f = 1 TO 3
FOR x = 1 TO 55
Inflow(x) = Miniflow + x * increment - increment
Flow(1) = Inflow(x)
FOR g = 1 TO max
```

```
Pressuredrp:
vel = Flow(g) * .408 / d(g) ^ 2

'k is a friction factor for PVC piping using very rough glue
       » lines which is typical.  For smoother piping, use a
       » proportionally smaller number for k.

k = .02024 - .01169 * d(g) + .002425 * d(g) ^ 2 + .0009241 *
       » d(g) ^ 3 - .0002939 * d(g) ^ 4
dp(g) = k * ntotl(g) * vel ^ 2 / d(g) * .263

usepointcalc:
IF f = 1 THEN Flow(g + 1) = Flow(g) - mnfl(g)
IF f = 2 THEN Flow(g + 1) = Flow(g) - avfl(g)
IF f = 3 THEN Flow(g + 1) = Flow(g) - pkfl(g)
mindptot(x) = mindptot(x) + dp(g)
NEXT g
Outflow(x) = Flow(max)
NEXT x
INPUT "Ready to print"; Q5$

printing:
LPRINT
LPRINT "      American Fluid Piping Pressure Drop Program"
IF f = 1 THEN LPRINT "      -  Minimum Flow Conditions -"
IF f = 2 THEN LPRINT "      -  Average Flow Conditions -"
IF f = 3 THEN LPRINT "      -  Peak Flow Conditions -"
LPRINT
LPRINT "Inlet FLOW - gpm Outlet Flow -gpm  Pressure Drop - psi"
FOR x = 1 TO 55
LPRINT Inflow(x); "                          "; Outflow(x);
       » ""; mindptot(x)

mindptot(x) = 0
NEXT x
NEXT f
END ■
```

Molecular Weight and $CaCO_3$ Conversion Factors for Ions in Water Treatment

Ion/Chemical	Chemical Formula	Molecular Weight	Substance to $CaCO_3$ Equiv.	$CaCO_3$ Equiv. to Substance Factor
Acetic Acid	$HC_2H_3O_2$	60.1	0.83	1.20
Aluminum	Al^{+++}	27.0	5.56	0.18
Ammonia	NH_3	17.0	2.94	0.34
Ammonium	NH_4^+	18.0	2.78	0.86
Barium	Ba^{++}	137.4	0.73	1.37
Bicarbonate	HCO_3^-	61.0	0.82	1.22
Calcium	Ca^{++}	40.1	2.50	0.4
Carbon Dioxide	CO_2	44.0	1.14	0.88
Carbonate	$CO_3^=$	60.0	0.83	1.20
Chloride	Cl^-	35.5	1.41	0.71
Chlorine	Cl_2	70.0	1.41	0.71
Copper	Cu^{++}	63.6	1.57	0.64
Iron (Ferric)	Fe^{+++}	55.8	2.69	0.37
Iron (Ferrous)	Fe^{++}	55.8	1.79	0.56
Fluoride	F^-	19.0	2.63	0.38
Hydrogen	H^+	1.0	50.0	0.02
Hydrogen Peroxide	H_2O_2	34.0	2.94	0.34
Hydrogen Sulfide	H_2S	34.1	2.93	0.34
Hydroxide	OH^-	17.0	2.94	0.34
Iodide	I^-	126.9	0.39	2.54
Iodine	I_2	253.8	0.39	2.54
Lead	Pb^{++}	207.	0.48	2.08
Magnesium	Mg^{++}	24.3	4.10	0.24
Manganese (Manganic)	Mn^{+++}	54.9	2.73	0.37
Manganese (Manganous)	Mn^{++}	54.9	1.82	0.55
Nitrate	NO_3^-	62.0	0.81	1.24
Peracetic Acid	$HC_2H_3O_3$	76.1	0.66	1.52
Phosphate	PO_4^{-3}	95.0	1.58	0.63
Potassium	K^+	39.1	1.28	0.78
Silica	SiO_2	60.1	0.83	1.20
Sodium	Na^+	23.0	2.18	0.46
Sulfate	SO_4	96.1	1.04	0.96
Sulfide	$S^=$	32.1	3.13	0.32
Water	H_2O	18.0	5.56	0.18

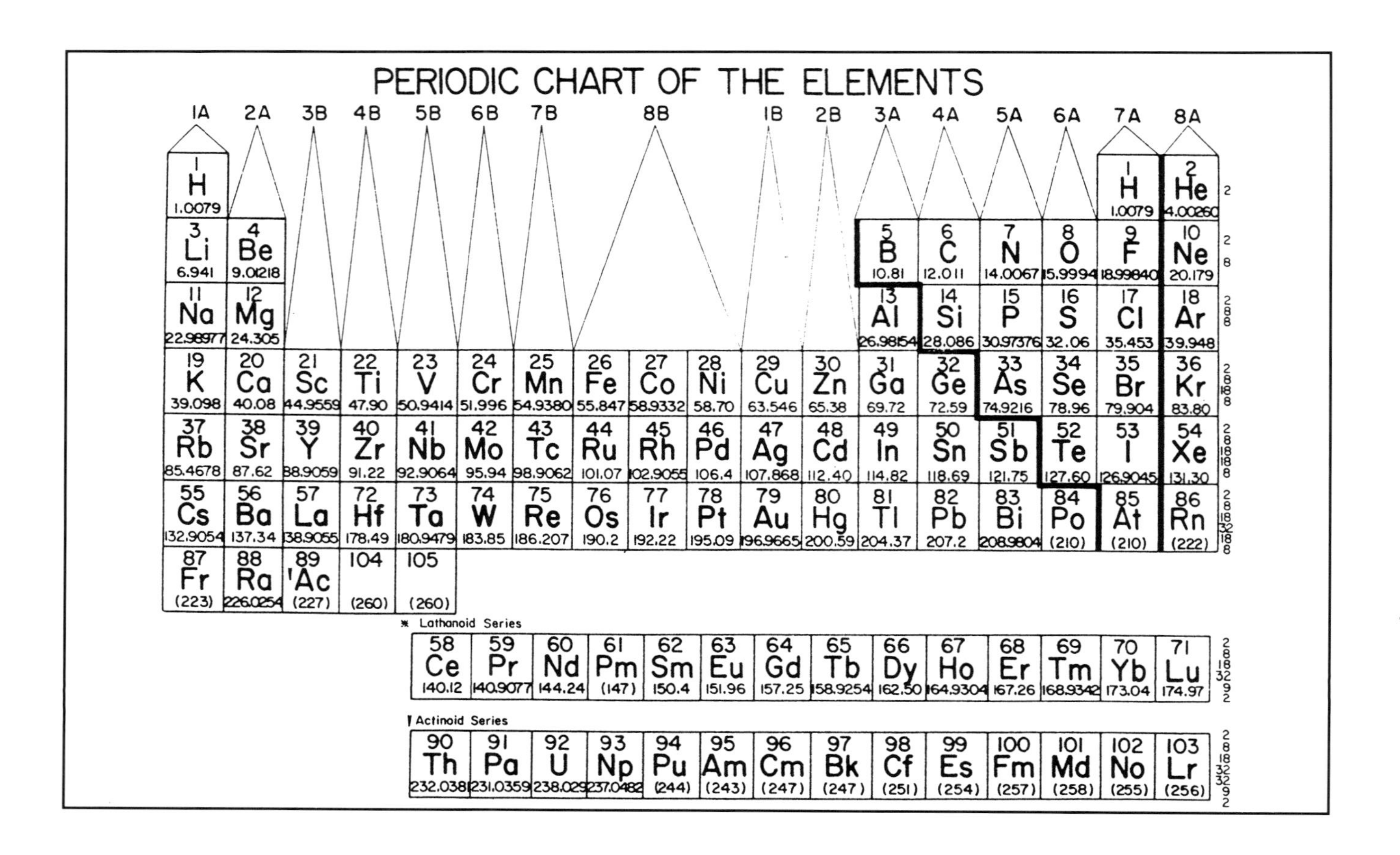
PERIODIC CHART OF THE ELEMENTS
1A 2A 3B 4B 5B 6B 7B 8B 1B 2B 3A 4A 5A 6A 7A 8A
1 H 1.0079
2 He 4.00260
3 Li 6.941
4 Be 9.01218
5 B 10.81
6 C 12.011
7 N 14.0067
8 O 15.9994
9 F 18.99840
10 Ne 20.179
11 Na 22.98977
12 Mg 24.305
13 Al 26.98154
14 Si 28.086
15 P 30.97376
16 S 32.06
17 Cl 35.453
18 Ar 39.948
19 K 39.098
20 Ca 40.08
21 Sc 44.9559
22 Ti 47.90
23 V 50.9414
24 Cr 51.996
25 Mn 54.9380
26 Fe 55.847
27 Co 58.9332
28 Ni 58.70
29 Cu 63.546
30 Zn 65.38
31 Ga 69.72
32 Ge 72.59
33 As 74.9216
34 Se 78.96
35 Br 79.904
36 Kr 83.80
37 Rb 85.4678
38 Sr 87.62
39 Y 88.9059
40 Zr 91.22
41 Nb 92.9064
42 Mo 95.94
43 Tc 98.9062
44 Ru 101.07
45 Rh 102.9055
46 Pd 106.4
47 Ag 107.868
48 Cd 112.40
49 In 114.82
50 Sn 118.69
51 Sb 121.75
52 Te 127.60
53 I 126.9045
54 Xe 131.30
55 Cs 132.9054
56 Ba 137.34
57 La 138.9055
72 Hf 178.49
73 Ta 180.9479
74 W 183.85
75 Re 186.207
76 Os 190.2
77 Ir 192.22
78 Pt 195.09
79 Au 196.9665
80 Hg 200.59
81 Tl 204.37
82 Pb 207.2
83 Bi 208.9804
84 Po (210)
85 At (210)
86 Rn (222)
87 Fr (223)
88 Ra 226.0254
89 'Ac (227)
104 (260)
105 (260)
* Lathanoid Series
58 Ce 140.12
59 Pr 140.9077
60 Nd 144.24
61 Pm (147)
62 Sm 150.4
63 Eu 151.96
64 Gd 157.25
65 Tb 158.9254
66 Dy 162.50
67 Ho 164.9304
68 Er 167.26
69 Tm 168.9342
70 Yb 173.04
71 Lu 174.97
' Actinoid Series
90 Th 232.0381
91 Pa 231.0359
92 U 238.029
93 Np 237.0482
94 Pu (244)
95 Am (243)
96 Cm (247)
97 Bk (247)
98 Cf (251)
99 Es (254)
100 Fm (257)
101 Md (258)
102 No (255)
103 Lr (256)

Temperature Correction Factors for RO Membrane Systems

Permeate Flow × Correction Factor = Permeate Flow @ 77 °F

°F	Divisor	Factor	°F	Divisor	Factor
35	.52	1.92	63	.82	1.22
36	.53	1.89	64	.83	1.2
37	.54	1.85	65	.85	1.18
38	.55	1.82	66	.86	1.16
39	.56	1.78	67	.87	1.15
40	.57	1.75	68	.88	1.14
41	.58	1.72	69	.89	1.12
42	.59	1.69	70	.9	1.11
43	.6	1.67	71	.91	1.1
44	.61	1.64	72	.93	1.08
45	.62	1.61	73	.94	1.06
46	.63	1.59	74	.95	1.05
47	.64	1.56	75	.97	1.03
48	.65	1.53	76	.98	1.02
49	.67	1.49	77	1	1
50	.68	1.47	78	1.01	.99
51	.69	1.45	79	1.02	.98
52	.7	1.43	80	1.04	.96
53	.71	1.41	81	1.05	.95
54	.72	1.39	82	1.06	.94
55	.73	1.37	83	1.07	.93
56	.74	1.35	84	1.08	.92
57	.75	1.33	85	1.1	.91
58	.76	1.31	86	1.11	.9
59	.77	1.29	87	1.12	.89
60	.78	1.27	88	1.14	.88
61	.8	1.25	89	1.15	.87
62	.81	1.23	90	1.17	.86

Conversion of U.S. to Other Units

Units	*Parts $CaCO_3$ per million (ppm)*	*Grains $CaCO_3$ per U.S. Gallon (grpg)*	*English degrees or Clark*	*French degrees*	*German degrees*	*Milliequivalents per liter meq/million*
1 part per million	1.	0.0583	0.07	0.1	0.0560	0.20
1 Grain per U.S. Gallon	17.1	1.	1.2	1.71	0.958	0.343
English or Clark degree	14.3	0.833	1.	1.43	0.800	0.286
1 French degree	10.	0.583	0.7	1.	0.560	0.20
1 German degree	17.9	1.04	1.24	1.79	1.	0.357
1 Milliequivalent/L	50.0	2.92	3.50		2.80	1.

Conversion Units and Equivalents

Water analysis units	*Parts per million (ppm)*	*Milligrams per liter (mg/L)*	*Grams per liter (g/L)*	*Grains U.S. gallon (grs/U.S. gal)*	*Grains British Imp. gallon*	*Kilograins per cubic foot (Kgr/ft^3)*
1 Part per million	1.	1.	0.001	0.583	0.07	0.0004
1 Milligram per liter	1.	1.	0.001	0.583	0.07	0.0004
1 Gram per liter	1000.	1000.	1.	68.3	70.0	0.436
1 Grain per U.S. gallon	17.1	17.1	0.017	1.	1.2	0.0075
1 Grain per British Imp. gal	14.3	14.3	0.014	0.833	1.	0.0062
1 Kilograin per cubic foot	2294.	2294.	229.4	134.	1.	

1.0 milligram/Liter (mg/L) = 1.0 part per million 10^{-3} g/L
1.0 nanogram/Liter (ng/L) = 1.0 part per trillion 10^{-9} g/L
1.0 microgram/Liter (#g/L) = 1.0 part per billion 10^{-6} g/L

Useful Calculations.

1 gallon of water weighs approximately 8.33 pounds.
1 gallon = 231 cubic inches
1 gallon = 3.785 liters
1 gallon = 3785 milliliters
1 cubic foot of water weighs approximately 62.4 pounds
1 cubic foot = 7.48 gallons
1 pound = 7000 grains
1 pound = 453.6 grams
1 gram = 15.43 grains
1 gram per liter = 58.41 grains per gallon
1 gram per liter = 1000 ppm
0.0171 grams per liter = 1 grain per gallon
0.0038 grams per gallon = 1 ppm
1 grain per gallon = 17.1 ppm
1 grain per gallon = 142.9# per million gallons
1 milligram per liter = 1 ppm
1 ppm = 0.058 grains per gallon
8.33 pounds/1,000,000 gallons = 1 ppm
10,000 ppm - 1%

To Convert from	to	Multiply by
Capacity		
Kgrs/ft.^{3} (as $CaCO_3$)	g CaO/l	1.28
Kgrs/ft.^{3} (as $CaCO_3$)	g $CaCO_3$/l	2.29
Kgrs/ft.^{3} (as $CaCO_3$)	eq/l	0.0458
g $CaCO_3$/l	Kgrs/ft^3 (as $CaCO_3$)	0.436
g CaO/l	Kgrs/ft^3 (as $CaCO_3$)	0.780
Flow Rate		
U.S. gpm/ft.^{3}	BV/hr	8.02
U.S. gpm/ft.^{2}	m/hr	2.45
U.S. gpm	m^3/hr	.227
BV/min	U.S..gpm/ft^3	7.46
Pressure Drop		
psi/ft.	mH_2O/m resin	2.30
	g/cm^2/m	230
Regenerant Concentration		
lbs/ft.^{3}	g/l	16.0
Density		
lbs/ft.^{3}	g/l	16.0
Rinse Requirements		
U.S. gal/ft.^{3}	BV	0.134

MEMBRANE MAINTENANCE

WHAT IS FOULING MY MEMBRANE SYSTEM?

Common Foulants

The most common membrane foulants and scale can be grouped into categories. These guidelines can be used to determine which cleaning solutions should be used in the initial attempt to remove the suspected foulant. If a particular cleaning solution is only minimally effective, it is likely that the foulant is not what was suspected. A different cleaning solution should be used.

Carbonate Scale

Calcium and magnesium carbonate solubilities are exceeded in most reverse osmosis systems. They can quickly precipitate if the pretreatment system fails at preventing this precipitation. Carbonate scale formation will tend to occur in the membrane elements that are exposed to higher salt concentrations prior to the water exiting in the concentrate stream. These elements will experience a decrease in their normalized permeate flow rate and a decrease in their salt rejection. This is most likely to occur in systems relying solely on the injection of a scale inhibitor to prevent carbonate formation, or in systems whose pretreatment softeners are malfunctioning.

Iron/Manganese

Iron and manganese can foul membrane systems if the dissolved metal is oxidized by chlorine or chloramine in the feed water, or the metal can be oxidized if the water comes into contact with air. Concentrations as low as 0.05 ppm can foul a membrane system, causing an increase in the normalized differential pressùre and possibly a decrease in the normalized permeate flow rate. The presence of oxidized iron will be visibly apparent as rust in the system prefilters. Usually if manganese is present, iron will also be present.

Sulfates or Silica

Some water sources are high in sulfate or silica concentrations. The potential for their precipitation can be predicted using a water analysis. If precipitation occurs, it will first affect the performance of the downstream end membrane elements where the concentration of salts are their highest. Heavy sulfate and silica fouling can be extremely difficult to clean. Specialized cleaning solutions are available from for cleaning sulfate and silica scale.

Suspended Solids and Silt

Organic suspended solids or silt should be suspected as the membrane foulant if a loss in normalized permeate flow rate occurs with a membrane system that is operating on a surface water source. A surface water originates in a lake, river, or reservoir. It typically contains a high concentration of organic suspended solids and silt that can readily foul a membrane system. A Silt Density Index (SDI) measurement greater than 3 or a turbidity

* Courtesy, American Fluid Technologies, Inc., Wayzata, MN.

reading greater than 1 would indicate a greater potential for fouling by suspended solids.

Biological Activity

Bacteria and other types of biological activity can readily foul a membrane system. Such activity is more likely if a biocide such as chlorine or chloramine is not present in the RO feed water. Since these must be removed upstream of PA thin -film systems, biological fouling is a common problem with these systems. Such fouling can cause an increase in the normalized differential pressure or a decrease in the normalized permeate flow rate. Standard (heterotrophic) bacteria counts can be used to check for high activity in the membrane system. Bacteria counts in excess of 100/m1 are considered excessive.

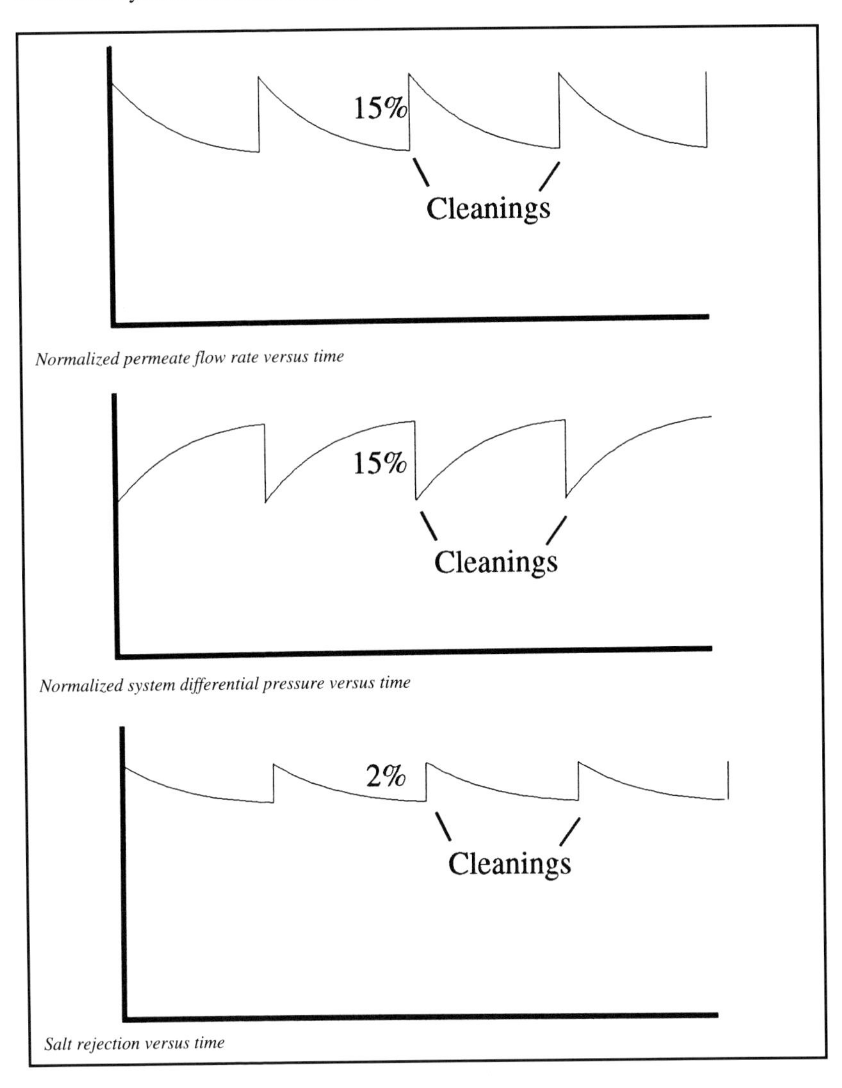

Normalized permeate flow rate versus time

Normalized system differential pressure versus time

Salt rejection versus time

Iron Bacteria

Iron bacteria consists of bacteria strains that live on the walls of iron or stainless steel pipe and fittings. They have the capability of dissolving the iron from the pipe or fittings using an acidic environment. They create a chemically-resistant outside layer around the colonies that can impede the removal of the bacteria. Heavy iron bacteria is visibly apparent as a brown slime that can quickly cause an increase in the system's normalized differential pressure. Since iron bacteria get their iron from pipe and fittings, it may be present even if incoming iron concentrations are low.

WHEN SHOULD I CLEAN?

Cleaning Frequency

There are certain types of scale that pose an immediate threat to the integrity of a membrane system. Key examples include calcium carbonate scale in a cellulose acetate (CA) membrane system and silica or sulfate scale in any type of membrane system. Calcium carbonate can create localized areas of high pH near the surface of the membrane which will result in an increased rate of CA membrane hydrolysis. It should be cleaned immediately if its presence is suspected. Silica and sulfate scale should also be cleaned immediately due to the difficulty in cleaning these particular types of scale.

As a general rule, if a CA membrane system's pretreatment does not include acid injection as a means of preventing calcium carbonate formation, calcium carbonate scale should be suspected if the membrane performance changes. It should be cleaned immediately.

If the solubility of silica or any of the sulfate scales are being exceeded within the membrane system, these scales should be suspected if the membrane performance changes. In this situation, the system should also be cleaned immediately.

Other types of fouling or scale formation do not pose an immediate threat to the membrane system. The following guidelines can be used with the normalized performance data to determine the maximum fouling to allow prior to cleaning the system:

10-15% Decrease in the Normalized Permeate Flow Rate

or

1-15% Increase in the Normalized System Differential Pressure

or

1-2% Decrease in the Salt rejection

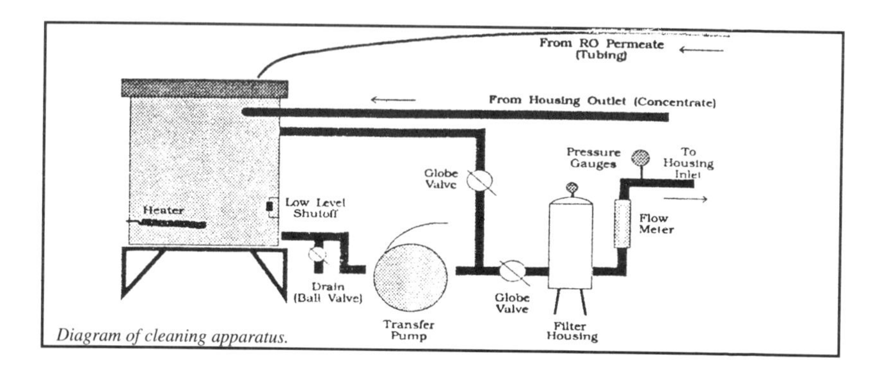

Diagram of cleaning apparatus.

HOW DO I CLEAN

Cleaning -in- Place (CIP)

1. Drain as much water from the RO/UF system as possible.

2. Fill cleaning tank with 5 gallons RO permeate water for each 4" x 40" membrane element to be cleaned, 20 gallons for each 8" x 40" element.

3. Add the recommendedcentration of cleaning chemical. Insure that the cleaner is completely dissolved by mixing the solution or by recirculating the solution around the cleaning tank with the transfer pump.

4. If possible, heat the solution to improve cleaning effectiveness. Do not exceed 110 °F or the membrane manufacturer's temperature guidelines.

5. Completely open the RO/UF concentrate throttling valve to minimize operating pressure during cleaning. Recirculate the solution in the normal direction of flow through the RO/UF system and back to the cleaning tank. Throttle the transfer pump to achieve the maximum individual housing flow rate without exceeding the membrane manufacturer's guidelines. If guidelines are unavailable, the following may be used (for spiral wound membrane systems):

Membrane Element Diameter	*Maximum Flow/Housing*
4"	10 GPM
8"	40 GPM

NOTE: Cleaning effectiveness can be improved by isolating membrane stages during cleaning. For example, cleaning the leading stage of a 2 -l array (the first two housings) and then cleaning the last housing in the array will make it possible to achieve maximum flow in all housings.

6. Recirculate for a minimum of 30 minutes. Additional recirculation and soaking can be used to improve cleaning effectiveness.

7. Discharge the cleaning solution to drain. Rinse the remaining solution out of the system at low pressure using the transfer pump or the normal system feed water. (Note: Neutralization of spent solutions may be required at your facility. Check with local environmental regulatory authorities.)

8. Return the concentrate throttling valve to its normal position but divert the RO/UF permeate water to drain until conductivity returns to normal.■

BIBLIOGRAPHY

AAMI (Association for the Advancement of Medical Instrumentation), *Standards--Hemodialysis Systems,* American National Standards Institute, March 16, 1992.

American Fluid Technologies, Wayzata, MN, not previously published.

American Iron and Steel Institute.

[Cited p 8-8. Complete info needed including year and title.]

Amjad, Z.; Workman, K.R.; Castete, D.R. "Considerations in Membrane Cleaning", pp 210-236. in *Reverse Osmosis: Membrane Technology, Water Chemistry, and Industrial Applications.* Amjad, Z., Ed., Van Nostrand Reinhold, New York (1993).

Ammerer, N.H. "Controlling TOC Effluent Levels from Newly Regenerated Resin", *Ultrapure Water 6(9),* pp. 34-40 (December 1989).

"Aquafine Owner's Manua,l" Aquafine Corp., , Valencia, CA (July 1992).

Aqua Media Ltd., Sunnyvale, CA (1981).

Argo, D.G. "Water Reuse: Where Are We Headed?", *Environmental Science and Technology 19,* pp 208-214 (1985).

Argo, D.G.; Ridgway, H.F. "Biological Fouling of Reverse Osmosis Membranes at Water Factory 21", *Proceedings of 10th Annual Conference and Trade Fair of the Water Supply Improvement Association,* Vol. 3, Honolulu, HI (July 25-29, 1982).

Aronovich, H. Personal communication (April 1993).

"Fourier Transform Infrared Specrtoscopy," AT&T Analytical Service Literature, Allentown, PA (1990).

Bauer, Robert C., and Snoeyink, Vernon L. "Reactions of Chloramines with Active Carbon," Journal WRCF, November, 1973.

Bernardin, F.E., Jr. "Granular Activated Carbon Operation: Problems and Solutions", QED Corporation, Ann Arbor, MI (1976).

Beseman, S. Personal communication (May 1993).

Betz, "Handbook of Industrial Water Conditioning," pp 55-56, (1982).

Borenstein, S.W. "Microbiologically Influenced Corrosion Failures of Austenitic Stainless Steel Welds", Corrosion/88, St. Louis, MO (August, 1988).

Brandt, D.C.; Leitner, G.F.; Leitner, W.E. "Reverse Osmosis Membranes State of the Art", pp 1-36 in *Reverse Osmosis: Membrane Technology, Water Chemistry, and Industrial Applications.* Amjad, Z., Ed., Van Nostrand Reinhold, New York (1993).

Brinda, P., Personal Communication (1993).

Bryers, J.D. "Bacterial Biofilms", *Curr. Opinion in Biotechnology, 4,* pp 197-204 (1993).

Byrne, W.; Bukay, M. "How to Monitor a Reverse Osmosis System", *Ultrapure Water 1(2),* pp 32-36 (September/October 1984).

Byrne, W.; Bukay, M."RO Troubleshooting, Part 1: Isolating the Location of a Decline in Salt Rejection", *Ultrapure Water 3(2),* pp 58-62 (March/April 1986).

Byrne, W.; Bukay, M. "RO Troubleshooting, Part 5: Causes and Prevention of an Increase in Differential Pressure", *Ultrapure Water 3(6),* pp 56-60 (November/ December 1986).

Byrne, W.; Bukay, M. "RO Troubleshooting, Part 2: Causes and Prevention of Front-End Salt Rejection Decline", *Ultrapure Water 3(3),* pp 59-62 (May/June 1986).

Byrne, W.; Bukay, M. "RO Troubleshooting, Part 3: Causes and Prevention of Tail-End Salt Rejection Declines", *Ultrapure Water 3(4),* pp 59-62 (July/August 1986).

Byrne, W.; Bukay, M. "RO Troubleshooting, Part 4: Causes and Prevention of Isolated and Uniform Salt Rejection Decline", *Ultrapure Water 3(5),* pp 58-62 (September/ October 1986).

Byrne, W. "Continuous Biocide Injection in Thin-Film RO", *Technology News* (December 1990), Volume III pp 1-2.

Byrne, W. "Tips from Our Readers: Calculations to Assist in Evaluating RO Performance Data", *Ultrapure Water 2(2),* pp 43-44 (March/April 1985).

Byrne, W.; Roesner, M. "An Alternative Biocide for Disinfection of Thin-Film Composite Membranes: Impact on the Pure Water Industry", *Proceedings, International Membrane Conference,* Boston (1990).

Carling, J.B. "Controlling Pretreatment Coagulants with Streaming Current Technology", *Ultrapure Water 6(6),* pp. 43-47 (September 1989).

Cloutier, M., Liquid Metronics Incorporated. Personal conversation (June 1993).

Coccagna, L. "Filtration: Theoretical Considerations, Practical Results", *Culligan Technology,* (House Publications), (summer and fall 1984).

Cohen, David, Personal communication (March, May, July and October 1993).

Collentro, W.V. "Contributions from Our Readers: Use of Progressive Piping", *Ultrapure Water 2(4),* pp 45-46 (July/August 1985).

Comb, L., Osmonics, Inc. Personal conversation (September 1986).

Comb, L., Osmonics, Inc. Personal conversation (May 1993).

Combs, R.F.; Veloz, T.M. "Fabrication of Stainless Steel for High Purity Water Systems", *Ultrapure Water 7(3),* pp. 54-57 (April 1990).

Connors, D. Personal communication (September 1993).

Continental Water Systems Corporation, San Antonio, TX.

Costerton, J.W.; Cheng, K.J.; Geesey, G.G.; Ladd, T.I.; Nickel, J.C.; Dasgupta, M.; Marrie, T.J. "Bacterial Biofilms in Nature and Disease", *Annual Review of Microbiology 41,* pp 435-464 (1987).

Costerton, J.W., Marrie, T.J.; Cheng, K.J. "Phenomena of Bacterial Adhesion", pp. 3-43 in *Bacterial Adhesion: Mechanisms and Physiological Significance.* Savage, D.C.;

Fletcher, M., Eds., Plenum Press, Inc., New York and London (1985).

Cowan, J.; Weintriff, D. *Water-Formed Scale Deposits.* Gulf Publishing Company (1976).

Crane Co., King of Prussia, PA.

Cubicciotti, D.; Licina, G.J. "Electrochemical Aspects of Microbially Induced Corrosion", *MP,* (January 1990).

DeJohn, P.B. "Dechlorination with Granular Activated Carbon", BASSA Symposium/ Workshop, Berkeley, CA (March 4, 1976).

Dorr-Oliver.

Dow Chemical Co., Filmtec Products.

DuPont Company, Permasep Products, Wilmington, DE. *Bulletin 502*, pp 15, 16.

DuPont Company, Permasep Engineering Manual, Wilmington, DE (December 1, 1982).

Durham, L. "Biological Growth Control Problem in RO Systems", *Ultrapure Water 6(6),* pp 30-37 (September 1989).

Filteau, Gerry, Personal communication (January, April 1993).

FilterCor, *Product Literature.*

Flemming, H.-C. "Biofouling in Water Treatment", pp 47-80 in *Biofouling and Biocorrosion in Industrial Water Systems.* Flemming, H.-C.; Geesey, G.G., Eds., Springer-Verlag, Berlin (1991).

Flemming, H.-C. "Mechanistic Aspects of Reverse Osmosis Membrane Biofouling and Prevention", pp 163-209 in *Reverse Osmosis: Membrane Technology, Water Chemistry, and Industrial Applications.* Amjad, Z., Ed., Van Nostrand Reinhold, New York (1993).

Flemming, H.-C.; Schaule, G. "Biofouling on Membranes - a Microbiological Approach", *Desalination 70,* pp 95-119 (1989).

Flemming, H.-C.; Schaule, G. "Investigations on Biofouling of Reverse Osmosis and Ultrafiltration Membranes: Part I: Initial Phase of Biofouling", *Vom Wasser 71*, pp 207-223 (1988b).

Flow Tech Industries.

Fluid Systems Corporation. Product Literature: "Reverse Osmosis Principles and Applications", p. 13 (October 1970).

Fulford, K. Personal communication (May 1990, March 1993).

Fulford, K. Personal communication (June 1993).

Gaso, Product Literature.

Geesey, G.G. "Microbial Exopolymers: Ecological and Economic Considerations", *ASM News 48,* pp 9-14 (1982.)

Gekas, V.; et al., "Food and Dairy Applications: the State of the Art", *Desalination 53,* pp 95-127 (1985).

Graber Olive House, not previously published.

Greenkorn, R.A.; Kessler, D.P. *Transfer Operations.* McGraw-Hill, New York (1972).

Hillestad, Jill, personal communications (1994).

Holland, D., Dalton Scientific, Inc. Personal conversation (July 1993).

Horizon Engineers, product Literature.

Hydranautics, Inc., Product Literature.

Isner, J.D.; Williams, R.C. "Analytical Techniques for Identifying Reverse Osmosis Foulants", B.F. Goodrich. pp 227-274 in *Reverse Osmosis: Membrane Technology, Water Chemistry, and Industrial Applications.* Amjad, Z., Ed., Van Nostrand Reinhold, New York (1993).

Iverson, D. Personal communication (April 1993).

Johnson, R., Minntech Corporation. Personal communication (May 1993).

Kates, P., DuPont Company, Personal communication (June 1993).

Koseoglu, S.S.; Guzman, G.J. "Applications of Reverse Osmosis Technology in the Food Industry", pp 300-333 in *Reverse Osmosis: Membrane Technology, Water Chemistry, and Industrial Applications.* Amjad, Z., Ed., Van Nostrand Reinhold, New York (1993).

Kraft, R.L. "Elimination of Degasifiers from High Purity Water Systems", *Ultrapure Water 2(2),* pp 20-27 (March/April 1985).

LeChevallier, M.W. "Biocides and the Current Status of Biofouling Control in Water Systems", pp 113-132 in *Biofouling and Biocorrosion in Industrial Water Systems.* Flemming, H.-C.; Geesey, G.G., Ed., Springer-Verlag, Berlin (1991).

Lepore, J.V.; Ahlert, R.C. "Fouling in Membrane Processes", pp 141-184 in *Reverse Osmosis Technology: Application for High-Purity Water Production.* Parekh, B.S., Ed., Marcel Dekker, Inc., New York and Basel (1988).

Light, William, Personal communication (March 1993).

Liquid Metronics Inc., Product Literature

Luss, G. Personal communication (April 1993).

Luss, G. Personal communication (May 1993).

Luss, G. Personal communication (June 1993).

Maltais, J.; Stern, T. "A Superior Biocide for Disinfecting Reverse Osmosis Systems", Minntech Corporation, Minneapolis, MN (1989).

Marshall, K.C. "Mechanisms of Bacterial Adhesion at Solid-Water Interfaces", pp 133-162 in *Bacterial Adhesion: Mechanisms and Physiological Significance.* Savage, D.C.; Fletcher, M., Eds., Plenum Press, New York and London (1985).

Marshall, K.C.; Blainey, B.L. "Role of Bacterial Adhesion in Biofilm Formation and Biocorrosion", pp 29-46 in *Biofouling and Biocorrosion in Industrial Water Systems.* Flemming, H.-C.; Geesey, G.G., Eds., Springer-Verlag, Berlin (1991).

Martinson, T., Advanced Structures Inc., Personal communication (August 1993).

McPherson, W.H. Personal communication (March, April, May 1981).

McPerson, W.H.; Bedford, D. "RO as a Final Filtration Process in High Purity Water Production", *Ultrapure Water 3(1),* pp 20-26 (January/February 1986).

McPherson, W.H., "The Control of Fouling in Reverse Osmosis Systems, Two Case Histories", Tenth Annual Water Supply Improvement Association (July 1982).

Means, E.G.; Hanami, L.; Ridgway, H.F.; Olson, B.H. "Evaluating Mediums and Plating Techniques for Enumerating Bacteria in Water Distribution Systems", *Journal of the American Water Works Association 73,* pp 119-124 (1981).

Michaud, C. Personal communication (March, June 1993).

Millipore Corp., Product Literature.

Mills, W.R. "Groundwater Recharge Success", *Water Environmental Technology 5,* pp 40-44 (1993).

Mohr, C.M.; Leeper, S.A.; Engelgau, D.E.; Charbonezu, B.L. "Membrane Applications and Research in Food Processing: An Assessment", EG&G Idaho Inc., DOE/ID/10210 (August 1988).

Mukhopadhyay, D. Personal communication (March 1993).

Nakagome, K.; Brady, M.F. "Membrane Modules for Producing High Purity Water", *Ultrapure Water 3(3),* pp 34-38 (May/June 1986).

National Electrical Code.

Nusbaum, I.; Argo, D.G. "Design, Operation, and Maintenance of a 5-mdg Wastewater Reclamation Reverse Osmosis Plant", pp 377-436 in *Synthetic Membrane Processes, Fundamental and Water Applications.* Belfort, G., Ed., Academic Press, Inc., Harcourt Brace Jovanovich, New York (1984).

Osmonics Inc., Minnetonka, MN, Product Literature.

Osmundson, P., Osmonics, Inc. Personal communication (August 1993).

Patterson, M.K.; Husted, G.R.; Rutkowski, A.; Mayette, D.C. "Isolation, Identification, and Microscopic Properties of Biofilms in High-Purity Water Distribution Systems", *Ultrapure Water 8(4),* pp 18-24 (May/June 1991).

Paul, J.H.; Jeffrey, W.H. "The Effect of Surfactants on the Attachment of Estuarine and Marine Bacteria to Surfaces", *Canadian Journal of Microbiology 31,* pp 224-228 (1984).

Paulson, D.J.; Wilson, R.L.; Spatz, D.D. "Reverse Osmosis and Ultrafiltration Applied to the Processing of Fruit Juices", ACS Symposium Series, 281:325-344 (1985).

Permasep Engineering Manual, Bulletin #304, p 4. DuPont Company, Polymer Products Department, Permasep Products, Wilmington, DE (1982).

Peterson, R.J. "The Expanding Roster of Commercial Reverse Osmosis Membranes", Filmtec Corporation (1986).

Ray, R.J., "Energy-Efficient Membrane Separations in the Sweetner Industry: Final Report", Bend Research, Inc. DOE/ID/1245-1 (February 14, 1986).

Reasoner; Geldreich, Personal communication. (1985).

Ridgway, H.F. "Microbial Fouling of Reverse Osmosis Membranes: Genesis and Control", pp 138-193 in *Biological Fouling of Industrial Water Systems: A Problem Solving Approach.* Mittelman, M.W.; Geesey, G.G., Eds., Water Micro Associates, San Diego, CA (1987).

Ridgway, H.F. "Microbial Adhesion and Biofouling of Reverse Osmosis Membranes", pp 429-481 in *Reverse Osmosis Technology: Application for High-Purity Water Production.* Parekh, B.S., Ed., Marcel Dekker, Inc., New York and Basel (1988).

Ridgway, H.F.; Safarik, J. "Biofouling of Reverse Osmosis Membranes", pp 81-111 in *Biofouling and Biocorrosion in Industrial Water Systems.* Flemming, H.-C.; Geesey, G.G., Eds., Springer-Verlag, Berlin (1991).

Ridgway, H.F.; Argo, D.G.; Olson, B.H. *Factors Influencing Biofouling of Reverse Osmosis Membranes at Water Factory 21: Chemical, Microbiological, and Ultrastructural Characterization of the Fouling Layer,* Vol. III-B. Final report prepared for U. S. Department of the Interior, Office of Water Research and Technology, contract No. 14-34-0001-8520, Washington, D.C. (1981).

Ridgway, H.F.; Justice, C.A.; Whittaker, C.; Argo, D.G.; Olson, B.H. "Biofilm Fouling of RO Membranes — Its Nature and Effect on Treatment of Water for Reuse", *Journal American Water Works Association 79,* pp 94-102 (1984a).

Ridgway, H.F.; Kelly, A.; Justice, C.A.; Olson, B.H. "Microbial Fouling of Reverse-Osmosis Membranes Used in Advanced Wastewater Treatment Technology: Chemical, Bacteriological, and Ultrastructural Analyses", *Applied and Environmental Microbiology 45,* pp 1066-1084 (1983).

Ridgway, H.F.; Rigby, M.G.; Argo, D.G. "Adhesion of a Mycobacterium Sp. to Cellulose Diacetate Membranes Used in Reverse Osmosis", *Applied and Environmental Microbiology 47,* pp 61-67 (1984b).

Ridgway, H.F.; Rigby, M.G.; Argo, D.G. "Biological Fouling Of Reverse Osmosis Membranes: The Mechanism Of Bacterial Adhesion", pp. 1314-1350 in *Proceedings of Water Reuse Symposium III, The Future of Water Reuse,* Volume 3, August 26-31, 1984, San Diego, CA, published by AWWA Research Foundation, Denver, CO (1984c).

Ridgway, H.F.; Rigby, M.G.; Argo, D.G. "Bacterial Adhesion and Fouling of Reverse Osmosis Membranes", *Journal of the American Water Works Association 77,* pp 97-106 (1985).

Ridgway, H.F.; Rodgers, D.M.; Argo, D.G. "Effect of Surfactants on the Adhesion of Mycobacteria to Reverse Osmosis Membranes", *Proceedings of the 5th Annual Semiconductor Pure Water Conference,* pp 133-164, San Francisco, CA (January 16-17, 1986).

Rohm and Haas Company, Philadelphia, PA. "Engineering Manual for the Amberlite® Ion Exchange Resins."

Safarik, J.; Williams, J.; Ridgway, H.F. "Analysis of Biofilm from Reverse Osmosis Membranes by Computer-Programmed Polyacrylamide Gel Electrophoresis", *Proceedings of the American Society for Microbiology,* New Orleans, LA (May 14-18, 1989).

Seelye Plastics, Inc., Product Literature

Sheu, M.J.; Wiley, R.C. "Preconcentration of Apple Juice by RO", *Journal of Food Science 48,* p 422 (1983).

Shurtleff, A. Personal communication (June 1993).

Snoeyink, V.L.; Suidan, M.T. "Dechlorination by Activated Carbon and Other Reducing Agents", Chapter 16 in *Disinfection: Water and Wastewater,* Johnson, J.D., Ed., Ann Arbor Science Publishers, Inc., Ann Arbor, MI (1975).

Stiff; Davis. (1952).

Stumm, W.; Morgan, J.J. *Aquatic Chemistry: An Introduction Emphasizing Chemical Equilibria in Natural Waters.* Wiley Interscience, New York (1981).

U.S. Department of Health and Human Services, Untitled Publication.

Victaulic Corp., Product Literature.

Vos, *Journal of Applied Polymer Science 10,* p 825 (1966).

Walsh, K. Personal communication (September 1993).

Whittaker, C.; Ridgway, H.F.; Olson, B.H. "Evaluation of Cleaning Strategies for Removal of Biofilms from Reverse-Osmosis Membranes", *Applied and Environmental Microbiology 48,* pp 395-403 (1984).

Wiens, J. "Information Exchange #8: pH Control in Reverse Osmosis Systems", *Ultrapure Water 2(6),* pp 49 (November/December 1985).

Wilf, M., Personal communications (1993).

Willets, C.O. et al., "Concentration by Reverse Osmosis of Maple Sap", *Food Technology 21,* p 24 (January 1967).

Zeiher, E.H.K.; Pierce, C.C.; Woods, D. "Biofouling of Reverse Osmosis Systems: Three Case Studies", *Ultrapure Water 8(7),* pp 50-64 (October 1991).■

SUBJECT INDEX